DEPARTMENT OF THE ARMY
U.S. Army Corps of Engineers
Washington, DC 20314-1000

CECW-CE

EM 1110-2-1100
(Change 2)

Manual
No. 1110-2-1100

1 April 2008

Engineering and Design
COASTAL ENGINEERING MANUAL

1. Purpose. The purpose of the *Coastal Engineering Manual* (CEM) is to provide a comprehensive technical coastal engineering document. It includes the basic principles of coastal processes, methods for computing coastal planning and design parameters, and guidance on how to formulate and conduct studies in support of coastal flooding, shore protection, and navigation projects. This Change 2 to EM 1110-2-1100, 1 April 2008, includes the following changes and updates:

    a. Part I-1. References were checked and some were deleted (Engineer Manuals that are no longer in the USACE inventory).

    b. Part I-4. Minor changes were made in the text to better reflect the contents of subsequent parts of the CEM.

    c. Part II-1. Figure II-1-9 has been revised; Equations II-1-128, II-1-160, and II-1-161 have been corrected.

    d. Part II-2. Equations II-2-4, II-2-5, and II-2-32 have been corrected along with other errors reported by various users.

    e. Part II-5. References were checked and some were deleted (Engineer Manuals that are no longer in the USACE inventory).

    f. Part II-6. The value of "e" used in Eq. II-6-28 has been corrected.

    g. Part II-7. The table of contents was corrected. A new section, II-7-11, Note to Users, Vessel Buoyancy, was added at the end of the chapter.

    h. Part III-3. Corrections have been made to format and spelling. Different plots were added to Figures III-3-24 and III-3-26.

    i. Part IV-1. Corrections have been made to references.

    j. Part V-1. Citation of an Engineer Regulation has been corrected.

    k. Part V-2. Citation of references has been changed, web pages with sources of wind and wave data have been added. Some minor text changes have also been made.

    l. Part V-3. Citations of unpublished reports or personal communications have been deleted, and links to other figures or parts of the CEM have been checked and corrected.

    m. Part V-4. Minor text changes, corrections to references and Figure V-4-1.

    n. Part V-5. Links to other parts of the CEM that were planned but never written have been deleted.

2. Applicability. This manual applies to all HQUSACE elements and all USACE commands having Civil Works and military design responsibilities.

3. Discussion. The CEM is divided into five parts in two major subdivisions: science-based and engineering-based. The first four parts of the CEM and Appendix A compose the science-based subdivision:

Part I, "Introduction"
Part II, "Coastal Hydrodynamics"
Part III, "Coastal Sediment Processes"
Part IV, "Coastal Geology"
Appendix A, "Glossary"

The engineering-based subdivision is oriented toward a project-type approach, Part V, "Coastal Project Planning and Design."

4. Distribution Statement. Approved for public release, distribution unlimited.

5. Note to Users. Revised chapters are dated 1 April 2008. Readers need to download the entire new chapters and discard earlier versions in their possession.

FOR THE COMMANDER:

STEPHEN L. HILL
Colonel, Corps of Engineers
Chief of Staff

Table of Contents

List of Tables

List of Figures

Chapter III-1
Coastal Sediment Properties

III-1-1. Introduction

a. Bases of sediment classification.

(1) Several properties of sediments are important in coastal engineering. Most of these properties can be placed into one of three groups: the size of the particles making up the sediment, the composition of the sediment, or bulk characteristics of the sediment mass.

(2) In some cases (in clay, for example) there are strong correlations among the three classification groups. A clay particle is, in the compositional sense, a mineral whose molecules are arranged in sheets that feature orderly arrays of silicon, oxygen, aluminum, and other elements (Lambe and Whitman 1969). Clay particles are small and platey. They are small in part because they originate from the chemical modification and disintegration of relatively small pre-existing mineral grains and because the sheet-like minerals are not strong enough to persist in large pieces. The geologist's size classification defines a particle as clay if it is less than 0.0039 mm. Because a clay particle is so small, it has a large surface area compared to its volume. This surface area is chemically active and, especially when wet, the aggregate of clay surfaces produces the cohesive, plastic, and slippery characteristics of its bulk form. Thus, the three classifications each identify the same material when describing "clay."

(3) On the other hand, most grains of beach sand are quartz, a simpler and chemically more inert material than clay minerals. In the geologist's size classification, sand grains are at least 16 times larger and may be more than 500 times larger in diameter than the largest clay particle (4,000 to more than 100 million times larger in volume). At this size, the force of gravity acting on individual sand grains dwarfs the surface forces exerted by those sand grains, so the surface properties of sand are far less important than surface properties of clay particles. Because sand grains do not stick together, a handful of pure dry sand cannot be picked up in one piece like a chunk of clay. Several differences between clay and sand are summarized on Table III-1-1. More inclusive discussions of sediment sizes, compositions, and bulk properties are given later in this chapter.

Table III-1-1
Relations Among Three Classifications for Two Types of Sediment

Name of Sediment	Bases of Classification		
	Usual Composition	Size Range, Wentworth	Bulk Properties
Clay	Clay Minerals (sheets of silicates)	Less than 0.0039 mm	Cohesive Plastic under stress Slippery Impermeable
Sand	Quartz (SiO_2)	Between 0.0625 mm and 2.00 mm	Noncohesive Rigid under stress Gritty Permeable

b. Sediment properties important for coastal engineering. Sediment properties of material existing at the project site, or that might be imported to the site, have important implications for the coastal engineering project. The following sections briefly discuss several examples of ways that sediment properties affect coastal engineering projects and illustrate their importance.

(1) Properties important in dredging.

(a) A hydraulic dredge needs to entrain sediment from the bottom and pump it through a pipe. The entrainment and the pumping are both affected by the properties of the sediment to be dredged. The subject is briefly treated in the following paragraphs, but more details on dredging practice can be obtained from books by Turner (1984) and Huston (1970). Other information is available through the Dredging Research Program of the U.S. Army Corps of Engineers.

(b) Sediment can be classified for entrainment by a hydraulic dredge as fluid, loose, firm, or hard. Fluid muds and loose silt or sand can be entrained relatively easily by dragheads. Firm sand, stiff clay, and organically bound sediment may require a cutterhead dredge to loosen the sediment. Usually, hard material such as rock or coral is not suitable for hydraulic dredging unless the material has previously been well broken.

(c) Sediment can be classified for pumping by a hydraulic dredge as cohesive, noncohesive, or mitigated (Turner 1984). Cohesive sediments get transported through the pipe as lumps and nodules whereas noncohesive sediments disperse as a slurry, which is more easily pumped through the pipe. Mitigated sediments consist mainly of noncohesive sediments with a small amount of clay, which increases the transport efficiency of the pipe.

(d) The diameter of the pipe and the size of the pump limit the size of the material which can be pumped. Usually, oversize material is prevented from entering the pipeline of a suction dredge by a grid placed across the draghead, or the cutterhead reduces material entering the pipeline to adequate size.

(e) Another property of the sediment important in dredging is the degree of its cohesiveness that allows the sediment to stand in near-vertical banks while being dredged. A dredge works more efficiently if the material will maintain such a steep "face" during the dredging process. Muds and loose sands that flow like liquids lack this property.

(f) The above statements apply to those dredging systems that remove material from the bottom through a pipeline by a pump. Such hydraulic dredges are not always the most feasible dredging system to use because of space constraints, navigation requirements, dredging depth, sediment properties, or disposal options. Under some conditions, mechanical dredges, which include a grab bucket operated from a derrick, or a dipper dredge (power shovel) on a barge (Huston 1970), may be more desirable. When mechanical dredges are used, looser sediments are usually dredged with a bucket and harder sediments with a dipper.

(2) Properties important in environmental questions.

(a) Recently, environmental problems associated with the handling and deposition of sediment have received increased attention. These concerns most frequently arise from dredging operations, but can occur anytime sediment is introduced into the marine environment. The usual issues involve the burial of bottom-dwelling organisms, the blockage of light to bottom-dwelling and water-column organisms, and the toxicity of the sediments.

(b) The sediment property of most environmental consequence is size. Turbidity in the water column depends on the fall velocity of the sediment particles, which is a strong function of the grain size. Turbid waters can be carried by currents away from the immediate project site, blocking the light to organisms over a wide area and, as the sediments settle out, blanketing the bottom at a rate faster than the organisms can accommodate. Fine sediments (silts and clays) get greater scrutiny under environmental regulation because they produce greater and longer-lasting turbidity, which will impact larger areas of the seafloor than will coarser, sand-sized material. The dredging of sand usually encounters less severe environmental objection, provided that there are few fines mixed with it and that the site has no prior toxic chemical history.

(c) Environmental regulation is changing, and many regulatory questions are outside the usual experience of coastal engineers. However, a basic coastal engineering contribution to facilitating the progress of a project through regulatory review is the early collection of relevant sediment samples from the site and obtaining accurate data on their size, composition, and toxicity.

(3) Properties important in beach fills.

(a) Beach fills have two primary functions: to provide temporary protection to upland property, and to increase temporarily the recreational space along the shore. Neither function can be well satisfied with sediment finer than sand. Because the recreational function is inhibited by material coarser than sand, and because fill coarser than sand is frequently less available, the primary beach fill material is usually sand.

(b) The source sediment for a beach fill is known as the borrow material, and the sediment on the beach prior to the fill is known as the native material. The sediment property most important for design is the size distribution of the borrow and native sands. The art of beach fill design consists of calculating the volume of borrow with a given size distribution that will produce a required volume of beach fill.

(c) Ideally, the median size of the borrow sand should not be less than the median size of the native sand, and the spread of the sizes in the borrow size distribution should not exceed the spread of sizes in the native sand. Often it is impossible to meet these ideal conditions because suitable borrow material does not exist in adequate volume at a reasonable cost. Further, on severely eroded beaches, the native sand may be skewed to coarser size ranges because the fines have eroded out, producing unrealistic requirements for borrow sand size distribution.

(d) Beach fill design aims to compensate for the differences between borrow sand and native sand, usually by overfilling with borrow sand and assuming preferential loss of the fine fractions. A favorable feature of beach fill technology is the accidental, partial loss of the fine fraction during the dredging and handling between borrow and beach. There have been cases (mostly anecdotal) where such handling losses have produced sand fill on the beach that is coarser than the borrow sand from which the fill was derived.

(e) The shore protection and the recreational qualities of a beach fill conflict when coarser sediment sizes are used. Usually, a beach provides more protection against erosion when its particles are coarser (also when they are more angular and more easily compacted). However, fill material larger than sand size (about 2.0 mm) will lessen the recreational value of the beach. Also odor and color of beach fill may be objectional to recreational users; but usually, if the grain size of the material is adequate, the objectional odor and color are temporary.

(4) Properties important in scour protection.

(a) Scour, the localized removal of bed material below its natural elevation, usually occurs near (and is usually caused by) marine structures such as jetties, seawalls, bridge pilings, etc. These types of structures can accelerate tidal currents, focus wave energy, and increase turbulence in the water column.

(b) To prevent scour, it is usual to place a layer of less erodible material on the surface of the sediment that is subject to erosion. Such a layer is called a revetment or scour blanket. A typical revetment consists of broken rock, known as riprap, which is essentially a sediment consisting of large rock particles. The sedimentary properties of riprap which are important as scour protection include its size distribution, density of the rock material, and the porosity and permeability of the material as placed. The riprap must be heavy enough to resist movement under design currents, the porosity and thickness of the riprap must be adequate to dissipate fluid energy before it reaches the underlying material being protected, and the permeability must be adequate to satisfactorily relieve pressure buildup at the seafloor-revetment interface.

(c) For long-term protection, riprap must consist of dense, durable material of blocky shape. Porous carbonate rocks such as coral and some limestones are not satisfactory, and thin slabs of material of any composition (such as shale) are usually more mobile than blocky shapes having the same weight. However, in any situation, economics and available materials may make it advantageous to use materials that depart from the ideal.

(d) It should be recognized that scour protection in coastal engineering differs from scour protection encountered in typical transportation projects both in the magnitude of the forces and in their reversing directions. Though scour protection design for highways is a well-developed art with extensive documentation, the direct transferal of highway riprap experience to coastal problems is usually unsatisfactory.

(5) Properties important in sediment transport studies.

(a) The underlying physics of how water moves sediment is not well understood. This is, perhaps, one reason for the large number of formulas (often conflicting) which have been proposed to predict transport rates. These formulas are usually functions of fluid properties, flow condition properties, and sediment properties. Sediment properties commonly used include: grain size, grain density, fall velocity, angle of repose, and volume concentration. Sediment size distribution and grain shape are also important.

(b) One method used to study sediment transport is to follow the movement of marked (tracer) particles in the nearshore environment. Ideally, tracers should react to coastal processes that move sediment in a manner identical to the native sand, yet provide some signal to the investigator that will distinguish the tracer from the native material. In the last two decades, the most common tracer has been dyed native sand grains. Typically representative samples of sand taken from the site are dyed with a fluorescent dye and then reintroduced to their environment. Care must be taken to ensure that the dying process does not significantly alter the sediment size or density. Transport of these tracers is then monitored by sampling. Because of their dilute distribution, tracers are a very labor-intensive means of studying sediment movement.

(c) Native sand tracers (trace minerals, heavy minerals) have been used to interpret sediment movement, but usually these tracers have a size and density that differ from the majority of grains on the beach. The problem becomes more complicated as beach fills become more common, and inadvertently introduce non-native tracer grains. See Galvin (1987) for a more thorough discussion of this topic.

(d) Recently, a few sensors have become available that can measure the concentration of moving sediment. These sensors are generally quite sensitive to grain size and require calibration using sediments obtained onsite. See, for example, a discussion of an optical backscatter sensor (Downing, Sternberg, and Lister 1981).

III-1-2. Classification of Sediment by Size

a. Particle diameter.

(1) One of the most important characteristics of sediment is the size of the particles. The range of grain sizes of practical interest to coastal engineers is enormous, covering about seven orders of magnitude, from clay particles to large breakwater armor stone blocks. A particle's size is usually defined in terms of its diameter. However, since grains are irregularly shaped, the term diameter can be ambiguous. Diameter is

normally determined by the mesh size of a sieve that will just allow the grain to pass. This is defined as a particle's sieve diameter. When performed in a standard manner, sieving provides repeatable results, although there is some uncertainty about how the size of a sieve opening relates to the physical size of the particle passing through the opening. (See page 58 of Blatt, Middleton, and Murray (1980) for further discussion.)

(2) Another way to define a grain's diameter is by its fall velocity. A grain's sedimentation diameter is the diameter of a sphere having the same density and fall velocity. This definition has the advantage of relating a grain's diameter to its fluid behavior, which is the usual ultimate reason for needing to determine the diameter. However, a settling tube analysis is somewhat less reproducible than a sieve analysis, and testing procedures have not been standardized. Other diameter definitions that have been occasionally used include the nominal diameter, the diameter of a sphere having the same volume as the particle; and the axial diameter, the length of one of the grain's principal axes, or some combination of these axes.

(3) For nearly spherical particles, as many sand grains are (but most shell and shell fragments are not), there is little difference in these definitions. When reporting the results of an analysis, it is always appropriate to define the diameter (or describe the measurement procedure), particularly if the sieve diameter is not being used.

(4) Usually, there is a need to characterize an appropriate diameter for an aggregation of particles, rather than the diameter of a single particle. Even the best-sorted natural sediments have a range of grain sizes. Most sediment samples found in nature have a few relatively large particles covering a wide range of diameters and many small particles within a small range of diameters. That is, most natural sediment samples have a highly skewed distribution, in an absolute sense. However, if size classes are based upon a logarithmic (power of 2) scale and if the contents of each size fraction are considered by weight (not by the number of particles), then most typical sediment samples will have a distribution that is near Gaussian (or normal). When such sediment samples are plotted as a percentage of the total weight of the sample being sieved, the sediment size distribution that results usually approximates a straight line on a log-normal graph. This line is known as the log-normal distribution. Meaningful descriptions of the distributions of this type of data can be made using standard statistical parameters.

(5) Sediment size data normally come from the weight of sediment that accumulates on each sieve in a nest of graduated sieves. This can be plotted on semilog paper as shown in Figure III-1-1 (ENG form 2087) or on log-normal paper as shown in Figure III-1-2.

b. Sediment size classifications.

(1) The division of sediment sizes into classes such as cobbles, sand, silt, etc., is arbitrary, and many schemes have been proposed. However, two classification systems are in general use today by coastal engineers. Both have been adopted from other fields.

(2) The first is the Modified Wentworth Classification, which is generally used in geologic work. Geologists have been particularly active in sediment size research because they have long been interested in interpreting slight differences in size as indicating particular processes or events in the geologic past. A summary of geological work applicable to coastal engineering is found in Chapter 3 of the book by Blatt, Middleton, and Murray (2nd ed., 1980).

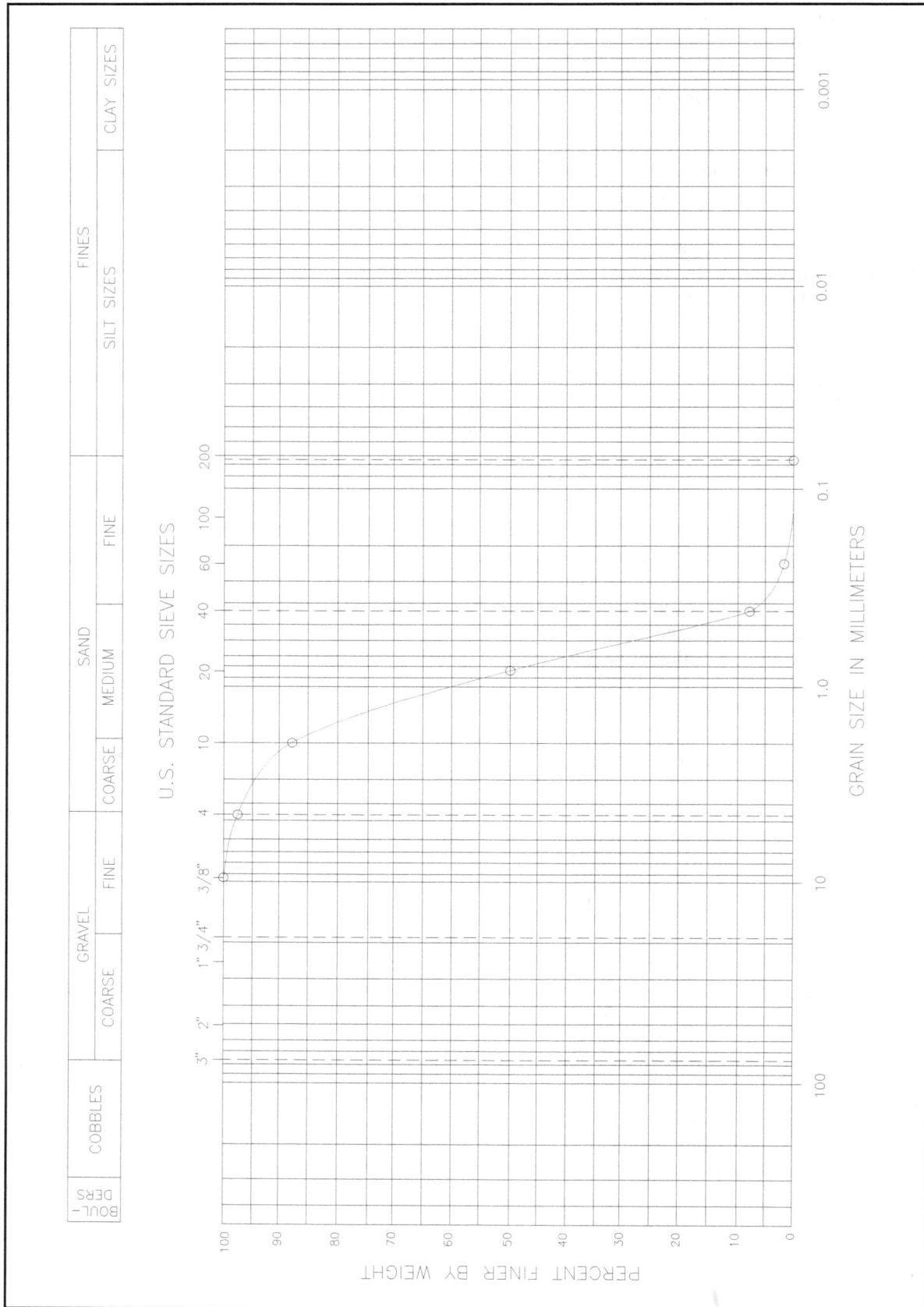

Figure III-1-1. Example of sediment distribution using semilog paper; sample 21c - foreshore at Virginia Beach

Coastal Sediment Properties

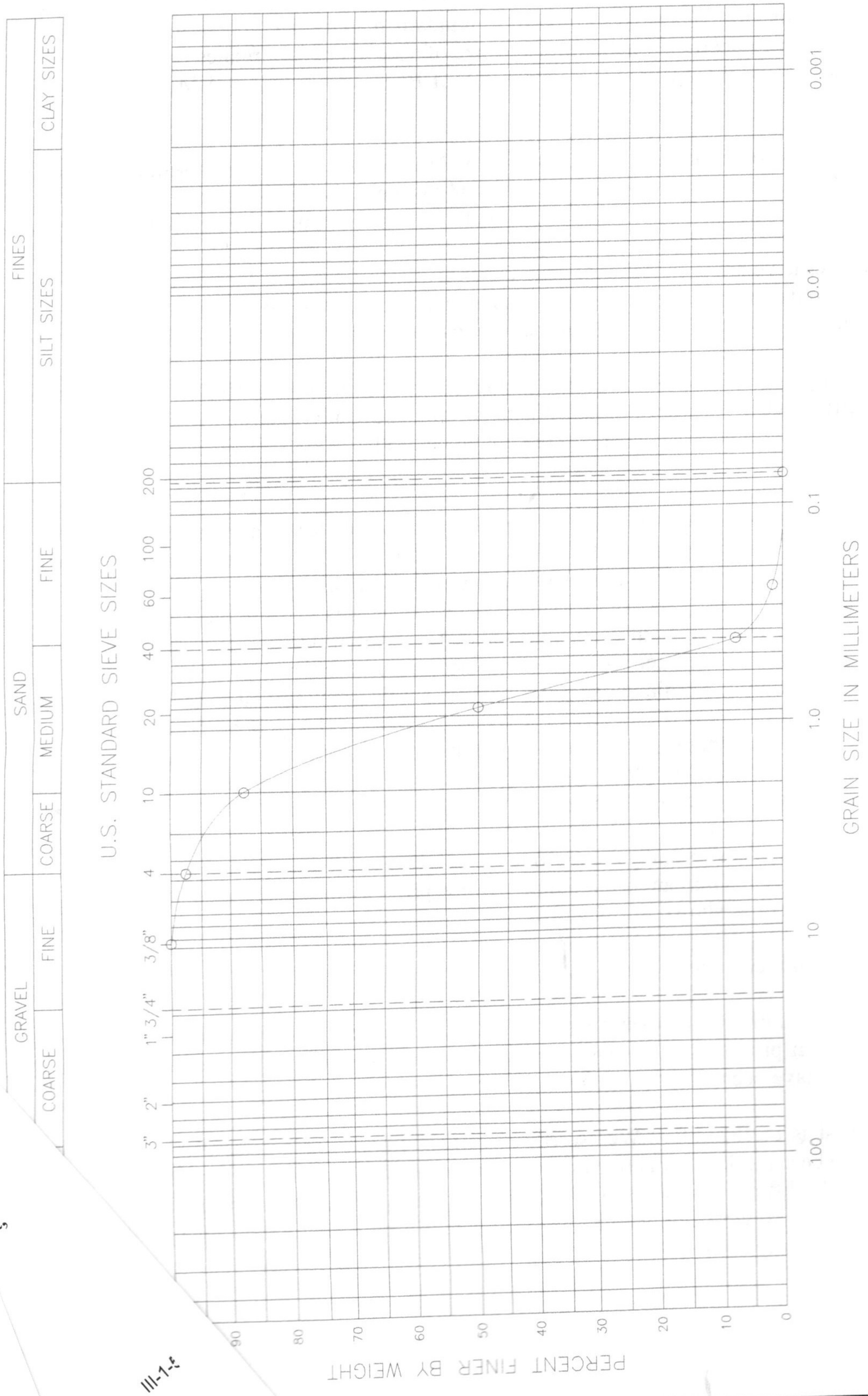

Figure III-1-1. Example of sediment distribution using semilog paper; sample 21c - foreshore at Virginia Beach

Coastal Sediment Properties

Table III-1-2
Sediment Particle Sizes

ASTM (Unified) Classification[1]	U.S. Std. Sieve[2]	Size in mm	Phi Size	Wentworth Classification[3]
Boulder		4096.	-12.0	
		1024.	-10.0	Boulder
	12 in. (300 mm)	256.	-8.0	
Cobble		128.	-7.0	Large Cobble
		107.64	-6.75	
		90.51	-6.5	Small Cobble
	3 in. (75 mm)	76.11	-6.25	
		64.00	-6.0	
		53.82	-5.75	
		45.26	-5.5	Very Large Pebble
Coarse Gravel		38.05	-5.25	
		32.00	-5.0	
		26.91	-4.75	
		22.63	-4.5	Large Pebble
	3/4 in. (19 mm)	19.03	-4.25	
		16.00	-4.0	
		13.45	-3.75	
		11.31	-3.5	Medium Pebble
Fine Gravel		9.51	-3.25	
	2.5	8.00	-3.0	
	3	6.73	-2.75	
	3.5	5.66	-2.5	Small Pebble
	4 (4.75 mm)	4.76	-2.25	
Coarse Sand	5	4.00	-2.0	
	6	3.36	-1.75	
	7	2.83	-1.5	Granule
	8	2.38	-1.25	
	10 (2.0 mm)	2.00	-1.0	
	12	1.68	-0.75	
	14	1.41	-0.5	Very Coarse Sand
	16	1.19	-0.25	
Medium Sand	18	1.00	0.0	
	20	0.84	0.25	
	25	0.71	0.5	Coarse Sand
	30	0.59	0.75	
	35	0.50	1.0	
	40 (0.425 mm)	0.420	1.25	
	45	0.354	1.5	Medium Sand
	50	0.297	1.75	
	60	0.250	2.0	
Fine Sand	70	0.210	2.25	
	80	0.177	2.5	Fine Sand
	100	0.149	2.75	
	120	0.125	3.0	
	140	0.105	3.25	
	170	0.088	3.5	Very Fine Sand
	200 (0.075 mm)	0.074	3.75	
Fine-grained Soil:	230	0.0625	4.0	
	270	0.0526	4.25	
Clay if PI ≥ 4 and plot of PI vs. LL is	325	0.0442	4.5	Coarse Silt
on or above "A" line and the presence	400	0.0372	4.75	
of organic matter does not influence		0.0312	5.0	
LL.		0.0156	6.0	Medium Silt
		0.0078	7.0	Fine Silt
		0.0039	8.0	Very Fine Silt
Silt if PI < 4 and plot of PI vs. LL is		0.00195	9.0	Coarse Clay
below "A" line and the presence of		0.00098	10.0	Medium Clay
organic matter does not influence LL.		0.00049	11.0	Fine Clay
		0.00024	12.0	
(PI = plasticity limit; LL = liquid limit)		0.00012	13.0	Colloids
		0.000061	14.0	

[1] ASTM Standard D 2487-92. This is the ASTM version of the Unified Soil Classification System. Both systems are similar (from ASTM (1994)).
[2] Note that British Standard, French, and German DIN mesh sizes and classifications are different.
[3] Wentworth sizes (in mm) cited in Krumbein and Sloss (1963).

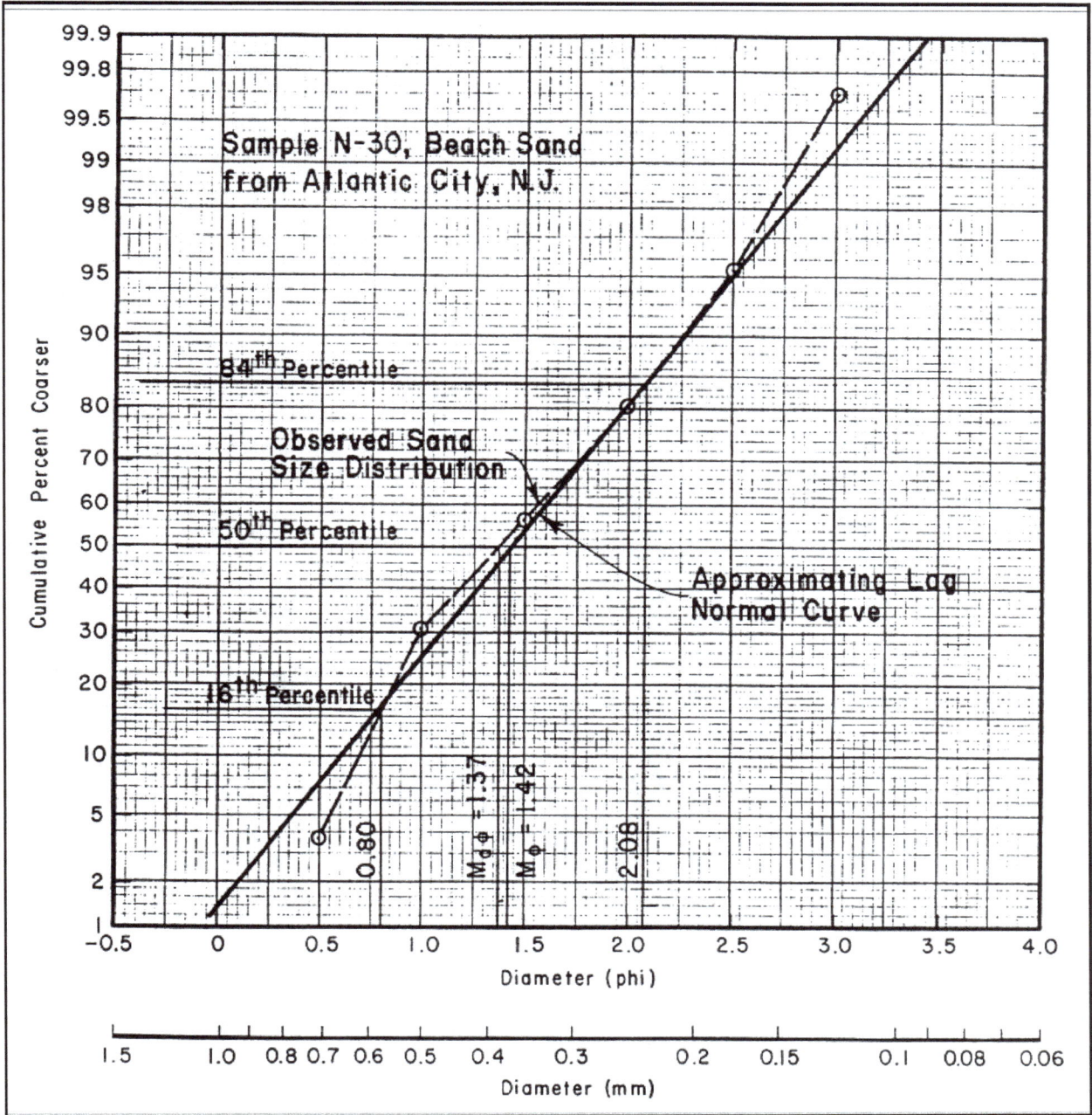

Figure III-1-2.   Example of sediment distribution using log-normal paper

(3) The other is the Unified Soils Classification or the ASTM Classification, which various American engineering groups have developed. These groups have been more concerned with standardizing ways to obtain repeatable results in the analyses of sediment. The definitive source for such information is Volume 4.08 of the Standards published by the American Society for Testing and Materials (ASTM). These standards are published annually, with revisions as needed. In what follows, reference to ASTM standards will be to the 1994 edition of Volume 4.08, which deals with soil, as defined by engineers.

(4) These two systems are compared in Table III-1-2. Note that there are differences in the category limits. For example, sand in the engineer's classification is between 0.074 and 4.76 mm, while in the

geologist's classification, sand is between 0.0625 and 2.0 mm. In describing a sediment, it is necessary to say which classification system is being utilized.

c. Units of sediment size.

(1) Table III-1-2 lists three ways to specify the size of a sediment particle: U.S. Standard sieve numbers, millimeters, and phi units. A sieve number is approximately the number of square openings per inch, measured along a wire in the wire screen cloth (Tyler 1991). Available sieve sizes can be obtained from catalogs on construction materials testing such as Soiltest (1983). The millimeter dimension is the length of the inside of the square opening in the screen cloth. This square side dimension is not necessarily the maximum dimension of the particle which can get through the opening, so these millimeter sizes must be understood as nominal approximations to sediment size. Table III-1-2 shows that the Wentworth scale has divisions that are whole powers of 2 mm. For example, medium sands are those with diameters between 2^{-2} mm and 2^{-1} mm. This property of powers of 2-mm class limits led Krumbein (1936) to propose a phi unit scale based on the definition:

$$\varphi = -\log_2 D \tag{III-1-1a}$$

where D is the grain diameter in millimeters. Phi diameters are indicated by writing φ after the numerical value. That is, a 2.0-φ sand grain has a diameter of 0.25 mm. To convert from phi units to millimeters, the inverse equation is used:

$$D = 2^{-\varphi} \tag{III-1-1b}$$

(2) The benefits of the phi unit include: (a) it has whole numbers at the limits of sediment classes in the Wentworth scale; and (b) it allows comparison of different size distributions because it is dimensionless. Disadvantages of this phi unit are: (a) the unit gets larger as the sediment size gets smaller, which is both counterintuitive and ambiguous; (b) it is difficult to physically interpret size in phi units without considerable experience; and (c) because it is a dimensionless unit, it cannot represent a unit of length in physical expressions such as fall velocity or Reynolds number.

d. Median and mean grain sizes.

(1) All natural sediment samples contain grains having a range of sizes. However, it is frequently necessary to characterize the sample using a single typical grain diameter as a measure of the central tendency of the distribution. The *median grain diameter M_d* is the sample characteristic most often chosen. The definition of M_d is that, by weight, half the particles in the sample will have a larger diameter and half will have a smaller. This quantity is easily obtained graphically, if the sample is sorted by sieving or other method, and the weight of the size fractions are plotted, as seen in Figures III-1-1 and III-1-2.

(2) The median diameter is also written as D_{50}. Other size fractions are similarly indicated. For example, D_{90} is the diameter for which 90 percent of the sediment, by weight, has a smaller diameter. An equivalent definition holds for the median of the phi-size distribution φ_{50} or for any other size fraction in the phi scale.

(3) Another measure of the central tendency of a sediment sample is the *mean grain size*. Several formulas have been proposed to compute this quantity, given a cumulative size distribution plot of the sample (Otto 1939, Inman 1952, Folk and Ward 1957, McCammon 1962). These formulas are averages of 2, 3, 5, or more symmetrically selected percentiles of the phi frequency distribution. Following Folk (1974):

$$M_\varphi = \frac{(\varphi_{16} + \varphi_{50} + \varphi_{84})}{3} \qquad \text{(III-1-2)}$$

where M_φ is the estimated mean grain size of the sample in phi units. This can be converted to a linear diameter using Equation 1-1b. The median and mean grain sizes are usually quite similar for most beach sediments. For example, a study of 465 sand samples from three New Jersey beaches, the mean averaged only 0.01 mm smaller than the median for sands having an average median diameter of 0.30 mm (1.74 phi) (Ramsey and Galvin 1977). If the grain sizes in a sample are log-normally distributed, the two measures are identical. Since the median is easier to determine and the mean does not have a universally accepted definition, the median is normally used in coastal engineering to characterize the central tendency of a sediment sample.

  e. *Higher order moments.*

  (1) Additional statistics of the sediment distribution can be used to describe how the sample varies from a log-normal distribution. The *standard deviation* is a measure of the degree to which the sample spreads out around the mean (i.e., its sorting). Following Folk (1974), the *standard deviation* can be approximated by:

$$\sigma_\varphi = \frac{(\varphi_{84} - \varphi_{16})}{4} + \frac{(\varphi_{95} - \varphi_5)}{6} \qquad \text{(III-1-3)}$$

where σ_φ is the estimated standard deviation of the sample in phi units. For a completely uniform sediment φ_{05}, φ_{16}, φ_{84}, and φ_{95} are all the same, and thus, the standard deviation is zero. There are also qualitative descriptions of the standard deviation. A sediment is described as well-sorted if all particles have sizes that are close to the typical size (small standard deviation). If the particle sizes are distributed evenly over a wide range of sizes, then the sample is said to be well-graded. A well-graded sample is poorly sorted; a well-sorted sample is poorly graded.

  (2) The degree by which the distribution departs from symmetry is measured by the *phi coefficient of skewness* α_φ defined in Folk (1974) as:

$$\alpha_\varphi = \frac{\varphi_{16} + \varphi_{84} - 2(\varphi_{50})}{2(\varphi_{84} - \varphi_{16})} + \frac{\varphi_5 + \varphi_{95} - 2(\varphi_{50})}{2(\varphi_{95} - \varphi_5)} \qquad \text{(III-1-4)}$$

  (3) For a perfectly symmetric distribution, the skewness is zero. A positive skewness indicates there is a tailing out toward the fine sediments, and conversely, a negative value indicates more outliers in the coarser sediments.

  (4) The *phi coefficient of kurtosis* β_φ is a measure of the peakedness of the distribution; that is, the proportion of the sediment in the middle of the distribution relative to the amount in both tails. Following Folk (1974), it is defined as:

$$\beta_\varphi = \frac{\varphi_{95} - \varphi_5}{2.44 \, (\varphi_{75} - \varphi_{25})} \qquad \text{(III-1-5)}$$

Values for the mean and median grain sizes are frequently converted from phi units to linear measures. However, the standard deviation, skewness, and kurtosis should remain in phi units because they have no corresponding dimensional equivalents. If these terms are used in equations, they are used in their dimensionless phi form. Relative relationships are given for ranges of standard deviation, skewness, and kurtosis in Table III-1-3.

Table III-1-3
Qualitative Sediment Distribution Ranges for Standard Deviation, Skewness, and Kurtosis

Standard Deviation	
Phi Range	**Description**
<0.35	Very well sorted
0.35-0.50	Well sorted
0.50-0.71	Moderately well sorted
0.71-1.00	Moderately sorted
1.00-2.00	Poorly sorted
2.00-4.00	Very poorly sorted
>4.00	Extremely poorly sorted
Coefficient of Skewness	
<-0.3	Very coarse-skewed
- 0.3 to - 0.1	Coarse-skewed
- 0.1 to +0.1	Near-symmetrical
+0.1 to +0.3	Fine-skewed
>+0.3	Very fine-skewed
Coefficient of Kurtosis	
<0.65	Very platykurtic (flat)
0.65-0.90	Platykurtic
0.90-1.11	Mesokurtic (normal peakedness)
1.11-1.50	Leptokurtic (peaked)
1.50-3.00	Very leptokurtic
>3.00	Extremely leptokurtic

f. Uses of distributions. The median grain size is the most commonly used sediment size characteristic, and it has wide application in coastal engineering practice. The standard deviation of sediment samples has been used in several ways, including beach-fill design (see Hobson (1977), Ch. 5, Sec. III,3) and sediment permeability (Krumbein and Monk 1942). When a set of samples are taken from a single project site, they will frequently show little or no consistent variation in median diameter. In this case, various higher order moments are usually used to distinguish different depositional environments. There is an extensive literature on the potential application of the measures of size distribution; see, for example, Inman (1957), Folk and Ward (1957), McCammon (1962), Folk (1965, 1966), Griffiths (1967), and Stauble and Hoel (1986).

g. Sediment sampling procedures.

(1) Although a beach can only be composed of the available sediments, grain size distributions change in time and space. In winter, beach distributions are typically coarser and more poorly sorted than in summer. Also, typically, there is more variability in the foreshore and the bar/trough regions than in the dunes and the nearshore.

(2) While a single sample is occasionally sufficient to grossly characterize the sediments at a site, usually a set of samples is obtained. Combining samples from across the beach can reduce the high variability in spacial grain size distributions on beaches (Hobson 1977). Composite samples are created by either physically combining several samples before sieving or by mathematically combining the individual sample

weights. The set of samples obtained can be fairly small if the intent is only to characterize the beach as a whole. However, if the intent is to compare and contrast different portions of the same beach, many more samples are needed. In this case it is usually necessary to develop a sampling scheme prior to fieldwork.

(3) In the cross-shore it is recommended that samples be collected at all major changes in morphology along the profile, such as dune base, mid-berm, mean high water, mid-tide, mean low water, trough, bar crest, and then at 3-m intervals to the depth of closure (Stauble and Hoel 1986). In the longshore direction, sediment sampling should coincide with survey profile lines so that the samples can be spatially located and related to morphology and hydrodynamic zones. Shoreline variability and engineering structures should be considered in choosing sampling locations. A suggested rule of thumb is that a sampling line be spaced every half mile, but engineering judgment is required to define adequate project coverage.

(4) Samples collected along profile sub-environments can be combined into composite groups of similar depositional energy levels as seen in Figure III-1-3. Intertidal and subaerial beach samples have been found to be the most usable composites to characterize the beach and nearshore environment. Stauble and Hoel (1986) found that a composite containing the mean high water, mid-tide, and mean low water gave the best representation of the foreshore beach.

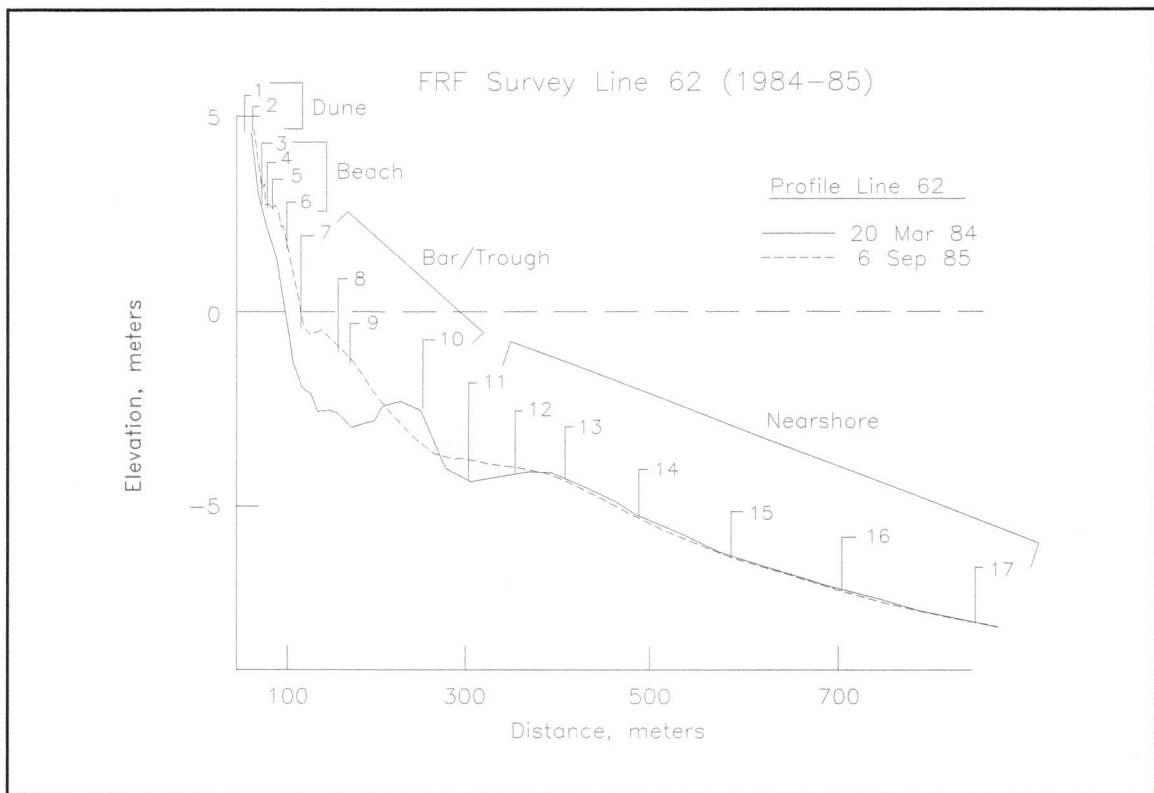

Figure III-1-3. Suggested composite sediment sample groups on a typical profile - example from Field Research Facility, Duck, NC

EXAMPLE PROBLEM III-1-1

FIND:
The statistics of the sediment size distribution shown in Figure III-1-2 and their qualitative descriptions.

GIVEN:
The needed phi values are: $\varphi_{05} = 0.56$, $\varphi_{16} = 0.80$, $\varphi_{25} = 0.93$, $\varphi_{50} = 1.37$, $\varphi_{75} = 1.87$, $\varphi_{84} = 2.08$, and $\varphi_{95} = 2.48$.

SOLUTION:
In phi units, the median grain size is given as:

$$\varphi_{50} = M_{d\varphi} = 1.37\varphi$$

From Equation 1-1b, the median grain size in millimeters is found as:

$$M_d = 2^{-1.37} = 0.39 \text{ mm}$$

From Equation 1-2, the mean grain size is found in phi units as:

$$M_\varphi = (0.80 + 1.37 + 2.08) / 3 = 1.42\varphi$$

From Equation 1-1b, this is converted to millimeters as:

$$D = 2^{-1.42} = 0.37 \text{ mm}$$

From Equation 1-3, the standard deviation is found as:

$$\sigma_\varphi = (2.08 - 0.80)/4 + (2.48 - 0.56)/6.0 = 0.32 + 0.32 = 0.64 \ \varphi$$

From Equation 1-4, the coefficient of skewness is found as:

$$\alpha_\varphi = 0.055 + 0.078 = 0.13$$

From Equation 1-5, the coefficient of kurtosis is found as:

$$\beta_\varphi = 1.92 / 2.29 = 0.84$$

Thus, using Tables III-1-2 and III-1-3, the sediment is a medium sand (Wentworth classification), it is moderately well sorted, and the distribution is fine-skewed and platykurtic.

h.   Laboratory procedures.

(1) Several techniques are available to analyze the size of beach materials, and each technique is restricted to a range of sediment sizes. Pebbles and coarser material are usually directly measured with calipers. However, this is not practical for sediments smaller than about 8 mm. Coarse sieves can also be used for material up to about 75 mm.

(2) Sand-sized particles (medium gravels through coarse silt) are usually analyzed using sieves. This requires an ordered stack of sieves of square-mesh woven-wire cloth. Each sieve in the stack differs from the adjacent sieves by having a nominal opening less than the opening of the sieve above it and greater than the opening of the sieve below it. A pan is placed below the bottom sieve. The sample is poured into the top sieve, a lid is placed on top, and the stack is placed on a shaker, usually for about 15 min. The different grains fall through the stack of sieves until each reaches a sieve that is too fine for it to pass. Then the amount of sediment in each sieve is weighed. ASTM Standard D422-63 (Reapproved 1990) is the basic standard for particle size analysis of soils, including sieving sedimentary materials of interest to coastal engineers. The application of D422 requires sample preparation described in Standard D421-85 (Reapproved 1993). The wire cloth sieves should meet ASTM Specification E-11.

(3) Sieves are graduated in size of opening according to the U.S. standard series or according to phi sizes. U.S. standard sieve sizes are listed in Table III-1-2, and the most commonly used sieves are shown in Figure III-1-1. Phi sized sieve openings vary by a factor of 1.19 from one sieve size to the next (by the fourth root of 2, or 0.25 phi units); e.g., 0.25, 0.30, 0.35, 0.42, and 0.50 mm.

(4) The range of sieve openings must span the range of sediment sizes to be sieved. Typically, about 6 full-height sieves or 13 half-height sieves plus a bottom pan are used in the analysis of a particular sediment. If 6 sieves are used, each usually varies in size from its adjacent neighbors by a half phi; if 13 sieves are used, each usually varies by a quarter phi. Normally, about 40 grams of sediment is sieved. More is needed if there are large size fractions (see ASTM Standard D2487-94).

(5) Determination of sediment size fractions for silts and clays is usually not necessary. Rather it is normally noted that a certain percentage of the material are fines with diameters smaller than the smallest sieve. When measurements are needed, the pipette method or the hygrometer method is usually used. Both of these methods are based upon determining the amount of time that different size fractions remain in suspension. These are discussed in Vanoni (1975). Coulter counters have also been used occasionally.

III-1-3. Compositional Properties

a. Minerals.

(1) Because of its resistance to physical and chemical changes and its common occurrence in terrestrial rocks, quartz is the mineral most commonly found in littoral materials. On temperate latitude beaches, quartz and feldspar grains commonly account for more than 90 percent of the material (Krumbein and Sloss 1963, p.134), and quartz, on average, accounts for about 70 percent of beach sand. However, it is important to realize that on individual beaches the percentage of quartz grains (or any other mineral) can range from essentially zero to 100 percent. Though feldspars are more common on the surface of the earth as a whole, comprising approximately 50 percent of all crustal rocks to quartz's approximately 12 percent (Ritter 1986), feldspars are more subject to chemical weathering, which converts them to clay minerals, quartz, and solutions. Since quartz is so inert, it accumulates during weathering processes. Thus, feldspars and related silicates are more commonly encountered in coastal sediment close to sources of igneous and metamorphic rocks, especially mountainous and glaciated coasts where streams and ice carry unweathered sediment

immediately to the shore. Quartz sand and clay are more common on shores far from mountains where weathering has had more time to reduce the relative proportions of feldspars and related silicates.

(2) Most carbonate sands owe their formation to organisms (both animal and vegetable) that precipitate calcium carbonate by modifying the very local chemical environment of the organism to favor carbonate deposition. Calcium carbonate may be deposited as calcite or aragonite, but aragonite is unstable and changes with time to either calcite or solutions. Calcite or limestone, the rock made from calcite, may, under some conditions, be altered to dolomite by the partial replacement of the calcium with magnesium. Carbonate sands can contribute up to 100 percent of the beach material, particularly in situations where it is produced in the local marine environment and there are limited terrestrial sediment supplies, such as reef-fringed, tropical island beaches. These sands are generally composed of a combination of shell and shell fragments, oolites, coral fragments, and algal fragments (Halimeda, foraminiferans, etc.). When carbonate shell is mixed with quartz sand, it may be necessary to dissolve out the shell to get a meaningful representation of each size fraction. ASTM Standard 4373 describes how to determine the calcium carbonate content of a beach.

(3) Other minerals that frequently form a small percentage of beach sands are normally referred to as heavy minerals, because their specific gravities are usually greater than 2.87. These minerals are frequently black or reddish and may, in sufficient concentrations, color the entire beach. The most common of these minerals (from Pettijohn (1957)) are andalusite, apatite, augite, biotite, chlorite, diopside, epidote, garnet, hornblende, hypersthene-enstatite, ilmenite, kyanite, leucoxene, magnetite, muscovite, rutile, sphene, staurolite, tourmaline, zircon, and zoisite. Their relative abundance is a function of their distribution in the source rocks of the littoral sediments and the weathering process. Heavy minerals have been occasionally used as natural tracers to identify sediment pathways from the parent rocks (Trask 1952; McMaster 1954; Giles and Pilkey 1965; Judge 1970).

In coastal engineering work, knowledge of the sediment composition is not normally important in its own right, but it is closely related to other important parameters such as sediment density and fall velocity.

b. Density.

(1) *Density* is the mass per unit volume of a material, which, in SI units, is measured in kilograms per cubic meter (kg/m^3). Sediment density is a function of its composition. Minerals commonly encountered in coastal engineering include quartz, feldspar, clay minerals, and carbonates. Densities for these minerals are given in Table III-1-4.

(2) Quartz is composed of the mineral silicon dioxide. Feldspar refers to a closely related group of metal aluminium silicate minerals. The most common clay minerals are illinite, montmorillonite, and kaolinite. Common carbonate minerals include calcite, aragonite, and dolomite. However, carbonate sands are usually not simple, dense solids; but rather the complex products of organisms which produce gaps, pores, and holes within the structure, all of which tend to lower the effective density of carbonate sand grains. Thus, carbonate sands frequently have densities less than quartz.

(3) The density of a sediment sample may be calculated by adding a known weight of dry sediment to a known volume of water. The change in volume is measured; this is the volume of the sediment. The sediment mass (= weight / acceleration of gravity) divided by its volume is the density. A complicating factor is that small pockets of air will stay in the pores and cling to the surfaces of almost all sediments. To obtain an accurate volume reading, this air must be removed by drawing a strong vacuum over the sand-water mixture. ASTM Volume 4.08 gives standards for measuring the density of soils and rocks, respectively (ASTM 1994).

Table III-1-4
Densities of Common Coastal Sediments

Mineral	Density kg/m³
Quartz	2,648
Feldspar	2,560 - 2,650
Illite	2,660
Montmorillonite	2,608
Kaolinite	2,594
Calcite	2,716
Aragonite	2,931
Dolomite	2,866

Above sand size, sediments encountered in coastal engineering are usually composites of several minerals, that is, rocks. Table III-1-5 lists densities of rocks commonly encountered. These rocks are also used for riprap, which are large sedimentary particles. Lines 5 and 6 of Table III-1-5 deal with carbonate rocks, the dolomites and limestones. Note 3 of Table III-1-5 suggests that typical dolomites are denser, less porous, and geologically older than typical limestones.

Table III-1-5
Average Densities of Rocks Commonly Encountered in Coastal Engineering

Number	Rock Type	Number of Samples	Mass Density, kg/m³
1	Basalt	323	2,740
2	Dolerite-Diabase	224	2,890
3	Granite	334	2,660
4	Sandstone	107	2,220
5	Dolomite	127	2,770
6	Limestone	182	2,540

Notes:
[1] Basalt and dolerite (or diabase) in lines 1 and 2 are fine-grained dark igneous rock often classed together as trap rock for engineering purposes.
[2] Lines 1, 2, 3, 4 from Table 5 of Johnson and Olhoeft (1984). Lines 5, 6 from Table 4-4 of Daly, Manger, and Clark (1966).
[3] Line 5 is the average of 127 samples from 3 dolomites or dolomitic limestones. Line 6 is the average of 182 samples from 5 limestones. The rocks in line 5 are denser and older than those in line 6. Average porosity of rocks in line 5 $\leq$ 3.0%; in line 6 average porosity is between 3% and 17%. Rocks in line 5 are older than Carboniferous.

c. Specific weight and specific gravity. The specific weight of a material is its density times the acceleration of gravity g. In SI units, the acceleration of gravity is 9.807 m/s², and thus, specific weight is measured in kilograms per meter squared per second squared (kg/(m² * s²)). The specific gravity of a material is its density divided by the density of water at 4^0 C, which is 1000.0 kg/m³. Specific gravity is a dimensionless quantity.

d. Strength.

(1) The material strength of a particle is the maximum stress which the particle can resist without failing, for a given type of loading. The SI unit of stress is the Pascal, and the convenient Pascal multiple is 1 million Pascals, or 1 mega Pascal (MPa). "Strength" in this section is unconfined ultimate strength in compression,

which is equivalent to crushing strength. Tensile strength, which is significantly less than compressive strength, is not discussed here, but tensile strength is usually proportional to compressive strength.

(2) The most commonly encountered material in coastal engineering is the quartz sand grain, which is very strong indeed. A single crystal of quartz has a strength on the order of 2,500 MPa. However, a sandstone, which is a composite of many sand grains, is surprisingly weak (top histogram on Figure III-1-4), being typically less than 100 MPa, or less than 4 percent of the strength of the single crystal. For this reason, sandstone is rarely used in coastal engineering construction.

(3) The difference in strength between quartz crystal and composite sandstone is due to weak intergranular cement and to flaws such as grain boundaries, bedding planes, cleavage, and joints that have a higher probability of being present in larger pieces. The data on Figure III-1-4 are excerpted from the extensive tabulation of Handin (1966), using all samples which were tested at room temperature and zero confining pressure.

(4) For calcium carbonate, strength varies with size in a direction opposite to that of quartz. Single crystals of calcite are weak (~14 MPa, depending on orientation) compared to single crystals of quartz (~2,500 MPa). But limestone rocks, made from interlocking calcium carbonate crystals, are much stronger than single crystals of calcium carbonate, and even somewhat stronger than sandstones, as shown by comparing the top two histograms on Figure III-1-4.

(5) The weaker rocks on the left end of the histograms of Figure III-1-4 are those with macroscopic flaws such as bedding planes, rather than flaws in single grains. The high-strength outliers on the right end of the sandstone and limestone histograms of Figure III-1-4 are special cases. The sandstone outlier is a quartzite, a metamorphosed sandstone recrystallized with silica cement. The limestone outlier is Solnhofen limestone, which is a dense, fine-grained, uniform limestone.

(6) Dolomite is a carbonate rock allied to limestone in which about half of the calcium of calcite has been replaced by magnesium (both the mineral and the rock are called dolomite). On average, dolomite and dolomitic limestones make better riprap than limestone and sandstone, as suggested by comparison of the histograms on Figure III-1-4.

(7) Where available, rocks classified commercially as trap rock (dense basalt, diorite, and related rocks) or granite (including rhyolite and dense gneiss) make even better riprap, with strength typically on the order of 140 to 200 MPa under conditions comparable to those in Figure III-1-4.

(8) A typical specification for rock used as riprap in coastal engineering, extracted from "Low Cost Shore Protection," Report on Section 54, U.S. Army Corps of Engineers (1981, p. 785), is as follows:

The stone shall be free of cracks, seams, and other defects that would tend to increase unduly its deterioration from natural causes or breakage in handling or dumping. The stone shall weigh, when dry, not less than 150 pounds per cubic foot.* The inclusion of objectionable quantities of sand, dirt, clay, and rock fines will not be permitted. Selected granite and quartzite, rhyolite, traprock and certain dolomitic limestones generally meet the requirements of these specifications.

*150 PCF is equivalent to 2,400 kg/m^3.

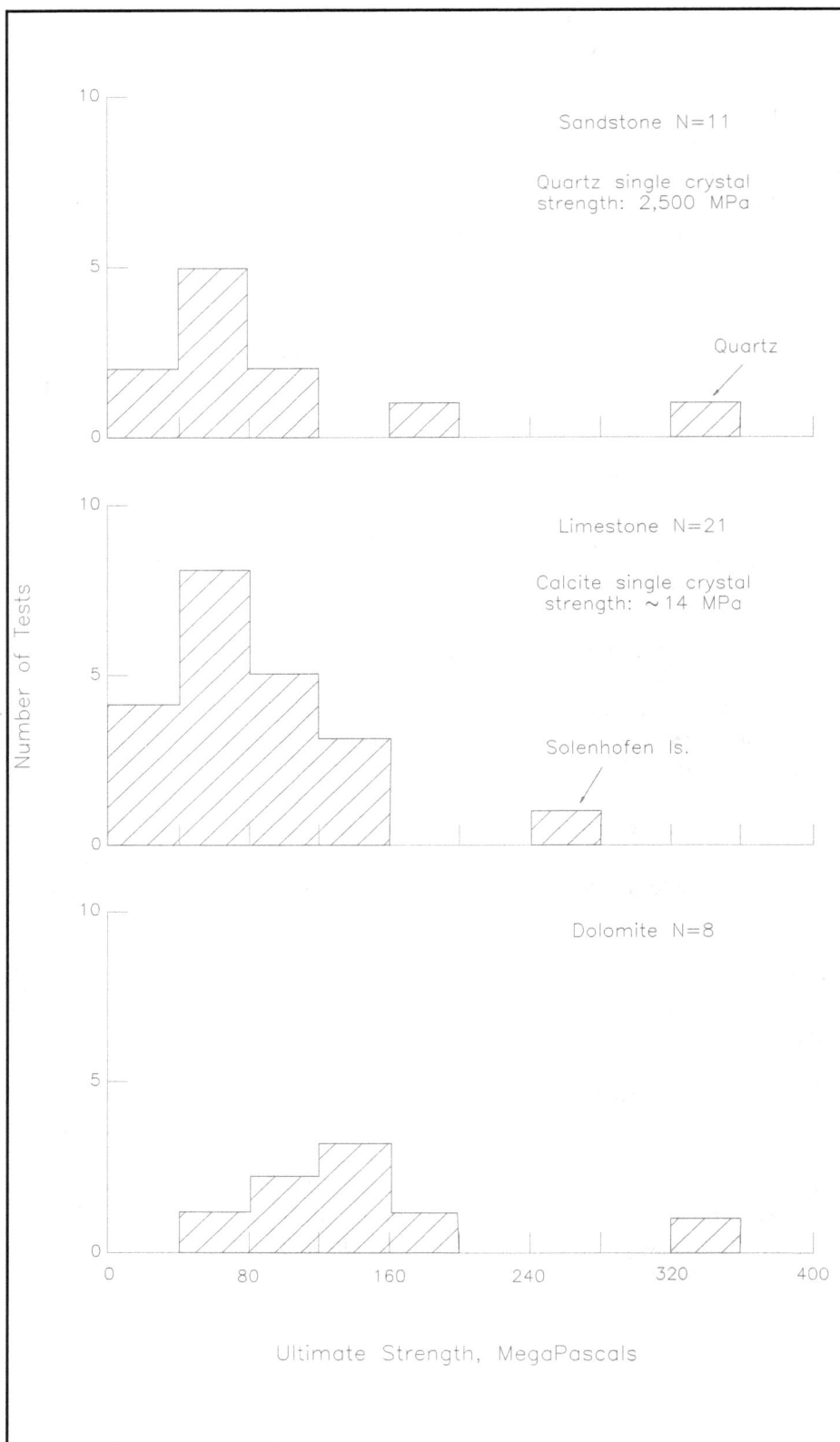

Figure III-1-4.   Unconfined ultimate strength of three rock types (after Handin (1966))

EXAMPLE PROBLEM III-1-2

FIND:
The density, specific weight, and specific gravity of a sediment sample.

GIVEN:
18.1 grams of the sample of dry beach sand exactly fills a small container having a volume of 10.0 cm³ after the filled container is strongly vibrated. When this amount of sand is poured into 50.0 ml of water which is subjected to a strong vacuum, the volume of the sand-water mixture is 56.8 ml.

SOLUTION:
It is important to recognize the difference in the volume of the grains themselves, which is 6.8 cm³ (= 56.8 - 50) (a milliliter is a cubic centimeter), and the volume of the aggregate (the grains plus the void spaces), which is 10 cm³.

Density is the sediment mass divided by its volume:

$$\rho_s = 18.1 \text{ gm} / 6.8 \text{ cm}^3 = 2.66 \text{ gm/cm}^3 = 2,660 \text{ kg/m}^3$$

If this problem were in English units rather than metric, the sediment weight (in pounds force) would probably be given, rather than the mass. To obtain the mass, the weight would need to be divided by the acceleration of gravity (32.2 ft/sec²) (mass = weight / g). After dividing by the volume (in ft³), the density would be obtained in slugs/ft³.

The specific weight of the sand grains themselves is the density times the acceleration of gravity:

$$\text{Sp wt of the grains} = 2.66 \text{ gm/cm}^3 * 980 \text{ cm/sec}^2$$
$$= 2,610 \text{ gm/(cm}^2 * \text{sec}^2) = 26,100 \text{ kg/(m}^2 * \text{sec}^2)$$

The specific gravity of a material has the same value as its density when measured in gm/cm³, because the density of water at 4⁰ C is 1.00 gm/cm³:

$$\text{Specific Gravity} = 2.66$$

e.   *Grain shape and abrasion.*

(1) Grain shape is primarily a function of grain composition, grain size, original shape, and weathering history. The shape of littoral material ranges from nearly spherical (e.g. quartz grains) to nearly disklike (e.g. shell fragments, mica flakes) to concave arcs (e.g. shells). Much of the early work on classifying sediment particle shape divided the problem into three size scales; the sphericity or overall shape of a particle, the roundness or the amount of abrasion of the corners, and the microtexture or the very fine scale roughness. These differences can be illustrated by noting that a dodecahedron has high sphericity but low roundness, while a thin oval has low sphericity but high roundness. A tennis ball has greater micro-texture than a baseball. More recent approaches to the quantification of grain shape have avoided the artificial division of shape into sphericity, roundness, and microtexture by characterizing all the wavelengths of the grain's

irregularities in one procedure using fractal geometry types of analysis. See Ehrlich and Weinberg (1970), and Frisch, Evans, Hudson, and Boon (1987).

(2) Grain shape is important to coastal engineers because it affects several other properties, particularly when the grains are far from spherical, which is the usual assumption. These include fall velocity, sieve analysis, initiation of motion, and also certain bulk properties, such as porosity and angle of repose. One particular area of interest to coastal engineers is in the design of man-made interlocking armor units on breakwaters that have high stability, even when stacked at a high angle of repose. Grain shape has also been used to indicate residence time in the littoral environment. See Krinsley and Doornkamp (1973) and Margolis (1969).

(3) However, most littoral grain shapes are close enough to spheres that a detailed study of their shape is not warranted. Frequently, a qualitative description of roundness is sufficient. This can be done by comparing the grains in a sample to photographs of standardized grains (see Krumbein 1941; Powers 1953; Shepard and Young 1961).

(4) Among the earliest research studies in coastal engineering were investigations of sand grain abrasion, done because of the worry that abrasion of beach sand contributes to beach erosion. These investigations found that abrasion of the typical quartz beach sand is rarely significant. In general, recent information lends further support to the conclusion of Mason (1942) that

On sandy beaches the loss of material ascribable to abrasion occurs at rates so low as to be of no practical importance in shore protection problems.

(5) To achieve the high stress required for quartz to abrade (to fail locally), very large impact forces are needed. These forces are developed by sudden changes in momentum, the product of mass and velocity. Given the small mass of a sand grain, large forces can be achieved only by grains moving at high velocities. But the drag on a sand grain moving in water increases as the square of its velocity, which limits sand grain velocity to a low multiple of its fall velocity. Because fall velocities are only a few centimeters per second, it is extremely difficult to achieve a stress between impacting grains that is anywhere near the strength of quartz. Thus, the rounding of the corners of angular quartz grains in riverine or littoral environments is a very lengthy process. Sands, silts, and clays found in the coastal environment can generally be considered as some of the stable end results of the weathering process of rocks so long as they remain on or near the surface of the earth.

(6) However, abrasion is common in large particles such as boulders and riprap subject to wave action. The mass of a particle increases with the cube of its diameter, so a minimum-size boulder (300 mm on Table III-1-2) compared to a minimum-size coarse sand grain (2 mm) (ASTM classification) will have $(300/2)^3 = 3.375$ million times more mass. If that boulder was a perfect sphere in shape, and this rock sphere rested with a point contact on the plane surface of another rock, then merely the weight of the boulder would crush the point contact of the sphere until the area of contact increased enough to reduce the pressure of the contact to below the crushing strength of the boulder (say 120 MPa, based on Figure III-1-4). The material crushed at this contact is abraded from the boulder. This process of stress concentration at points of contact is considered quantitatively by Galvin and Alexander (1981). If the boulder moves, say, with the rocking motion imparted by the arrival of a wave crest, the slight velocity of the boulder mass provides a momentum which can produce impact forces in excess of crushing strength at points of contact between a boulder and its neighbors, thus abrading the rock.

(7) Because it is probable that a large rock will break along surfaces of weakness, the resulting pieces after breakage will usually be stronger than the rock from which the pieces are broken. Thus, abraded gravel

pieces on a wave-washed shingle beach are apt to have greater strength than the bulk strength of the rock from which the gravel was derived.

III-1-4. Fall Velocity

When a particle falls through water (or air), it accelerates until it reaches its fall or settling velocity. This is the terminal velocity that a particle reaches when the (retarding) drag force on the particle just equals the (downward) gravitational force. This quantity figures prominently in many coastal engineering problems. While simple in concept, its precise calculation is usually not. A particle's fall velocity is a function of its size, shape, and density; as well as the fluid density, and viscosity, and several other parameters.

a. General equation.

(1) For a single sphere falling in an infinite still fluid, the balance between the drag force and the gravitational force is:

$$C_D \; \frac{\pi \, D^2}{4} \; \frac{\rho \, W_f^2}{2} \; = \; \frac{\pi \, D^3}{6} \; (\rho_s \; - \; \rho) \; g \qquad\qquad \text{(III-1-6)}$$

or, solving for the velocity:

$$W_f \; = \; \left(\frac{4}{3} \; \frac{gD}{C_D} \left[\frac{\rho_s}{\rho} \; - \; 1 \right] \right)^{\frac{1}{2}} \qquad\qquad \text{(III-1-7)}$$

where

W_f = fall velocity

C_D = dimensionless drag coefficient

D = grain diameter

ρ = density of water

ρ_s = density of the sediment

(2) The units of the fall velocity will be the same as the units of $(gD)^{\frac{1}{2}}$. The problem now usually becomes one of determining the appropriate drag coefficient. Figure III-1-5, which is based upon extensive laboratory data of Rouse and many others, shows how the drag coefficient C_D varies as a function of the Reynolds number ($Re = W_f \, D/v$, where v is the kinematic viscosity) for spherical particles. Re is dimensionless, but W_f, D, and v must have common units of length and time.

(3) The plot in Figure III-1-5 can be divided into three regions. In the first region, Re is less than about 0.5, and the drag coefficient decreases linearly with Reynolds number. This is the region of small, light grains gently falling at slow velocities. The drag on the grain is dominated by viscous forces, rather than inertia forces, and the fluid flow past the particle is entirely laminar. The intermediate range is from about

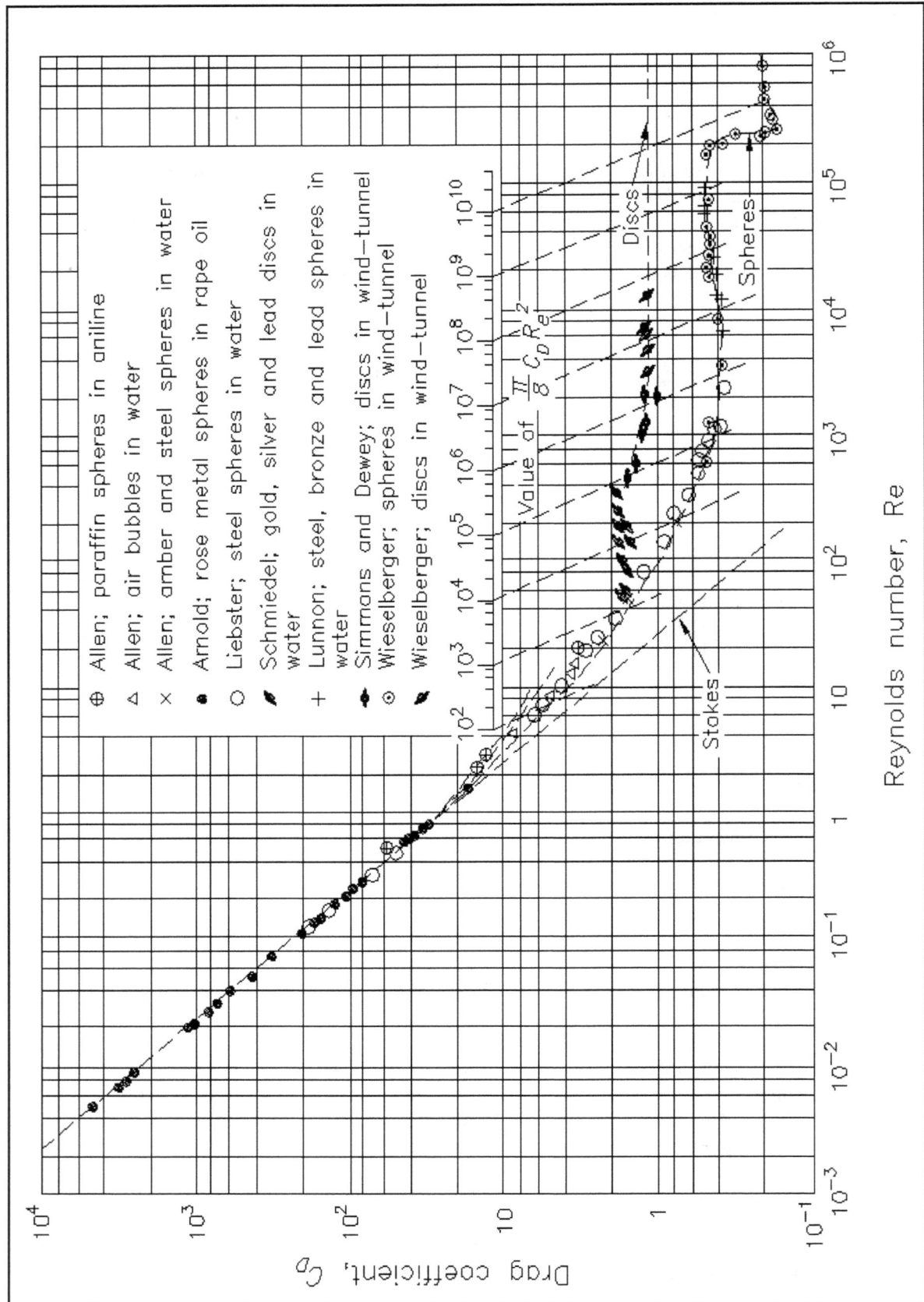

Figure III-1-5.   Drag coefficient as a function of Reynolds Number (Vanoni 1975)

$Re > 400$ to $Re < 200,000$. Here the drag coefficient has the approximately constant value of 0.4 to 0.6. In this range the particles are larger and denser, and the fall velocity is faster. The physical reason for this change in the behavior of C_D is that inertial drag forces have become predominant over the viscous forces, and the wake behind the particle has become turbulent. At about $Re = 200,000$ the drag coefficient decreases abruptly. This is the region of very large particles at high fall velocities. Here, not only is the wake turbulent, but the flow in the boundary layer around the particle is turbulent as well.

(4) In the first region, Stokes found the analytical solution for C_D as:

$$C_D = \frac{24}{Re} = \frac{24 \, \nu}{W_f \, D} \qquad \text{(III-1-8)}$$

(5) This line is labeled "Stokes" in Figure III-1-5. Substituting Equation 1-8 into Equation 1-7 gives the fall velocity in this region:

$$W_f = \frac{g \, D^2}{18 \, \nu} \left(\frac{\rho_s}{\rho} - 1 \right) \qquad \text{(III-1-9)}$$

(6) Note that in this region the velocity increases as the square of the grain diameter, and is dependent upon the kinematic viscosity.

(7) For the region of $400 < Re < 200,000$, the approximation $C_D \sim 0.5$ is used in Equation 1-7 to obtain:

$$W_f = 1.6 \left(g \, D \left[\frac{\rho_s}{\rho} - 1 \right] \right)^{\frac{1}{2}} \qquad \text{(III-1-10)}$$

(8) Here it is seen that the fall velocity varies as the square root of the grain diameter and is independent of the kinematic viscosity.

(9) Similarly, in the region $Re > 200,000$, the approximation $C_D \sim 0.2$ is used in Equation 1-7 to obtain:

$$W_f = 2.6 \left(g \, D \left[\frac{\rho_s}{\rho} - 1 \right] \right)^{\frac{1}{2}} \qquad \text{(III-1-11)}$$

(10) There is a large transition region between the first two regimes (between $0.5 < Re < 400$). For quartz spheres falling in water, these Reynolds numbers correspond to grain sizes between about 0.08 mm and 1.9 mm. Unfortunately, this closely corresponds to all sand particles, as seen in Table III-1-2. Thus, for very small particles (silts and clays), the fall velocity is proportional to D^2 and can be calculated from Equation 1-9. For gravel size particles the fall velocity is proportional to $D^{1/2}$ and can be calculated from Equation 1-10. However, for sand, the size of most interest to coastal engineers, no simple formula is available. The fall velocity is in a transition region between a D^2 dependence and a $D^{1/2}$ dependence. In this size range, it is easiest to obtain a fall velocity value from plots such as Figure III-1-6, which show the fall velocity as a function of grain diameter and water temperature for quartz spheres falling in both water and air. The vertical and horizontal axes are grain diameter and fall velocity, in millimeters and centimeters per second, respectively. The short straight lines crossing the curves obliquely are various values of Re.

(11) The transition between the second and third regime corresponds to approximately 90-mm quartz spheres (cobbles, as shown on Table III-1-2) falling in water.

(12) Generally for grain sizes outside the range shown in Figure III-1-6 or for spheres with a density other than quartz, or fluids other than air or water, Equation 1-7 can be used with a value of C_D obtained from Figure III-1-5. However, this is an iterative procedure involving repeated calculations of fall velocity and Reynolds number. Instead, Equation 1-7 can be rearranged to yield:

$$\frac{\pi}{8} C_D Re^2 = \frac{\pi D^3 \left(\frac{\rho_s}{\rho} - 1 \right) g}{6 \, \nu^2}$$

(III-1-12)

This quantity, $\pi/8 \, C_D Re^2$, can be used in Figure III-1-5 to obtain a value of C_D or Re, either of which can then be used to calculate the fall velocity.

b. Effect of density. Equation 1-7 shows that fall velocity depends on $((\rho_s/\rho)-1)^{1/2}$. For quartz sand grains in fresh water, this factor is about 1.28. For quartz grains in ocean water, this factor reduces to about 1.25 because of the slight increase in density of salt water. For quartz grains in very turbid (muddy) water, the factor will decrease somewhat more. If the suspended mud is present in a concentration of 10 percent by mass, $((\rho_s/\rho)-1)^{1/2}$ becomes 1.19. Thus, natural increases in water density encountered in coastal engineering will decrease the fall velocity of quartz by not much more than (1.28 - 1.19) / 1.28 = 0.07, or on the order of 7 percent.

c. Effect of temperature. Temperature has an effect on the density of water; however, this is very small compared to its effect on the coefficient of viscosity. Changes in the fluid viscosity affect the fall velocity for small particles but not for large. (Equation 1-8 contains a viscosity term while Equations 1-9 and 1-10 do not.) Figure III-1-6 has separate lines labeled with temperatures of $0°$, $10°$, $20°$, $30°$, and $40 °C$, and these lines are reversed for the two fluids. A grain will fall faster in warmer water, but slower in warmer air, compared to its fall velocities at lower temperatures. This difference is the direct result of how viscosity varies with temperature in the two fluids.

d. Effect of particle shape.

(1) Grain shape affects the fall velocity of large particles (those larger than Re about 10 or D about 0.125 mm in water) but has a negligible effect on small particles. For large grains of a given nominal diameter, the less spherical the particle, the slower the fall velocity. In Figure III-1-5 it is seen that at large Reynolds numbers, discs have higher C_D values, and thus lower fall velocities than spheres.

(2) Mehta, Lee, and Christensen (1980) investigated the fall velocity of natural shells (unbroken bivalve halves). For Re in the range of 1,000-5,000, they found that if the shells rocked but did not spin as they fell, C_D values ranged from about 0.7 to 1.0 depending upon a shape factor. The less spherical, the greater the drag. If the shells spun while falling, the drag coefficient increased to the range of 1.0 to 1.5. For large particles that are far from spherical, it is usually best to determine their fall velocity experimentally.

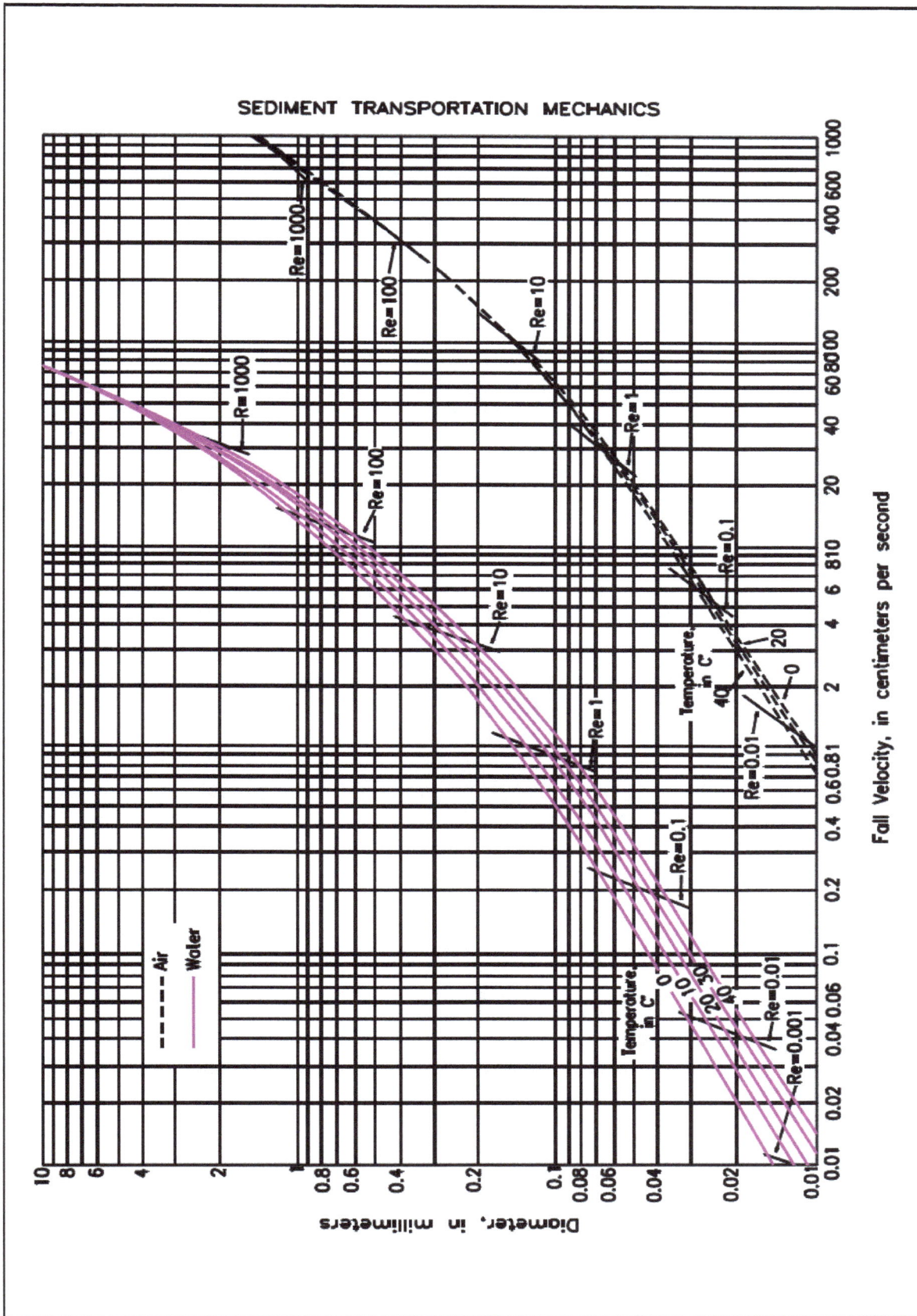

Figure III-1-6. Fall velocity of quartz spheres in air and water (Vanoni 1975)

e. Other factors.

(1) Several other factors affect the fall velocity. A tight clump of grains in an otherwise clear fluid will fall faster than a single grain because the adjacent fluid is partially entrained and thus the drag on each particle decreases. However, if the grains are uniformly distributed in the fluid, each will fall slower because, as each grain falls, replacement fluid must flow upward and this flow impedes the other grains. Likewise, an adjacent wall will decrease the fall velocity. These effects must be considered in the design and calibration of a settling tube used to measure the relationship between grain diameter and fall velocity. Theoretical calculations have shown that an increase in turbulence levels in the fluid should reduce the fall velocity for large particles. However, this has not been shown experimentally. These and other effects are hard to measure and parameterize. However, taken together, they are probably of minor importance and only serve to somewhat decrease the accuracy of fall velocity predictions.

(2) There are no ASTM standards for estimating sand size from fall velocity (see ASTM volume 4.08), and equipment to measure such fall velocity is not usually offered by the construction materials testing industry (Soiltest 1983). However, the fall velocity of grains is an essential parameter in coastal engineering research on sediment transport. Where large grain sizes are of importance and where shell material makes up a large percentage of the sample, fall velocity may be a preferable way to characterize the material as opposed to sieve size.

III-1-5. Bulk Properties

a. Porosity.

(1) Porosity, bulk density, and permeability are related bulk properties that arise from the fact that aggregations of sediments have void spaces around each grain. The *porosity P* is defined as the ratio of pore space, or voids, to the whole volume. It is related to the *volume concentration N*, which is the ratio of the solid volume to the whole volume; and to the *voids ratio e*, which is the ratio of pore space to solid space; by the equation:

$$e = \frac{P}{N}$$

(III-1-13a)

$$N = 1 - P = \frac{1}{1 + e}$$

(III-1-13b)

(2) Porosity is a function of how tightly the grains are packed together, and thus, is not a constant for a given sediment. As a grain settles to the bed, the greater the effects of gravity relative to the effects of the lateral fluid stresses over the bed, the lower the volume concentration. That is, the grains have less opportunity to roll around and find a position of maximum stability (= most tightly packed position). Thus, grains in the surf zone are typically compacted to near their maximum volume concentration while this is not the case in many quiet estuaries.

(3) In natural sands, volume concentration is essentially independent of grain size within the sand size range. However, the volume concentration is complicated by the irregular shape and nonuniform size of the grains. In general, an increase in nonuniformity of grain sizes increases the volume concentration (decreases the porosity) because small grains can fit into the pore spaces of the large grains. In engineering terms,

porosity decreases as the sand becomes better graded. An increase in the irregularity of the grain shapes tends to decrease the volume concentration. That is, an irregular grain of longest dimension D usually occupies almost as much space as a sphere of the same diameter D.

(4) For uniform spheres the loosest packing arrangement is cubic, yielding $P = 0.476$, while the densest is rhombohedral, yielding $P = 0.260$. Natural sands have porosities in the range $0.25 < P < .0.50$. Blatt, Middleton, and Murray (1980, Chapter 12) and Terzaghi and Peck (1967) showed that for laboratory studies of sands with gaussian size distributions, porosity decreased from 42.4 percent for extremely well-sorted sands to 27.9 percent for very poorly sorted sands, with no clay matrix. Chamberlain (1960) and Dill (1964) measured volume concentration of the fine sand on the beach and around the head of Scripps Canyon in Southern California. When compacted by vibration to its densest packing, the sand had a porosity of 0.40. In situ measurements from the beach yielded $P = 0.42$, while measurements from offshore near the canyon head yielded $P = 0.50$, and some micaceous (very irregular) sands from the canyon yielded $P = 0.62$.

(5) Good porosity data are not often available. The standard assumption in longshore transport computations is that sand has a porosity of 0.40, although there are likely to be significant variations from that figure, as discussed by Galvin (1979).

b. Bulk density. Density was defined in Part III-1-3-b as relating to a particle itself. Bulk density refers to a group of particles. *Dry bulk density* is the mass of an aggregation of grains divided by the volume of the grains (solids) plus the volume of the pore spaces. That is:

$$Dry \ Bulk \ Density \ = \ N \ \rho_s \tag{III-1-14}$$

Saturated bulk density is the mass of an aggregation of grains plus the mass of the interstitial water divided by the volume of the sample. That is:

$$Saturated \ Bulk \ Density \ = \ (N \ \rho_s) \ + \ (P \ \rho) \tag{III-1-15}$$

The dry bulk density is never greater than the grain density, and the saturated bulk density is only greater than the grain density if the interstitial fluid is more dense than the grains (if the grains float). Table III-1-6 lists typical bulk quantities for several sediments that are useful for coastal engineering computations. Table III-1-6 contains three parts: A – typical engineering data; B – saturated densities of naturally occurring surficial soils, along with porosity information; and C – dry densities of synthetic laboratory soils. Comparison of the two columns of data in Part C of Table III-1-6 gives an idea of the consolidation to be expected from settling, and a minimum estimate of the "bulking" of newly placed dry material.

c. Permeability. Permeability is the ability of water to flow through a sediment bed, and is largely a function of the size and shape of the pore spaces. Several aspects of this flow are of interest to coastal engineers. Flow into and out of the bed is one source of energy dissipation for waves traveling in shallow water (see Reid and Kajiura (1957), Packwood and Peregrine (1980)). Permeability is also a major factor in determining the steepness of the foreshore. Sediment is carried shoreward during the wave uprush in the swash zone. The permeability of the swash zone helps control how much of this water returns to the sea on the surface (above the bed) and how much returns through the bed. The surface backrush will transport sediment seaward, decreasing the equilibrium foreshore slope (Savage 1958; McLean and Kirk 1969; Packwood 1983). Just seaward of the breaker zone recent studies have suggested that even small amounts of wave-induced flow into and out of the bed have a major effect on the bottom boundary layer and the resulting sediment transport (Conley and Inman 1992).

EXAMPLE PROBLEM III-1-3

FIND:
Estimate W_f, Re, and C_D for the following particles falling through 20° C fluids.

a) 0.2-mm (=0.02-cm) quartz grain in fresh water
b) 0.2-mm quartz grain in salt water
c) 0.2-mm quartz grain in air
d) 0.001-mm-(=0.0001-cm-) diam kaolinite clay particle (unflocculated) in fresh water
e) 4.5-mm (=0.45-cm) steel BB pellet in air
f) 6-1/2-(=16.5-cm-) diam pine sphere in fresh water

GIVEN:

$$g = 980 \text{ cm/sec}^2$$
$$v_{water} = 0.011 \text{ cm}^2/\text{sec}$$
$$v_{air} = 0.15 \text{ cm}^2/\text{sec}$$
$$\rho_{fresh\ water} = 1.00 \text{ gm/cm}^3$$
$$\rho_{salt\ water} = 1.03 \text{ gm/cm}^3$$
$$\rho_{air} = 0.0012 \text{ gm/cm}^3$$
$$\rho_{quartz} = 2648 \text{ kg/m}^3 \text{ from Table III-1-4 } (=2.648 \text{ gm/cm}^3)$$
$$\rho_{kaolinite} = 2594 \text{ kg/m}^3 \text{ from Table III-1-4 } (=2.594 \text{ gm/cm}^3)$$
$$\rho_{steel} = 7800 \text{ kg/m}^3 \ (=7.8 \text{ gm/cm}^3)$$
$$\rho_{pine} = 480 \text{ kg/m}^3 \ (=0.48 \text{ gm/cm}^3)$$

SOLUTION:

For convenience, all calculations will be done in the cgs (centimeters, grams, seconds) system.

a) The fall velocity is read from Figure III-1-6 as 2.3 cm/sec. Re is calculated from its definition as:
$$Re = (2.3 * 0.02)/0.011 = 4.2$$

C_D can be found from Equation III-1-7 when rearranged as:
$$C_D = (4/3) * 980*0.02/(2.3)^2 * ((2.65/1) - 1) = 8.1$$

This agrees reasonably well with Figure III-1-5.

b) From section III-1-4-b, the fall velocity should be decreased from its value in part (a) by the factor $1.25/1.28 = 0.977$, thus:
$$W_f = 2.3 * 0.977 = 2.2 \text{ cm/sec}$$

$$Re = (2.2 * 0.02)/0.011 = 4.0$$

$$C_D = 4/3 * 980*0.02/(2.2)^2 * (2.65/1-1) = 8.9$$

(Continued)

Example Problem III-1-3 (Concluded)

c) The fall velocity is read from Figure III-1-6 as 150 cm/sec. Re is calculated from its definition as:

$$Re = (150 * 0.02)/0.15 = 20$$

C_D is calculated from 1-7 as:

$$C_D = 4/3 * 980*0.02/(150)^2 * ((2.65/0.0012)-1) = 2.5$$

This agrees reasonably well with Figure III-1-5.

d) Expecting to find Re < 0.5, Equation 1-9 can be used to find W_f,

$$W_f = 980/18 \ (0.0001)^2/0.011 \ (2.594/1-1) = 7.9 \times 10^{-5} \text{ cm/sec}$$

From its definition:

$$Re = 7.9 \times 10^{-5} * 0.0001 / 0.011 = 7.2 \times 10^{-7}$$

Thus, use of Equation 1-9 is well justified. From Equation 1-8:

$$C_D = 24/7.2 \times 10^{-7} = 3.3 \times 10^{7}$$

e) Estimating that Re will be in the range $400 < Re < 200,000$, Equation 1-10 can be used to find an estimate of W_f,

$$W_f = 1.6 \ (980 * 0.45 \ ((7.8/0.0012)-1))^{1/2} = 2,700 \text{ cm/sec}$$
$$Re = .45 * 2700 / 0.15 = 8100$$

Thus, use of Equation 1-10 is justified. From Figure III-1-5, it is seen that a C_D value of 0.4 is better than the value of 0.5 used in Equation III-1-10. Iterating using this value in Equation 1-7 yields:

$$W_f = (4/3 * 980 * 0.45/0.4 * (7.8/0.0012-1))^{1/2} = 3,100 \text{ cm/sec}$$
$$Re = .45 * 3,100 / 0.15 = 9,300$$

Alternatively, Equation III-1-12 could have been used to obtain:

$$\pi/8 \ C_D \ Re^2 = 3.14*(0.45)^3*(7.8/0.0012-1)*980/(6 * (0.15)^2) = 1.3 \times 10^{7}$$

Using Figure III-1-5, this yields similar values of 12,000 and 0.4 for Re and C_D, respectively.

f) It is important to realize that this particle is lighter than water and its direction of motion will be upward. This is implied in the result by the fall velocity having a negative value. Estimating that Re will be in the range $Re > 200,000$, Equation 1-11 can be used to find an estimate of W_f,

$$W_f = 2.6 \ (980 * 16.5 \ ((0.48/1)-1))^{1/2} = -240 \text{ cm/sec}$$
$$Re = 240 * 15.2 / 0.011 = 360,000$$

Thus, use of Equation 1-11 is justified. Alternatively, Equation 1-12 could have been used to obtain:

$$\pi/8 \ C_D \ Re^2 = 3.14*(16.5)^3*(0.48/1-1)*980/(6 * (0.011)^2) = (-)9.9 \times 10^{9}$$

Using Figure III-1-5, this yields similar values of 4×10^5 and 0.2 for Re and C_D, respectively.

Table III-1-6
Soil Densities Useful for Coastal Engineering Computations

A. Typical Engineering Values (from Terzaghi and Peck (1967), Table 6.3)

Material	Dry Bulk Density, kg/m³	Saturated Bulk Density, kg/m³
Uniform Sand		
Loose	1,430	1,890
Dense	1,750	2,090
Mixed Sand		
Loose	1,590	1,990
Dense	1,860	2,160
Clay		
Stiff glacial		2,070
Soft, very organic		1,430

B. Natural Surface Soils (from Daly, Manger, and Clark (1966), Table 4-4)

Material	No. of Samples	Mean Porosity, %	Saturated Bulk Density, kg/m³	Locality
Sand	12	38.9	1930	Cape May sand spits
Loess	3	61.2	1610	Idaho
Fine Sand	54	46.2	1930	CA seafloor
Very Fine Sand	15	47.7	1920	CA seafloor
Sand-Silt-Clay	3	74.7	1440	CA seafloor

C. Laboratory Soils (from Johnson and Olhoeft (1984), Table 4)

Material	Dry Bulk Density "fluffed" kg/m³	Dry Bulk Density "tapped", kg/m³
Gravelly Soil	1,660	1,770
Sandy Soil	1,440	1,560
Dune Sand	1,610	1,760
Loess	990	1,090
Peat	270	320
Muck	800	850

Note: Data for loess in Parts B and C are representative of silty material.

d. Angle of repose.

(1) When a dry sand-size sediment is poured onto a flat surface, it will form a cone-shaped mound. The slope of the surface of the cone (or dune) at the moment of avalanching is called the angle of repose. The angle of repose concept is of interest to coastal engineers for several reasons, including the stability of rubble-mound breakwaters and the modeling of sediment transport.

(2) The angle of repose is a function of grain shape; it increases with increasing grain irregularity. This is easy to see in the extreme case: a pile of jacks (or interlocking breakwater armor units) will stack at a much steeper angle than a pile of ball bearings. Several workers, including Lane (1955), Simons and Albertson (1960), Allen (1970), Cornforth (1973), Statham (1974), and Simons and Senturk (1977) have looked at the variation of the angle of repose in relationship to various sediment characteristics.

(3) Bagnold used the relationship between the shear forces, the normal forces, and the angle of repose to develop his energetics-based sediment transport model. He reasoned that the amount of sediment by weight at the top of a bed that can be supported in an elevated state and then be transported is related to the applied shear stress of the overlying (moving) fluid by the tangent of the angle of repose. Bagnold's concept forms the basis of many of the sediment transport models most frequently used by coastal engineers. For further discussion, see Bagnold (1963), (1966), Bailard and Inman (1979), and Bailard (1981).

e. Bulk properties of different sediments.

(1) Clays, silts, and muds.

(a) Coastal engineers typically encounter clay as a foundation material or as a material to be dredged. The flat topography of coastal plains and the quiet waters of bays and lagoons are often underlain by clay. Some older clays are consolidated and can stand with near-vertical slopes when eroded. For example, the deepest parts of tidal inlets may have steep sides cut into stiff clay. Many eroding coastal flats contain much clay. Revetments laid on these clays, or sheet piles driven into them, need particular attention; such design may require the advice of a geotechnical engineer.

(b) A silt-sized particle is intermediate between sand and clay. Most silt is produced by the gradual chemical weathering of rocks, but some silt is rock flour ground out by glaciers. Sediments consisting mostly of silt are common in deltas, estuaries, and glacial lakes, but are relatively uncommon on beaches where waves drive the dominant processes. Silt remains in suspension far longer than sand grains, so it is easily removed from any shore where wave action is moderate or severe. A mass of pure silt differs from a mass of clay in that, when dry, the silt has very little cohesion and will easily fall apart, whereas clay will cohere like a brick.

(c) Muds are watery mixtures of clay and silt, typically in approximately equal proportions, often with minor amounts of sand and organic material. Muds act more as a viscous fluid than as a cohesive solid. They have coastal engineering importance because muds often accumulate in dredged channels where the upper and lower boundaries of the mud layer can be difficult to determine. Maintenance dredging of channels that accumulate mud requires particular attention, especially a clear definition of the material to be dredged.

(d) Some deposits of mud lie offshore of the coast, where the mud is believed to modify the transmission of water waves. See, for example, Jiang and Zhao (1989); Shen, Isobe, and Watanabe (1994); and deWit, Kranenburg, and Batljes (1994) for introduction to recent work. Such mud deposits have been reported at the mouth of most large Asian rivers, off the Gulf Coast of Louisiana, northeast of the mouth of the Amazon River, and seaward of the Kerala district in India. These mud deposits are believed to oscillate with the passage of water waves above them, absorbing energy from the waves and reducing their height.

(2) Organically bound sediment.

(a) Marsh grasses growing in back bays and other tidal wetlands bind sands, silts, and clays with their root systems to form an organically bound sediment, sometimes called *peat* (typically, organic silt in the engineer's soil classification). These sediments are very compressible; when overlain by a barrier island, field evidence indicates that the compression of organic matter by the weight of the sand results in subsidence of the barrier island. Cores illustrating the vertical sequence of organic sediment overlain by barrier island sands have been shown in many geological studies (e.g., Figures 9 and 15 of Kraft et al. (1979)). Shore erosion can expose the organically bound sediment to ocean waves. Erosion by ocean waves produces pillow- or cobble-shaped fragments of organically bound sediment, often found on barrier island coasts after storms.

EXAMPLE PROBLEM III-1-4

FIND:

The volume concentration, the porosity, the voids ratio, the dry bulk density, and the saturated bulk density of the densely packed aggregate given in example problem III-1-2.

GIVEN:

18.1 grams of dry beach sand exactly fills a small container having a volume of 10.0 cm^3 after the filled container is strongly vibrated. When this amount of sand is poured into 50.0 ml of water that is subjected to a strong vacuum, the volume of the sand-water mixture is 56.8 ml.

SOLUTION:

It is important to recognize the difference in the volume of the grains themselves, 6.8 cm^3 (= 56.8 - 50) (a milliliter is a cubic centimeter), and the volume of the aggregate (the grains plus the void spaces), 10 cm^3.

The volume concentration is the ratio of solid volume to total volume:
$$N = 6.8 \text{ cm}^3 / 10.0 \text{ cm}^3 = 0.68$$

The porosity is the ratio of void space to total volume, which, from Equation 1-13 is:
$$P = 1 - N = 1 - 0.68 = 0.32$$

The voids ratio is the ratio of pore space to solid space, which, from Equation 1-13 is:
$$e = P/N = 0.32/0.68 = 0.47$$

The dry bulk density from Equation 1-14 is:

$$\text{Dry Bulk Density} = 0.68 * 2{,}660 \text{ kg/m}^3 = 1{,}800 \text{ kg/m}^3$$

The saturated bulk density from Equation 1-15 is:

$$\text{Saturated Bulk Density} = (0.68*2{,}660) +(0.32*1{,}000) \text{ kg/m}^3 = 2{,}100 \text{ kg/m}^3$$
These bulk density values are within the expected range as shown in Table III-1-6A.

(b) Where they are present at coastal engineering projects, such organic sediments must be considered. Pile foundations bearing significant loads must penetrate the organic material. It may be required to remove the compressible sediment before construction.

(3) Sand and gravel.

(a) Ocean beaches of the world consist of sand and some gravel. The typical temperate zone ocean beach is made of quartz sand whose median diameter D_{50} is in the range $0.15 \leq D_{50} \leq 0.40$ mm. In colder latitudes where geologically recent glacial action has affected sediment supply and the wave climate is more severe, the composition of the beach material becomes more varied and size tends to increase. See, for example, Figure 89 of Davies (1980). In these glacier-affected latitudes, sand grains include silicate and oxide compositions (heavy minerals) as noticeable components along with the quartz, and coarse pebbles (gravel or "shingle") may locally dominate the sediment.

(b) In tropical latitudes, quartz beach sand may give way to calcium carbonate sand produced from shelly organisms, algae, or coral. See, for example, Figure 92 of Davies (1980) derived from earlier work of Hayes. Carbonate sands differ from quartz sand in two important ways: carbonates are chemically more active than quartz, making carbonate grains subject to local cementation or sometimes dissolution, and individual carbonate grains are structurally weaker than single quartz grains so that they may be crushed by traffic and often have sharper edges.

(c) Oolites are a particular variety of carbonate sands locally abundant in the Bahamas and other semi-tropical lands. Oolites are sands whose grains are ellipsoidal in shape, consisting of thin smooth concretionary layers of calcium carbonate. There is a very large literature on this subject, mainly by geologists; section 13.3 of Blatt, Middleton, and Murray (1980) is a good starting point.

(d) Beaches in bays and estuaries often differ from ocean beaches at the same latitudes. Often the beach material along interior shores is coarser and much more limited in volume than along ocean shores, possibly because limited wave action on interior waters has not eroded enough material to produce extensive sand beaches.

(4) Cobbles, boulders, and bedrock. Glaciated lands, tectonically active areas, and hilly or mountainous shorelines often have shores with abundant cobbles, boulders, and bedrock. Such shores result from enhanced supplies of rock produced by glaciers, by stream erosion of mountainous terrain, or by enhanced wave erosion at the downwind end of long fetches.

III-1-6. References

Allen 1970
Allen, J. R. L. 1970. "The Avalanching of Granular Solids on Dune and Similar Slopes," *Journal of Geology*, Vol 78, pp 326-351.

American Society for Testing Materials 1994
American Society for Testing Materials. 1994. Volume 04.08, Soil and Rock (1): D420 – D4914, American Society for Testing and Materials, Philadelphia, PA.

Bagnold 1963
Bagnold, R. A. 1963. "Beach and Nearshore Processes; Part I: Mechanics of Marine Sedimentation," *The Sea: Ideas and Observations*, M. N. Hill, ed., Interscience, New York, Vol 3, pp 507-528.

Bagnold 1966
Bagnold, R. A. 1966. "An Approach to the Sediment Transport Problem from General Physics," U. S. Geological Survey Professional Paper 422-I, U. S. Department of the Interior.

Bailard 1981
Bailard, J. A. 1981. "An Energetics Total Load Sediment Transport Model for a Plane Sloping Beach," *Journal of Geophysical Research*, Vol 86, No. C11, pp 10938-10954.

Bailard and Inman 1979
Bailard, J. A., and Inman, D. L. 1979. "A Reexamination of Bagnold's Granular Fluid Model and Bed Load Transport," *Journal of Geophysical Research*, Vol 84, No. C12, pp 7827-7833.

Blatt, Middleton, and Murray 1980
Blatt, H., Middleton, G., and Murray, R. 1980. *Origin of Sedimentary Rocks*, 2nd ed., Prentice–Hall, Englewood Cliffs, NJ.

Chamberlain 1960
Chamberlain, T. K. 1960. "Mechanics of Mass Sediment Transport in Scripps Submarine Canyon," Ph.D. diss., Scripps Institution of Oceanography, Univ. California, San Diego.

Conley and Inman 1992
Conley, D. C., and Inman, D. L. 1992. "Field Observations of the Fluid-Granular Boundary Layer Under Near-Breaking Waves," *Journal of Geophysical Research*, Vol 97, No. C6, pp 9631-9643.

Cornforth 1973
Cornforth, D. H. 1973. "Prediction of Drained Strength of Sands from Relative Density Measurements," American Society for Testing and Materials, Spec. Tech. Pub. 523, pp 281-303.

Daly, Manger, and Clark 1966
Daly, R. A., Manger, G. E., and Clark, S. P., Jr. 1966. "Density of Rocks," *Handbook of Physical Constants*, S. P. Clark, Jr., ed., revised edition, pp 19-26.

Davies 1980
Davies, J. L. 1980. *Geographical Variation in Coastal Development*, 2nd ed., Longman, London.

deWit, Kranenburg, and Batljes 1994
deWit, P. J., Kranenburg, C., and Batljes, J. A. 1994. "Liquefaction and erosion of mud due to waves and current," *International Conference on Coastal Engineering*, Book of Abstracts, Paper 115, Kobe, pp 278-9.

Dill 1964
Dill, R.F. 1964. "Sedimentation and Erosion in Scripps Submarine Canyon Head," *Papers in Marine Geology*, Shepard Commemorative Volume, R.I. Miller, ed., Macmillian & Co., NY, pp 23-41.

Downing, Sternberg, and Lister 1981
Downing, J. P., Sternberg, R. W., and Lister, C. R. B. 1981. "New Instrumentation for the Investigation of Sediment Suspension Processes in the Shallow Marine Environment," *Marine Geology*, Vol 42, pp 19-34.

Ehrlich and Weinberg 1970
Ehrlich, R., and Weinberg, B. 1970. "An Exact Method for Characterization of Grain Shape," *Journal Sed. Pet.*, Vol 40, pp 205-212.

Folk 1965
Folk, R. L. 1965. *Petrology of Sedimentary Rocks*, Hemphill Publishing Company, Austin, TX.

Folk 1966
Folk, R. L. 1966. "A Review of Grain-size Parameters," *Sedimentology*, Vol 6, pp 73-93.

Folk 1974
Folk, R. L. 1974. *Petrology of Sedimentary Rock,* Hemphill Publishing Company, Austin, TX.

Folk and Ward 1957
Folk, R. L. and Ward, W. C. 1957. "Brazos River Bar. A Study in the Significances of Grain Size Parameters," *Journal of Sedimentary Petrology*, Vol 27, pp 3-26.

Frisch, Evans, Hudson, and Boon 1987
Frisch, A. A., Evans, D. A., Hudson, J. P., and Boon, J. 1987. "Shape Discrimination of Sand Samples using the Fractal Dimension," *Proc. Coastal Sediments '87*, American Society of Civil Engineers, New Orleans, Louisiana, May 12-14, pp 138-153.

Galvin 1979
Galvin, C. 1979. "Relation Between Immersed Weight and Volume Rates of Longshore Transport," Technical Paper 79-1, Coastal Engineering Research Center, U.S. Army Engineer Waterways Experiment Station, Vicksburg, MS.

Galvin 1987
Galvin, C. 1987. "Vertical Profile of Littoral Sand Tracers from a Distribution of Waiting Times," *Coastal Sediment '87*, Vol 1, pp 436–451.

Galvin and Alexander 1981
Galvin, C., and Alexander, D. F. 1981. "Armor Unit Abrasion and Dolos Breakage by Wave-Induced Stress Concentrations," American Society of Civil Engineers Preprint 81-172.

Giles and Pilkey 1965
Giles, G. T. and Pilkey, O. H. 1965. "Atlantic Beach and Dune Sediments of the Southern United States," *Journal of Sedimentary Petrology*, Vol 35, No. 4, pp 900-910.

Griffiths 1967
Griffiths, J. C. 1967. *Scientific Method in Analysis of Sediments*, McGraw-Hill, New York.

Handin 1966
Handin, J. 1966. "Strength and Ductility," *Handbook of Physical Constants*, S. P. Clark, Jr., ed., Geological Society of America, Memoir 97, pp 223–289.

Hobson 1977
Hobson, R. D. 1977. "Review of Design Elements for Beach-Fill Evaluation," TP 77-6, Coastal Engineering Research Center, U.S. Army Engineer Waterways Experiment Station, Vicksburg, MS.

Huston 1970
Huston, J. 1970. *Hydraulic Dredging.* Cornell Maritime Press, Cambridge, MD.

Inman 1952
Inman, D. L. 1952. "Measures for Describing the Size Distribution," pp 125-145.

Inman 1957
Inman, D. L. 1957. "Wave Generated Ripples in Nearshore Sands," Technical Memorandum 100, Beach Erosion Board, U.S. Army Corps of Engineers.

Jiang and Zhao 1989
Jiang, L., and Zhao, Z. 1989. "Viscous Damping of Solitary Waves over Fluid-Mud Seabeds," *Journal Waterway, Port, Coastal, and Ocean Engineering*, American Society of Civil Engineers, Vol 115, No. 3, pp 345-362.

Johnson and Olhoeft 1984
Johnson, G. R., and Olhoeft, G. R. 1984. "Density of Rocks and Minerals," *Handbook of Physical Properties of Rocks*, Vol III, R. S. Carmichael, ed., CRC Press, Inc., pp 1-28.

Judge 1970
Judge, C. W. 1970. "Heavy Minerals in Beach and Stream Sediments as Indicators of Shore Processes between Monterey and Los Angeles, California," TM-33, Coastal Engineering Research Center, U.S. Army Engineer Waterways Experiment Station, Vicksburg, MS.

Kraft, Allen, Belknap, John, and Maurmeyer 1979
Kraft, J. C., Allen, E. A., Belknap, D. F., John, D. J., and Maurmeyer, E. M. 1979. "Processes and Morphologic Evolution of an Estuarine and Coastal Barrier System," *Barrier Islands*, S. P. Leatherman, ed., Academic Press, New York, pp 149-183.

Krinsley and Doornkamp 1973
Krinsley, D. H., and Doornkamp, J. C. 1973. "Atlas of Quartz Sand Surface Textures," Cambridge University Press, Cambridge.

Krumbein 1936
Krumbein, W. C. 1936. "Application of Logarithmic Moments to Size Frequency Distribution of Sediments," *Journal of Sedimentary Petrology*, Vol 6, No. 1, pp 35-47.

Krumbein 1941
Krumbein, W. C. 1941. "Measurement and Geological Significance of Shape and Roundness of Sedimentary Particles," *Journal of Sedimentary Petrology*, Vol 11, No. 2, pp 64-72.

Krumbein and Monk 1942
Krumbein, W. C., and Monk, G. D. "Permeability as a Function of the Size Parameters of Unconsolidated Sand, American Institute Mining and Metallurgy Engineering," Technical Publication No. 1492, Petroleum Technology, pp1-11.

Krumbein and Sloss 1963
Krumbein, W. C., and Sloss, L. L. 1963. "Stratigraphy and Sedimentation," Ch. 4, *Properties of Sedimentary Rocks*, W. H. Freeman & Company, pp 93-149.

Lambe and Whitman 1969
Lambe, T. W., and Whitman, R. V. 1969. *Soil Mechanics*, John Wiley, New York.

Lane 1955
Lane, E. W. 1955. "Design of Stable Channels," *Transactions, American Society of Civil Engineers,* Vol 120, pp 1234-1279.

Margolis 1969
Margolis, S. V. 1969. "Electron Microscopy of Chemical Solution and Mechanical Abrasion Features on Quartz Sand Grains," *Sedimentary Geology*, Vol 2, pp 243-256.

Mason 1942
Mason, M. A. 1942. "Abrasion of Beach Sands," TM-2, U.S. Army Corps of Engineers, Beach Erosion Board, Washington, DC.

McCammon 1962
McCammon, R. B. 1962. "Efficiencies of Percentile Measures for Describing the Mean Size and Sorting of Sedimentary Particles," *Journal of Geology*, Vol 70, pp 453-465.

McLean and Kirk 1969
McLean, R. F., and Kirk, R. M. 1969. "Relationship Between Grain Size, Size-sorting, and Foreshore Slope on Mixed Sandy-Shingle Beaches," *N. Z. J. Geol. and Geophys.*, Vol 12, pp 128-155.

McMaster 1954
McMaster, R. L. 1954. "Petrography and Genesis of the N.J. Beach Sands," *Geology Series*, Vol 63, New Jersey Department of Conservation & Economic Development.

Mehta, Lee, and Christensen 1980
Mehta, A. J., Lee, J., and Christensen, B. A. 1980. "Fall Velocity of Shells as Coastal Sediment," *Journal of the Hydraulics Division, American Society of Civil Engineers,* Vol 106, No. HY 11, pp 1727-1744.

Otto 1939
Otto, G. H. 1939. "A Modified Logarithmic Probability Graph for the Interpretation of Mechanical Analyses of Sediments," *Journal of Sedimentary Petrology*, Vol 9, pp 62-76.

Packwood 1983
Packwood, A. R. 1983. "The Influence of Beach Porosity on Wave Uprush and Backwash," *Coastal Engineering*, Vol 7, pp 29-40.

Packwood and Peregrine 1980
Packwood, A. R., and Peregrine, D. H. 1980. "The Propagation of Solitary Waves and Bores over a Porous Bed," *Coastal Engineering*, Vol 3, pp 221-242.

Pettijohn 1957
Pettijohn, F. J. 1957. *Sedimentary Rocks*, Harper and Brothers, New York, p 117.

Powers 1953
Powers, M. C. 1953. "A New Roundness Scale for Sedimentary Particles," *Journal Sed. Pet.*, Vol 23, pp 117-119.

Ramsey and Galvin 1977
Ramsey, M. D., and Galvin, C. 1977. "Size Analyses of Sand Samples from Southern New Jersey Beaches," Miscellaneous Paper 77-3, Coastal Engineering Research Center, U.S. Army Engineer Waterways Experiment Station, Vicksburg, MS.

Reid and Kajiura 1957
Reid, R. O., and Kajiura, K. 1957. "On the Damping of Gravity Waves Over a Permeable Sea Bed," *Trans. Amer. Geophys. Union*, Vol 38, No. 5 (October), pp 662-666.

Ritter 1986
Ritter, D. F. 1986. *Process Geomorphology*, 2nd ed., Wm. C. Brown, Dubuque, IA.

Savage 1958
Savage, R. P. 1958. "Wave Run-Up on Roughened and Permeable Slopes," *Journal of the Waterways and Harbors Division, American Society of Civil Engineers,* Vol 84, No. WW3, Paper 1640.

Shen, Isobe, and Watanabe 1994
Shen, D., Isobe, M., and Watanabe, A. 1994. "Mud Transport and Muddy Bottom Deformation Due to Waves," Book of Abstracts, Paper 30, *International Conference on Coastal Engineering*, Kobe, pp 106-107.

Shepard and Young 1961
Shepard, F. P., and Young, R. 1961. "Distinguishing Between Beach and Dune Sands," *Journal Sed. Pet.*, Vol 31, No. 2, pp 196-214.

Simons and Albertson 1960
Simons, D. B., and Albertson, M. L. 1960. "Uniform Water Conveyance Channels in Alluvial Material," *Journal Hydraulic Division, American Society of Civil Engineers,* Vol 86, No. HY5, pp 33-99.

Simons and Senturk 1977
Simons, D. B., and Senturk, F. 1977. *Sediment Transport Technology,* Water Res. Pub., Fort Collins, CO.

Soiltest 1983
Soiltest. 1983. *Soiltest Catalog,* Evanston, IL.

Statham 1974
Statham, I. 1974. *Ground Water Hydrology*, John Wiley & Sons, New York.

Stauble and Hoel 1986
Stauble, D. K., and Hoel, J. 1986. "Guidelines for Beach Restoration Projects, Part II -Engineering," Report SGR-77, Florida Sea Grant, University of Florida, Gainesville.

Terzaghi and Peck 1967
Terzaghi, K., and Peck, R. B. 1967. *Soil Mechanics in Engineering Practice*, 2nd ed., John Wiley, New York.

Trask 1952
Trask, P. D. 1952. "Source of Beach Sand at Santa Barbara, California, as Indicated by Mineral Grain Studies," Tech. Memo. No. 28, U. S. Army Corps of Engineers, Beach Erosion Board.

Turner 1984
Turner, T. M. 1984. *Fundamentals of Hydraulic Dredging*, Cornell Maritime Press, Centreville, MD.

Tyler 1991
Tyler, W. S. 1991. *Testing Sieves and Their Uses*, Handbook 53, LP9-91.

U. S. Army Corps of Engineers 1981
U. S. Army Corps of Engineers. 1981. *Low-cost Shore Protection*, Final Rept on Shoreline Erosion Control Demonstration Program (Section 54).

Vanoni 1975
Vanoni, V. A. 1975. *Sedimentation Engineering*, American Society of Civil Engineers, Manual No. 54.

III-1-7. Definition of Symbols

α_φ	Phi coefficient of skewness (Equation III-1-4)
β_φ	Phi coefficient of kurtosis (Equation III-1-5)
ν	Kinematic viscosity [length²/time]
ρ	Mass density of water (salt water = 1,025 kg/m³ or 2.0 slugs/ft³; fresh water = 1,000kg/m³ or 1.94 slugs/ft³) [force-time²/length⁴]
ρ_s	Mass density of sediment grains [force-time²/length⁴]
σ_φ	Standard deviation of a sediment sample (Equation III-1-3) [phi units]
φ	Sediment grain diameter in phi units ($\varphi = -\log_2 D$, where D is the sediment grain diameter in millimeters)
C_D	Drag coefficient [dimensionless]
D	Sediment grain diameter [length - generally millimeters]
D_x	Sediment grain diameter for which x-percent of the sediment, by weight, has a smaller diameter
e	Voids ratio of a sediment sample - ratio of pore space to solid space (Equation III-1-13a)
g	Gravitational acceleration (32.17 ft/sec², 9.807m/sec²) [length/time²]
M_d	Median grain diameter - half the particles in the sample will have a larger diameter and half will have a smaller diameter [length]
M_φ	Estimated mean grain size of a sediment sample (Equation (III-1-2) [phi units]
N	Volume concentration of a sediment sample - ratio of the solid volume to the whole volume (Equation III-1-13b)
P	Porosity - ratio of pore space, or voids, to the whole volume
Re	Reynolds number [dimensionless]
W_f	Sediment fall velocity [length/time]

III-1-8. Acknowledgments

Authors of Chapter III-1, "Coastal Sediment Properties:"

David B. King, Ph.D., Coastal and Hydraulics Laboratory (CHL), Engineer Research and Development
    Center, Vicksburg, Mississippi.
Cyril J. Galvin, Coastal Engineer, Springfield, Virginia.

Reviewers:

Andrew Morang, Ph.D., CHL
Joan Pope, CHL
Todd L. Walton, Ph.D., CHL

Table of Contents

Page

List of Tables

List of Figures

Chapter III-2
Longshore Sediment Transport

III-2-1. Introduction

a. Overview.

(1) The breaking waves and surf in the nearshore combine with various horizontal and vertical patterns of nearshore currents to transport beach sediments. Sometimes this transport results only in a local rearrangement of sand into bars and troughs, or into a series of rhythmic embayments cut into the beach. At other times there are extensive longshore displacements of sediments, possibly moving hundreds of thousands of cubic meters of sand along the coast each year. The objective of this chapter is to examine techniques that have been developed to evaluate the longshore sediment transport rate, which is defined to occur primarily within the surf zone, directed parallel to the coast. This transport is among the most important nearshore processes that control the beach morphology, and determines in large part whether shores erode, accrete, or remain stable. An understanding of longshore sediment transport is essential to sound coastal engineering design practice.

(2) Currents associated with nearshore cell circulation generally act to produce only a local rearrangement of beach sediments. The rip currents of the circulation can be important in the cross-shore transport of sand, but there is minimal net displacement of beach sediments along the coast. More important to the longshore movement of sediments are waves breaking obliquely to the coast and the longshore currents they generate, which may flow along an extended length of beach (Part II-4). The resulting movement of beach sediment along the coast is referred to as littoral transport or longshore sediment transport, whereas the actual volumes of sand involved in the transport are termed the littoral drift. This longshore movement of beach sediments is of particular importance in that the transport can either be interrupted by the construction of jetties and breakwaters (structures which block all or a portion of the longshore sediment transport), or can be captured by inlets and submarine canyons. In the case of a jetty, the result is a buildup of the beach along the updrift side of the structure and an erosion of the beach downdrift of the structure. The impacts pose problems to the adjacent beach communities, as well as threaten the usefulness of the adjacent navigable waterways (channels, harbors, etc.) (Figure III-2-1).

(3) Littoral transport can also result from the currents generated by alongshore gradients in breaking wave height, commonly called diffraction currents (Part II-4). This transport is manifest as a movement of beach sediments toward the structures which create these diffraction currents (such as jetties, long groins, and headlands). The result is transport in the "upwave" direction on the downdrift side of the structure. This, in turn, can create a buildup of sediment on the immediate, downdrift side of the structure or contribute to the creation of a crenulate-shaped shoreline on the downdrift side of a headland.

b. Scope of chapter. This chapter defines terms associated with the longshore transport of littoral material, presents relationships for the longshore sediment transport rate as a function of breaking waves and longshore currents, discusses the dependence of longshore transport relationships on sediment grain size, presents a method for calculating the cross-shore distribution of longshore sand transport, and overviews analytical and numerical models for shoreline changes which include longshore sediment transport relationships.

III-2-2. Longshore Sediment Transport Processes

a. Definitions. On most coasts, waves reach the beach from different quadrants, producing day-to-day and seasonal reversals in transport direction. At a particular beach site, transport may be to the right (looking

Figure III-2-1.   Impoundment of longshore sediment transport at Indian River Inlet, Delaware

seaward) during part of the year and to the left during the remainder of the year. If the left and right transports are denoted respectively $Q_{\ell L}$ and $Q_{\ell R}$, with $Q_{\ell R}$ being assigned a positive quantity and $Q_{\ell L}$ assigned a negative value for transport direction clarification purposes, then the net annual transport is defined as $Q_{\ell NET}$ $= Q_{\ell R} + Q_{\ell L}$. The net longshore sediment transport rate is therefore directed right and positive if $Q_{\ell R} > Q_{\ell L}$, and to the left and negative if $Q_{\ell R} < /Q_{\ell L}/$. The net annual transport can range from essentially zero to a large magnitude, estimated at a million cubic meters of sand per year for some coastal sites. The gross annual longshore transport is defined as $Q_{\ell GROSS} = Q_{\ell R} + /Q_{\ell L}/$, the sum of the temporal magnitudes of littoral transport irrespective of direction. It is possible to have a very large gross longshore transport at a beach site while the net transport is effectively zero. These two contrasting assessments of longshore sediment movements have different engineering applications. For example, the gross longshore transport may be utilized in predicting shoaling rates in navigation channels and uncontrolled inlets, whereas the net longshore transport more often relates to the deposition versus erosion rates of beaches on opposite sides of jetties or breakwaters. (It is noted, however, that the latter may capture the gross transport rate in some cases.)

b. Modes of sediment transport.

(1) A distinction is made between two modes of sediment transport: suspended sediment transport, in which sediment is carried above the bottom by the turbulent eddies of the water, and bed-load sediment transport, in which the grains remain close to the bed and move by rolling and saltating. Although this distinction may be made conceptually, it is difficult to separately measure these two modes of transport on prototype beaches. Considerable uncertainty remains and differences of opinion exist on their relative contributions to the total transport rate.

(2) Because it is more readily measured than the bed-load transport, suspended load transport has been the subject of considerable study. It has been demonstrated that suspension concentrations decrease with height above the bottom (Kraus, Gingerich, and Rosati 1988, 1989). The highest concentrations typically are found in the breaker and swash zones, with lower concentrations at midsurf positions. On reflective beaches, at which a portion of the wave energy is reflected back to sea, individual suspension events are correlated with the incident breaking wave period. In contrast, on dissipative beaches, at which effectively all of the arriving wave energy is dissipated in the nearshore, long-period water motions have been found to account for significant sediment suspension. For dissipative beaches, the suspension concentrations due to long-period (low-frequency) waves have been measured as 3 to 4 times larger than those associated with the short-period high-frequency incident waves (Beach and Sternberg 1987).

c. Field identification of longshore sediment transport. Emphasis is placed herein upon field, or proto-type, measurement and identification of littoral transport. Laboratory measurement of longshore sediment transport is generally thought to underestimate prototype transport rates, primarily because of scale effects. Laboratory measurements are also complicated by the need to establish a continuous updrift sediment supply in the model.

(1) Experimental measurement.

(a) Longshore sediment transport relationships are typically based on data measured by surveying impoundments of littoral drift at a jetty, breakwater, spit, or deposition basin; bypassing impounded material (e.g., at an inlet); or measuring short-term sand tracer transport rates. Other techniques focus upon measurement of only the suspended load transport. Longshore sediment transport estimates using impoundment are believed to come closest to yielding total quantities (i.e., the bed load plus suspended load transport), and typically represent longer-term measures (i.e., weeks to years). It is these longer-term, total transport quantities that are of central importance to practical coastal engineering design. Impoundment techniques are discussed below in Part III-2-2.c.(2).

(b) Measurement of sand tracer transport rates involves tagging the natural beach sediment with a coating of fluorescent dye or low-level radioactivity. Tracers are injected into the surf zone, and the beach material is sampled on a grid to determine the subsequent tracer distribution. The longshore displacement of the center of mass of the tracer on the beach between injection and sampling provides a measure of the mean transport distance. The sand advection velocity is obtained by dividing this distance by the elapsed time. The time between tracer injection and sampling is usually an hour to a few hours, so the measurement is basically the instantaneous longshore sand transport under a fixed set of wave conditions. The technique, therefore, provides measurements that are particularly suitable for correlations with causative waves and longshore currents, as in time-dependent numerical models of longshore transport rates and beach change, but is not particularly useful in determining long-term net transport rates or directions. Identification of the appropriate sediment mixing depth also limits the technique's quantitative accuracy. Numerous studies have used sand tracers to determine sand transport rates; examples include Komar and Inman (1970), Knoth and Nummedal (1977), Duane and James (1980), Inman et al. (1980), Kraus et al. (1982), White (1987), and many others. Tracer techniques can also be used in conjunction with geologically distinct materials, such as naturally occurring mineral sediments. These methods, which generally involve larger quantities of material and larger scales of measurement, are potentially more useful for determining longer-term rates and directions and are discussed below.

(c) Measurements of suspended load transport have been the focus of numerous studies. One approach has been to pump water containing suspended sand from the surf zone, a technique which has the advantage that large quantities can be processed, leading to some confidence that the samples are representative of sediment concentrations found in the surf (Watts 1953b; Fairchild 1972, 1977; Coakley and Skafel 1982). These measurements, combined with simultaneous measurements of alongshore current velocity, yield estimates of the longshore sediment transport suspended load. The disadvantages of the approach are that one cannot investigate time variations in sediment concentrations at different phases of wave motions, and the sampling has often been undertaken from piers which may disturb the water and sediment motion. Recent studies of suspended load transport have employed optical and acoustic techniques (Brenninkmeyer 1976; Thornton and Morris 1977; Katoh, Tanaka, and Irie 1984; Downing 1984; Sternberg, Shi, and Downing 1984; Beach and Sternberg 1987; Hanes et al. 1988). These approaches yield continuous measurements of the instantaneous suspended sediment concentrations. Arrays of instruments can be employed to document variations across the surf zone and vertically through the water column.

(d) Another method for measuring suspension concentrations is with traps, usually consisting of a vertical array of sample bins which collect sediment but allow water to pass, and so can be used to examine the vertical distribution of suspended sediments and can be positioned at any location across the surf zone (e.g., Hom-ma, Horikawa, and Kajima 1965; Kana 1977, 1978; Inman et al. 1980; Kraus, Gingerich, and Rosati 1988). Figure III-2-2 shows vertical distributions of the longshore sediment flux (transport rate per unit area) through the water column obtained in a 5-min sampling interval by traps arranged across the surf zone at Duck, North Carolina (Kraus, Gingerich, and Rosati 1989). The steep decrease in transport with elevation above the bed is apparent; such considerations are important in groin and weir design. The sharp decrease in transport at the trap located seaward of the breaker line indicates that the main portion of longshore sediment transport takes place in the surf and swash zones.

(e) The only method presently suitable for distinct measurement of bed load transport is bed-load traps. These are bins which are open-ended or dug into the seabed into which the bed-load transport is to settle. There are questions as to sampling efficiency when used in the nearshore because of the potential for scour (Thornton 1972; Walton, Thomas, and Dickey 1985; Rosati and Kraus 1989).

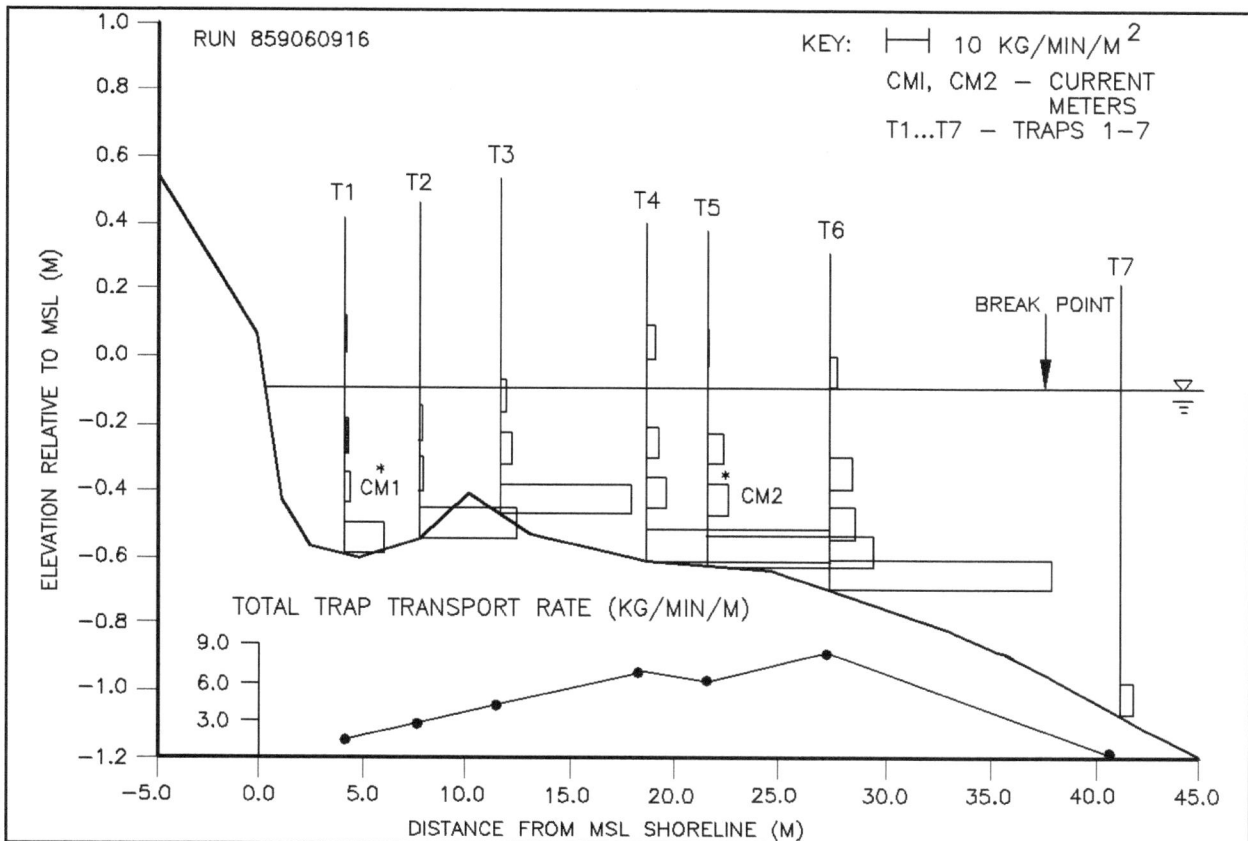

Figure III-2-2.   Cross-shore distribution of the longshore sand transport rate measured with traps at Duck, North Carolina (Kraus, Gingerich, and Rosati 1989)

(2) Qualitative indicators of longshore transport magnitude and direction.

(a) The ability to assess directions of longshore sediment transport is central to successful studies of coastal erosion and the design of harbor structures and shore protection projects. It is equally important to evaluate quantities of that transport as a function of wave and current conditions.

(b) Multiple lines of evidence have been used to discern directions of longshore sediment transport. Most of these are related to the net transport; i.e., the long-term resultant of many individual transport events. Blockage by major structures such as jetties can provide the clearest indication of the long-term net transport direction (see Figure III-2-1). Sand entrapment by groins is similar, but generally involves smaller volumes and responds to the shorter-term reversals in transport directions; therefore, groins do not always provide a definitive indication of the net annual transport direction. Other geomorphologic indicators of transport direction include the deflection of streams or tidal inlets by the longshore sand movements, shoreline displacements at headlands similar to that at jetties, the longshore growth of sand spits and barrier islands, and the pattern of upland depositional features such as beach ridges. Grain sizes and composition of the beach sediments have also been used to determine transport direction as well as sources of the sediments. It is often believed that a longshore decrease in the grain size of beach sediment provides an indication of the direction of the net transport. This is sometimes the case, but grain size changes can also be the product of alongcoast variations in wave energy levels and/or sediment sources and sinks that have no relation to sediment transport directions.

(c) Identification of unique minerals within the sand has also been used to deduce transport paths. By using the heavy mineral augite as a tracer of longshore sand movements, Trask (1952, 1955) demonstrated that the sand filling the harbor at Santa Barbara, California, originates more than 160 km up the coast. The augite was derived from volcanic rocks in the Morro Bay area north of Santa Barbara. Likewise, Bowen and Inman (1966) estimated the directions and magnitudes of the net transport along the California coast by studying the progressive dilution of beach sand augite by addition of sand from other sources. In other examples, Meisburger (1989) utilized oolitic carbonate as a natural mineral tracer to deduce transport paths along Florida's east coast. Johnson (1992) noted the net transport paths along parts of the Lake Michigan coastline by examining the movement of gravel placed as beach restoration.

(d) Many of the geomorphic and sedimentological indicators of longshore sediment transport directions are not absolute, and too strong a reliance upon them can lead to misinterpretations. It is best to examine all potential evidence that might relate to transport direction, and consider the relative reliabilities of the indicators.

(3) Quantitative indicators of longshore transport magnitude.

(a) Some of the qualitative indicators of longshore transport discussed above can also be used to estimate the quantities involved in the process. Repeated surveys over a number of years and analyses of aerial photographs of the longshore growth of sand spits have been used to establish approximate rates of sediment transport. For the estimates to be reasonable, it is typically necessary that such surveys span a decade or longer, so the results represent a long-term net sediment transport rate.

(b) The blockage of longshore sediment transport by jetties and breakwaters and the resulting growth and erosion patterns of the adjacent beaches have yielded reasonable evaluations of the net (and sometimes gross) transport rates at many coastal sites. Sand bypassing plants have been constructed at some jetties and breakwaters as a practical measure to reduce the accretion/erosion patterns adjacent to the structures. The first measurements obtained relating sand transport rates to causative wave conditions were collected by Watts (1953a) at Lake Worth Inlet, Florida, using measured quantities of sand pumped past the jetties. The best correlation was obtained using month-long net sand volumes. A number of subsequent studies have similarly employed sand blockage by jetties and breakwaters to obtain data relating transport rates to wave conditions. Caldwell (1956) estimated the longshore sand transport from erosion rates of the beach downdrift of the jetties at Anaheim Bay, California. Bruno and Gable (1976); Bruno, Dean, and Gable (1980); Bruno et al. (1981); and Walton and Bruno (1989) measured transport rates by repeatedly surveying the accumulating blocked sand at Channel Islands Harbor, California; Dean et al. (1982) measured sand accumulations in the spit growing across the breakwater opening at Santa Barbara, California; and Dean et al. (1987) collected data from the weir jetty at Rudee Inlet, Virginia. All of these studies yielded measurements of longshore sediment transport rates that are used in correlations with wave energy flux estimates. Errors are introduced with the use of jetties and breakwaters to measure sediment transport rates, the foremost being the local effects of the structure on waves and currents, and the long-term nature of the determinations. In some cases it takes a month or longer of sand accumulation to have a volume that is large enough to be outside the range of possible survey error, during which time the waves and currents are continuously changing. Therefore, a correlation between a given wave condition and the resulting sediment transport rate due solely to that particular wave condition cannot be obtained; rather, wave parameters occurring during the interval of interest are integrated and then correlated with the total accumulated volume. Nevertheless, rates determined by impoundment and erosion are very valuable as they are closely related to gross quantities involved in the design of projects, such as the amount of sediment required to be bypassed at an inlet.

(c) As discussed above, sand tracer has been used to make short-term estimates of longshore sand transport. Other techniques that have been used to measure sediment movements on beaches include various sand traps, pumps, and optical devices. However, such sampling schemes are also short-term measures and may be related to specific modes of transport, either the bed load or suspended load, rather than yielding total quantities of the long-term littoral sediment transport. Generally, it is the latter quantity which is of importance to engineering design.

(4) Longshore sediment transport estimation in the United States.

(a) Early attempts to estimate the direction and magnitude of net longshore sediment transport along the U.S. coastline centered chiefly upon examination of sand impoundment and bypassing volumes at jetties and breakwaters. Johnson (1956, 1957) compiled data of this type for many shorelines and found net transport rates ranging up to approximately 1 million m^3 (MCM) of sand per year in some locations. Based on Johnson's work, estimated patterns of littoral drift for a portion of the east coast of the United States are shown in Figure III-2-3. Table III-2-1 lists representative net longshore transport rates for selected U.S. coasts (SPM 1984). Transport rate magnitudes are clearly related to the general wave climate as energetic (west coast), intermediate (east coast and parts of the Gulf of Mexico shoreline), and low (Great Lakes and parts of the Gulf of Mexico shoreline).

(b) Since these early estimates of the transport rates, numerous investigations have yielded refined values at specific sites along the U.S. coastlines. These values are not generally archived in a single source, and must therefore be found through a literature search of various site-specific reports and articles. Estimates of both the net and gross transport rates along any particular portion of the coastline may also be developed using hindcast wave data (see Part III-2-2.d.(3) of this chapter below).

III-2-3. Predicting Potential Longshore Sediment Transport

In engineering applications, the longshore sediment transport rate is expressed as the volume transport rate Q_ℓ having units such as cubic meters per day or cubic yards per year. This is the total volume as would be measured by survey of an impoundment at a jetty and includes about 40 percent void space between the particles as well as the 60-percent solid grains. Another representation of the longshore sediment transport rate is an immersed weight transport rate I_ℓ related to the volume transport rate by

$$I_\ell = (\rho_s - \rho) \, g \, (1 - n) \, Q_\ell \tag{III-2-1a}$$

or

$$Q_\ell = \frac{I_\ell}{(\rho_s - \rho) \, g \, (1 - n)} \tag{III-2-1b}$$

where

ρ_s = mass density of the sediment grains

ρ = mass density of water

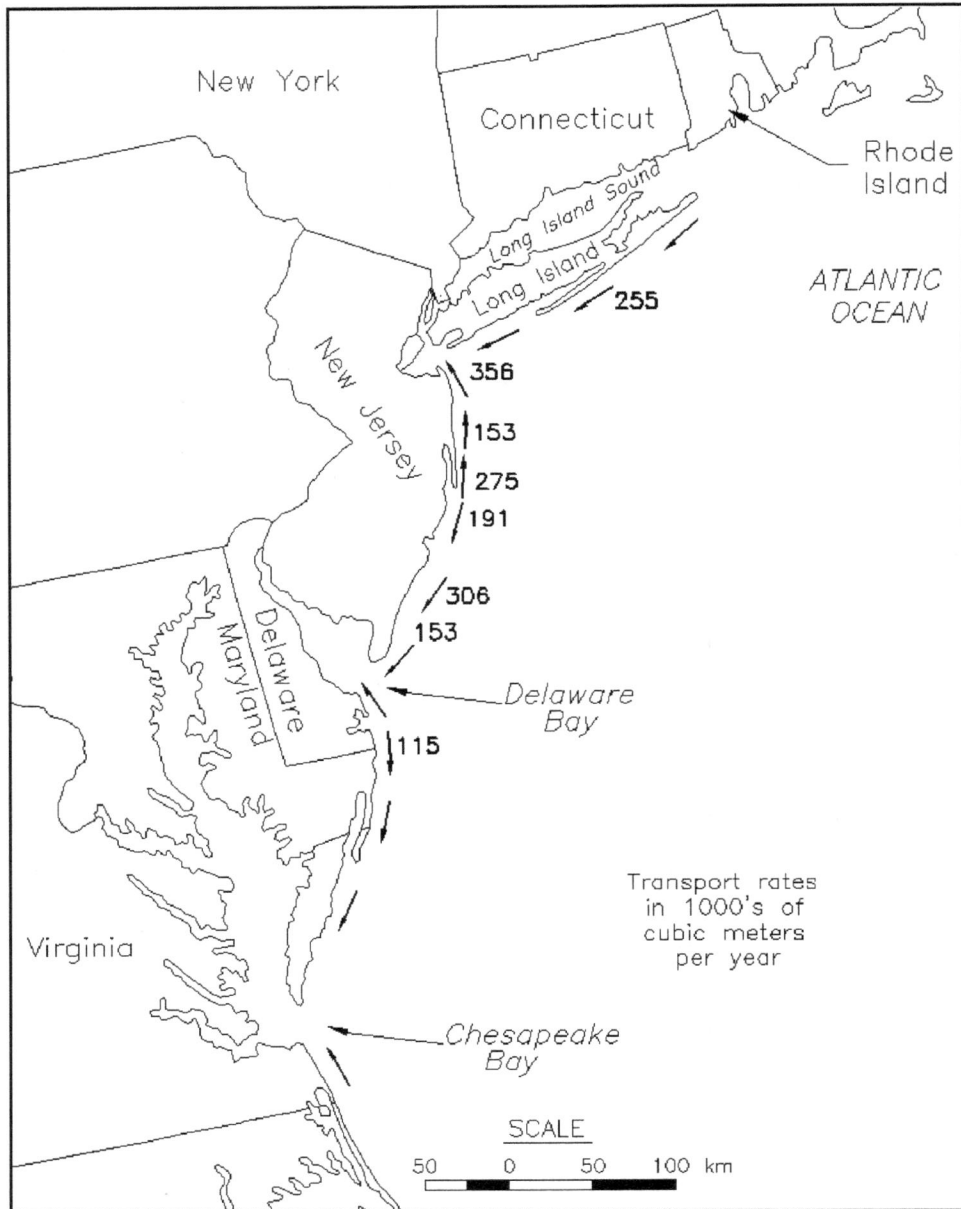

Figure III-2-3.   Estimated annual net longshore transport rates and directions along the east coast of the United States based on data from Johnson (1956, 1957) and Komar (1976)

Table III-2-1
Longshore Transport Rates from U.S. Coasts (SPM 1984)[1]

Location	Predominant Direction of Transport	Longshore[2] Transport, cu m/yr	Date of Record	Reference
Atlantic Coast				
Suffo k County, NY	W	153,000	1946-55	New York District (1955)
Sandy Hook, NJ	N	377,000	1885-1933	New York District (1954)
Sandy Hook, NJ	N	333,000	1933-51	New York District (1954)
Asbury Park, NJ	N	153,000	1922-25	New York District (1954)
Shark River, NJ	N	229,000	1947-53	New York District (1954)
Manasquan, NJ	N	275,000	1930-31	New York District (1954)
Barnegat Inlet, NJ	S	191,000	1939-41	New York District (1954)
Absecon Inlet, NJ	S	306,000	1935-46	New York District (1954)
Ocean City, NJ	S	306,000	1935-46	U.S. Congress (1953a)
Cold Spring Inlet, NJ	S	153,000	---	U.S. Congress (1953b)
Ocean City, MD	S	115,000	1934-36	Baltimore District (1948)
Atlantic Beach, NC	E	22,500	1850-1908	U.S. Congress (1948)
Hillsboro Inlet, FL	S	57,000	1850-1908	U.S. Army (1955b)
Palm Beach, FL	S	115,000 to 175,000	1925-30	BEB (1947)
Gulf of Mexico				
Pinellas County, FL	S	38,000	1922-50	U.S. Congress (1954a)
Perdido Pass, AL	W	153,000	1934-53	Mobile District (1954)
Pacific Coast				
Santa Barbara, CA	E	214,000	1932-51	Johnson (1953)
Oxnard Plain Shore, CA	S	765,000	1938-48	U.S. Congress (1953c)
Port Hueneme, CA	S	382,000	---	U.S. Congress (1954b)
Santa Monica, CA	S	206,000	1936-40	U.S. Army (1948b)
El Segundo, CA	S	124,000	1936-40	U.S. Army (1948b)
Redondo Beach, CA	S	23,000	---	U.S. Army (1948b)
Anaheim Bay, CA	E	115,000	1937-48	U.S. Congress (1954c)
Camp Pendleton, CA	S	76,000	1950-52	Los Angeles District (1953)
Great Lakes				
Milwaukee County, WI	S	6,000	1894-1912	U.S. Congress (1946)
Racine County, WI	S	31,000	1912-49	U.S. Congress (1953d)
Kenosha, WI	S	11,000	1872-1909	U.S. Army (1953b)
IL State Line to Waukegan	S	69,000	---	U.S. Congress (1953e)
Waukegan to Evanston, IL	S	44,000	---	U.S. Congress (1953e)
South of Evanston, IL	S	31,000	---	U.S. Congress (1953e)
Hawaii				
Waikiki Beach	--	8,000	---	U.S. Congress (1953f)

[1] Method of measurement is by accretion except for Absecon Inlet and Ocean City, NJ, and Anaheim Bay, CA, which were measured by erosion, and Wa kiki Beach, HI, which was measured according to suspended load samples.
[2] Transport rates are estimated net transport rates Q_N. In some cases, these approximate the gross transport rates Q_g.

g = acceleration due to gravity

n = in-place sediment porosity ($n \approx 0.4$)

The parameter n is a pore-space factor such that $(1 - n) Q_\ell$ is the volume transport of solids alone. One advantage of using I_ℓ is that this immersed weight transport rate incorporates effects of the density of the sediment grains. The factor $(\rho_s - \rho)$ accounts for the buoyancy of the particles in water. The term "potential" sediment transport rate is used, because calculations of the quantity imply that sediment is available in sufficient quantity for transport, and that obstructions (such as groins, jetties, breakwaters, submarine canyons, etc.) do not slow or stop transport of sediment alongshore.

a. *Energy flux method.*

(1) Historical background. An extensive discussion of the evolution of energy-based longshore transport formulae is presented by Sayao (1982), in his dissertation. The following is a summary of Sayao's discussion, focussing on evolution of the so-called "CERC" formula.

Munch-Peterson, a Danish engineer, first related the rate of littoral sand transport to deepwater wave energy in conjunction with harbor studies on the Danish coast (Munch-Peterson 1938). Because of a lack of wave data, Munch-Peterson used wind data in practical applications, which gave preliminary estimations of the littoral drift direction. In the United States, use of a formula to predict longshore sediment transport based on wave energy was suggested by the Scripps Institute of Oceanography (1947), and applied by the U.S. Army Corps of Engineers, Los Angeles District to the California coast (Eaton 1950). Watts (1953a) and Caldwell (1956) made the earliest documented measurements of longshore sediment transport (at South Lake Worth, Florida, and Anaheim Bay, California, respectively) and related transport rates to wave energy, resulting in modifications to the existing formulae. Savage (1962) summarized the available data from field and laboratory studies and developed an equation which was later adopted by the U.S. Army Corps of Engineers in a 1966 coastal design manual (U.S. Army Corps of Engineers 1966), which became known as the "CERC formula." Inman and Bagnold (1963), based on Bagnold's earlier work on wind-blown sand transport and on sand transport in rivers, suggested use of an immersed weight longshore transport rate, rather than a volumetric rate. An immersed weight sediment transport equation was calibrated by Komar and Inman (1970) based on the available field data including their tracer-based measurements at Silver Strand, California, and El Moreno, Mexico. Based on Komar and Inman's (1970) transport relationship and other available field data, the CERC formula for littoral sand transport was updated from its 1966 version, and has been presented as such in the previous editions of the *Shore Protection Manual* (1977, 1984).

(2) Description.

(a) The potential longshore sediment transport rate, dependent on an available quantity of littoral material, is most commonly correlated with the so-called longshore component of wave energy flux or power,

$$P_\ell = (EC_g)_b \, \sin\alpha_b \, \cos\alpha_b \qquad\qquad (\text{III-2-2})$$

where E_b is the wave energy evaluated at the breaker line,

$$E_b = \frac{\rho g H_b^2}{8} \qquad\qquad (\text{III-2-3})$$

and C_{gb} is the wave group speed at the breaker line,

$$C_{gb} = \sqrt{gd_b} = \left(g \, \frac{H_b}{\kappa} \right)^{\frac{1}{2}} \qquad\qquad (\text{III-2-4})$$

where κ is the breaker index H_b / d_b . The term $(EC_g)_b$ is the "wave energy flux" evaluated at the breaker zone, and α_b is the wave breaker angle relative to the shoreline. The immersed weight transport rate I_ℓ has the same units as P_ℓ (i.e., N/sec or lbf/sec), so that the relationship

$$I_\ell = KP_\ell \qquad\qquad (\text{III-2-5})$$

is homogeneous, that is, the empirical proportionality coefficient K is dimensionless. This is another advantage in using I_ℓ rather than the Q_ℓ volume transport rate. Equation 2-5 is commonly referred to as the "CERC formula."

(b) Equation 2-5 may be written

$$I_\ell = K P_\ell = K(E C_g)_b \sin\alpha_b \cos\alpha_b \tag{III-2-6a}$$

which, on assuming shallow water breaking, gives

$$I_\ell = K \left(\frac{\rho g H_b^2}{8} \right) \left(\frac{g H_b}{\kappa} \right)^{\frac{1}{2}} \sin\alpha_b \cos\alpha_b \tag{III-2-6b}$$

$$I_\ell = K \left(\frac{\rho g^{3/2}}{8 \kappa^{1/2}} \right) H_b^{5/2} \sin\alpha_b \cos\alpha_b \tag{III-2-6c}$$

$$I_\ell = K \left(\frac{\rho g^{\frac{3}{2}}}{16 \kappa^{\frac{1}{2}}} \right) H_b^{\frac{5}{2}} \sin(2\alpha_b) \tag{III-2-6d}$$

(c) By using Equation 2-1b, the relationships for I_ℓ can be converted to a volume transport rate:

$$Q_\ell = \frac{K}{(\rho_s - \rho) g (1 - n)} P_\ell \tag{III-2-7a}$$

$$Q_\ell = K \left(\frac{\rho \sqrt{g}}{16 \kappa^{\frac{1}{2}} (\rho_s - \rho) (1 - n)} \right) H_b^{\frac{5}{2}} \sin(2\alpha_b) \tag{III-2-7b}$$

(d) Field data relating I_ℓ and P_ℓ are plotted in Figure III-2-4, for which the calculations of the wave power are based on the root-mean-square (rms) wave height at breaking $H_{b\ rms}$. Data presented in Figure III-2-4 include those measured by: (1) sand accumulation at jetties and breakwaters (South Lake Worth Inlet, Florida (Watts 1953a); Anaheim Bay (Caldwell 1956), Santa Barbara (Dean et al. 1987), and Channel Islands (Bruno et al. 1981, Walton and Bruno 1989), California; Rudee Inlet, Virginia (Dean et al. 1987); Cape Thompson, Alaska (Moore and Cole 1960); and Point Sapin, Canada (Kamphuis 1991)); (2) sand tracer at Silver Strand, California (Komar and Inman 1970); El Moreno, Mexico (Komar and Inman 1970); Torrey Pines, California (Inman et al. 1980); and Ajiguara, Japan (Kraus et al. 1982)); and (3) sediment traps at Kewaunee County, Wisconsin (Lee 1975); and Duck, North Carolina (Kana and Ward 1980)). Because of questions in methodologies and trapping efficiencies, probably the data sets most appropriate for engineering application are those based upon category 1 above, sand accumulation (impoundment) at jetties and breakwaters.

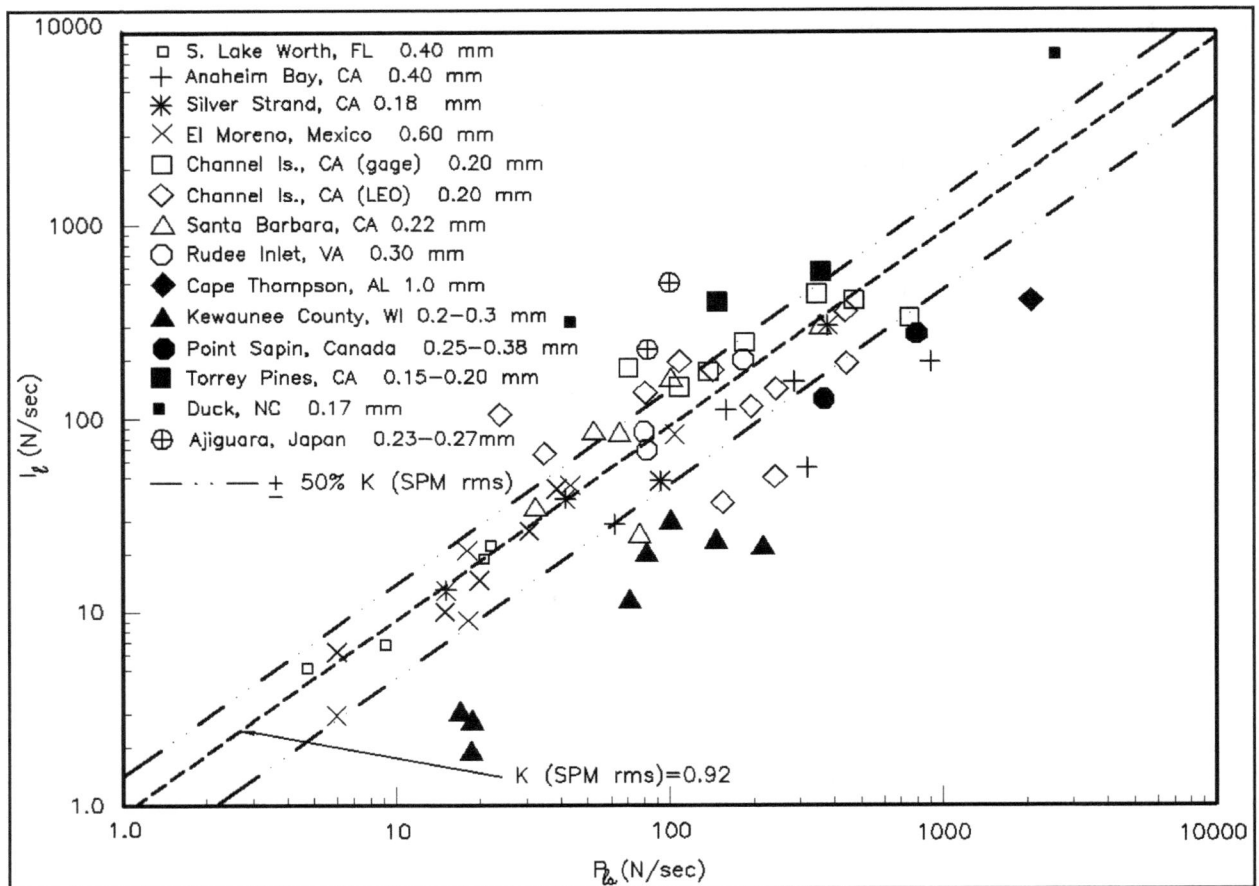

Figure III-2-4. Field data relating I_ℓ and P_ℓ

(e) The K coefficient defined here is based on utilizing the rms breaking wave height $H_{b\,rms}$. The *Shore Protection Manual* (1984) presented a dimensionless coefficient $K_{SPM\,sig} = (0.39)$ based on computations utilizing the significant wave height. The value of this SPM coefficient corresponding to the rms wave height $H_{b\,rms}$ is $K_{SPM\,rms} = 0.92$, which is indicated in Figure III-2-4 with a dashed line for reference. Judgement is required in applying Equation 2-7. Although the data follow a definite trend, the scatter is obvious, even on the log-log plot. The dash-double-dot lines represent a ±50 percent interval around the SPM reference line ($K_{SPM\,rms} = 0.92$).

(f) An early design value of the K coefficient was introduced for use with rms breaking wave height by Komar and Inman (1970); $K_{K\&I\,rms} = 0.77$. This value is commonly seen in many longshore transport rate computations.

(g) Values of the other parameters for use in the sediment transport equations are: $\rho_s = 2,650$ kg/m^3 (5.14 slugs/ft^3) for quartz-density sand; $\rho = 1,025$ kg/m^3 (1.99 slug/ft^3) for 33 parts per thousand (ppt) salt water; and $\rho = 1,000$ kg/m^3 (1.94 slug/ft^3) for fresh water; $g = 9.81$ m/sec^2 (32.2 ft/sec^2); and $n = 0.4$. The breaker index (κ) is ≈ 0.78 for flat beaches and increases to more than 1.0 depending on beach slope (Weggel 1972).

(3) Variation of K with median grain size.

(a) Longshore sand transport data presented in Figure III-2-4 represent beaches with quartz-density grain sizes ranging from ~0.2 mm to 1.0 mm, and wave heights ranging from 0.5 to 2.0 m. Bailard (1981, 1984) developed an energy-based model, which presents K as a function of the breaker angle and the ratio of the orbital velocity magnitude and the sediment fall speed, also based on the rms wave height at breaking. Bailard calibrated the model using eight field and two laboratory data sets, and developed the following equation:

$$K = 0.05 + 2.6 \sin^2 (2\alpha_b) + 0.007 \frac{u_{mb}}{w_f} \tag{III-2-8}$$

where u_{mb} is the maximum oscillatory velocity magnitude, obtained from shallow-water wave theory as

$$u_{mb} = \frac{\kappa}{2} \sqrt{g d_b} \tag{III-2-9}$$

and w_f is the fall speed of the sediment, either calculated using the relationships described in Part III-1, or, if the spherical grain assumption cannot be applied to the material, measured experimentally. Bailard developed his relationship using the following data ranges: $2.5 \le w_f \le 20.5$ cm/sec; $0.2° \le \alpha_b \le 15°$; and $33 \le u_{mb} \le 283$ cm/sec. A comparison of observed and predicted K coefficients using Bailard's Equation 2-8 is presented in Figure III-2-5, using the data sets on which Bailard based his calibration. Because Bailard's relationship is based on a limited data set, predicted K coefficients may be highly variable. Bailard's relationship for K is similar to a relationship presented by Walton (1979) and Walton and Chiu (1979), which was compared to limited laboratory data.

(b) Others have proposed empirically based relationships for increasing K with decreasing grain size (or equivalently, fall speed) (Bruno, Dean, and Gable 1980; Dean et al. 1982; Kamphuis et al. 1986; Dean 1987). Komar (1988), after reexamining available field data, suggested that the previous relationships resulted from two data sets with K values based on erroneous or questionable field data. Revising these K values, Komar (1988) concluded that existing data suggests little dependence of the empirical K coefficient on sediment grain sizes, at least for the range of sediments in the data set. Data from shingle beaches, however, indicated a smaller K, but the data were too limited to establish a correlation. Komar stressed that K should depend on sediment grain size, and the absence of such a trend in his analysis must result from the imperfect quality of the data.

(c) Recently, del Valle, Medina, and Losada (1993) have presented an empirically based relationship for the K parameter, adding sediment transport data representing a range in median sediment grain sizes (0.40 mm to 1.5 mm) from the Adra River Delta, Spain to the available database as modified by Komar (1988). Del Valle, Medina, and Losada obtained wave parameters from buoy and visual observations, and sediment transport rates were evaluated from aerial photographs documenting a 30-year period of shoreline evolution for five locations along the delta. Results of their analysis reinforce a decreasing trend in the empirical coefficient K with sediment grain size. Their empirical fit based on the corrections to the database as suggested by Komar (1988) is shown in Figure III-2-6 and given in Equation III-2-10. The empirically based relationship is to be applied with rms breaking wave height,

$$K = 1.4 \ e^{(-2.5 \ D_{50})} \tag{III-2-10}$$

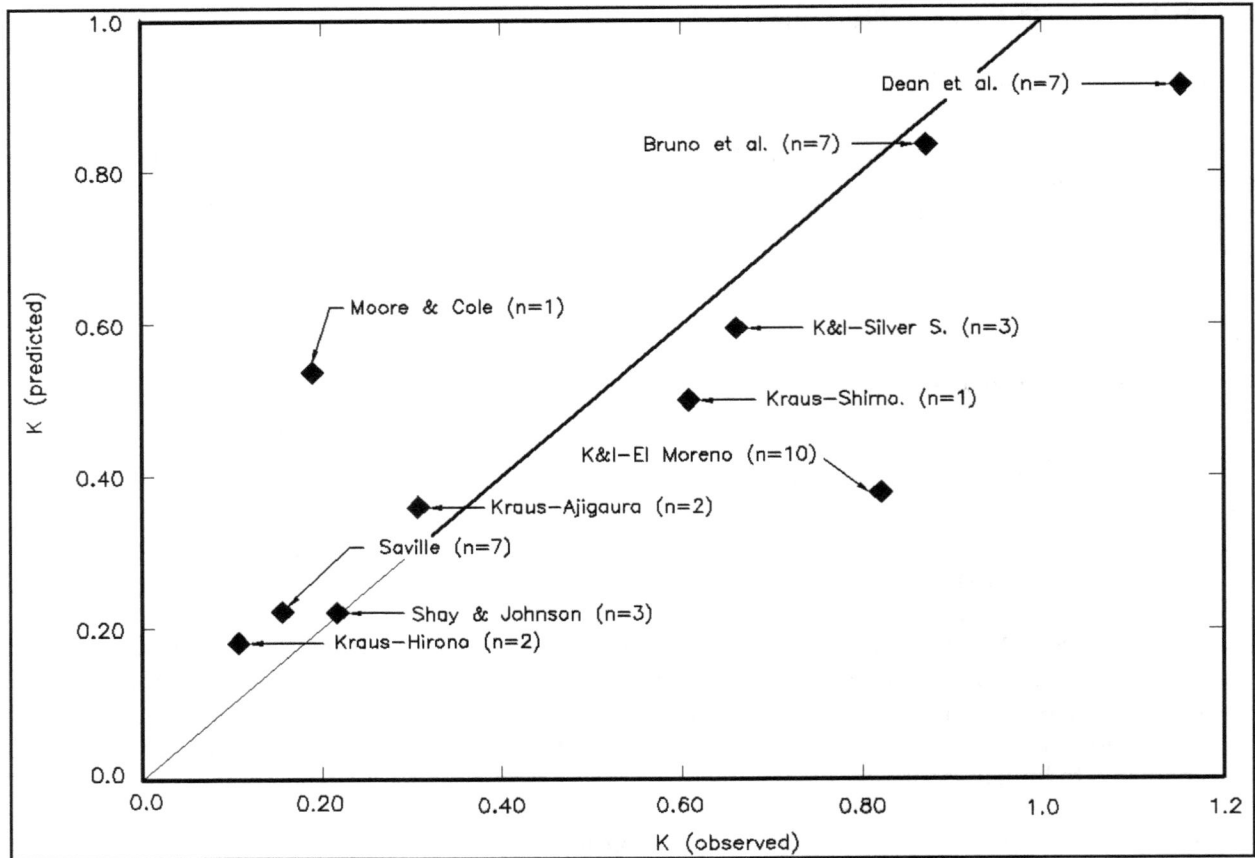

Figure III-2-5.   Measured and predicted K coefficients using Bailard's (1984) equation

where D_{50} is the median grain size of the beach sediment in millimeters. This relationship is based on limited data and depends strongly on the data from the Adra River Delta.

(d) While it is generally thought that the K coefficient should decrease with increasing grain size, the nature of this relationship is not well understood at present. Again, because of the limited data set and inherent variability in measuring longshore sediment transport rates, predicted K coefficients may vary considerably from appropriate values for any particular site.

(4) Variation of K with surf similarity.

(a) From laboratory data, a relationship between K and the surf similarity parameter $\xi_b = m / (H_b/L_o)^{1/2}$ has also been observed (Kamphuis and Readshaw 1978). These data suggest that the value of K increases with increasing value of the surf similarity parameter (i.e. as the breaking waves tend from spilling to collapsing condition).

(b) Numerous other relationships for predicting longshore sediment transport rates exist (e.g., Watts (1953a), Kamphuis and Rea shaw (1978), Kraus et al. (1982), Kamphuis et al. (1986), Kamphuis (1991)). A promising empirical relationship based principally upon laboratory results is that developed by Kamphuis (1991) which correlates well with laboratory and field data sets. However, for application to field studies, the use of a physically based relationship based solely on field data, such as the one presented herein (often called the CERC formula) is preferred.

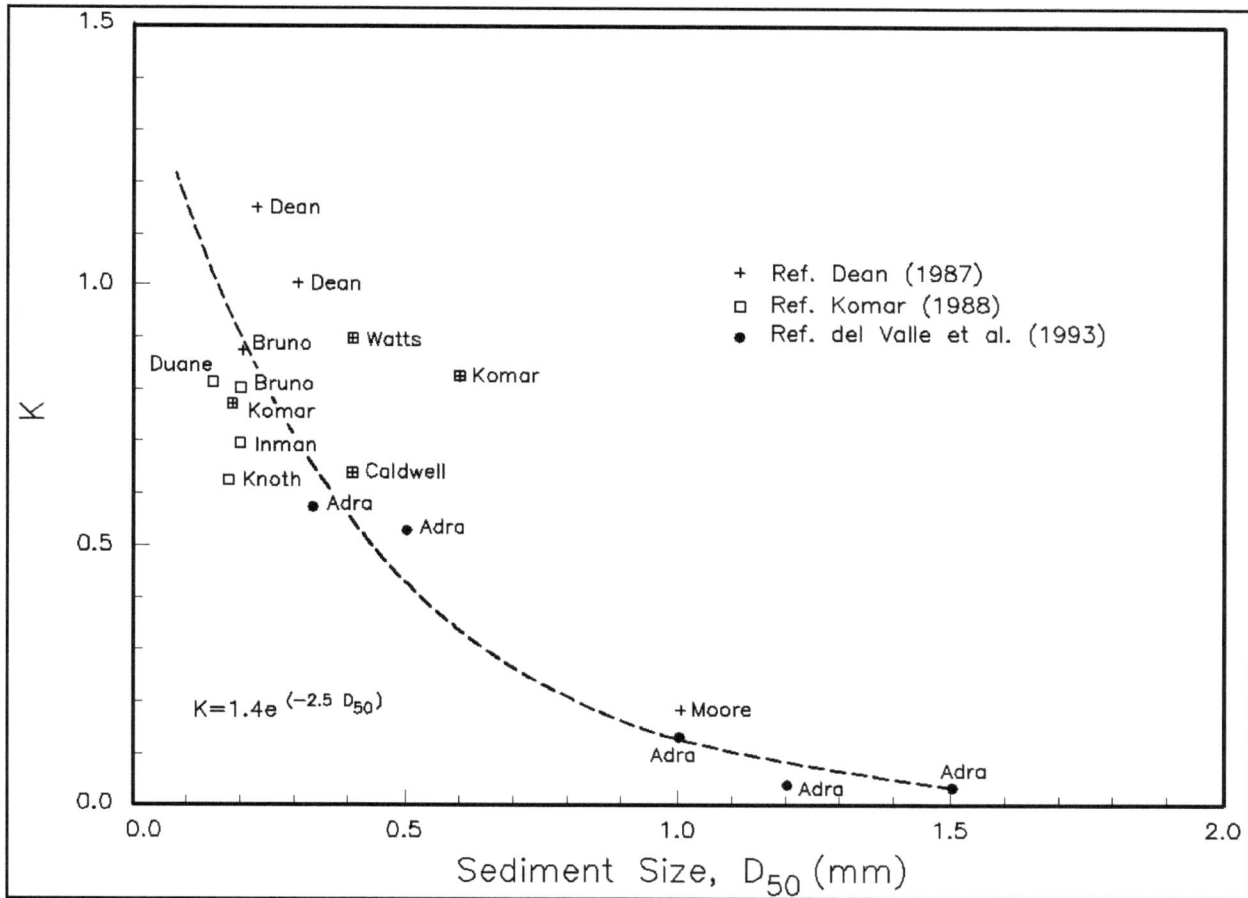

Figure III-2-6. Coefficient K versus median grain size D_{50} (del Valle, Medina, and Losada 1993)

(c) Researchers at CHL (unpublished) used sand tracer data from Santa Barbara and Torrey Pines, California, and profile and dredging records from Santa Barbara Harbor, California, to compare 20 longshore sediment transport models. They concluded that the CERC equation performed as well or better than the other models. Of the six models including effects of grain size evaluated (they did not evaluate del Valle, Medina, and Losada's (1993) relationship), they concluded that Bailard's (1984) relationship (given in Equation 2-8) performed the best. However, the data set used for comparison only represented grain sizes in the range 0.15 to 0.25 mm, and therefore was not particularly suited for testing the dependence of longshore sediment transport relationships on grain size outside this limited range.

b. Longshore current method. Early workers such as Grant (1943) stressed that sand transport in the nearshore results from the combined effects of waves and currents; i.e., the waves placing sand in motion and the longshore currents producing a net sand advection. Walton (1980, 1982) proposed a longshore sediment transport calculation method using the breaking-wave-driven longshore current model of Longuet-Higgins (1970) from which the longshore energy flux factor becomes

$$P_\ell = \frac{\rho \ g \ H_b \ W \ V_\ell \ C_f}{\left(\dfrac{5\pi}{2}\right)\left(\dfrac{V}{V_o}\right)}$$

(III-2-11)

EXAMPLE PROBLEM III-2-1

FIND:

Determine the K parameter for use at a site using Bailard's (1984) and del Valle, Medina, and Losada's (1993) relationships, then calculate the potential volumetric longshore transport rate along the beach using each K value.

GIVEN:

Waves having an rms wave height of 2.0 m (6.7 ft) break on a quartz sand beach with median grain size, $D_{50} = 1.0$ mm at a wave breaking angle $\alpha_b = 4.5°$. The temperature of the seawater is 20° C and w_f = fall velocity = 13.1 cm/sec. Assume $\kappa = 1.0$ for this problem.

SOLUTION:

w_f = 13.1 cm/sec (0.43 ft/sec), which is within Bailard's data range.

The maximum oscillatory velocity magnitude is calculated from Equation 2-9 as

$$u_{mb} = \frac{\kappa (g \, d_b)^{1/2}}{2} = \frac{1 (9.81 \times 2.0)^{1/2}}{2}$$

$u_{mb} = 2.21$ m/sec, which is within Bailard's (1984) data range.

Therefore, Equation 2-8 may be used, and the coefficient to apply in the longshore sediment transport equation is,

$$K = 0.05 + 2.6 \sin^2 (2\alpha_b) + 0.007 \, u_{mb}/w_f = 0.05 + 2.6 \sin^2 (2(4.5°)) + 0.007 (2.21/0.131)$$

$$K = 0.23$$

Applying del Valle, Medina, and Losada's relationship, Equation 2-10,

$$K = 1.4 \, e^{(-2.5 \, D_{50})} = 0.12$$

(Continued)

Example Problem III-2-1 (Concluded)

Using Equation 2-7b with $K = 0.23$,

$$Q_\ell = K \left(\frac{\rho\sqrt{g}}{16 \; \kappa^{1/2} \; (\rho_s - \rho)(1 - n)} \right) H_{b \; rms}^{5/2} \; \sin(2\alpha_b)$$

$$Q_\ell = 0.23 \; \frac{(1025) \; (9.81)^{1/2}}{16 \; (1) \; (2650 - 1025) \; (1 - 0.4)} \; (2.0)^{5/2} \; \sin \; (2 \times 4.5°)$$

$$Q_\ell = (0.042 \; \text{m}^3/\text{sec}) \; (3600 \; \text{sec/hr}) \; (24 \; \text{hr/day})$$

$$Q_\ell = 3.6 \times 10^3 \; \text{m}^3/\text{day} \; (4.7 \times 10^3 \; \text{yd}^3/\text{day}) \; \text{using Bailard's methodology.}$$

Using del Valle, Medina, and Losada's relationship ($K = 0.12$)

$$Q_\ell = 0.12 \; \frac{(1025) \; (9.81)^{1/2}}{16 \; (2650 - 1025) \; (1 - 0.4)} \; (2.0)^{5/2} \; \sin \; (2 \times 4.5°)$$

$$Q_\ell = (0.021 \; \text{m}^3/\text{sec}) \; (3600 \; \text{sec/hr}) \; (24 \; \text{hr/day})$$

$$Q_\ell = 1.9 \times 10^3 \; \text{m}^3/\text{day} \; (2.5 \times 10^3 \; \text{yd}^3/\text{day})$$

It is noted that the estimates of the transport rate developed from the two relationships for K, Equations 2-8 and 2-10, differ by a factor of almost 1.9. This serves to highlight the uncertainty associated with selection of the appropriate K coefficient for a given site, and the resulting uncertainty in the prediction of the longshore transport rate's value. The grain size used in this problem is much larger than that typically found on U.S. beaches, and the K coefficient found is correspondingly smaller than that recommended in *Shore Protection Manual* (1984).

EXAMPLE PROBLEM III-2-2

FIND:
Calculate the potential immersed-weight and volumetric longshore sand transport rates.

GIVEN:
The same conditions as in Example Problem III-2-1, except assume a median grain size D_{50} corresponding to a K coefficient = 0.60. Assume $\kappa = 1$ for simplicity.

SOLUTION:

With $H_{b\,rms}$ = 2.0 m (6.6 ft), ρ = 1,025 kg/m³ (1.99 slug/ft³), ρ_s = 2,650 kg/m³ (5.14 slug/ft³), g = 9.8 m/sec² (32.2 ft/sec²) and α_b = 4.5°, Equation 2-6d gives

$$I_\ell = K \left(\frac{\rho g^{\frac{3}{2}}}{16\,\kappa^{\frac{1}{2}}} \right) H_{b\,rms}^{\frac{5}{2}} \sin(2\alpha_b)$$

$$0.60 \; \frac{(1025)\,(9.81)^{\frac{3}{2}}}{16\,(1)} \; (2.0)^{\frac{5}{2}} \; \sin(2\times4.5^\circ) \cong 1050 \; N/sec \; (236 \; lbf/sec)$$

This potential immersed-weight transport rate may be converted to a potential volumetric rate by using Equation 2-7b

$$Q_\ell = K \left(\frac{\rho\sqrt{g}}{16\,\kappa^{\frac{1}{2}}\,(\rho_s - \rho)(1 - n)} \right) H_{b\,rms}^{\frac{5}{2}} \sin(2\alpha_b)$$

$$Q_\ell = 0.60 \; \frac{(1025)\,(9.81)^{\frac{1}{2}}}{16\,(1)\,(2650 - 1025)\,(1 - 0.4)} \; (2.0)^{\frac{5}{2}} \; \sin(2\times4.5^\circ)$$

$$Q_\ell = (0.109 \; m^3/sec)\,(3600 \; sec/hr)\,(24 \; hr/day)$$

$$Q_\ell = 9.4 \times 10^3 \; m^3/day \; (12.3 \times 10^3 \; yd^3/day)$$

EXAMPLE PROBLEM III-2-3

FIND:

Calculate the potential volumetric longshore sand transport rate along the beach.

GIVEN:

Spectral analysis of wave measurements at an offshore buoy in deep water yields a wave energy density E_o of 2.1 x 10^3 N/m (144 lbf/ft), with a single peak centered at a period $T = 9.4$ sec. At the measurement site, the waves make an angle of $\alpha_o = 7.5°$ with the trend of the coast, but after undergoing refraction, the waves break on a sandy beach with an angle of $\alpha_b = 3.0°$. Assume that the K coefficient is 0.60.

SOLUTION:

The group speed of the waves in deep water is given in Part II-1 as

$$C_{go} = gT/4\pi = 9.8\,(9.4)\,/\,(4\pi) = 7.3 \text{ m/sec (24.0 ft/sec)}$$

The energy flux per unit shoreline length in deep water is

$$(EC_g)_o \cos\alpha_o = (2.1 \times 10^3)(7.3)\cos(7.5°) = 1.5 \times 10^4 \text{ N/sec (3.4 x } 10^3 \text{ lbf/sec)}$$

The conservation of wave energy flux allows the substitution

$$(EC_g)_b \cos\alpha_b = (EC_g)_o \cos\alpha_o$$

where bottom friction and other energy losses are assumed to be negligible. Equation 2-2 for the longshore component of the energy flux at the shoreline then becomes

$$P_\ell = (EC_g)_b \sin\alpha_b \cos\alpha_b = [(EC_g)_o \cos\alpha_o]\sin\alpha_b = (1.5 \times 10^4)\sin(3.0°)$$

$$P_\ell = 800 \text{ N/sec (180 lbf/sec)}$$

Spectra yield wave parameters equivalent to rms conditions, and therefore $K = 0.60$ may be used in Equation 2-7a to calculate the potential volumetric sand transport rate. This gives

$$Q_\ell = \frac{K}{(\rho_s - \rho)\,g\,(1 - n)}\,P_\ell$$

$$Q_\ell = \frac{0.60}{(2650 - 1025)\,(9.81)\,(1 - 0.4)}\,800$$

$$Q_\ell = 0.050 \text{ m}^3/\text{sec x 3600 sec/hr x 24 hr/day}$$

$$Q_\ell = 4.3 \times 10^3 \text{ m}^3/\text{day (5.7 x } 10^3 \text{ yd}^3/\text{day)}$$

EXAMPLE PROBLEM III-2-4

FIND:
Calculate the resulting volumetric longshore transport of sand using measured surf zone velocity.

GIVEN:
Breaking waves have an rms height of 1.8 m (5.9 ft) and there is a persistent longshore current in the surf zone with mean velocity 0.25 m/sec (0.82 ft/sec) as measured at approximately the mid-surf position. The width of the surf zone is approximately 75 m (246 ft). Assume the K coefficient was calculated as 0.60.

SOLUTION:

With $H_{b\ rms}$ = 1.8 m (5.9 ft) and V_ℓ = 0.25 m/sec (0.82 ft/sec), Equation 2-12 gives

$$V/V_o = 0.2\ (37.5/75) - 0.714\ (37.5/75)\ \ln\ (37.5/75) = 0.35$$

and Equation 2-11 gives

$$P_\ell = [(1025)(9.81)(1.8)(75)(0.25)(0.01)]/[(5\pi/2)(0.35)] = 1234\ \text{N/sec}\ (277\ \text{lbf/sec})$$

From Equation 2-7a,

$$Q_\ell = (0.60)(1234)/[(2650-1025)(9.81)(1-0.4)]$$

$$Q_\ell = 0.077\ \text{m}^3/\text{sec} = 6.7 \times 10^3\ \text{m}^3/\text{day}\ (8.8 \times 10^3\ \text{yd}^3/\text{day})$$

in which W is the width of the surf zone, V_ℓ is the measured longshore current at a point in the surf zone, C_f is a friction coefficient dependent on Reynolds' number and bottom roughness, and V_o is the theoretical longshore velocity at breaking for the no-lateral-mixing case. A theoretical velocity distribution for a linear beach profile that best fits Longuet-Higgins nondimensionalized data is chosen

$$\left(\frac{V}{V_o}\right) = 0.2\left(\frac{Y}{W}\right) - 0.714\left(\frac{Y}{W}\right)\ln\left(\frac{Y}{W}\right) \tag{III-2-12}$$

in which Y equals the distance to the measured current from the shoreline, and V/V_o equals Longuet-Higgins dimensionless longshore current velocity for an assumed mixing coefficient equal to 0.4, which agrees reasonably well with laboratory data. Values of the friction factor C_f in Equation 2-11 were shown by Longuet-Higgins to be approximately 0.01, based on laboratory data. Thornton and Guza (1981) calculated the friction factor using field data measured at Torrey Pines Beach, San Diego, California, and a mean value of the parameter, averaged over four selected days, was 0.01 with a standard deviation of 0.01. Using Equations 2-11 and 2-12 with knowledge of breaking wave height, width of the surf zone, longshore velocity (at some point within the surf zone), distance to the measured longshore velocity, and an assumed friction factor, the longshore sand transport rate may then be calculated. From a practical standpoint, it is often easier and more accurate to measure the longshore current V_ℓ than it is to determine the breaker angle α_b needed in the P_ℓ formulation. In either method, calculations of longshore sediment transport may be very different than actual values. Formulations for the longshore current distribution across nonplanar, concave-up beach profiles are presented by McDougal and Hudspeth (1983a, 1983b, and 1989) and by Bodge (1988).

   c. Using hindcast wave data. Potential longshore sand transport rates can be calculated using Wave Information Study (WIS) hindcast wave estimates (see Part II-2). First, refraction and shoaling of incident linear waves are calculated using Snell's law and the conservation of wave energy flux. The shallow-water wave breaking criterion then defines wave properties at the break point, and potential longshore sand transport rates are calculated by means of the energy flux method.

   (1) Wave transformation procedure. To calculate the potential longshore sand transport rate using Equation 2-7b, the breaking wave height and incident angle with respect to the shoreline are required. WIS hindcast estimates, however, are given for intermediate to deepwater depths (Hubertz et al. 1993). Refraction and shoaling transformation of the WIS hindcast wave estimates to breaking conditions are therefore necessary and can be accomplished using linear wave theory for coastlines having reasonably straight and parallel bottom contour lines. Assuming that offshore depth contours are straight and parallel to the trend of the shoreline and neglecting energy dissipation prior to breaking, the wave height and angle at breaking can be computed from the coupled equations

$$H_b = H_1^{\frac{4}{5}} \left[\frac{C_{g1} \cos\alpha_1}{\sqrt{\frac{g}{\kappa}} \cos\alpha_b} \right]^{\frac{2}{5}}$$

(III-2-13)

$$\sin\alpha_b = \sqrt{g \frac{H_b}{\kappa} \frac{\sin\alpha_1}{C_1}}$$

(III-2-14)

where the subscript "1" refers to offshore (WIS) conditions. Equations 2-13 and 2-14 are derived from energy conservation and Snell's Law. Where it is assumed that wave breaking occurs for shallow-water wave conditions; i.e.,

$$C_b = C_{gb} = \sqrt{g\,d_b} = \sqrt{g\frac{H_b}{\kappa}}$$

(III-2-15)

Employing the identity $sin^2\alpha_b = 1 - cos^2\alpha_b$, Equations 2-13 and 2-14 can be combined as

$$H_b = H_1^{\frac{4}{5}}(C_{g1} \cos \alpha_1)^{\frac{2}{5}} \left[\frac{g}{\kappa} - \frac{H_b\,g^2\,\sin^2(\alpha_1)}{\kappa^2\,C_1^2} \right]^{-\frac{1}{5}}$$

(III-2-16)

and solved iteratively for H_b. The angle α_b can then be found using Equation 2-14. In application of Equations 2-14 and 2-16, the offshore wave angle α_1 should be transformed relative to the shoreline-perpendicular as described below.

(2) Wave conditions.

(a) As discussed in Part II-2, WIS hindcast wave estimates are compiled in intermediate to deepwater depths. The examples presented herein use Revised Atlantic Level 2 (RAL2) WIS data, which are presented in 45-deg wave angle bands. Angles reported for RAL2 WIS Phase III stations α_{WIS} are measured in degrees "from" in a compass sense, for which due north equals zero and other angles are measured clockwise from north. For calculation of longshore sand transport, a right-handed coordinate system is more convenient, in which waves approaching normal to the shoreline are given an angle of 0 deg. Looking seaward, waves approaching from the right are associated with negative angles, and waves approaching from the left are associated with positive angles such that transport directed to the right is given a positive value and transport directed to the left is given a negative value in accord with sign convention discussed earlier. WIS angles may be converted to angles associated with transport calculations α by means of the following relationship:

$$\alpha = \theta_n - WIS \ angle \tag{III-2-17}$$

where θ_n is the azimuth angle of the outward normal to the shoreline. Angles of α over $\pm 90^0$ would be excluded from calculations. The WIS angle is the azimuth angle the waves are coming from.

(b) RAL2 WIS wave data for one particular wave-angle band are summarized in Table III-2-2. The number of occurrences are listed for specific wave height and period bands, with the total number of occurrences listed in the right column (for each wave band) and bottom row (for each period band). The header gives the time period for the data record (1956 - 1975), location of the station in latitude and longitude, water depth, and angle band represented by the tabulated data.

(c) Data given in WIS statistical tables may be used in several ways to calculate the potential longshore sand transport rate. Two examples using the data in Table III-2-2 are given below. In Example III-2-5, the potential longshore sand transport rate is estimated with the average significant wave height for the given

Table III-2-2

Occurrence of Wave Height and Period for Direction Band 112.5° - 157.49° RAL2 Station 72

WIS ATLANTIC REVISION 1956 - 1975

LAT: 40.25 N LONG: 73.75 W, DEPTH = 27 M

OCCURRENCES OF WAVE HEIGHT AND PEAK PERIOD

FOR 45-DEG DIRECTION BAND

STATION 72

$(112.50^0$ to $157.49^0)$ mean = 135.0^0

Hmo(m)	T_p, seconds								
	3.0-4.9	5.0-6.9	7.0-8.9	9.0-10.9	11.0-12.9	13.0-14.9	15.0-16.9	17.0-18.9	Total
0.00-0.99	1,022	1,509	6,018	3,261	918	120	10	.	12,858
1.00-1.99	124	1,380	1,437	1,095	610	127	4	.	4,777
2.00-2.99	.	66	440	384	183	60		.	1,133
3.00-3.99	.	.	56	140	76	16	.	.	288
4.00-4.99	.	.	.	17	22	6	.	.	45
5.00-5.99	.	.	.	1	2	.	.	.	3
6.00-6.99	.	.	.	.	1	1	.	.	2
7.00-7.99	.	.	.	.	.	.	.	.	0
8.00-8.99	.	.	.	.	.	.	.	.	0
9.00-Greater	.	.	.	.	.	.	.	.	0
Total	1,146	2,955	7,951	4,898	1,812	330	14	0	19,106

EXAMPLE PROBLEM III-2-5

FIND:

Calculate (a) the average (over the directional band) wave height H_{mo} and determine an equivalent significant wave height H_{sig}, (b) wave approach angle in the sediment transport coordinate system, (c) percentage of the total wave data represented by this wave directional angle band, (d) average peak spectral wave period T_p, (e) breaking wave height H_b and angle α_b, and (f) potential longshore sand transport rate.

GIVEN:

Wave statistics presented in Table III-2-2. The water depth at which the wave statistics were developed is 27 m. The azimuth angle (from north) of the outward normal to the beach of interest is approximately 102°. The sediment is quartz-density sand, and $K_{sig} = 0.39$. Assume $\kappa = 1.0$ for simplicity.

SOLUTION:

(a) Weighting the average wave height in each wave height band with the total number of observations in that band, calculate an average wave height, $H_{mo} = 0.93$ m; that is:

$$H_{mo} = [(0.5)(12858) + (1.5)(4777) + (2.5)(1133) + (3.5)(288) + (4.5)(45) + (5.5)(3) + (6.5)(2)] / 19106$$
$$= 0.93 \text{ m}$$

As discussed in Part II-1, H_s and H_{mo} are approximately equal when irregular wave profiles are sinusoidal in shape. However, as depth decreases and waves shoal prior to breaking, they become nonlinear and peaked in shape rather than sinusoidal. According to Part II-1, the two parameters are within 10 percent of each other if the depth (in meters) is greater than or equal to $0.0975\ T_p^2$. In the example, 27 m is greater than $0.0975\ (8.5\ \text{sec})^2 = 7.1$ m; therefore, the average significant wave H_{sig} may be said to be approximately equal to H_{mo}.

(b) The data in Table III-2-2 represent the angle band that is centered around WIS angle = 135°. However, the shoreline outward normal angle is 102° (from north), which means that the angle associated with transport calculations $\alpha = 102° - 135° = -33°$.

(c) WIS RAL2 data are presented every 3 hr for a 20-year time period, totalling 58,440 data points for each station (in all wave direction angle bands). The total number of occurrences listed in Table III-2-2 for the 45° direction band centered around 135° equals 19,106, which means that the percentage of the total data represented by this angle band equals 19,106/58,440 x 100 percent = 32.7 percent. Scanning the other data for this station (Hubertz et al. 1993, p. A-288 - A-289), this angle band constitutes the dominant wave direction in terms of number of occurrences for station 72.

(Continued)

Example Problem III-2-5 (Concluded)

(d) A weighting procedure using wave period bands similar to that used for the wave height bands can be used to calculate an average peak spectral wave period $T_p = 8.4$ sec; that is,

$$T_p = [(4)(1146) + (6)(2955) + (8)(7951) + (10)(4898) + (12)(1812) + (14)(330) + (16)(14)] / 19106$$
$$= 8.4$$

(e) Compute the wave celerity C_1 and group celerity C_{g1} for the offshore wave data in water depth $d = 27$ m and for the wave period $T_p = 8.4$ sec; and find $C_1 = 12.2$ m/s and $C_{g1} = 7.6$ m/s. Use Newton-Raphson iteration (or a similar technique) to solve Equation 2-16 for the breaking wave height H_b using the offshore wave height value $H_1 = 0.93$ m and angle $\alpha_1 = -33°$, and find

$$H_b \cong 1.2 \text{ m (3.9 ft)}$$

The breaking wave angle is then computed from Equation 2-14:

$$\alpha_b = -8.8 \text{ deg}$$

(f) Compute the potential longshore sediment transport rate for $K_{sig} = 0.39$ using Equation 2-7b:

$$Q = (0.39) [(1025)(9.81)^{0.5} / (16)(2650-1025)(1-0.4)] (1.2)^{5/2} \sin(2(-8.8)) = -0.038 \text{ m}^3/\text{s}$$

Convert the result to annual equivalent transport and multiply by the percent annual occurrence of this event, 32.7%:

$$Q = (-0.038 \text{ m}^3/\text{sec}) (3600 \text{ sec/hr}) (24 \text{ hr/day}) (365.25 \text{ day/yr}) (0.327)$$
$$= -393,000 \text{ m}^3/\text{yr} (-514,000 \text{ cy/yr}) \text{ (directed to the left)}.$$

Note that a similar approach can be utilized to find an answer using the CEDRS database (see Part II, Chapter 8) which provides "percent" occurrences (rather than number of observations) in 22.5° energy bands.

EXAMPLE PROBLEM III-2-6

FIND:

Calculate the potential longshore transport rate using each of the seven wave height bands given in Table III-2-2.

GIVEN:

Data in Table III-2-2, and information in Example III-2-5.

SOLUTION:

For each wave height band in Table III-2-2, the associated weighted average peak wave period T_p is calculated. Using the criteria discussed previously, the wave data are checked to ensure that $H_{sig} \sim H_{mo}$. The percent occurrence for that wave condition is determined by dividing the total number of occurrences in the wave height band in the given 45^0 directional band by the total number of records (58,440) in all directional bands. The results are shown in the table below as the "input" conditions. The breaking conditions may be computed directly by the program WISTRT or by Equations 2-14 and 2-16 in the same manner as described in Example 5, above. The potential longshore sediment transport from the 45° direction band centered approximately 33° to the right of shore-normal (looking seaward) was calculated to be -1,014,200 m³/year (-1,325,500 yd³/year) (directed to the left) for RAL2 Station 72.

Data for Example Problem III-2-6								
Input Conditions						Program Output		
Wave Condition	H_s (m)	T (sec)	α (deg)	Depth (m)	Percent Occurrence	H_b (m)	α_b (deg)	Q_I (m³/year)
1	0.5	8.3	-33	27	22.0	0.73	-7.8	-66,500
2	1.5	8.5	-33	27	8.17	1.8	-12.0	-355,000
3	2.5	9.5	-33	27	1.94	2.9	-14.2	-320,000
4	3.5	10.4	-33	27	0.493	4.0	-16.0	-198,800
5	4.5	11.5	-33	27	0.077	5.1	-17.4	-61,200
6	5.5	11.3	-33	27	0.005	6.0	-19.0	-6,500
7	6.5	13.0	-33	27	0.003	7.1	-20.0	-6,200
Total (directed to the left)								-1,014,200

Note that a similar approach can be utilized to find an answer using the CEDRS database (see Part II, Chapter 8) which provides "percent" occurrences (rather than number of observations) in 22.5° energy bands. Note that engineering judgment must be utilized when assessing actual transport rates. Values of longshore sand transport calculated with Equations 2-14 and 2-16 are based on a limitless supply of sand available for transport, which is often not the case (e.g., if structures are present, geologic features control sand movement, or sediments other than sand are in the transport region). In addition, longshore transport rates calculated using WIS data have been found to be larger than accepted rates (Bodge and Kraus 1991).

For this example calculated using waves offshore of the New Jersey coast, generally-accepted transport rates might be used to adjust the calculated values. Inspection of Figure III-2-3 indicates that net longshore transport rates are between 153,000 and 275,000 cu m/year for this portion of the New Jersey shore. This "calibration" implies that longshore sand transport rates calculated with WIS data and Equations 2-14 and 2-16 might be reduced from 15 to 27 percent of their calculated values to have application for this part of the coastline. Future calculations using WIS data for this portion of the coast could use this calibration for adjustment.

directional band. In Example III-2-6, wave data are more accurately represented by calculating a representative wave period for each of the given wave height bands for the given directional band. Each example requires the transformation of offshore wave data to breaking conditions, and subsequent computation of the associated longshore transport rate. The former can be accomplished using Equations 2-14 and 2-16 or using the program WISTRT (Gravens 1989). Both require knowledge or input of the offshore wave height H_l (WISTRT requires the significant wave height H_{sig}), associated period T, angle relative to the shoreline α, and water depth associated with wave data. The longshore transport rate can then be computed directly by Equation 2-7b. The program WISTRT uses a K value and a breaker wave height-to-depth ratio different than those used here. It also requires the percent occurrence associated with the given wave condition. The ACES program "Longshore Sediment Transport" (Leenknecht, Szuwalski, and Sherlock 1992) also provides a method for calculation of potential longshore sediment transport rates under the action of waves. Again, different constants K and κ are utilized than those presented here. Both WISTRT and the ACES programs use individual wave events as input, rather than an extended time series of wave information. A program for processing the WIS time series to obtain values of sediment transport is presented in Gravens, Kraus, and Hanson (1991).

(d) Note that both Examples III-2-5 and III-2-6 employed the same wave data, but Example III-2-6 computed the transport for discrete wave height bands whereas Example III-2-5 computed the transport for a single, band-averaged wave height. The transport computed in Example III-2-6 is more than double that in Example III-2-5. This difference is due to the nonlinear dependence of the transport equation on breaking wave height. If, for example, wave heights are Rayleigh distributed and the waves are all of uniform period, the transport rate computed using the distribution of wave heights will be about 1.53 times larger than that computed using only the band-averaged wave height.

(e) Bodge and Kraus (1991) and others (e.g., Kraus and Harikai 1983; Gravens, Scheffner, and Hubertz 1989; Gravens 1990a) have observed that use of the CERC formula (with the K_{SPM} coefficient) and WIS hindcast wave data have yielded potential longshore sand transport magnitudes that are two to five times larger than values for the region as estimated from dredging records, bypassing rates, or volumetric change. The longshore sand transport rate determined in Example III-2-6 represents the potential longshore transport rate, which depends on an available supply of littoral material. Consideration of the availability of littoral material; location, type, and condition of coastal structures; and sheltering specific to the project shoreline may contribute to a lower actual longshore transport rate. It is recommended when using hindcast wave data to predict potential longshore sand transport rates that other independent measures or estimates of longshore transport be used to supplement the potential transport estimate.

d. Deviation from potential longshore sediment transport rates.

(1) Temporal variations and persistence.

(a) Longshore sediment transport is a fluctuating quantity which can be depicted as shown in Figure III-2-7 where positive sediment transport is defined as positive in value if toward the right for an observer looking seaward from the beach, and negative in value if sediment transport is toward the left as noted previously and consistent with notation utilized by Walton (1972), Walton and Dean (1973), Dean (1987), and others. In terms of "Q_ℓ," on Figure III-2-7 the net longshore sediment transport rate is the "time average" transport given by

$$Q_{\ell NET} \equiv \overline{Q_\ell} = \frac{1}{T_o} \int_o^{T_o} Q_\ell(t)\,dt \qquad \qquad \text{(III-2-18)}$$

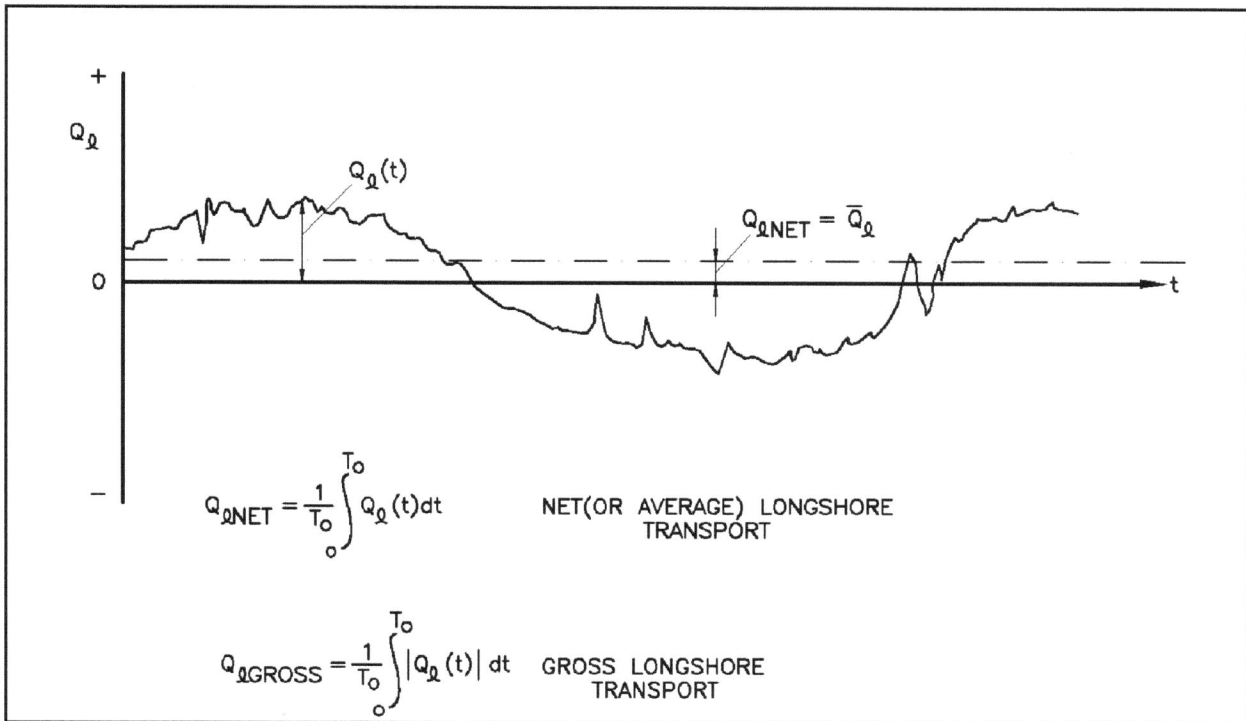

Figure III-2-7.　Longshore transport definitions

(b) The gross longshore sediment transport rate is given by

$$Q_{\ell GROSS} = \frac{1}{T_o} \int_o^{T_o} |Q_\ell(t)| \, dt \qquad\qquad \text{(III-2-19)}$$

where T_o is the length of record, often taken to be greater than 1 year, and $|Q_\ell|$ is the absolute magnitude of the longshore sediment transport rate. The gross longshore sediment transport rate is always defined positive. When the gross longshore sediment transport rate is multiplied by the time period, the derived quantity represents the total volume of sediment passing through a plane perpendicular to the shoreline regardless of the direction. The net longshore sediment transport rate may be either positive or negative.

(c) Although net and gross longshore sediment transport rates are often the most meaningful quantities for use in engineering design, the variability of the longshore sediment transport on much shorter time intervals is often of critical concern. As an example, if a channel is to be maintained clear of sediment, the pump(s) necessary to dredge the channel would need to be sufficiently large to handle the instantaneous maximum rate of longshore rate of sediment transport (that is projected to reach the channel). Typically, sizing for the maximum instantaneous rate of sediment transport would be economically unfeasible, and therefore some type of temporary sediment storage would be provided to allow for reduced pump sizing. Optimizing of the pump size and the provided sediment storage structures (possibly a groin, a breakwater, or some combination of both) will require knowledge of the fluctuating longshore sediment transport rates. Figure III-2-8 provides an example of the variability of annual rates of (net) longshore energy flux factor (which is proportional to the longshore sediment transport rate as given in Equation 2-2) for three locations on the East Coast of the United States as computed using 20 years of hindcast wave climate.

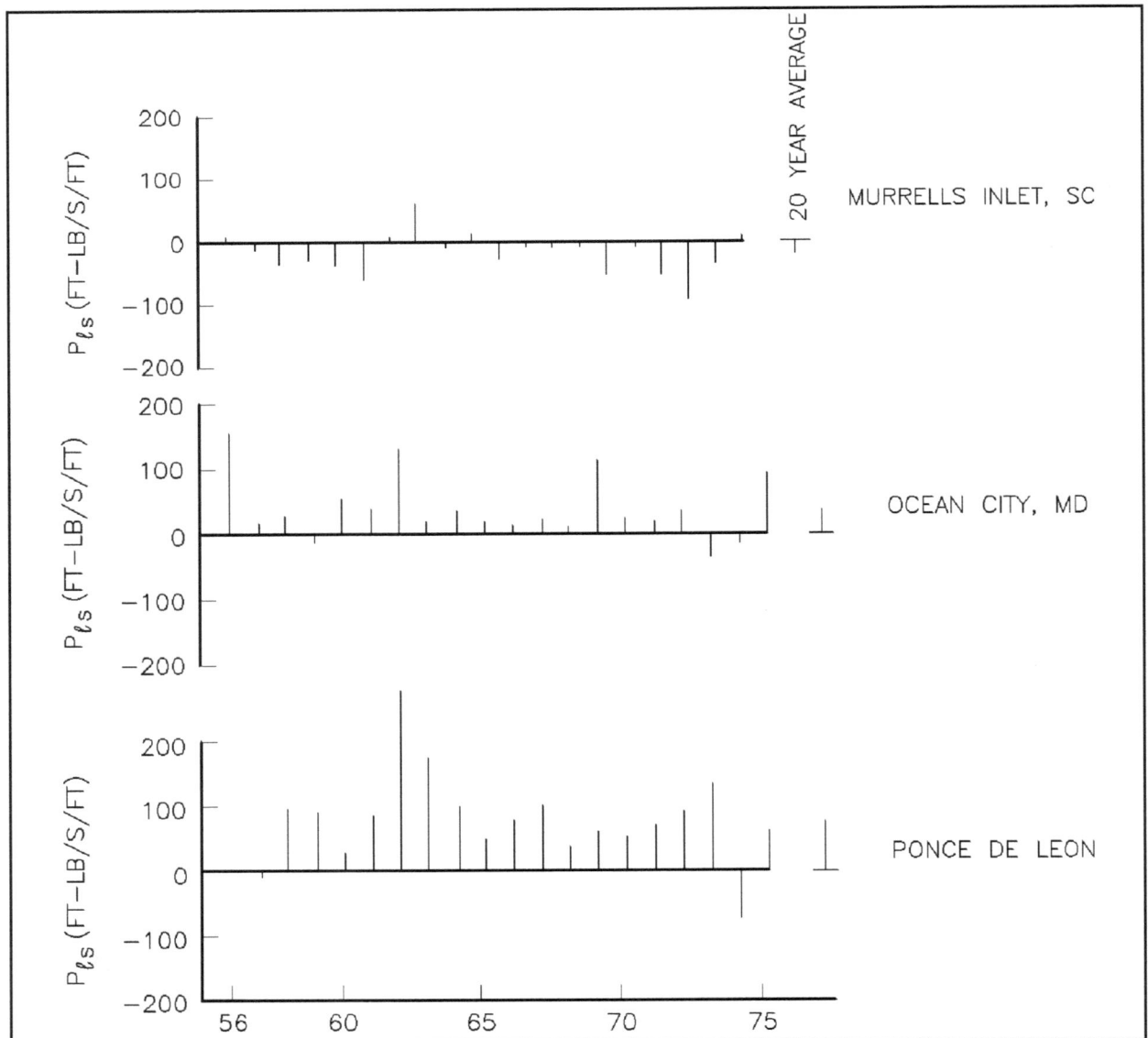

Figure III-2-8.   Time plot of annual longshore energy flux factor at three east coast sites (after Douglass (1985))

Figure III-2-9 provides an example of the variability of the monthly (net) longshore energy flux factor for the same three locations. Figures III-2-8 and III-2-9 show that as the averaging period gets shorter, the variability of the sediment transport rate increases due to the fact that integration is a smoothing process and reduces the variance in the data. For design purposes, an assessment of the uncertainty in the sediment transport climate can be addressed via simulation of the sediment transport rates. Walton and Douglass (1985) have provided an approach to such simulation for the case of monthly sand transport rates which, for the locations assessed, appeared to be in reasonable conformance with a normal distribution assumption. Figure III-2-10 from Walton and Douglass (1985) shows the distribution of monthly averages of longshore energy flux factor for a coastal location in South Carolina. The data are reasonably represented by a normal distribution, as shown by the fit of the solid line to the data. Figure III-2-11, also from Walton and Douglass (1985), is for the same location but with weekly averages. In this case, the weekly averages vary considerably from a normal probability distribution.

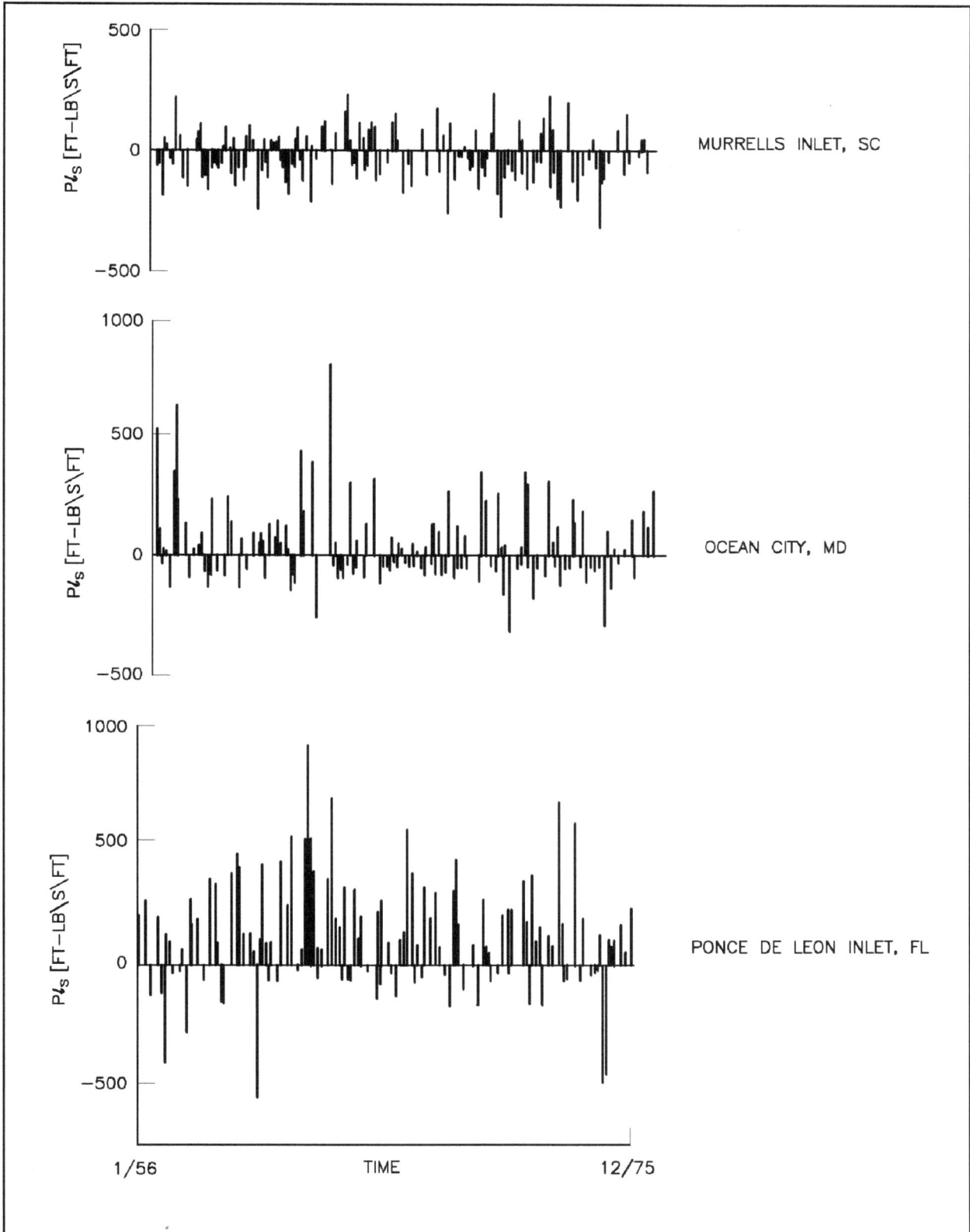

Figure III-2-9.   Time plot of monthly longshore energy flux factor time series for 1956-1975 at three east coast sites (after Douglass (1985))

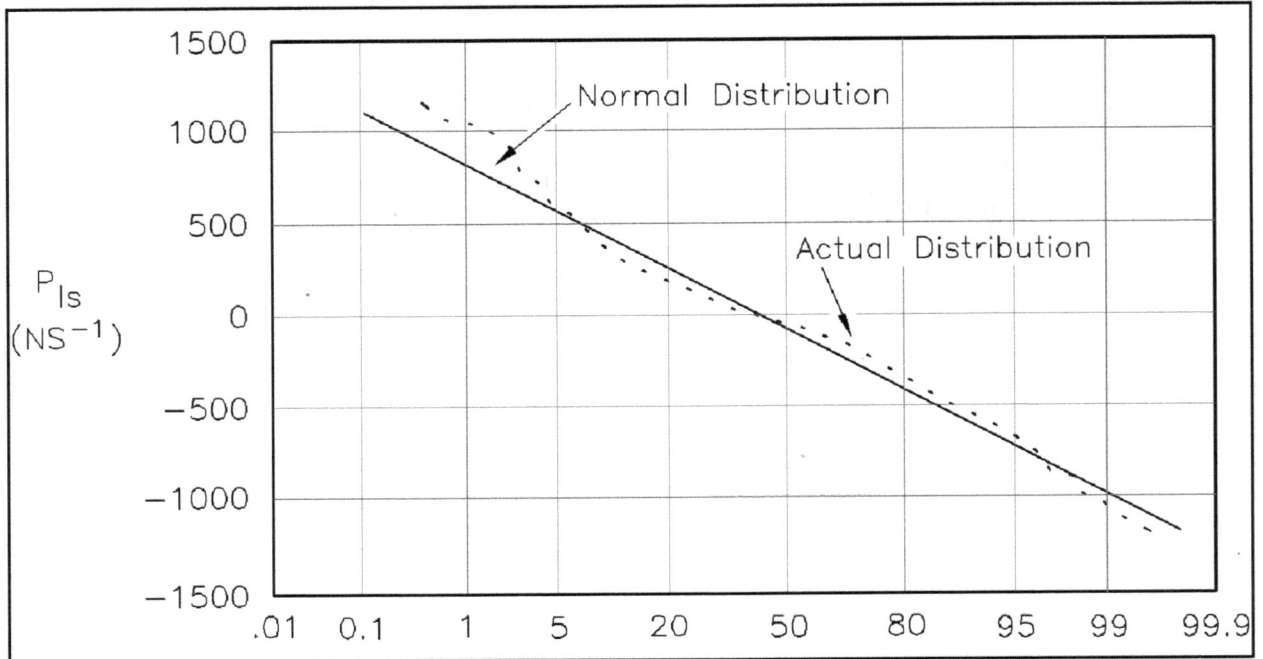

Figure III-2-10. Probability plot for monthly average P_{ls} series (Walton and Douglass 1985)

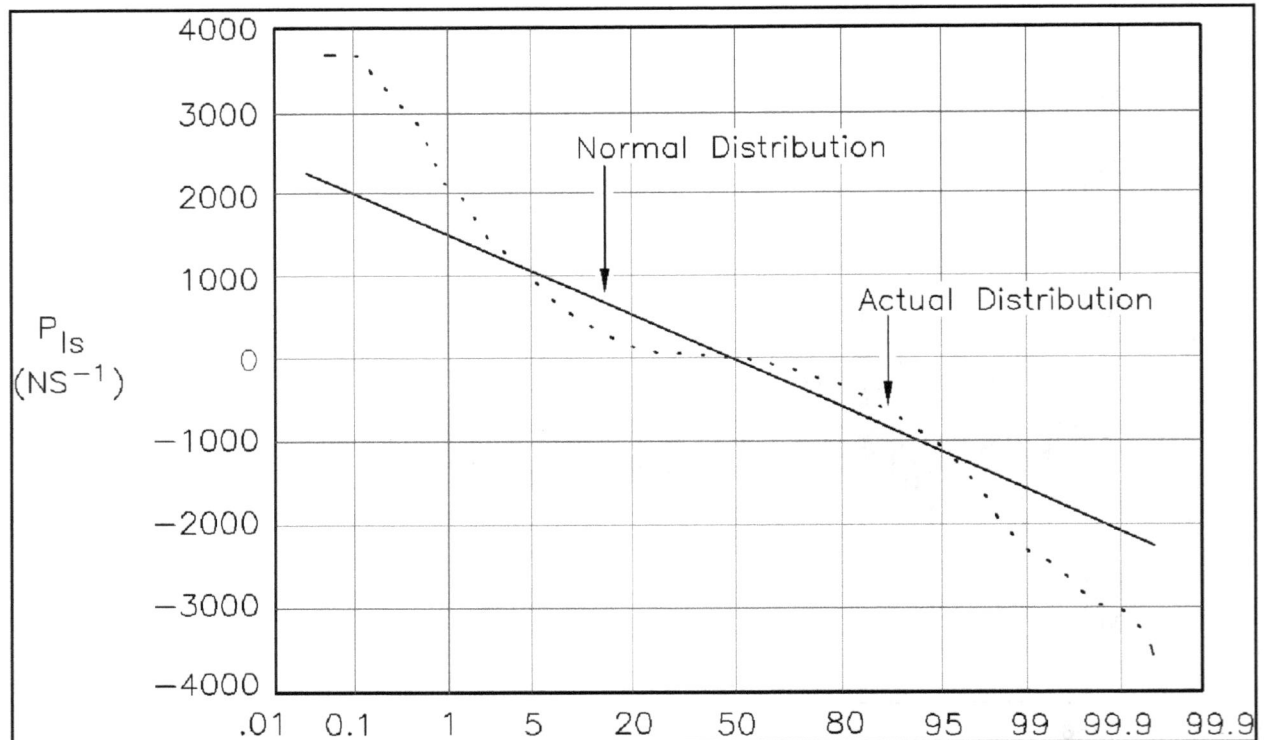

Figure III-2-11. Probability plot for weekly average P_{ls} series (Walton and Douglass 1985)

Longshore Sediment Transport

(d) Weggel and Perlin (1988) demonstrated that unidirectional annual transport variability may be described by a log-normal distribution. Thieke and Harris (1993) described similar results and noted that shorter-term (daily) dominant transports were better described by a Weibull distribution. These results apply only to statistics of transport in a single (positive or negative) direction and bidirectional distributions of transport are necessary to assess the true nature of the transport statistics for proper simulation (Walton 1989).

(e) When assessing various design alternatives, the proper distribution of the sediment transport rate must often be modeled as well as the natural persistence inherent in the data. Persistence is the measure of correlation for data closely spaced in time, i.e., large persistence means high sediment transport rates follow high sediment transport rates and low sediment transport rates follow low sediment transport rates in adjacent timeaveraging intervals. Persistence is typically measured via the autocorrelation coefficient (see, for example, Box and Jenkins (1976)). Autocorrelations for monthly and weekly longshore energy flux factors at the same South Carolina site are shown in Figures III-2-12 and III-2-13 (Walton and Douglass (1985)). Data exceeding the dotted lines suggest that persistence is evident in the data. Nonstationary and non-normal types of data may be simulated via an approach provided in Walton and Borgman (1991).

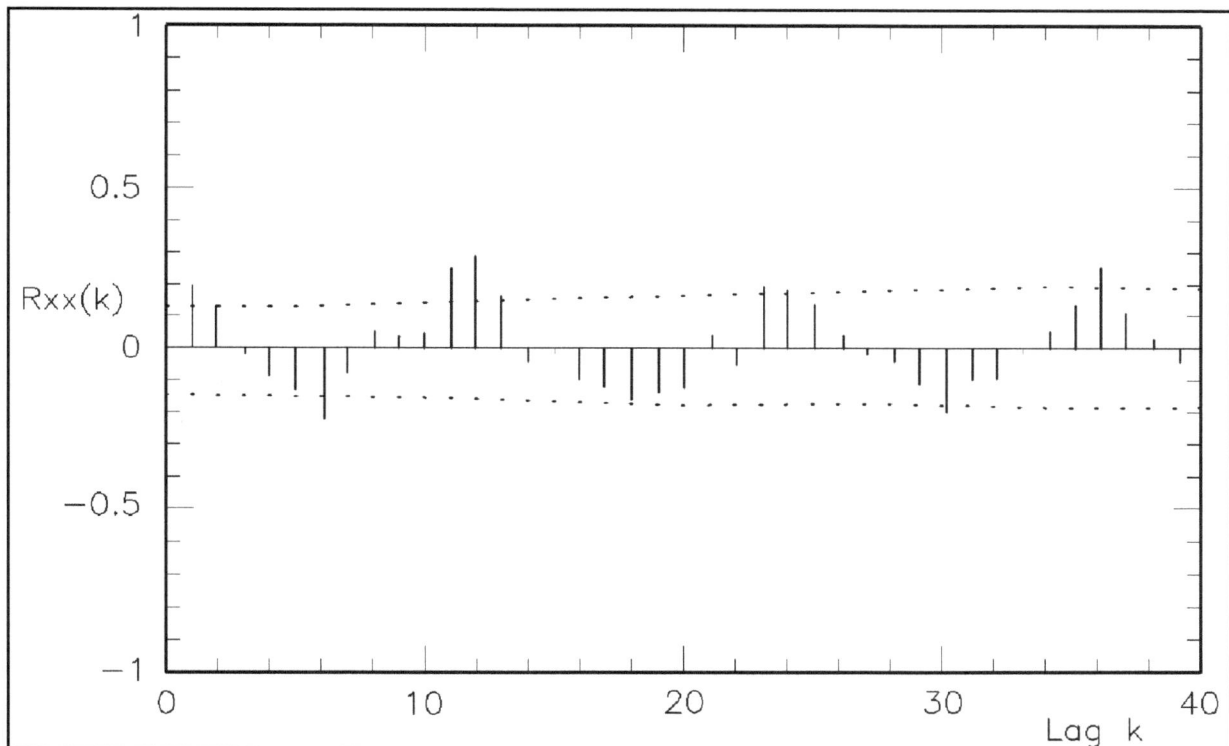

Figure III-2-12.   Autocorrelation of monthly P_{ls} series (Walton and Douglass 1985)

(f) The significance of "episodic" transport to annual conditions is not well understood. It is thought that storm-related or other episodic events may cause the bulk of longshore sediment transport observed at some locations in the long term. Douglass (1985) suggested that 70 percent of the gross transport occurred during only 10 percent of the time at many sites via analysis of wave hindcast data.

(2) Wave data accuracy. The accuracy of wave data used to calculate potential transport rates also leads to uncertainty in predictions. Wave measurements and observations have associated uncertainties based on

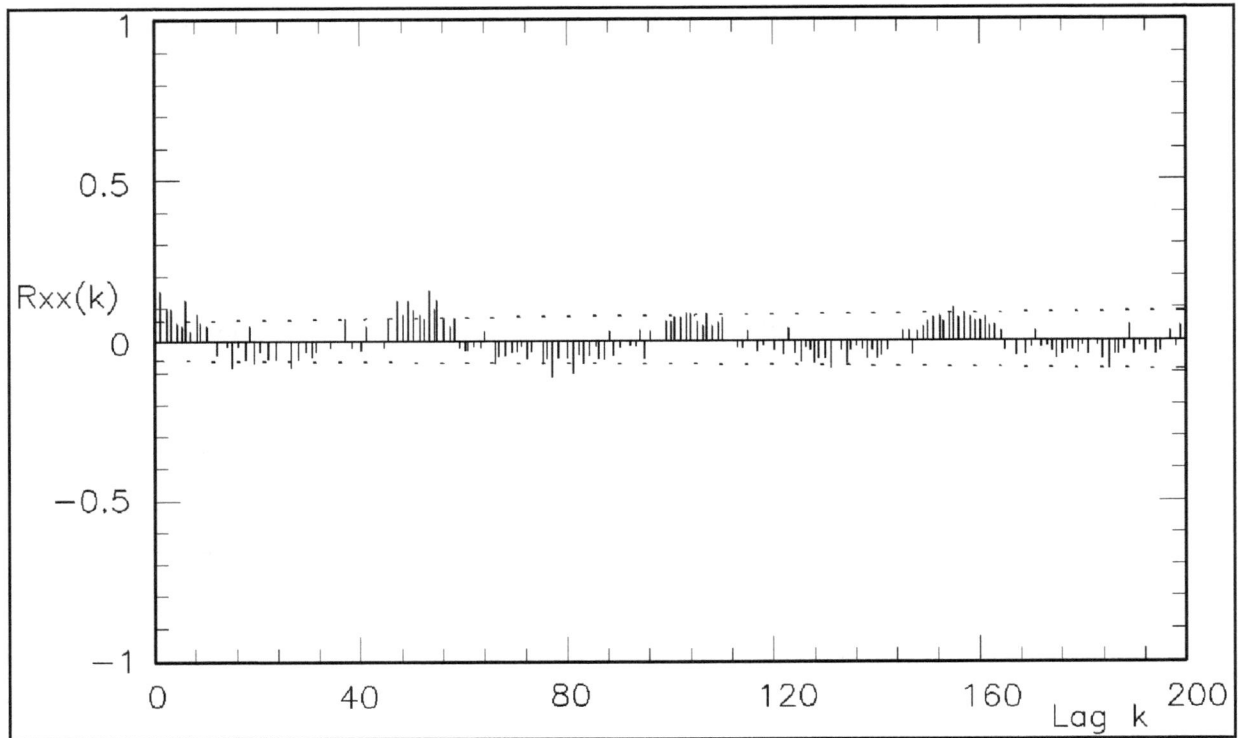

Figure III-2-13. Autocorrelation of weekly average P_{ls} series (Walton and Douglass 1985)

instrumentation accuracy and observer bias. Given that there are breaking wave height and wave angle uncertainty values ΔH_b and $\Delta \alpha_b$, respectively, an associated longshore transport uncertainty ΔQ_ℓ can be calculated. From Equation 2-7b, we see that $Q_\ell \sim H_b^{5/2} \sin 2\alpha_b$. Uncertainty in the longshore transport rate can be estimated by including the uncertainties in breaking wave height and angle:

$$Q_\ell \pm \Delta Q_\ell \sim (H_b \pm \Delta H_b)^{\frac{5}{2}} \, sin2(\alpha_b \pm \Delta\alpha_b) \tag{III-2-20}$$

Assuming that the wave angle at breaking is small, and using the first two terms of a Taylor series expansion of Equation 2-20, the uncertainty in the longshore transport rate is estimated as

$$\Delta Q_\ell \sim \pm Q_\ell \left(\frac{\Delta\alpha_b}{\alpha_b} + \frac{5}{2}\frac{\Delta H_b}{H_b} \right) \tag{III-2-21}$$

Thus, a 15-percent uncertainty in wave height and 15 percent uncertainty in wave angle result in 37.5- and 15-percent uncertainty contributions for height and angle, respectively, totaling a 52.5-percent uncertainty in Q_ℓ.

   (3) Sand supply availability. Application of the potential longshore sand transport equations presented herein results in an estimate of transport that implies the availability of an unlimited supply of sand. If sand availability is limited, such as on a rocky or reef coastline, or interrupted such as in the vicinity of groins, jetties, or breakwaters, the actual longshore sand transport rate will be less than the calculated rate.

e. Littoral drift roses.

(1) The littoral drift rose is a potentially useful tool for interpreting littoral drift trends along a section of shoreline where the shoreline curvature is mild and bottom contours are reasonably parallel to the shoreline (Walton 1972, Walton and Dean 1973). The littoral drift rose shows how the littoral drift changes with change in shoreline orientation relative to prevailing wave climatology. The drift rose is constructed using standard prediction techniques for calculating littoral drift given that the shoreline is oriented in a direction Θ_n, where Θ_n is the azimuth angle of the perpendicular to the shoreline in the seaward direction (see Figure III-2-14 for an example measurement of Θ_n at Ponte Vedra Beach, FL). The principle behind the littoral drift rose is that a range of shoreline orientations is considered. These orientations correspond to the range that exists at the study site. For each possible shoreline orientation, as described by Θ_n, the total positive and negative littoral drifts along the shoreline are calculated for a given time-averaging interval (i.e., 20 years, annual, monthly, etc.). These calculated drift values are plotted in a polar plot, as shown in Figure III-2-15 for the Ponte Vedra Beach, FL area. From the plots, the littoral transport rate for any given shoreline orientation can be determined by entering the plot with the seaward directed normal of the shoreline orientation "Θ_n" and reading off the total positive, total negative, and net littoral drift values. As the littoral drift for a given wave angle is proportional to $\sin(2\alpha_b)$, the net drift rose average for a real wave climatology has lobes that cause the magnitude to vary in a similar manner as $\sin(2\Theta_n)$.

(2) Littoral drift roses so constructed may be utilized in helping to identify tendencies of the shoreline toward stability or instability. As an example, for a long shoreline with variations of shoreline orientation, it is possible that there is a null (also termed nodal) point along the shoreline, which is defined as the location for which the averaged positive and negative littoral transport have the same magnitudes, yielding zero net drift. The littoral drift rose can be utilized in helping to identify this null (nodal) point. On the littoral drift rose, this null (nodal) point is reflected as a crossing of the positive and negative littoral drifts on the total littoral drift rose, or, as the point at which the net drift is zero on a net drift rose. Walton (1972) and Walton and Dean (1976) identified potential null points on barrier islands along the Florida coastline using the littoral drift rose and by integrating the longshore sand transport equation using changes in historical shorelines along with a known boundary condition of sediment accumulation at the end of the islands. Along the East Coast of the United States, there are several well-documented null points, such as just north of the Delaware-Maryland state boundaries and in New Jersey near Barnegat Inlet. At both of these locations the shoreline orientation changes significantly near the sites of the null (nodal) points and the net drift is to the south, south of the null (nodal) points, and to the north, north of the nodal points. Mann and Dalrymple (1986) examined the location of a null point along the Delaware coastline and concluded that there was a significant variation in its annual location, corresponding to the variation in the annual littoral drift rates. Such a shifting should be reflected in littoral drift roses for different averaging periods.

(3) Walton (1972) and Walton and Dean (1973) examined the stability of many shorelines using the littoral drift rose concept. Figure III-2-16 shows an "unstable" littoral drift rose and a barrier island orientation such that the initial island is oriented at the angle of the "unstable" null (nodal) point. Applying this rose to the island in the figure, it can be seen that if a negative perturbation (recession of shoreline due to mining, barrier overwash, etc.) is initiated on the island (away from the ends of the island where the transport scenario would not conform to the assumptions of the calculated littoral drift), the induced transport is away from the null point resulting in an erosional feature growth and potential island breakthrough (hence the term "unstable"). In a similar manner, for the "unstable" littoral drift rose an induced transport response to a positive shoreline perturbation in the same location (perhaps due to dredge spoil placement) would induce further transport of sand toward the perturbation, causing the perturbation to grow (a self-sustaining "positive" feedback mechanism). Walton (1972) has postulated that cuspate features such as shown in Figure III-2-17 (often found in bays, rivers, etc.) may be the result of "unstable" littoral drift roses for those sites due to the wave climatology experienced in long narrow fetch-enclosed locations.

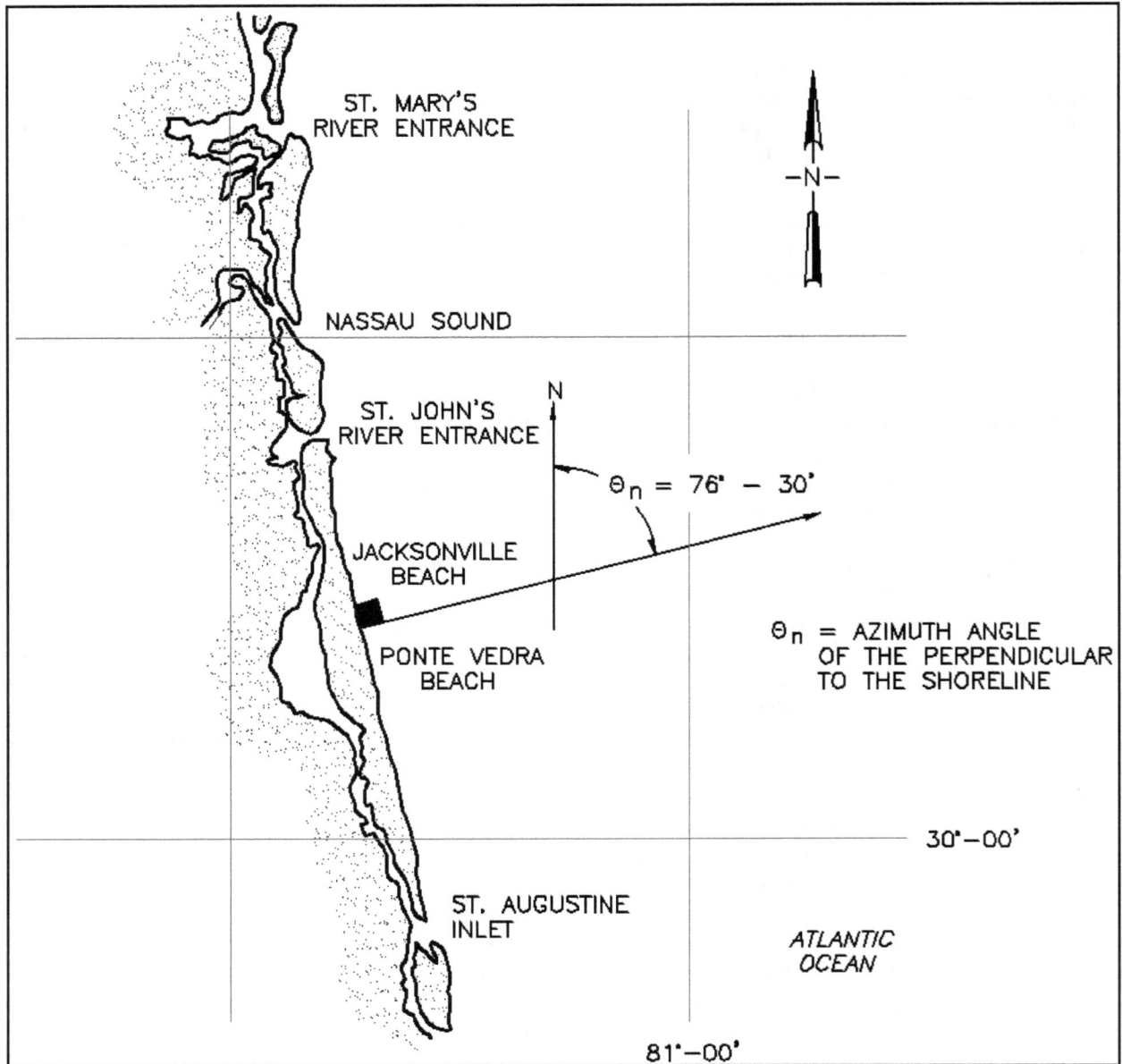

Figure III-2-14. Azimuth of normal to shoreline at Ponte Vedra Beach, Florida (Walton 1972)

(4) Another possibility is that the littoral drift rose along a shoreline is as shown in Figure III-2-18. In this case the null (nodal) point is stable. Perturbations in the shoreline, positive (accretions) or negative (recessions), now result in transport which tends to reduce rather than accentuate the perturbation (a "negative" feedback mechanism). In this case there is not a self-sustaining tendency toward island breaching Walton (1972) noted that the east coast of Florida is characterized by this type of littoral drift rose and, in fact, there are relatively few natural inlets along this coastline and a historical tendency of those inlets that are opened by man to close (unless closure is prevented via coastal structures such as jetties).

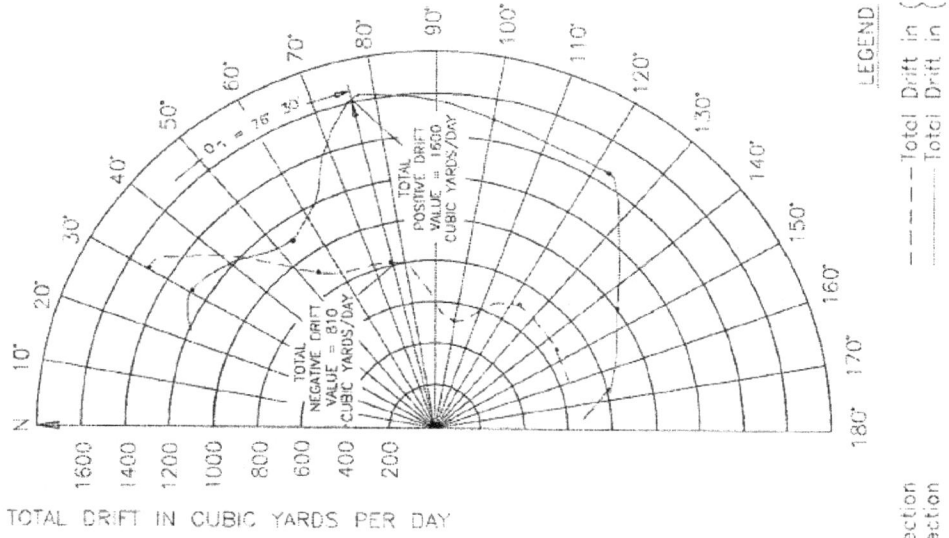

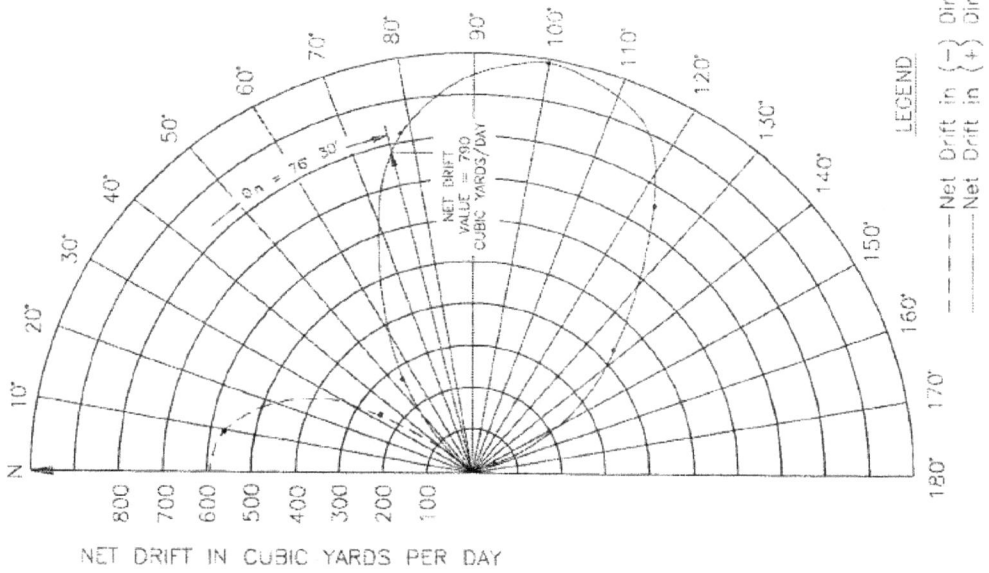

Figure III-2-15. Net and total littoral drift at Ponte Vedra Beach, Florida (Walton 1972)

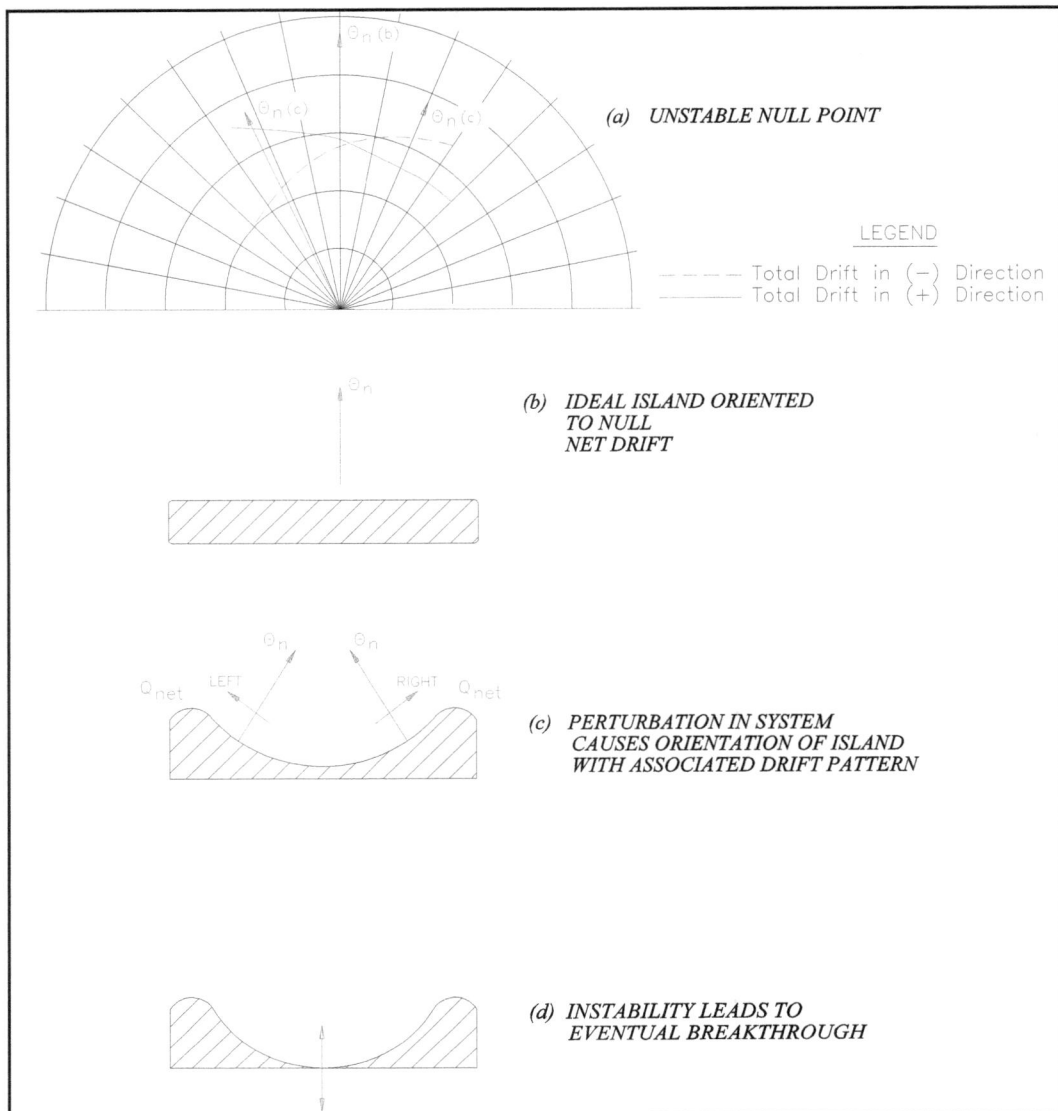

Figure III-2-16.   Ideal case of an unstable null point (Walton 1972)

f.   Cross-shore distribution of longshore sediment transport.

(1) The relationships given herein yield rates of total longshore sediment transport. Some applications require evaluations of the cross-shore distribution of the transport; e.g., for the effective design of groins, jetties, and sand weirs for weir jetties. In addition, a description of the cross-shore transport distribution is required in many computer simulation models.

(2) Bodge (1986) presents a literature review of the data and models which describe the cross-shore distribution of the longshore sediment transport. The collection of field data regarding this topic is difficult. The earliest approach was to use sand tracers (e.g., Zenkovitch 1960; Ingle 1966; Inman, Komar, and Bowen 1968; Inman, Tait, and Nordstrom 1971; Inman et al. 1980; Kraus et al. 1982, White 1987). As an example, Zenkovitch (1960) determined distributions at a coastal site by averaging a large number of tracer observations and found three maxima for the sand transport: two over longshore bars and a third in the swash

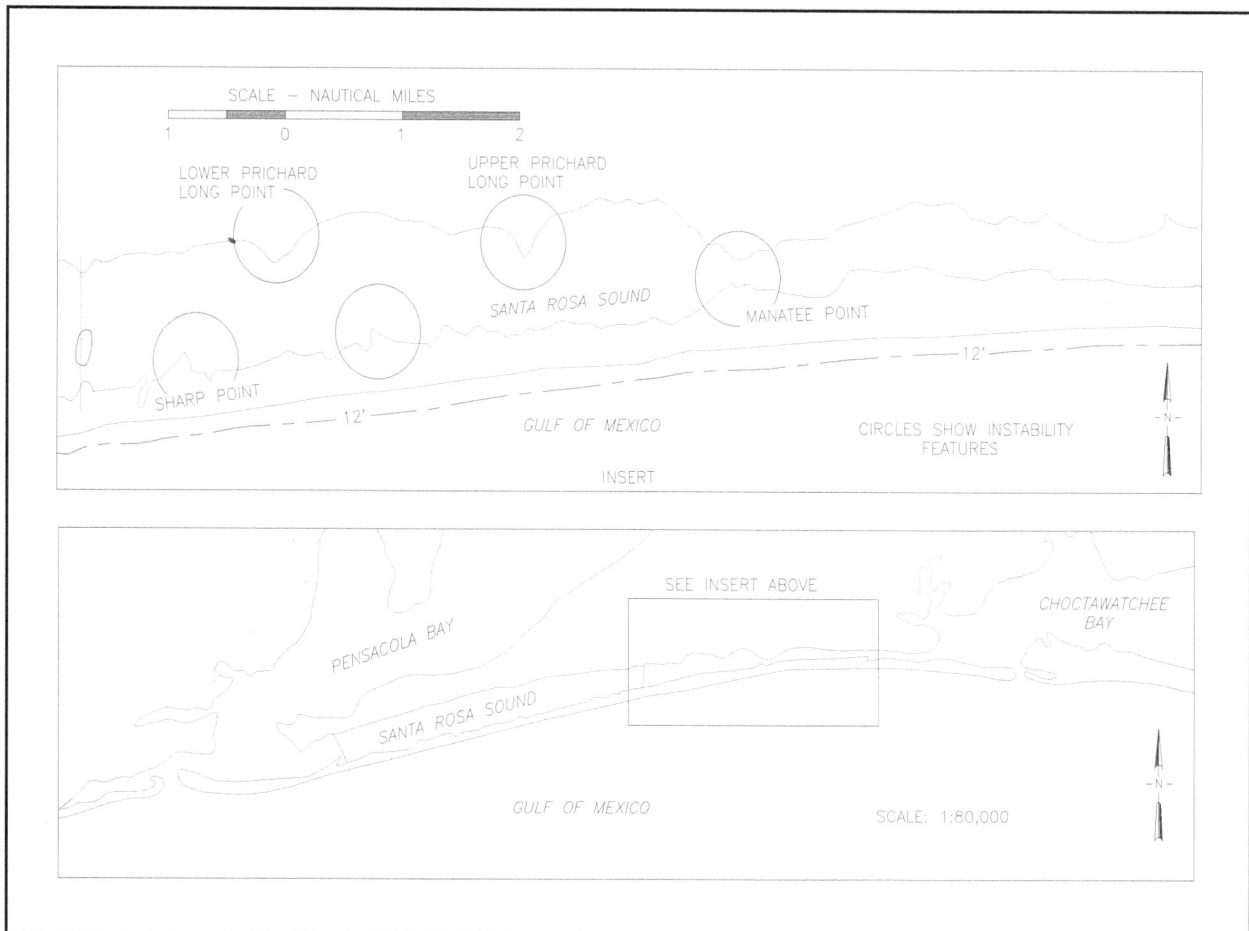

Figure III-2-17.   Instability-formed capes in Santa Rosa Sound (Walton 1972)

zone. Ingle (1966) used fluorescent-dyed sediment at five test sites along the California coast to investigate the longshore rate of tracer dispersion, as well as its cross-shore distribution. An example of the cross-shore distribution of tracer density measured at Goleta Point in October 1961 is shown at three measurement intervals for two surf zone cross-sections in Figure III-2-19. Cross-section A represents conditions 46 m (150 ft) downdrift of the release line, and cross-section B is 46 m (150 ft) downdrift of cross-section A. The distributions indicate that the most rapid tracer dispersal took place at the edge of the swash zone, while the greatest tracer transport took place at the breaker zone.

(3) Using sediment traps operated from a Florida pier across the outer portion of the surf zone, Thornton (1972) found that longshore transport was a maximum on the seaward side of the longshore bar where waves were breaking. Sawaragi and Deguchi (1978) placed round sediment traps divided into pie-shaped sections into the beach and found three basic longshore transport distribution profiles: (1) maximum in the swash zone or near the shoreline, (2) maximum at the breaker line, and (3) bimodal with maxima at the shore and breaker lines. From a field study using fluorescent tracers, Kraus et al. (1982) likewise observed similar results in addition to a generally uniform cross-shore distribution. Berek and Dean (1982) inferred longshore transport distribution from measurements of contour rotation in a pocket beach at Santa Barbara, California, and found from their data that the transport was greatest across the inner surf zone. Kraus and Dean (1987) and Kraus, Gingerich, and Rosati (1989) give general examples of vertical and cross-shore distributions of longshore sediment flux measured with portable traps (see Figure III-2-2).

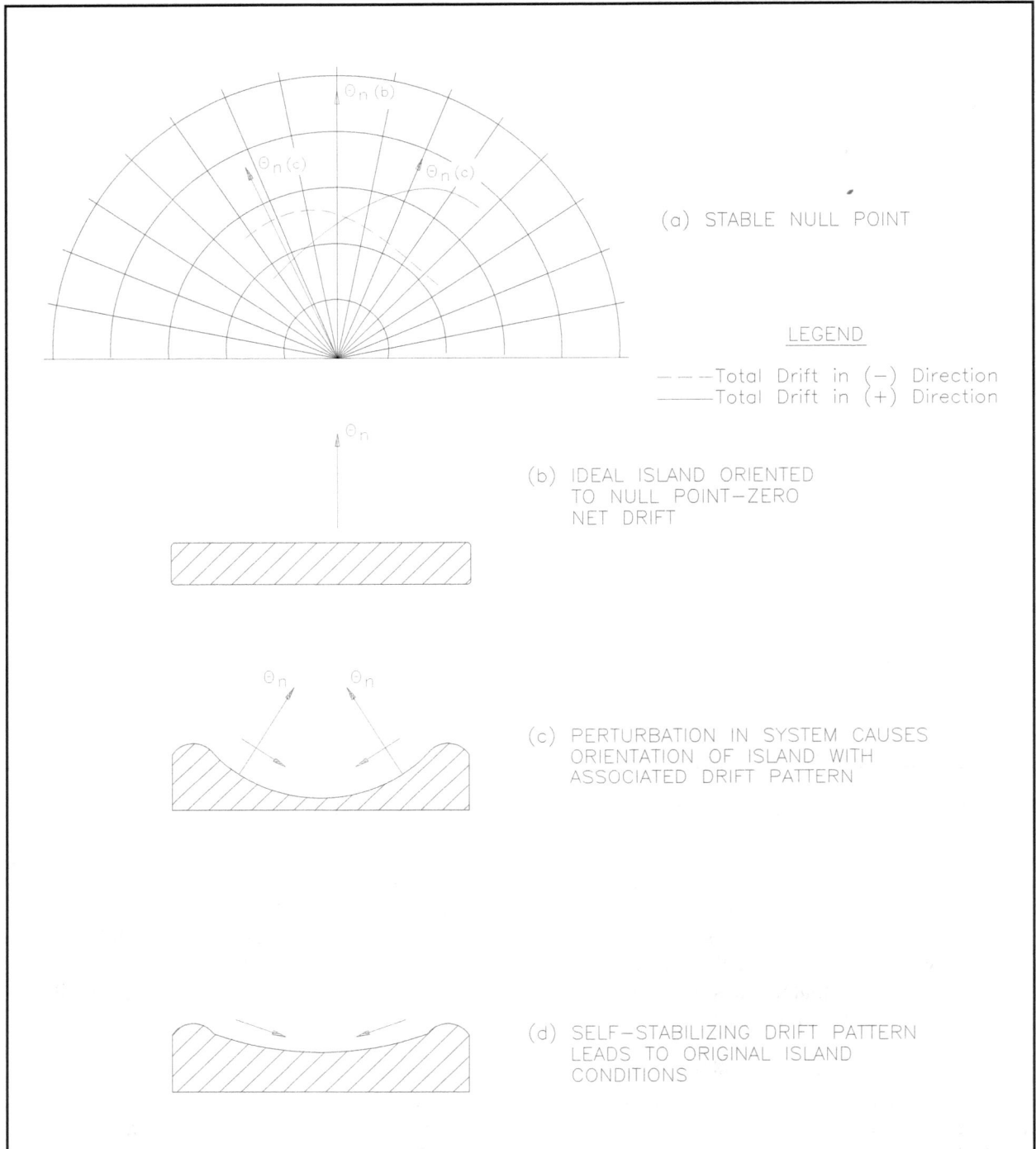

Figure III-2-18 shows a half-circle radial diagram with labels $\Theta_n(b)$, $\Theta_n(c)$, $\Theta_n(c)$.

(a) STABLE NULL POINT

LEGEND

– – –Total Drift in (−) Direction
———Total Drift in (+) Direction

Θ_n

(b) IDEAL ISLAND ORIENTED
TO NULL POINT−ZERO
NET DRIFT

Θ_n Θ_n

(c) PERTURBATION IN SYSTEM CAUSES
ORIENTATION OF ISLAND WITH
ASSOCIATED DRIFT PATTERN

(d) SELF−STABILIZING DRIFT PATTERN
LEADS TO ORIGINAL ISLAND
CONDITIONS

Figure III-2-18.   Ideal case of a stable null point (Walton 1972)

(4) Kana (1978) measured the vertical suspended sediment concentration at locations across the surf zone, and longshore current velocity at a mid-surf position. His measurements indicated that, for spilling waves, sediment concentration rapidly increased inside the breakpoint, then remained relatively constant under the bore as it propagated towards the beach. For plunging waves, sediment concentration peaked

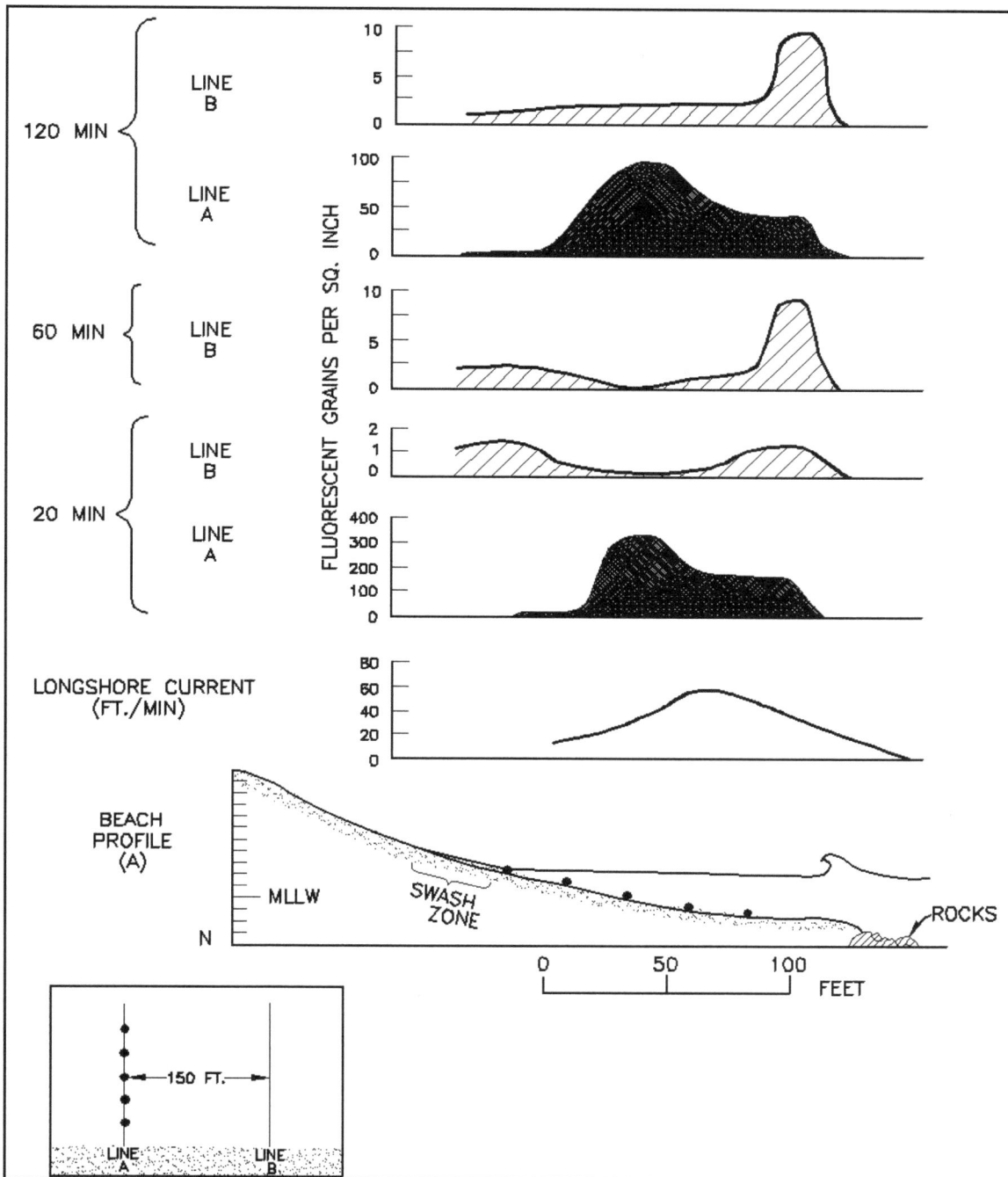

Figure III-2-19.   Distribution of tracer density across the surf zone, 20 October 1961 Goleta Point, California, experiment (Ingle 1966)

within a few meters of the breakpoint, then decreased gradually towards shore (see Figure III-2-20). Downing (1984) and Sternberg, Shi, and Downing (1984) likewise measured vertical sediment concentration profiles simultaneously with the longshore current, and found maximum transport at about the mid-surf-zone location. The measurements did not include the inshore portion of the surf zone and did not account for bed-load transport.

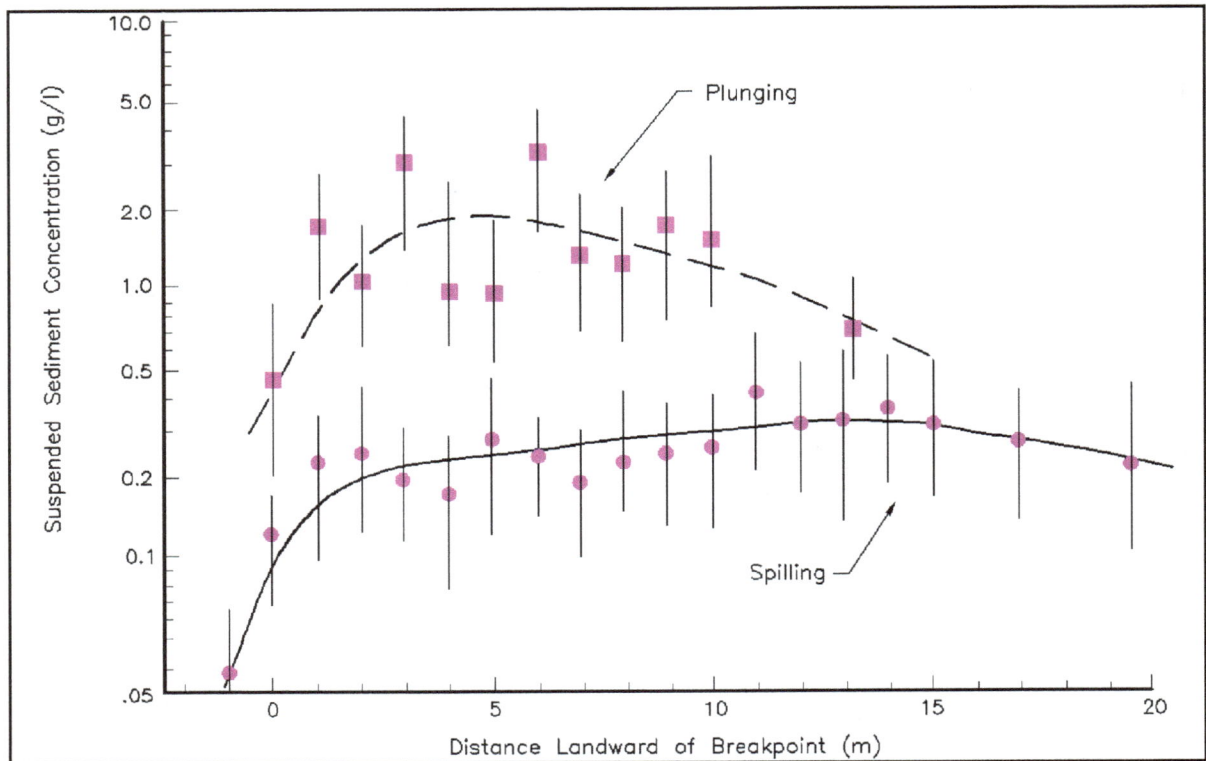

Figure III-2-20. Distribution of mean sediment concentration at 10 cm above the bed, relative to wave breakpoint (Kana 1978)

Bodge and Dean (Bodge 1986; Bodge and Dean 1987a, 1987b) utilized a short-term sediment impoundment scheme in the field and laboratory, consisting of the rapid deployment of a low profile, shore-perpendicular barrier. Beach profile changes in the vicinity of the barrier were determined from repeated surveys over short intervals of time, simultaneously with measurements of surf zone wave heights and currents. Four separate field experiments were undertaken at the CHL Field Research Facility at Duck, North Carolina. Two cross-shore distribution profiles from the field experiments are shown by the light lines in Figures III-2-21a and III-2-21b, indicating the presence of a maxima in the outer surf zone just shoreward of the breaker zone, and a second maxima in the swash zone. Light lines in Figures III-2-21c through III-2-21g show laboratory results for wave types including spilling, plunging, and collapsing. The plunging/spilling laboratory conditions (Figure III-2-21d) were modeled after the surf zone conditions of the field experiment in Figure III-2-21a. The error bars on the light lines indicate the most probable range of the local longshore transport contribution at each location across the surf zone, reflecting uncertainties in the assessed local magnitude of cross-shore transport, updrift limit of impoundment, and/or the degree of groin bypassing. The data suggest that the transport distribution is generally bimodal with peaks at the shoreline and at the mid-outer surf zone. The relative significance of the peaks was seen to shift from the near-breakpoint peak to the near-shoreline peak as the breaking wave condition varied from spilling to collapsing. Longshore transport seaward of the breakpoint represented about 10 to 20 percent of the total. Swash zone transport accounted for at least 5 to 60 percent of the total for spilling to collapsing conditions, respectively.

(5) In general, the field (and laboratory) studies of longshore transport indicate that (1) significant levels of transport may occur at and above the shoreline, (2) about 10 to 30 percent of the total transport occurs seaward of the breaker line, (3) maximum local transport has been noted within the shoreward half of the surf zone as often as within the seaward half, and (4) greater transport is often associated with shallower depths

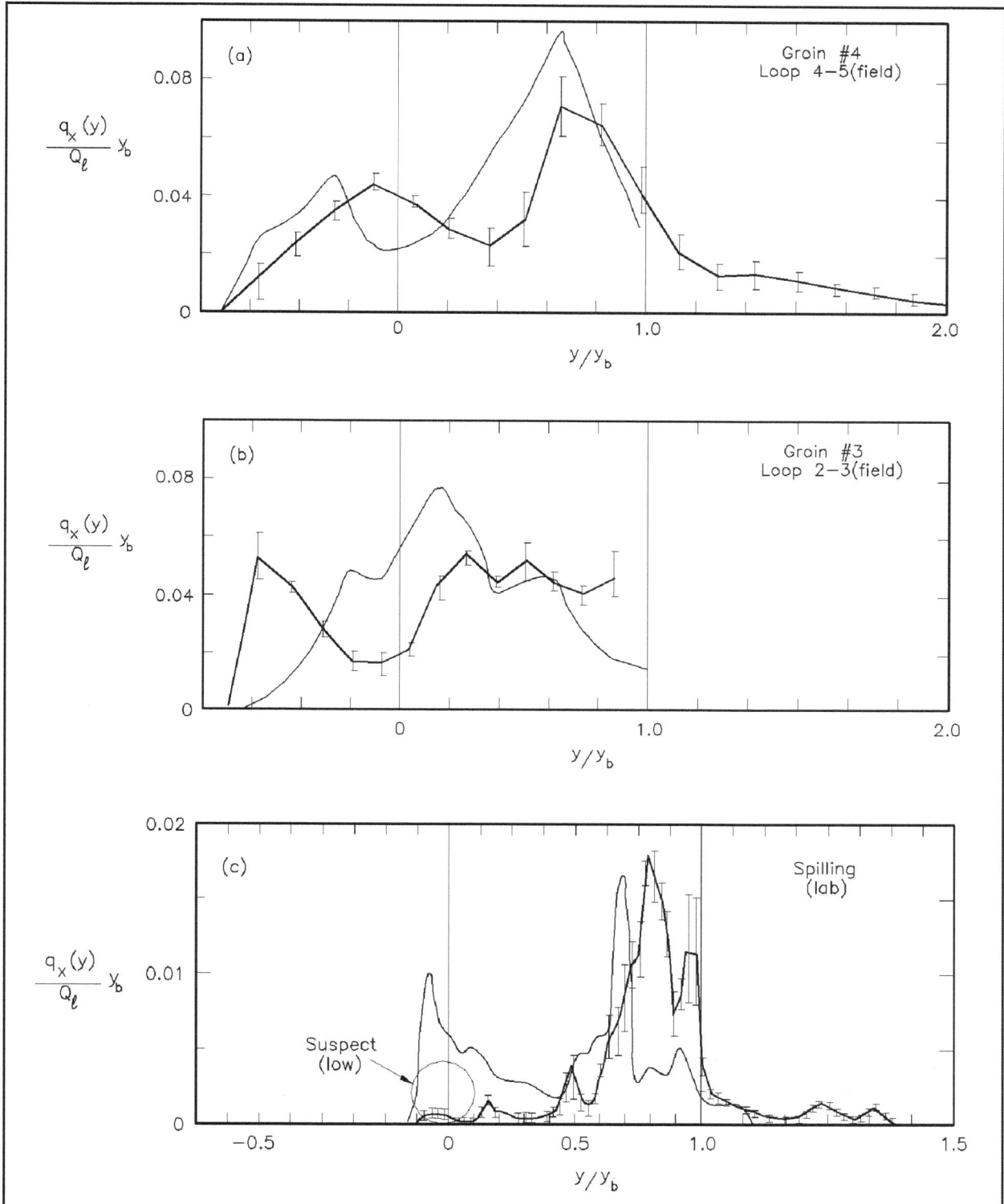

Figure III-2-21. Cross-shore distribution of longshore sediment transport as measured by Bodge and Dean (Bodge 1986, Bodge and Dean 1987a, 1987b) at Duck, North Carolina, and in the laboratory (Continued)

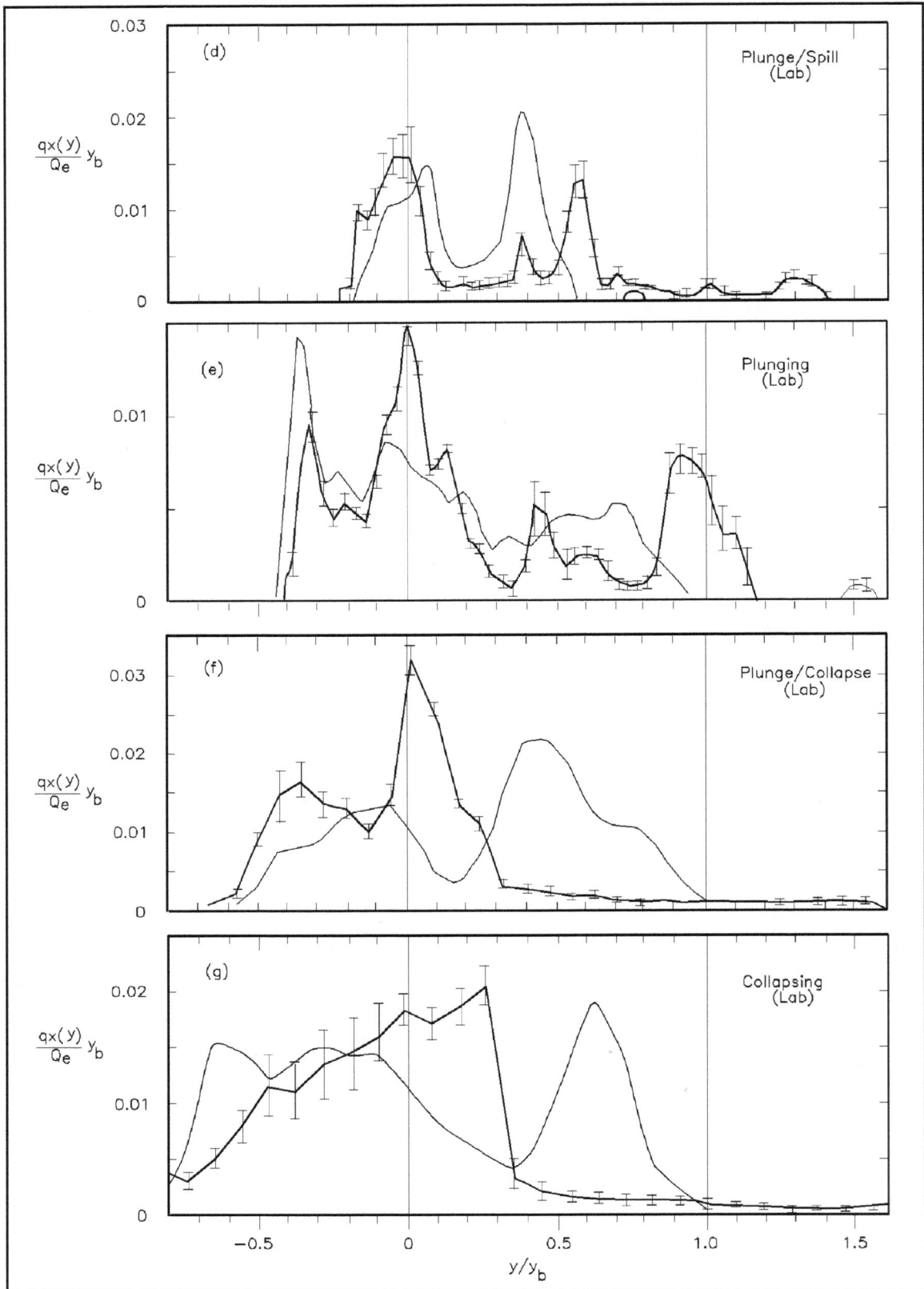

Figure III-2-21. (Concluded)

and breaking waves (i.e., breakpoint bars and the shoreline). Overall, field measurements demonstrate great variability in the shape of the longshore transport distribution profile across shore.

(6) Stresses exerted by waves vary in the cross-shore direction, generally decreasing from the breaker zone to the shoreline, but not necessarily in a uniform manner due to the presence of bars and troughs on the beach profile. The longshore current also has a characteristic profile, and because sand transport is a result of the combined waves and currents, its distribution will be related to their distributions. Theoretical relationships for the cross-shore distribution of longshore sediment transport have been postulated (e.g., Bagnold 1963, Komar 1977, Walton 1979, McDougal and Hudspeth 1981, Bailard and Inman 1981); however, they have not been shown to reproduce field measurements well. Using data from their field and laboratory experiments, Bodge and Dean tested five existing cross-shore distribution relationships, which were concluded to give from fair to poor correlation with measurements. Bodge and Dean based on the work of Bagnold (1963) also proposed a relationship for the cross-shore distribution of longshore sediment transport which assumes that sediment is mobilized in proportion to the local rate of wave energy dissipation per unit volume, and transported alongshore by an ongoing current

$$q_x(y) = k_q \frac{1}{d} \frac{\partial}{\partial x} (E \, C_g) \, V_\ell \qquad\qquad \text{(III-2-22)}$$

where $q_x(y)$ is the local longshore transport per unit width offshore, y represents the cross-shore coordinate, k_q is a dimensional normalizing constant, d is the local water depth in the surf zone (including wave-induced setup), E represents the local wave energy density, C_g is the local wave group celerity, and V_ℓ is the local longshore current speed. Equation 2-22 can be expanded by assuming shallow water conditions, small angles of wave incidence, and assuming a nonlinear value for $C_g = (g(H+d))^{1/2}$

$$q_x(y) = k_q \frac{1}{8} \rho g^{\frac{3}{2}} \frac{H}{\sqrt{H+d}} \left[2\frac{dH}{dy} + \frac{H}{2(H+d)} \frac{d}{dy} (H+d) \right] V_\ell \qquad\qquad \text{(III-2-23)}$$

in which H is the local wave height in the surf zone. Equation 2-23 represents conditions landward of the breakpoint; seaward of the breakpoint, $q_x(y) = 0$ under the assumption that no energy dissipation occurs. In application of Bodge and Dean's relationship, the dimensional constant k_q may be determined by integrating the distribution $q_x(y)$ across the surf zone, and equating this quantity to the total longshore sand transport rate Q_ℓ

(7) The model (solid lines in Figure III-2-21) was compared with field data, and predicted the general trend of the measured transport distribution fairly well for one case (Figure III-2-21a), but shifted the cross-shore distribution slightly shoreward relative to the measured data for the second case (Figure III-2-21b). Comparison of the model with laboratory data indicated that the model generally overpredicted transport in the mid-surf zone (especially for the plunging/collapsing and collapsing cases) and modeled the near-shoreline transport distribution to a more reasonable degree than previous approaches.

g. Application of longshore sediment transport calculations.

(1) Littoral budgets.

(a) A littoral sediment budget reflects an application of the principle of continuity or conservation of mass to coastal sediment. The time rate of change of sediment within a system is dependent upon the rate at which material is brought into a control volume versus the rate at which sediment leaves the same volume. The budget involves assessing the sedimentary contributions and losses and equating these to the net balance of sediment in a coastal compartment. Any process that results in a net increase in sediment in a control

EXAMPLE PROBLEM III-2-7

FIND:
Cross-shore distribution of longshore sediment transport using Equation 2-23.

GIVEN:
Waves in 10-m (32.8-ft) depth have an rms wave height of 2.0 m (6.6 ft), angle of 10 deg to the shoreline, and wave period 8.5 sec. The beach profile is as given below. Assume the wave height to water depth ratio for incipient breaking $\kappa = 1.0$, the stable wave height to water depth ratio for wave re-formation $\Gamma_{stb} = 0.40$, the energy flux dissipation rate $\gamma = 0.15$, the lateral mixing coefficient $\Lambda_{mix} = 0.30$, and bottom friction coefficient $C_f = 0.01$. The K parameter (for use in longshore transport relationships) was calculated as 0.60.

No.	Distance Offshore (m)	Depth (m)
1	100.0	-0.60
2	111.0	0.10
3	132.0	1.00
4	145.0	1.50
5	156.0	1.70
6	169.0	1.90
7	173.0	1.95
8	186.0	1.85
9	190.0	1.80
10	195.0	1.70
11	199.0	1.75
12	207.0	1.81
13	214.0	2.00
14	246.0	3.10
15	340.0	4.00

Beach profile for Example III-2-7

SOLUTION:
Part II-3 presents relationships for nearshore wave transformation and Part II-4 discusses the cross-shore distribution of nearshore currents. Alternately, the PC-based numerical model NMLONG (see Part II-4 for a complete description) may be used to calculate the cross-shore distribution of total water depth and longshore current speed over an irregular bottom profile (Kraus and Larson 1991), which can be used in application of Equation 2-23. Entering the given data set, with 100 computation points, a cross-shore spacing of 2.0 m (6.6 ft), no wind, and a tidal reference elevation of 0.0, the cross-shore distribution of waves and currents as shown in Figure III-2-22b is obtained. A reduced listing of the NMLONG output, chosen to represent the peaks and minima of the wave height and longshore current distributions, is presented in the first four columns of the following table. The predicted $q_x(y)/k_q$ is shown in the last column and in Figure III-2-22a.

↓↓Example Problem III-2-7 (Sheet 1 of 4)

Example Problem III-2-7 (Continued)

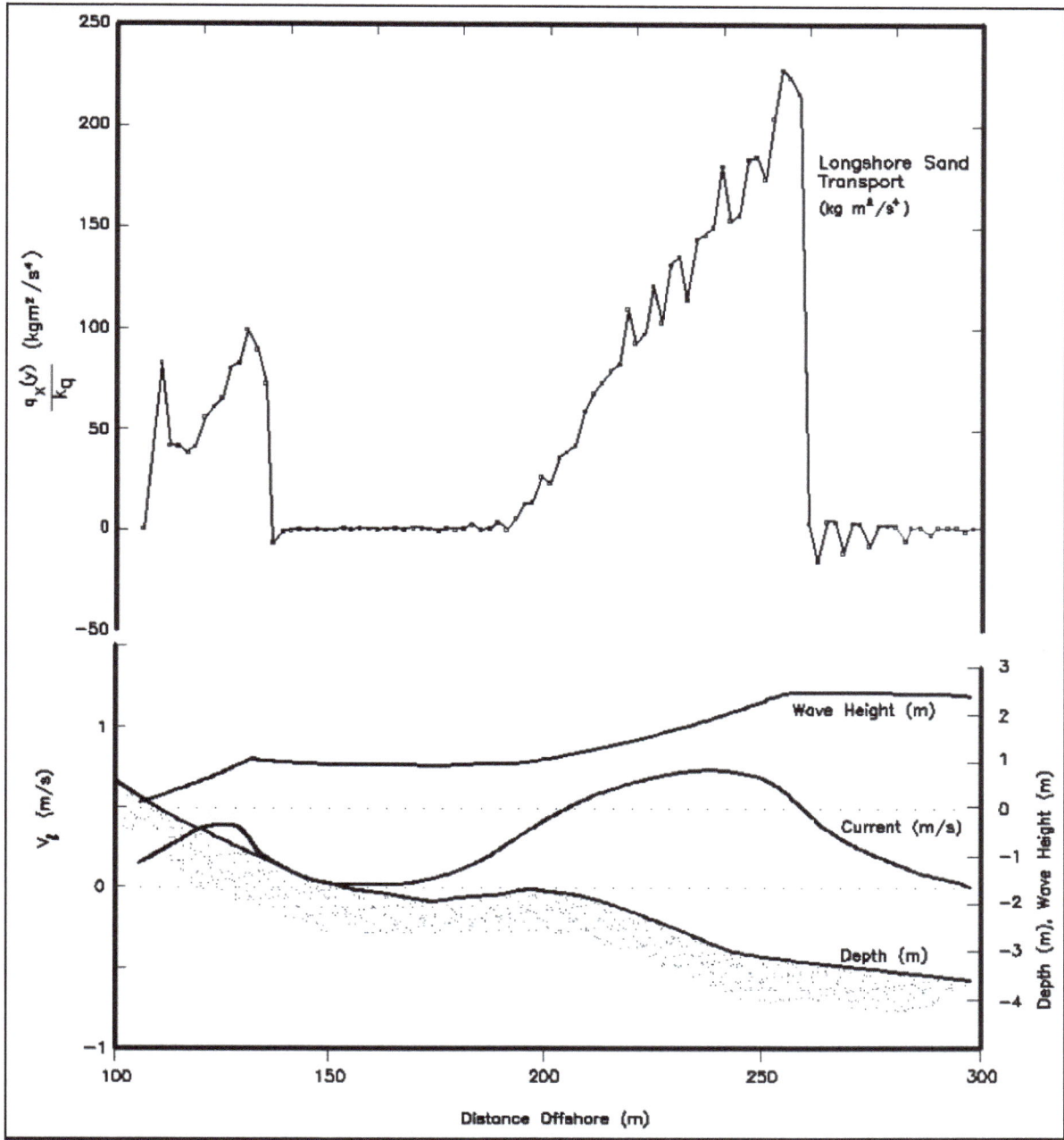

Figure III-2-22. Example III-2-7, predicted cross-shore distribution of longshore sand transport, wave height, and longshore current speed

Example Problem III-2-7 (Sheet 2 of 4)

Example Problem III-2-7 (Continued)

Output from NMLONG and Calculation of Cross-shore Distribution of Longshore Transport for Example Problem III-2-7

Cross-shore Coordinate (m)	Depth d (m)	Wave Height H (m)	Longshore current speed (m/sec)	$q_x(y)/k_q$ (kg m²/sec⁴)
110	0.09	0.20	0.20	83.3
112	-0.04	0.26	0.22	41.6
114	-0.14	0.32	0.25	40.8
126	-0.66	0.67	0.39	80.4
128	-0.74	0.74	0.39	82.9
130	-0.83	0.82	0.39	99.1
132	-0.91	0.90	0.34	89.4
134	-1.00	0.98	0.26	72.0
136	-1.08	0.96	0.20	- 6.9
148	-1.52	0.90	0.05	- 0.6
150	-1.55	0.89	0.04	- 0.7
152	-1.59	0.89	0.03	0.2
160	-1.73	0.87	0.02	- 0.4
162	-1.76	0.87	0.02	0.10
164	-1.79	0.87	0.02	0.10
178	-1.93	0.85	0.07	- 0.4
180	-1.91	0.85	0.08	- 0.3
182	-1.90	0.86	0.10	2.0
204	-1.80	1.10	0.51	37.5
210	-1.84	1.14	0.54	67.0
212	-1.89	1.18	0.56	72.2
238	-2.76	1.77	0.74	149.5
240	-2.83	1.83	0.74	179.7
242	-2.89	1.88	0.74	152.6
256	-3.18	2.34	0.63	223.8
258	-3.20	2.42	0.59	215.2
260	-3.21	2.42	0.53	2.3
278	-3.39	2.39	0.19	1.5
280	-3.41	2.39	0.17	1.4
282	-3.43	2.38	0.15	- 5.2

Example Problem III-2-7 (Sheet 3 of 4)

Example Problem III-2-7 (Concluded)

For some applications, such as weir design, only the percentage of longshore transport in a given region of the surf zone is of interest, and determination of the dimensional normalizing constant k_q may not be necessary. However, k_q may be calculated by integrating $q_x(y)/k_q$ across the surf zone, determined here by calculating the area below the points in Figure III-2-22b

$$\int_{y_s}^{y_b} \frac{q_x(y)}{k_q}\,dy = \frac{1}{k_q}\int_{106}^{260} q_x(y)\,dy \approx \frac{9300}{k_q}\ \frac{kg\ m^2}{sec^4}$$

where y_s is the cross-shore coordinate at the start of the surf zone, (approximately 106 m) and y_b is the cross-shore coordinate at the breaker line (the breaking wave height of 2.42 m occurs at 260 m). Next, the total longshore sand transport rate is calculated. The offshore wave celerity may be calculated using linear wave theory as $C_{gl} = 9.0$ m/sec. Breaking wave angle may be calculated using Snell's Law (Equation 6-13) with input wave conditions ($\alpha_1 = 10°$, $d_1 = 10$ m) and a depth at breaking = 3.2 m

$$\frac{\sin \alpha_1}{C_{gl}} = \frac{\sin \alpha_b}{C_{gb}}$$

$$\frac{\sin 10°}{9.0} = \frac{\sin \alpha_b}{(9.81\ (3.2))^{\frac{1}{2}}}$$

$$\alpha_b = 6.2°$$

The total longshore transport rate may be calculated using Equation 2-7b where $K = 0.60$

$$Q_\ell = K\left(\frac{\rho\sqrt{g}}{16\ \kappa\ (\rho_s - \rho)(1 - n)}\right) H_{b\ rms}^{\frac{5}{2}}\ \sin(2\alpha_b)$$

$$Q_\ell = 0.60\ \frac{(1025)\ (9.81)^{\frac{1}{2}}}{16\ (1)\ (2650 - 1025)\ (1 - 0.4)}\ (2.42)^{\frac{5}{2}}\ \sin(2(6.2°))$$

$$Q_\ell = (0.242\ m^3/sec)\ (3600\ sec/hr)\ (24\ hr/day)$$

$$Q_\ell = 20.9 \times 10^3\ m^3/day\ (27.3 \times 10^3\ yd^3/day)$$

The value of k_q may be calculated

$$\frac{9300}{k_q}\ \frac{kg\ m^2}{sec^4} = 0.242\ \frac{m^3}{sec}$$

Therefore, $k_q \sim 38,000$ sec³/kg, and the transport rates in Figure III-2-22b and the table can be converted to values of $q_x(y)$ (units m³/sec/m), if desired.

Example Problem III-2-7 (Sheet 4 of 4)

volume is called a *source*. Alternately, any process that results in a net loss of sediment from a control volume is considered a *sink*. Some processes can function as sources and sinks for the same control volume (e.g., longshore sediment transport). The balance of sediment between losses and gains is reflected in localized erosion and deposition. In general, longshore movement of sediment into a coastal compartment, onshore transport of sediment, additions from fluvial transport, and dune/bluff/cliff erosion provide the major sources of sediment. Longshore movement of sediment out of a coastal compartment, offshore transport of sediment, and aeolian transport and washover that increase beach/island elevation produce losses from a control volume. Sediment budgets, including the types and importance of sources and sinks, are discussed in detail in Part IV-5.

(b) The appropriate level of detail for a littoral budget is a function of the intended uses of the littoral budget, and the available resources to complete the project. The essential components of a littoral budget include a site description, background, and examination of previous analyses. Past and present conditions, and the results of other studies, must be examined before initiating a new budget analysis.

(c) The longshore sediment transport rate must be determined next. This requires data on wave conditions over as long a time period as possible. These waves are propagated to and transformed in the surf zone. Appropriate sediment transport equations must be applied, ideally using historical shoreline positions and wave conditions for the same time period to, in effect, "calibrate" the transport equations for the study site. Very often, shoreline change models, which use the sediment transport equations, are applied. Boundary conditions defined at the start of the analysis are changed, so that sensitivity of the budget to these conditions may be evaluated.

(d) The actual sediment budget may then be determined. Usually there are poorly quantified components remaining in the analysis, such as offshore gains and losses. These must be estimated using any available data, engineering judgment, and the requirement that the budget close. Although a significant effort goes into the development of a littoral budget, it must be remembered that it is an estimate and may easily be in error. In addition, the budget is usually calibrated with shoreline positions over a number of years, and therefore indicates long-term average rates of change. It may not be indicative of the changes in any one year.

(2) Variations in longshore sediment transport along the coast.

(a) As described in the next section, noncohesive shorelines are not typically straight. The shoreline orientation and the degree to which waves refract, shoal, and converge or diverge along the shoreline determine variations in the potential transport rate along the coast. These variations are important determinants to shoreline change along the coast. For a nonuniform coastline, the potential longshore sediment transport rate is computed at discrete points along the coastline using values of the local breaking wave height and angle, where the latter is expressed relative to the local shoreline orientation. A ray-tracing or grid-based wave refraction analysis is typically employed to determine these values.

(b) Shoreline change may be related to the computed gradients in transport rate along the coast. For example, areas of convergent transport may correspond to a sediment sink (or deposition). Areas of divergent transport may correspond to a sediment source (or erosion if the area is not a source). As long as there is an unlimited sediment source, a shoreline's response to longshore sediment transport should be dependent on gradients in transport along the coast rather than magnitudes of transport.

h. Three-dimensionality of shoreline features.

(1) The three-dimensionality of noncohesive shoreline shape and its corresponding underwater expression are important to various aspects of engineering design. Dunes are more susceptible to breakthrough where the beach width fronting the dune is narrow due to the diminished protection afforded by the berm. Noncohesive shorelines are typically neither straight nor of smooth curvature. They often have isolated "bumps" or more regularly spaced shoreline features. Migration of such shoreline features can undercut or flank the landward ends of coastal structures such as seawalls or groins. For this reason any regularly spaced (rhythmic) topographic features inherent in the beach/shoreline structure as reflected in its planform shape may be of importance to the engineer for consideration of risk factors in evaluation of projects. Little is known about these features from a quantitative standpoint, although considerable qualitative descriptions of the features are documented.

(2) Typically, shoreline rhythmic features are classified by their planform (longshore) spacing or approximate wavelength if they are of reasonably regular form and hence are often referred to as "sand waves" by many authors. The term "sand wave" in this context should not be confused with underwater sand waves that are ubiquitous in most marine environments. Planform amplitude or height of the shoreline features (defined as cross-shore distance from embayment to cusp point) often is correlated with the wavelength. That is, longer shoreline planform wavelength (or alongshore spacing) suggests larger planform amplitudes (Sonu 1969). Along most shorelines, a spectrum of rhythmic shapes and sizes is present, making the underlying shoreline planform characteristics somewhat confusing. Due to the continuous range of scale in observed shoreline rhythmic features it is impossible to completely separate discussion of the various features by size since the physics governing the various length scales may be similar.

(3) At the small end of the scale of rhythmic shoreline features (and consequently of lesser engineering significance) are "beach cusps." Figure III-2-23 is an example of a beach with developed beach cusps. Russell and McIntire (1965) compile observational statistics on beach cusps from a number of ocean beaches and show cusp planform wavelengths (or longshore spacing) from 6 to 67 m. Numerous authors have postulated theories for the conditions of formation as well as spacing and amplitude of these smaller-scale features.

(4) At present, though, an explanation that would encompass all the numerous small-scale rhythmic features noted along the shoreline is lacking. Extended discussion on these smaller-scale shoreline features can be found in Komar (1976).

(5) As an example of larger-scale rhythmic topography, Figure III-2-24 shows a shoreline from Tokai Beach, Japan (Mogi 1960), in which two predominant wavelength scales of rhythmic sinuous topography dominate over a 3-month period of study. The shorter planform wavelengths in this example are on the order of 250 m while the longer planform wavelengths are on the order of 2.5 km. Although phasing changes are evident, the rhythmic feature length scales appear to have prevailed throughout the different survey periods. Lippmann and Holman (1990) have documented the conditions for rhythmic bars along one section of the North Carolina shoreline (Duck, NC). They found that rhythmic bars were a predominant feature observed in 68 percent of video imaging records and noted that during the strongest wave activity the rhythmic features were destroyed but that the features returned 5-16 days following peak wave events.

(6) Along beaches in Japan, Hom-ma and Sonu (1963) observed that under certain conditions crescentic bars with a regular alongshore spacing would weld to the shoreline with consequent large cusps formed at the attachment points (see Figure III-2-25). Sonu (1973) notes a second type of rhythmic topography in which rhythmic shoreline features are associated with the presence of rip current cell circulation (see Figure III-2-26). Sonu (1973) noted that both types of rhythmic topography could be present independently

Figure III-2-23.   Beach cusps on a sandy beach in Mexico (photograph courtesy of Paul Komar)

Table III-2-3
List of Authors Postulating Theories for Cusp Development

Johnson (1910, 1919)	Dalrymple and Lanan (1976)
Dolan and Ferm (1968)	Dubois (1978)
Escher (1937)	Shepard (1973)
Longuet-Higgins and Parkin (1962)	Dolan, Vincent, and Hayden (1974)
Russell and McIntire (1965)	Flemming (1964)
Bagnold (1940)	Krumbein (1944b)
Williams (1973)	Komar (1973)
Kuenen (1948)	Zenkovitch (1964)
Sonu (1972, 1973)	Otvos (1964)
Sonu and Russell (1967)	Sonu, McCloy, and McArthur (1967)
Bowen (1973)	Bowen and Inman (1969)
Guza and Bowen (1981)	Guza and Inman (1975)
Sallanger (1979)	Seymour and Aubrey (1985)
Holman and Bowen (1979, 1982)	Darbyshire (1977)
Darbyshire and Pritchard (1978)	Dean and Maurmeyer (1980)

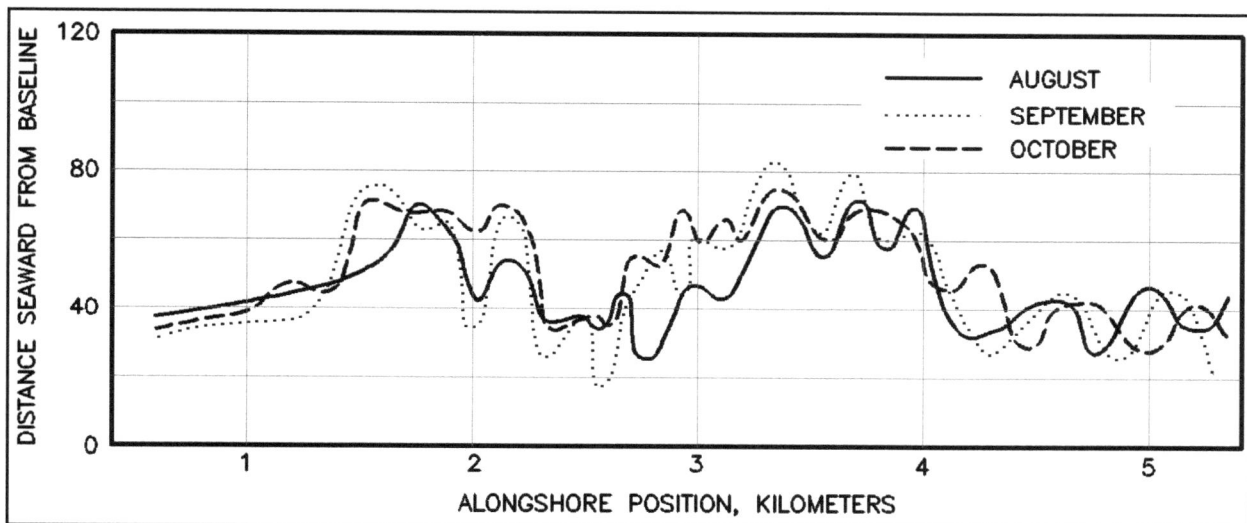

Figure III-2-24.   Shoreline fluctuations in plan view at Tokai Beach, Japan (after Mogi (1960))

along a shoreline, and that the crescentic bar type of feature is typically at larger scales than the features associated with rip current cell circulation.

   (7) Sonu (1973) details numerous examples of world coastlines with rhythmic shoreline/nearshore features, and notes the ubiquitous nature of these features via their existence on long uninterrupted coastlines as well as embayment shorelines between headlands, in tideless seas as well as coasts with tidal ranges up to 4 m, and on beaches with grain size ranging from sand to gravel. Observations of the planform wavelength (or alongshore spacing) of the features varies markedly, including 100 to 300 m along the east coast of Florida (Bruun and Manohar 1963); 64 to 218 m between transverse bars on the low-energy sheltered coast

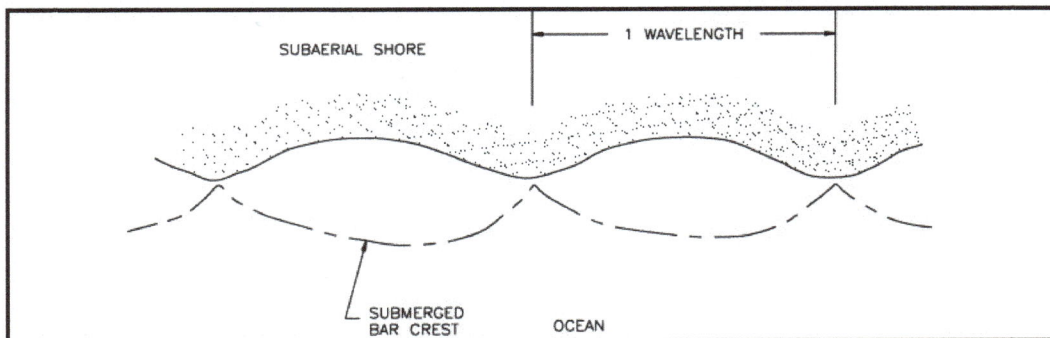

Figure III-2-25. Rhythmic shoreline features associated with the presence of crescentic bars welded to the shoreline

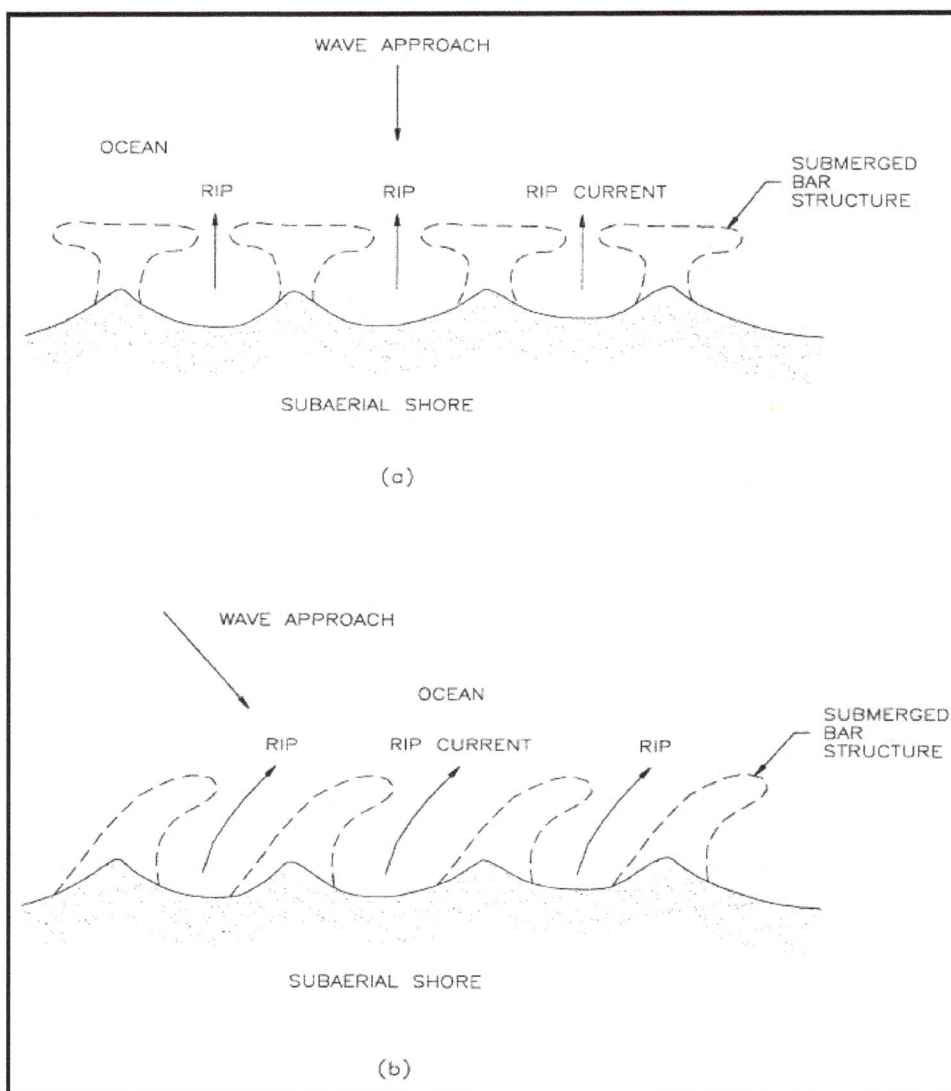

Figure III-2-26. Cusp rhythmic features in conjunction with rip currents under wave action approaching shoreline obliquely (adapted from Sonu (1973))

of St. James Island, Florida (Niedoroda and Tanner 1970); 100 to several thousand meters in the case of the example shown in Figure III-2-25 (Sonu 1973); up to 1,500 m for giant cusps noted along various beaches (Shepard 1952, 1973); and on the order of hundreds of meters along the Atlantic bluff shoreline of Cape Cod, Massachusetts (Aubrey 1980). Sonu and Russell (1967) noted, and later Dolan (1971) measured shoreline rhythmic features along the North Carolina coast with alongshore spacing ranging from 150 to 1,000 m with the predominant spacing about 500 to 600 m. In the same study, Dolan (1971) measure planform amplitudes from 15 to 25 m with large sand waves reaching amplitudes of 40 m. Numerous, well-documented surveys of sand wave/giant cusp rhythmic beach planforms also exist along the Danish and Dutch coasts (van Bendegom 1949, Brunn 1954, Verhagen 1989).

(8) Migration rates of rhythmic features vary widely with some studies reporting short-term fluctuations in position but no net long-term migration. Van Bendegom (1949) documents the monitoring over an 80-year period of large sand waves with amplitudes up to 200 m along the Dutch coast, and discusses the corresponding cycle of beach erosion and accretion as these waves migrate along the coastline with an average speed of 200 m/year. Dolan (1971) noted migration velocities of large sand waves along the North Carolina coast ranging from 100 to 200 m/month during heavy weather seasons. Verhagen (1989) has documented sand waves along the Dutch coast with amplitudes from 25 to 2,500 m and longshore speeds ranging from 45 to 310 m/year. Sonu (1969) has suggested that migration velocities of such shoreline features are inversely proportional to some power of the feature's alongshore spacing (i.e., the larger the feature, the slower the movement).

(9) Edge waves and rip currents are often cited as the main contributing forcing functions to the formation of rhythmic topography at various scales. A discussion of edge wave generation and hypothesized effects on beaches can be found in Guza and Inman (1975), Huntley and Bowen (1973, 1975a, 1975b, 1979), Huntley (1976), Holman and Bowen (1979, 1982), Wright et al. (1979), Guza and Bowen (1981), and Bowen and Inman (1971). A discussion of rip current formation and its effects on beaches can be found in Bowen (1969), Bowen and Inman (1969), Hino (1974), Dalrymple (1975), Dalrymple and Lanan (1976), Dolan (1971), Komar (1971), Komar and Rea (1976), and Komar (1978). Conclusive evidence proving the mechanisms for the formation of the many types of rhythmic topography is lacking.

(10) From an engineering standpoint the importance of rhythmic shoreline features (especially larger ones) and their potential for migration should not be overlooked in planning engineering structures or in analysis of design dune width for storm protection. For example, van Bendegom (1949) documents the structural failure of a groin due to the erosion produced by a large, migrating sand wave along the Dutch coast. Brunn (1954) described migrating sand waves along the Danish North Sea coast with observed planform spacings on the order of 300 to 2,000 m and amplitudes on the order of 60 to 80 m in areas where seasonal beach change was only 20 m/year and long-term shoreline recession only 2 m/year. In this regard, Brunn (1954) also cites a case of a sand wave of 900 m wavelength and 60 m amplitude with a migration speed of 700 m/year, and notes the difficulty of drawing definitive conclusions on average shoreline movements in such areas. Dolan (1971) noted that the regular spacing of dune breaching on Bodie Island, North Carolina, during the Ash Wednesday storm of 7 March 1962 correlated well with the rhythmic topography seen in the shoreline. Dolan (1971) also documented erosion along the Cape Hatteras, North Carolina, shoreline corresponding to embayments of rhythmic topography and suggested that when analyzing beach variability for specific sites, in addition to the seasonal recession-progradation cycle, additional variation (about 20 percent along the Outer Banks) should be considered to account for migration of the rhythmic topographic features.

(11) In practice, aerial photography at a reasonable scale (1 in. = 100 m or larger) or shoreline surveys are necessary to document the existence of rhythmic shoreline features. Sets of such aerial photographs/shoreline surveys with common control points and interspersed over long periods of time should

be useful in detailing both the potential existence and characteristics of such features for consideration in engineering planning.

i. Empirical shoreline models.

(1) In nature, many sections of coastline which are situated in the lee of a natural or artificial headland feature a curved shoreline geometry. Where sections of coastline are situated between two headlands, and particularly when there is a single, dominant wave direction, the shoreline may likewise assume a curved or "scalloped" shape (see Figure III-2-27a). In both cases, the curved portion of the shoreline related to the headland(s) is termed a crenulate or "spiral bay." Because of their geometries, these shorelines are also sometimes termed "parabolic," "zeta-bay," or "log-spiral" shorelines. The shape results from longshore transport processes which move sediment in the downdrift direction along the down-wave section of the shoreline, and from processes associated with wave diffraction which move sediment in the opposite direction in the immediate lee of the up-wave headland.

(2) Krumbein (1944b) and Yasso (1965) were among the first investigators to suggest that many "static" shorelines in the vicinity of rocky or erosion-resistant headlands could be fit to a log-spiral curve. Silvester (1970); Silvester and Ho (1972); Silvester, Tsuchiya, and Shibano (1980); and Hsu and Evans (1989) utilized the concept to develop empirical guidance for maximum coastal indentation between two headlands or coastal structures (such as seawalls or breakwaters) based on one dominant wave direction. Practical application of the approach requires identification of a predominant wave direction and the proper origin of the log-spiral curve. In a more theoretical effort, LeBlond (1972, 1979) derived equations for an equilibrium shoreline shape in the shadow zone of an upcoast headland based upon many simplifying assumptions concerning refraction and diffraction and found the resulting shoreline to be very similar to the log spiral shape. Rea and Komar (1975), Parker and Quigley (1980), and Finkelstein (1982) have also noted the similarity of bay shoreline shapes to log spiral curves. Walton (1977) and Walton and Chiu (1977) demonstrated that the log spiral curve is robust in the sense that most smooth curves found in nature can be fit to a log spiral if fortuitous values of its parameters are chosen. Walton (1977) presents a simplified procedure for evaluating a dynamic progression of static equilibrium shorelines downcoast from headland-type features using the concept of the littoral energy rose.

(3) Using shoreline data from prototype bays considered to be in static equilibrium and from physical models, Hsu, Silvester, and Xia (1987, 1989a, 1989b) presented an alternate expression to approximate the shoreline in the lee of headland-type features:

$$\frac{R}{R_o} = C_o + C_1 \left(\frac{\beta}{\theta} \right) + C_2 \left(\frac{\beta}{\theta} \right)^2 \qquad\qquad (III-2-24)$$

where the geometric parameters R, R_o, β, and θ are as shown in Figure III-2-27a, and values for the coefficients C_0, C_1, and C_2 are shown in Figure III-2-27b. The distance R_o corresponds to a control line drawn between the ends of the headlands that define a given section of shoreline. In the case of a single, upcoast headland, the distance R_o is the length of a control line drawn from the end of the headland to the nearest point on the downcoast shoreline at which the shoreline is parallel with the predominant wave crest. The distance R, measured from the end of the upcoast headland, defines the location of the shoreline at angles θ measured from the predominant wave crest. The angle β is that between the predominant wave direction and the control line R_o.

(4) The tidal shoreline which Equation 2-24 represents is not clear, but might be interpreted to represent the mean water shoreline. The data upon which Equation 2-24 is based are principally limited to $\beta > 22°$.

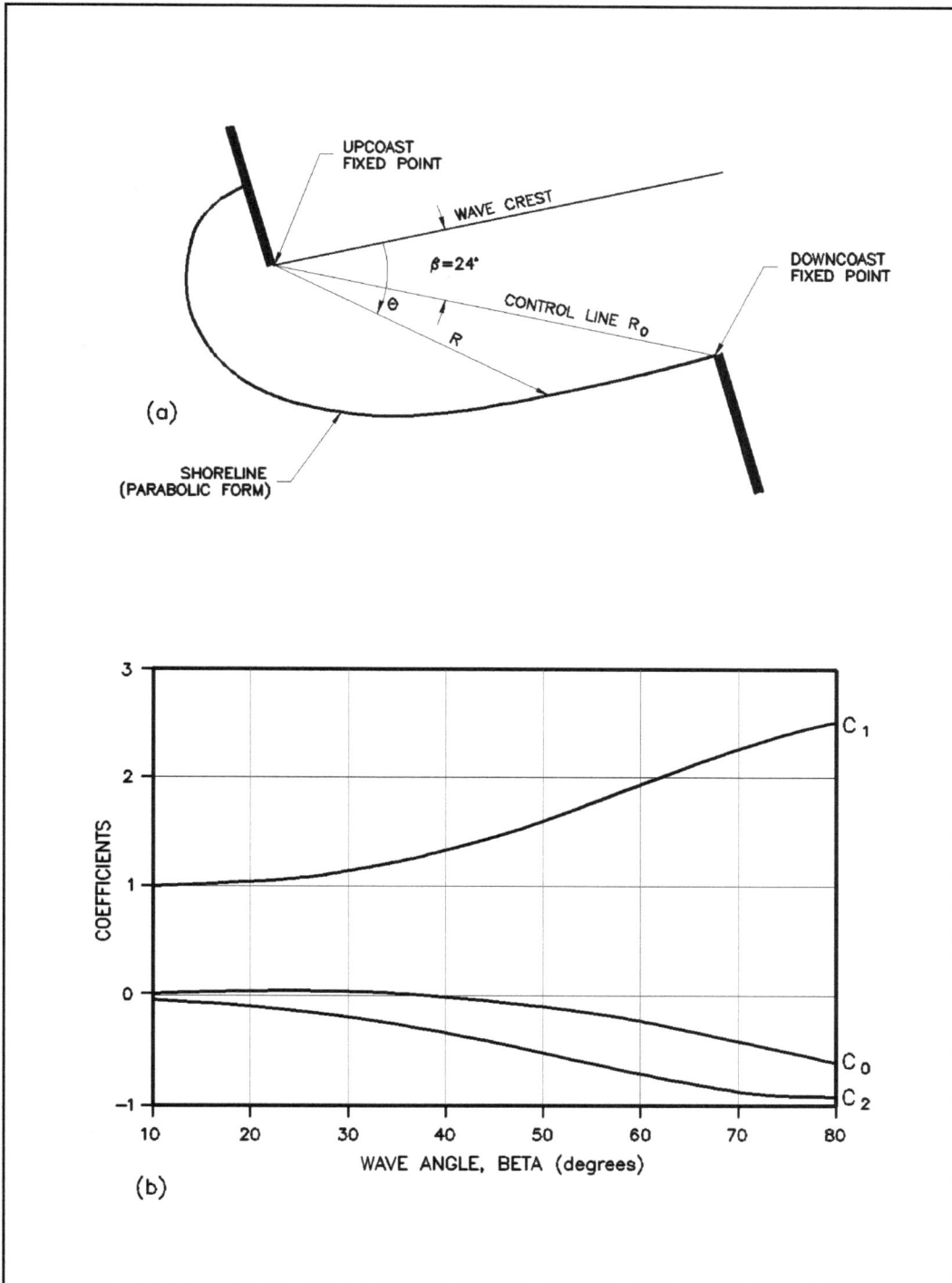

Figure III-2-27.   Spiral bay geometry: (a) definition sketch and (b) coefficients describing parabolic shoreline shape (Silvester and Hsu 1993)

Additionally, Equation 2-24 is intended for application for $\beta \le \theta \le 180°$, and assumes that a predominant wave direction exists at the site of interest. The latter is often not the case in nature, and so engineering judgement must be utilized in practical application of this method. For $\theta > 180°$, the distance R may be assumed to be constant and equal to the value of R computed at $\theta = 180°$.

(5) Additional empirical guidance on shoreline change at seawalls is provided in Walton and Sensabaugh (1979) where additional localized recession at a seawall under a storm condition (hurricane Eloise along the Florida panhandle), is provided. Similar guidance for other storms and other locations is not available, although McDougal, Sturtevant, and Komar (1987) and Komar and McDougal (1988), have reported similar findings at laboratory scales.

(6) The approach(es) outlined above may be useful for rough, preliminary calculations and estimates of "static" shoreline equilibriums when the assumptions necessary for application of the approaches are fulfilled, where detailed dynamics of the changing shoreline are not sought, and where time and/or budget constraints preclude a more detailed approach. For detailed prediction of shoreline change due to longshore gradients in sand transport or otherwise complicated geometries, a preferred approach would be to utilize a physical model and/or a numerical model, as appropriate to the scale of the study area.

j.   Analytical longshore sand transport shoreline change models.

(1) If the angle of the shoreline is small with respect to the x axis and simple relationships describe the waves, analytical solutions for shoreline change may be developed. As an example, utilizing the expression provided in Equation 2-7b for longshore sediment transport along with the assumption that the breaking wave angle α_b is small, the following planform shoreline change equation can be derived utilizing the coordinate system given in Figures III-2-28 and III-2-29:

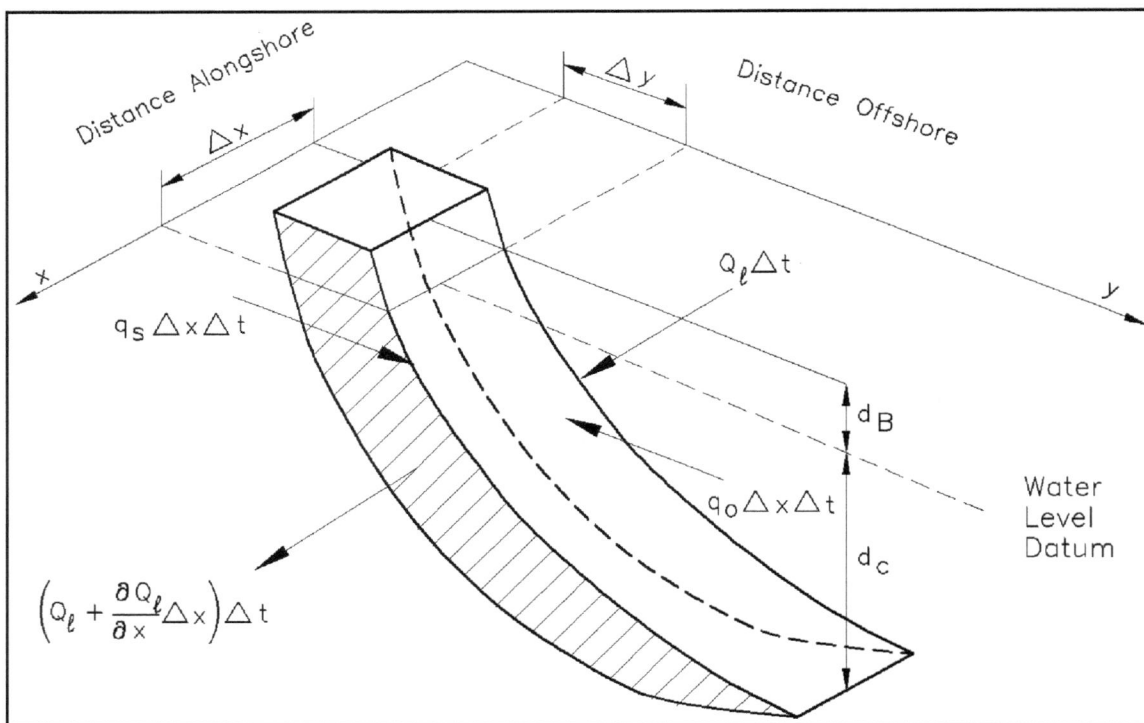

Figure III-2-28.   Elemental volume on equilibrium beach profile

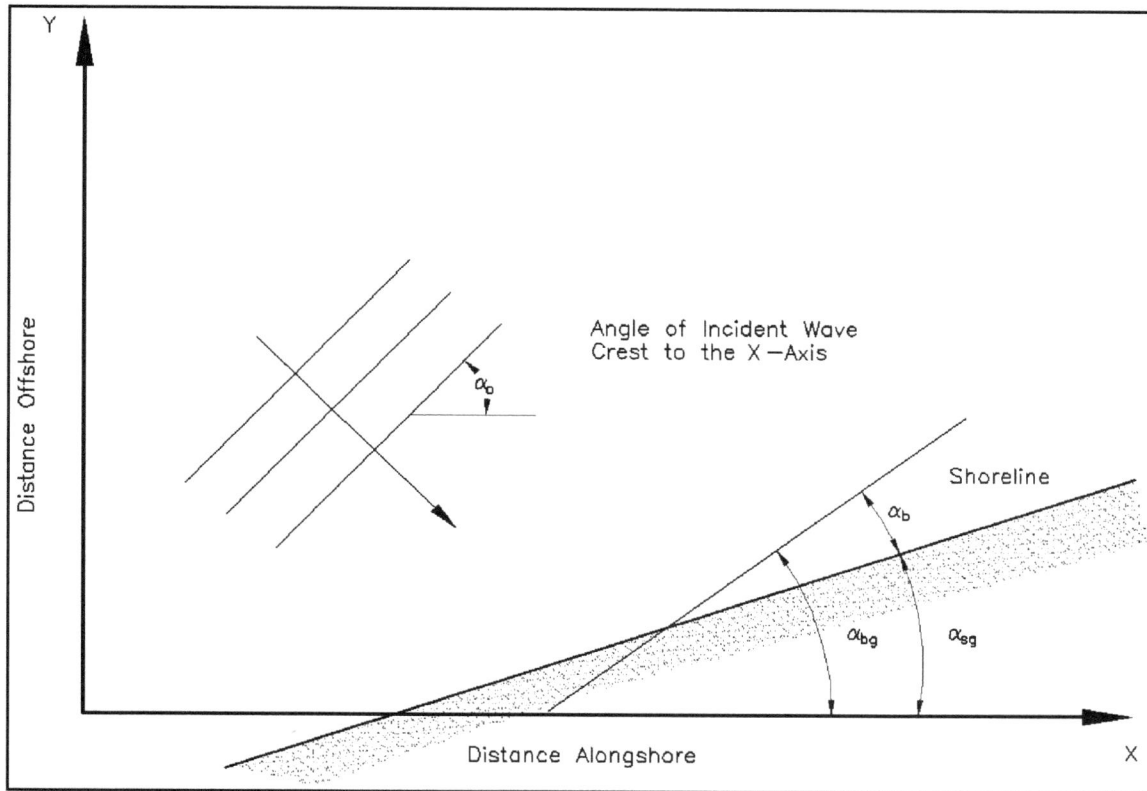

Figure III-2-29. **Definition of local breaker angle**

$$\varepsilon \, \frac{\partial^2 y}{\partial x^2} = \frac{\partial y}{\partial t}$$

(III-2-25)

where

$$\varepsilon = \frac{K H_b^2 C_{gb}}{8} \left(\frac{\rho}{\rho_s - \rho} \right) \left(\frac{1}{1 - n} \right) \left(\frac{1}{d_B + d_c} \right)$$

(III-2-26)

and where d_B = beach berm height above still-water level; d_c = depth of appreciable sand transport as measured from still-water level. Equation 2-25 is a partial differential equation, as it is dependent on both space (variable x) and time (variable t). A number of researchers have employed this equation or slight variations of it to provide analytical solutions to shoreline change under certain assumptions (the boundary conditions and initial conditions of the problem). Pelnard-Considére (1956) first presented an analytical solution to this simplified shoreline change equation for the case of an impermeable groin or jetty impounding the longshore sand transport on the updrift side of the structure under a stationary (constant) wave climate. Pelnard-Considére also verified its applicability with laboratory experiments and derived analytical solutions of the linearized shoreline change equation for two other boundary conditions: shoreline evolution updrift of an impermeable groin (with bypassing) and release of an instantaneous plane source of sand on the beach.

EXAMPLE PROBLEM III-2-8

FIND:
Compute the shoreline geometry of a crenulate bay located between two rock headlands for a shoreline where one dominant wave direction exists.

GIVEN:
The distance between the ends of the headlands is 175 m. The incident wave crests make an angle of 30 deg with a line drawn between the two headlands.

SOLUTION:

From Figure III-2-27b, the values of the coefficients for the wave angle β = 30 deg are approximately C_0 = 0.05, C_1 = 1.14, and C_2 = -0.19. The location of the shoreline may be predicted by plotting the distance R, measured from the end of the upwave headland, at angles θ measured from the line drawn between the headlands. The values R/R_o for various arbitrary angles between the wave angle, 30 deg, and a maximum angle, 180 deg, are computed from Equation 2-24. The corresponding dimensional values of R are then computed by multiplying R/R_o by the distance between the headlands R_o = 175 m. Representative examples are given below:

For θ = 30 deg: R = [0.05 + 1.14(30/30) - 0.19(30/30)^2] (175 m) = 175 m

For θ = 75 deg: R = [0.05 + 1.14(30/75) - 0.19(30/75)^2] (175 m) = 83 m

For θ = 180 deg: R = [0.05 + 1.14(30/180) - 0.19(30/180)^2] (175 m) = 41 m

For θ > 180°, the distance R may be assumed to be constant and equal to the value of R computed at θ = 180°.

(2) Le Méhauté and Brebner (1961) discuss solutions for shoreline change at groins, with and without bypassing of sand, and the effect of sudden dumping of material at a given point. They also present solutions for the decay of an undulating shoreline, and the equilibrium shape of the shoreline between two headlands.

(3) Bakker and Edelman (1965) modified the longshore sand transport rate equation to allow for an analytical treatment without linearization. The sand transport rate is divided into two different cases:

$$Q_\ell = Q_o K_a \tan(\alpha_b) \quad \text{for} \quad 0 \le \tan\alpha_b \le 1.23 \tag{III-2-27a}$$

and

$$Q_\ell = Q_o \frac{K_b}{\tan(\alpha_b)} \quad \text{for} \quad 1.23 < \tan(\alpha_b) \tag{III-2-27b}$$

where K_a and K_b are constants. The growth of river deltas was studied with these equations.

(4) Bakker (1968) extended a one-line shoreline change theory to include the shoreline and an additional offshore depth contour to describe beach planform change. Bakker hypothesized that the two-line theory provides a better description of sand movement downdrift of a long groin since it describes representative changes in the contours seaward of the groin head. Near structures such as groins, offshore contours may have a different shape from the shoreline. The two lines in the model are represented by a system of two differential equations which are coupled through a term describing cross-shore transport. According to Bakker (1968), the cross-shore transport rate depends on the steepness of the beach profile; a steep profile implies offshore sand transport; and a gently sloping profile implies onshore sand transport. Additional complex solutions of cases with groins under very simplistic assumptions are discussed in Bakker, Klein-Breteler, and Roos (1971). Le Méhauté and Soldate (1977) provide an analytical solution of the linearized shoreline change equation for the spread of a rectangular beach fill. Walton (1994) has extended this case to the fill case with tapered ends.

(5) Walton and Chiu (1979) present two derivations of the linearized shoreline change equation. The difference between the two approaches, which both arrive at the same partial differential equation, is that one uses the so-called "CERC Formula" (see Equation 2-5) for describing the longshore sand transport rate by wave action and the other a formula derived by Dean (1973) based on the assumption that the major sand transport occurs as suspended load. Walton and Chiu (1979) also present solutions for beach fill in a triangular shape, a rectangular gap in a beach, and a semi-infinite rectangular fill, and present previous analytical solutions in the literature in a nondimensionalized graphical solution form.

(6) Dean (1984) gives a brief survey of some analytical shoreline change solutions applicable to beach nourishment calculations, especially in the form of characteristic quantities describing loss percentages. One solution describes the shoreline change between two groins initially filled with sand. Larson, Hanson, and Kraus (1987) provide a review of a number of analytical solutions to the one-line model as well as additional solutions where the amplitude of the longshore sand transport rate is a discontinuous function of x, the shoreline coordinate in the longshore direction.

(7) Analytical solutions presented here are in the nondimensionalized form of Walton and Chiu (1979) and easily adaptable to solving simple scenarios. More difficult scenarios are best handled by a numerical model. In arriving at all solutions, it is tacitly assumed that sand is always available for transport unless explicitly restricted by boundary and/or initial conditions.

(8) The first case to be considered is that of a structure trapping sediment. This formulation can be applied to the prediction of the shoreline updrift and downdrift of a littoral barrier extending perpendicular to the initially straight and uniform shoreline. At the barrier, all sediment is assumed to be trapped by the barrier (no bypassing). This boundary condition requires that the shoreline at the structure be parallel to the incoming wave crests. Figure III-2-30 shows the resulting shoreline evolution with increasing time updrift (accretion) and downdrift (erosion). Figures III-2-31a, III-2-31b, and III-2-31c are nondimensionalized solution graphs (at different scales) for the condition of no sand transport at the structure location ($x = 0$),

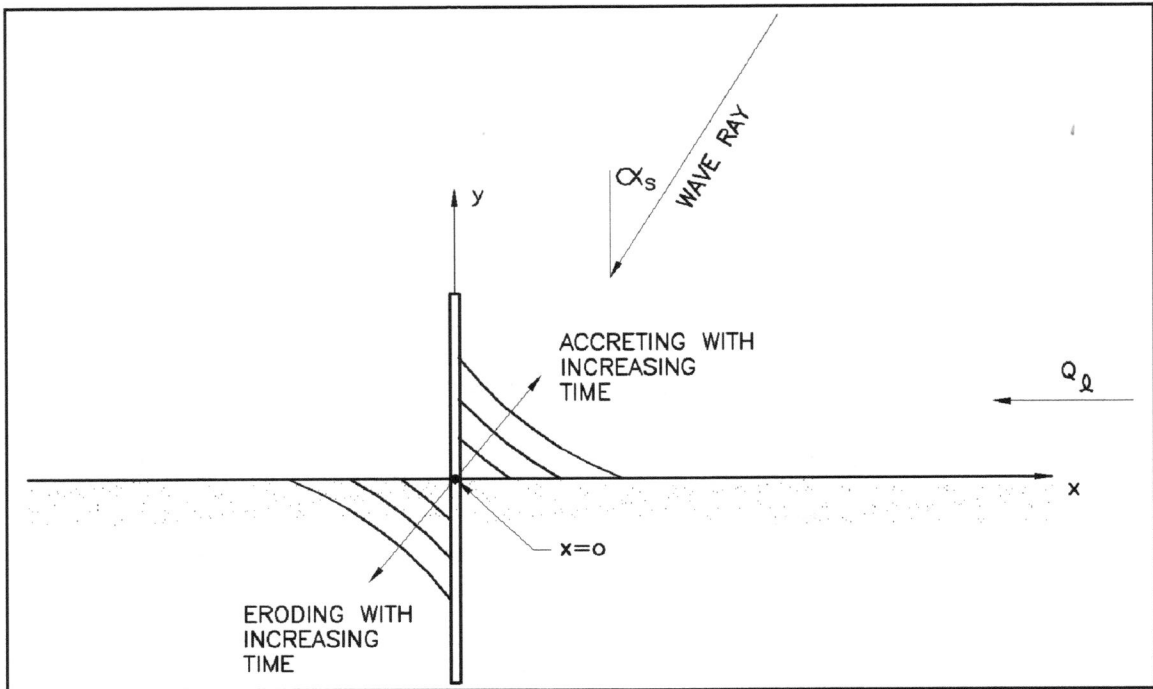

Figure III-2-30.   Structure placed perpendicular to shore

with the boundary condition at the structure being tan α_b = dy/dx, and the boundary condition at $x = \infty$ being $y = 0$ for all times. The initial condition for this solution is that $y = 0$ for $t = 0$. This particular solution graph can be utilized to estimate planform shapes on the updrift side of coastal structures where bypassing does not take place.

(9)   The dimensionalized solution for these conditions is given by Pelnard-Considére (1956) to be

$$y = 2\sqrt{\varepsilon t}\ \tan(\alpha_b) \left\{ \frac{1}{\sqrt{\pi}}\ \exp\left[-\left(\frac{x}{2\sqrt{\varepsilon t}} \right)^2 \right] - \frac{x}{2\sqrt{\varepsilon t}}\ erfc\left(\frac{x}{2\sqrt{\varepsilon t}} \right) \right\}\ ,\ \text{for}\ t < t_f \qquad \text{(III-2-28)}$$

where t_f is the time at which the structure fills to its capacity via this solution and where $erfc(\ )$ is the complementary error function defined as $erfc(\ ) = 1 - erf(\ )$ where $erf(\ )$ is the error function and

$$\text{erf}\ (\ ) = \frac{2}{\sqrt{\pi}} \int_0^{(\ )} e^{-z^2}\ dz$$

Both $erfc(\ )$ and $erf(\ )$ are tabulated in various mathematical handbooks. Figures III-2-32 and III-2-33 provide graphs of $erfc(\ )$ and $erf(\ )$.

(10) The time required for the structure to fill to capacity $t = t_f$ can be found from the previous solution with ordinate $x = 0$; i.e.,

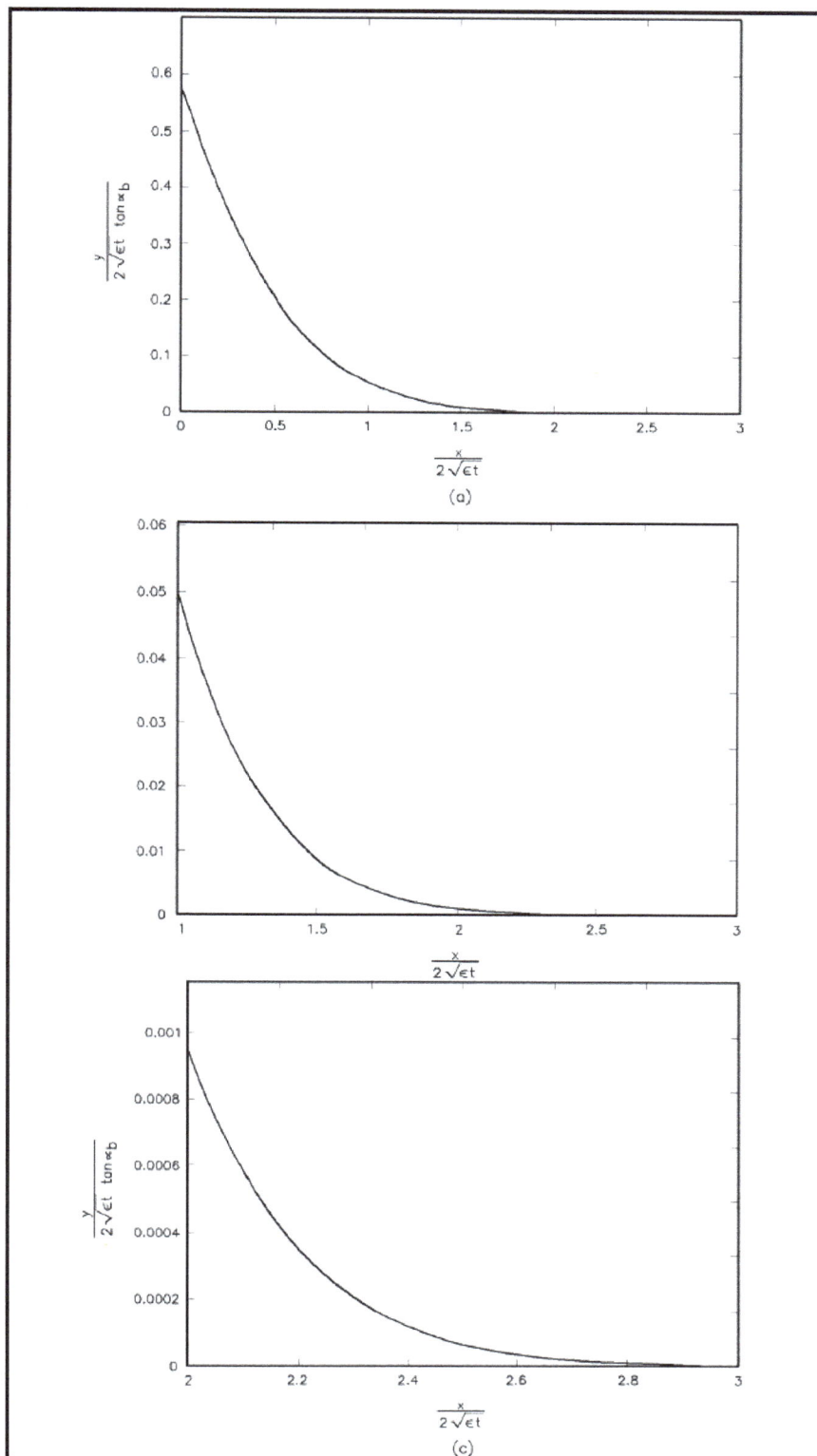

Figure III-2-31. Nondimensionalized solution graphs (at different scales) for the condition of no sand transport at the structure location

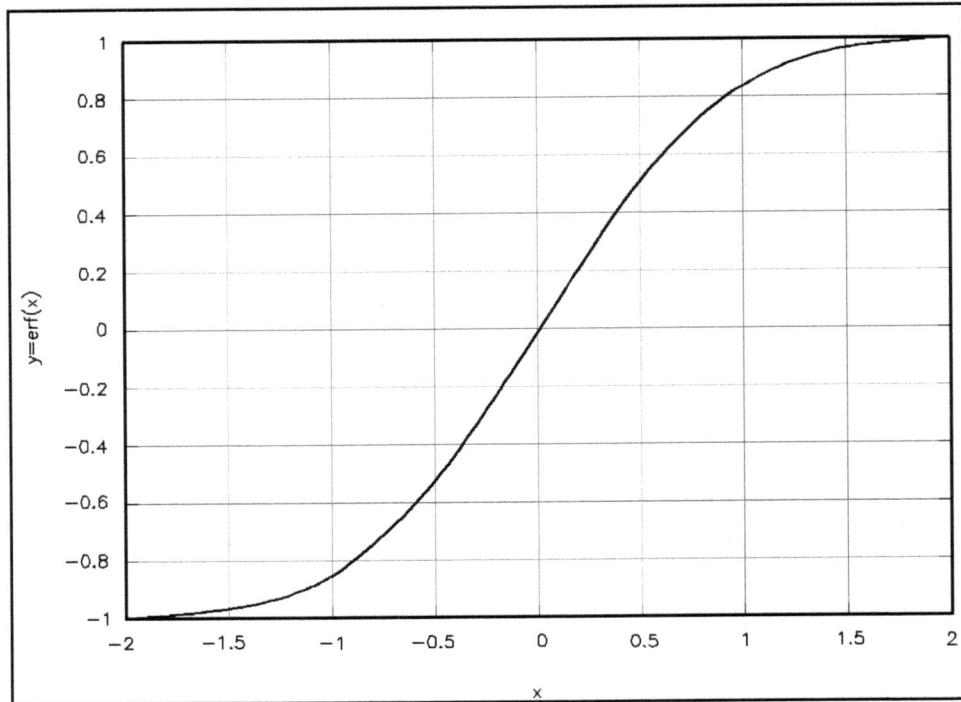

Figure III-2-32.   Error function

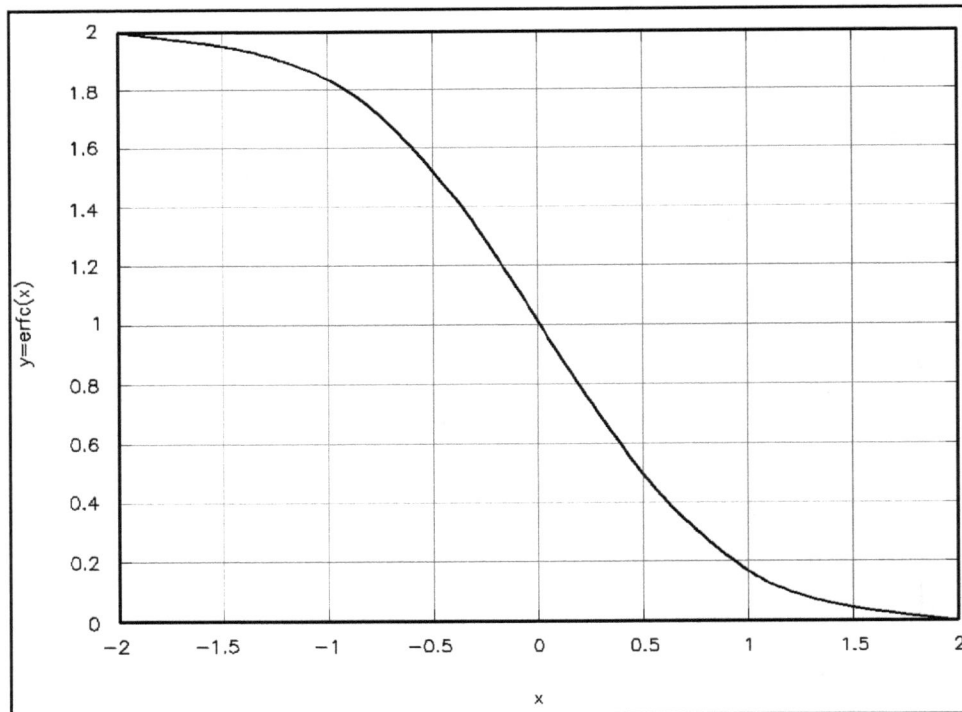

Figure III-2-33.   Complementary error function

$$t_f = \frac{Y^2 \pi}{4 \, \varepsilon \, \tan^2 (\alpha_b)} \qquad\qquad\text{(III-2-29)}$$

where Y = length of structure.

(11) Pelnard-Considére (1956) also provides a second solution for times after the structure has filled to capacity and bypassing of sediment begins to occur naturally. The boundary conditions for his second solution are that $y = Y$ at $x = 0$ and $y = 0$ at $x = \infty$ for all $t > 0$. The initial conditions are as in the previous solution $y = 0$ at $t = 0$ for $x > 0$. The solution to these specific boundary conditions is as follows:

$$y = Y \, erfc \left(\frac{x}{2 \sqrt{\varepsilon \, t_2}} \right) \quad , t > t_f \qquad\qquad\text{(III-2-30)}$$

which can be made dimensionless by dividing the above equation by the length of structure Y. The dimensionless solution is presented graphically in Figures III-2-34a, III-2-34b, and III-2-34c (at different scales). Pelnard-Considére (1956) used a time = t_2 in Equation 2-30 such that areas of shoreline above the x axis would be equal at the time $t = t_f$ when the structure is just filled to capacity (Equation 2-29), i.e., matched solution plan areas. In this manner t_2 was found to be $t_2 = t - 0.38 t_f$, where t is the initial solution time at which the structure begins to trap sand. Although the planforms do not match at time $t = t_f$ for the two solutions of Pelnard-Considére, the formulations are still useful for conceptual preliminary design and evolution of projects.

(12) The solution of Pelnard-Considére prior to bypassing (as given by Equation 2-28 and Figure III-2-31) may also be utilized for the situation in which erosion occurs on the sand-starved beach downdrift of an impermeable coastal structure that has no bypassing (natural or man-made). In this specific instance the solution would provide shoreline recession values as opposed to shoreline progradation values. In this scenario the solution should only be utilized far enough downdrift of the structure (i.e., beyond the immediate "shadow" of the structure) such that diffraction and refraction effects due to the structure do not influence the wave field and shoreline geometry.

(13) When applied in this scenario, the solution of Pelnard-Considére suggests that the ultimate downdrift extent of erosion caused by the structure is infinite. In practice, Equations III-2-28 and III-2-29 may be applied to estimate the theoretical downdrift extent of erosion, prior to bypassing, in terms of the structure length Y. That is, the distance downdrift of a structure at which the shoreline recession is less than or equal to some fraction of the structure's length (i.e., y / Y) can be expressed as a multiple of the structure's length (i.e., x / Y). Example III-2-10, below, illustrates this application. It is important to note that this solution is idealized and assumes that the breaking wave angle α_b can be approximated as an average, quasi-steady value. At present, the actual downdrift extent of erosion associated with a structure or other sediment sink is not well understood.

(14) The second case to be considered is that of a rectangular beach fill as shown in Figure III-2-35. Figure III-2-36 is a nondimensionalized solution graph that can be utilized in estimating plan area change for the rectangular beach nourishment fill on an initially straight reach of beach. Fill exists from $-a < x < +a$ and extends Y distance seaward from the original beach. The solution for this specific case is as follows:

$$y = \frac{Y}{2} \left\{ erf \left[\left(\frac{a}{2 \sqrt{\varepsilon t}} \right) \left(1 - \frac{x}{a} \right) \right] + erf \left[\left(\frac{a}{2 \sqrt{\varepsilon t}} \right) \left(1 + \frac{x}{a} \right) \right] \right\} \qquad\text{(III-2-31)}$$

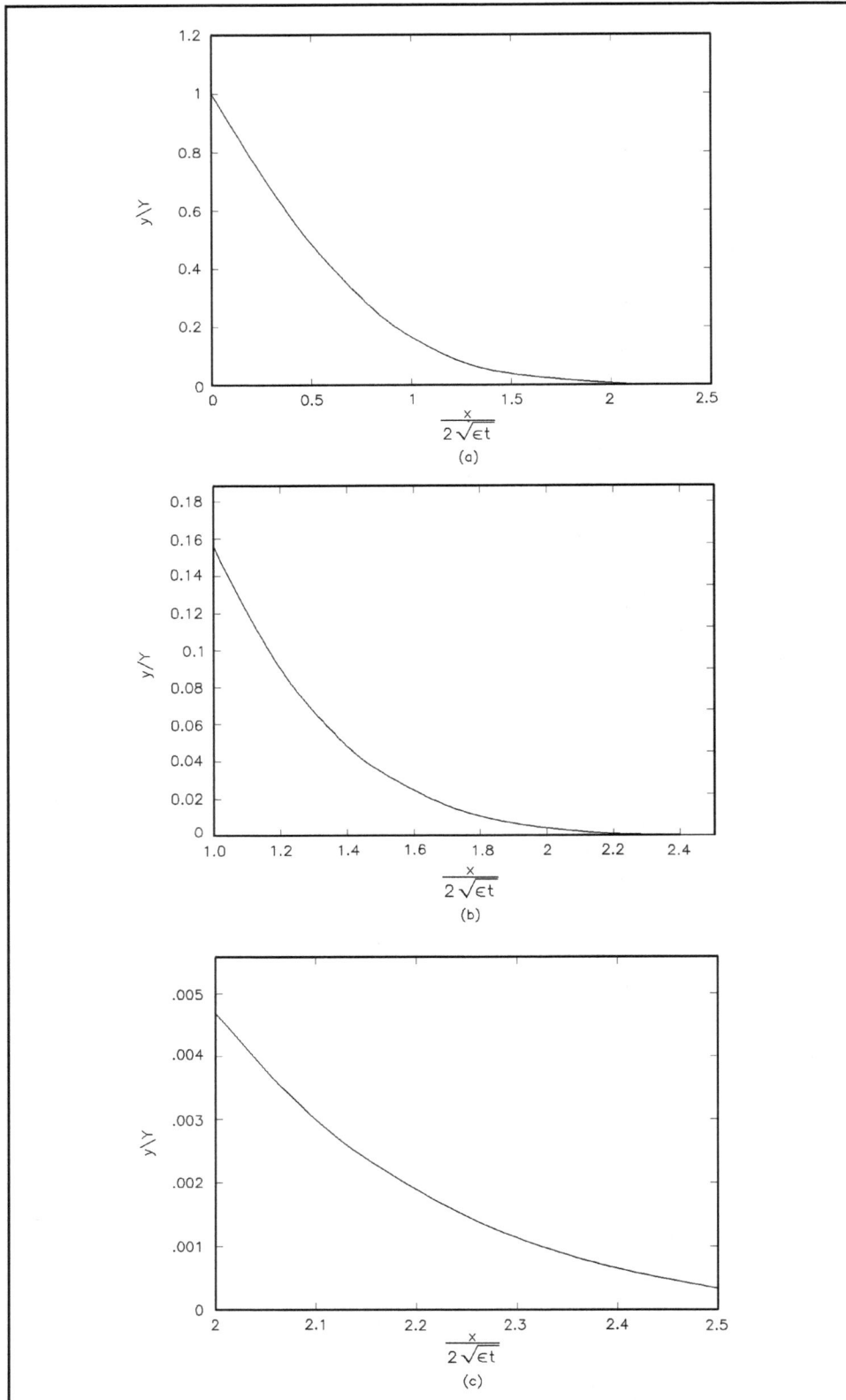

Figure III-2-34. Nondimensional solution curve for plan view of sediment accumulation at a coastal structure after natural bypassing initiated

EXAMPLE PROBLEM III-2-9

FIND:

    (a) The time it will take for the structure to fill to half its length, if its length is 600 m.

    (b) The distance seaward the shoreline will extend at the structure (y at $x = 0$) after 1 week ($t = 604,800$ sec) of continuous wave activity, and;

    (c) the distance seaward the shoreline will extend at 500 m updrift from the structure (y at $x = 500$ m) after 1 week ($t = 604,800$ sec) of continuous wave climate.

GIVEN:

    A long terminal groin extending a few surf zone widths at the end of a project reach adjacent to an inlet has been built to prevent sand from being lost from the beach into the inlet shoal system. Waves approach the inlet from the updrift side with a breaking angle $\alpha_b = 5$ deg. Breaking wave height $H_b = 2m$, and wave period $T = 10$ sec. The structure is initially expected to block all sediment (i.e., no bypassing). Sediment density to water density ratio $\rho_s/\rho = 2.65$, and porosity $n = 0.4$. Assume $d_B + d_c = 6$ m, $K = 0.77$, and $K = 1$.

SOLUTION:

    Equation 2-26 gives

$$\varepsilon = \frac{0.77\,(2)^2\,\sqrt{9.81 \cdot 2}}{8} \cdot \left(\frac{1}{2.65-1}\right) \cdot \frac{1}{(1-0.4)}\left(\frac{1}{6}\right) = 0.287\,\frac{m^2}{\text{sec}}$$

    (a) Express Equation 2-28 for the shoreline location y at the structure $x = 0$:

$$\frac{y}{2\sqrt{\varepsilon\,t}\,\tan(\alpha_b)} = \frac{1}{\sqrt{\pi}} \cong 0.564$$

and solve for $y = 1/2$ the length of the structure = 300 m, where $\alpha_b = 5$ deg:

$$\frac{300}{2\sqrt{0.287\,t}\,\tan(5^0)} \cong 0.564$$

Solve this expression for time t and find $t = 3.22 \times 10^7$ sec (372 days).

    (b) Use the expression for Equation 2-28 at the shoreline $x = 0$, and compute y for $t = 1$ week:

$$\frac{y}{2\sqrt{0.287\,(604800)}\,\tan(5^0)} = \frac{1}{\sqrt{\pi}}$$

and find $y = 41.1$ m (at x = 0).

    (c) Solve Equation 2-28 for $x = 500$ m and $t = 1$ week.

$$\frac{x}{2\sqrt{\varepsilon\,t}} = \frac{500}{2\sqrt{0.287\,(604800)}} = 0.60 \quad ; \quad 2\sqrt{\varepsilon\,t}\,\tan(\alpha_b) = 72.9\ \text{m}$$

From Figure III-2-33, $erfc(0.60) \approx 0.42$, so that Equation 2-28 is solved directly (or by Figure III-2-31) as

$$\frac{y}{2\sqrt{\varepsilon\,t}\,\tan(\alpha_b)} \cong \left[\frac{1}{\sqrt{\pi}}\exp(-0.6^2) - 0.6\,(0.42)\right] = 0.14$$

and

$$y = (0.14)\,(72.9\ \text{m}) = 10.2\ \text{m}$$

EXAMPLE PROBLEM III-2-10

FIND:

The distance downdrift of a structure at which the shoreline recession is less than or equal to 10 percent of the structure's length before sand begins to naturally bypass the structure.

GIVEN:

A long groin extending a few surf zone widths has been built with no artificial sand fill on either side. The groin's length, measured from the original shoreline, is Y. Assume that the wave activity is continuous with breaking angle $\alpha_b = 5$ deg.

SOLUTION:

The structure becomes filled to capacity at time t_f. Substitution of Equation 2-29 for t_f into Equation 2-28 yields:

$$\frac{y}{Y} = [\exp(-u^2) - \sqrt{\pi}\, u\, erfc(u)]$$

where

$$u = \frac{x}{2\sqrt{\varepsilon\, t_f}} = \frac{x}{Y}\frac{1}{\sqrt{\pi}}\, \tan(\alpha_b)$$

Determine the value u (graphically or by iteration), for which $y/Y = 0.10$. Find $u \approx 0.96$. Determine the downdrift distance x (relative to the structure's length Y) using this value for u and using $\alpha_b = 5$ deg:

$$\frac{x}{Y} = u\, \frac{\sqrt{\pi}}{\tan(\alpha_b)} = 0.96\, \frac{\sqrt{\pi}}{\tan(5^0)} = 19.4$$

That is, the shoreline recession is equal to or less than 10 percent of the structure's length beyond approximately 19.4 structure-lengths downdrift. If, for instance, the structure's length was $Y = 200$ m, the downdrift location at which the shoreline recession is less than 20 m (at the time the structure is filled to capacity) is (19.4)(200) = 3,880 m downdrift of the structure.

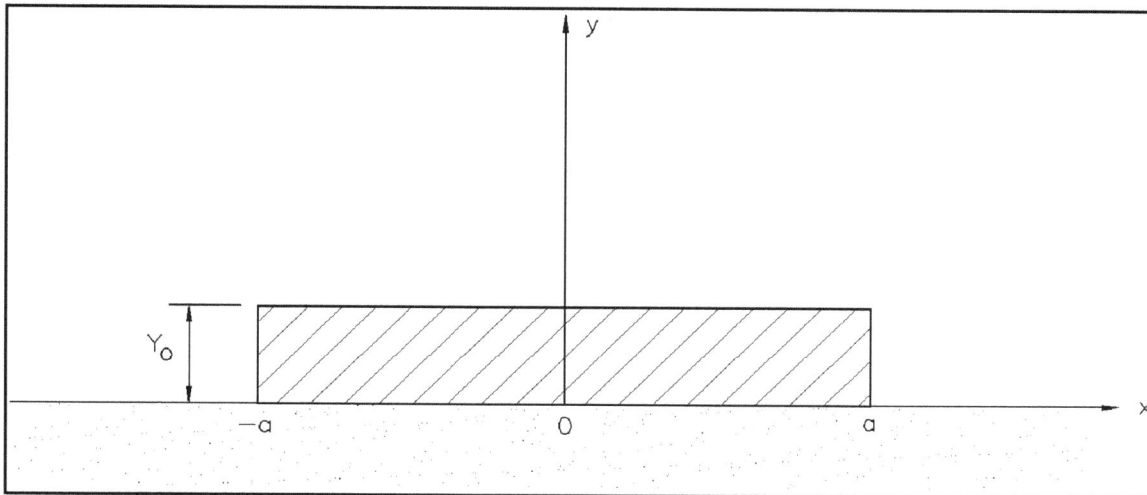

Figure III-2-35.   Rectangular beach fill (t=0)

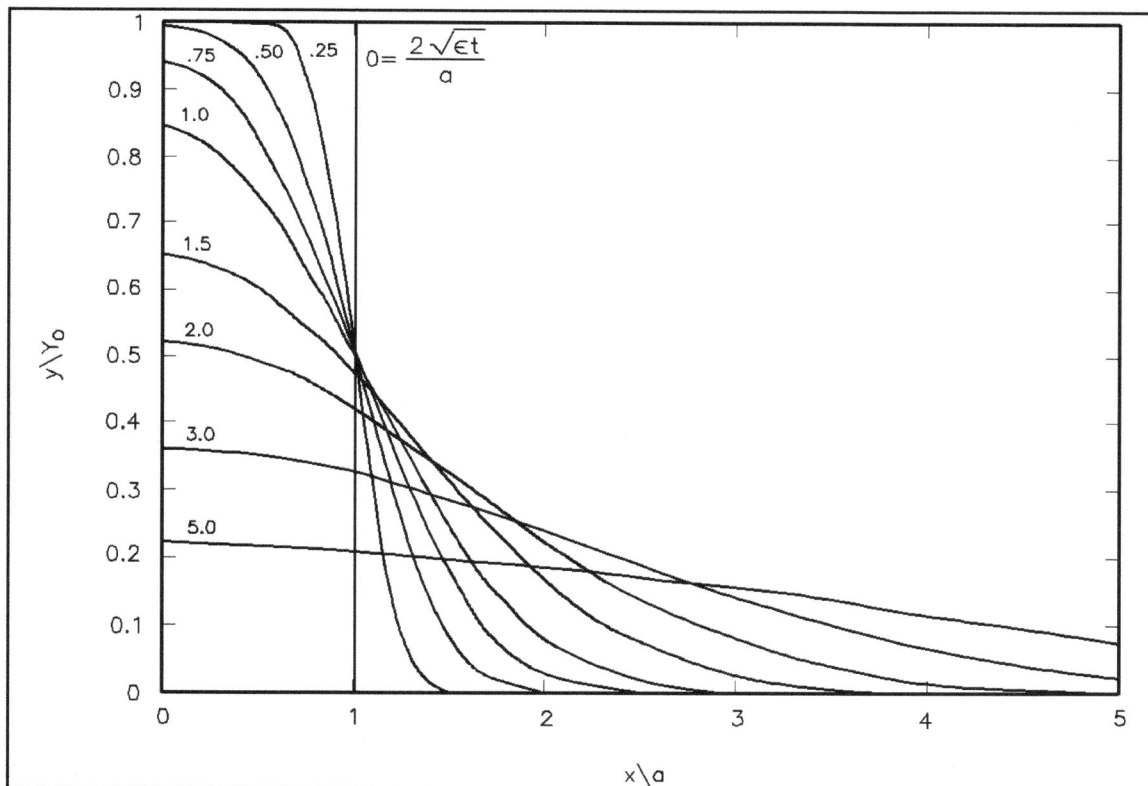

Figure III-2-36.   Nondimensional solution curves for rectangular initial plan view fill area

Only the portion of the solution graph for $x \geq 0$ is presented, as the solution is symmetric for values of x less than zero. The situation where tapered sections are included at the ends of the project is presented in Walton (1994).

(15) This equation may be integrated over the project limits to allow estimation of the proportion $p(t)$ of fill left within the project boundaries at a given time after project initiation to give:

$$p(t) = \frac{1}{\sqrt{\pi}} \left(\frac{\sqrt{\varepsilon t}}{a} \right) \left\{ \exp \left(- \left(\frac{a}{\sqrt{\varepsilon t}} \right)^2 \right) - 1 \right\} + erf \left(\frac{a}{\sqrt{\varepsilon t}} \right) \qquad \text{(III-2-32)}$$

which is plotted in Figure III-2-37 as a function of the inverse of the nondimensionalized beach half length parameter $\frac{a}{2\sqrt{\varepsilon t}}$

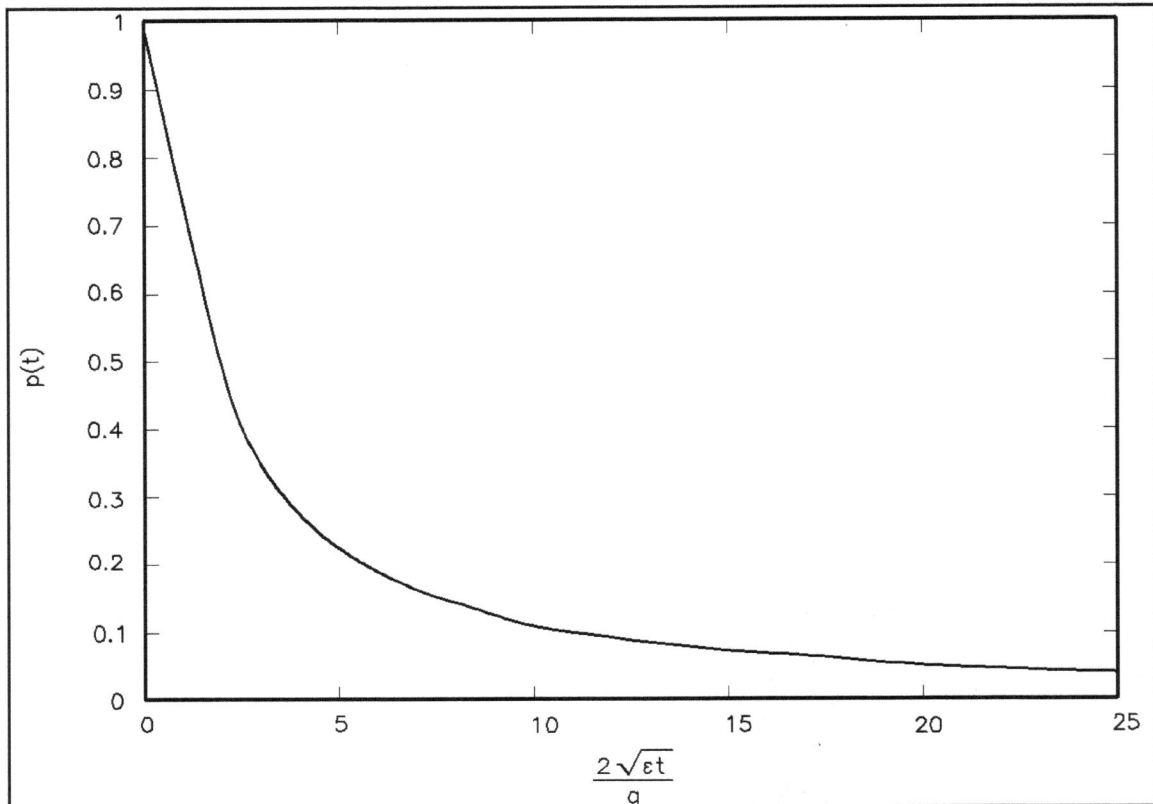

Figure III-2-37.   Proportion of fill p(t) remaining within limits of rectangular plan view fill area

(16) A third case to be considered is that of a triangular beach fill as shown in Figure III-2-38. A triangular planform beach nourishment might exist at a location where either a truck dumping of sand occurs or where drag scraping of sand from the offshore occurs. The solution for this specific case (where the original fill has been assumed triangular in shape) is:

$$y = \frac{Y_o}{2} \left\{ (1-X)\ erf\ (U(1-X)) + (1+X)\ erf\ (U\ (1+X)) -2\ Xerf\ (UX) \right.$$
$$\left. + \frac{1}{\sqrt{\pi}U} (e^{-U^2\ (1+X)^2} + e^{-U^2\ (1-X)^2} - 2e^{-(UX)^2}) \right\} \qquad \text{(III-2-33)}$$

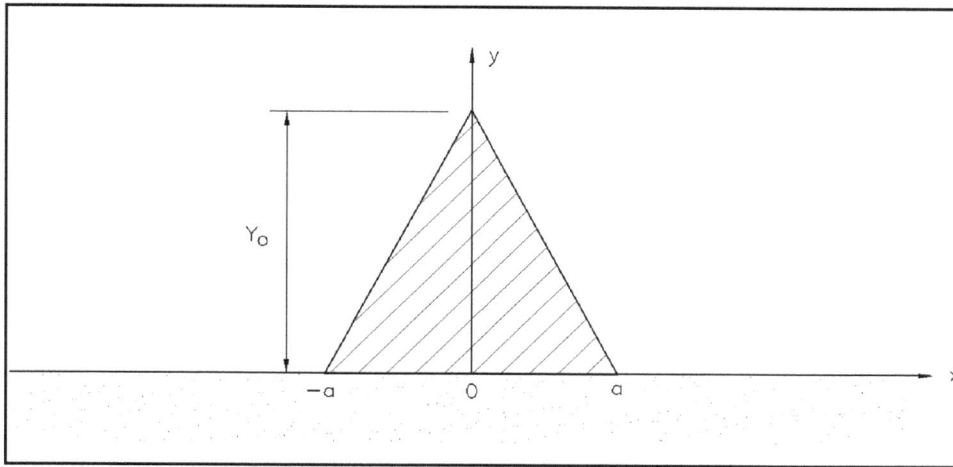

Figure III-2-38.   Triangular beach fill (*t*=0)

where

$$X = \frac{x}{a} \quad and \quad U = \frac{a}{2\sqrt{\varepsilon t}}$$

(17) Figure III-2-39 is a nondimensionalized solution graph for the case of a triangular planform beach nourishment. Only the portion of the solution graph for $x \geq 0$ is presented, as the solution is symmetric for values of x less than zero.

(18) A fourth case to be considered is that of a long fill project with a gap as shown in Figure III-2-40. For the case in which a beach nourishment has been placed on an existing beach but a gap has been left in the beach nourishment project (such as occurred in a beach nourishment project on Jupiter Island, Florida, in 1974), the following solution would apply:

$$y = \frac{Y}{2}\left\{ erfc\left[\left(\frac{a}{2\sqrt{\varepsilon t}}\right)\left(1 - \frac{x}{a}\right)\right] + erfc\left[\left(\frac{a}{2\sqrt{\varepsilon t}}\right)\left(1 + \frac{x}{a}\right)\right]\right\} \qquad \text{(III-2-34)}$$

The nondimensionalized solution graph for this particular planform of beach nourishment is provided in Figure III-2-41. Only the portion of the solution graph for $x \geq 0$ is presented, as the solution is symmetric for values of x less than 0.

(19) A fifth case to be considered is that of the end of a rectangular fill on a long beach nourishment project as shown in Figure III-2-42. The nondimensional solution graph for the case of a semi-infinite beach fill where the fill is contained within the area $x > 0$ is given in Figure III-2-43. This semi-infinite beach fill case can be utilized to provide the shape of the end of a long beach fill and the extent of the fill as it progresses down the coast with time. The solution for this specific case is

$$y = \frac{Y_o}{2}\left\{ 1 + erf\left(\frac{x}{2\sqrt{\varepsilon t}}\right)\right\} \qquad \text{(III-2-35)}$$

EXAMPLE PROBLEM III-2-11

FIND:

The width of the beach at the ends of the fill after 3 days and after 12 weeks of design wave action on the beach fill.

GIVEN:

A beach nourishment 5,000 m in length is placed on a reasonably stable shoreline to widen the beach as a protective measure for upland construction during storms. The new beach width is to be 50 m and is to be placed at the same berm elevation as the natural beach $d_B = 3$ m. Depth of appreciable sand transport as estimated from historical profiles in the region is assumed $d_c = 7$ m. A design wave for the area is estimated to be $H_b = 3$ m, $T = 12$ sec, $\alpha_b = 10$ deg. The sediment size of the natural beach is the same size as the fill sediment size and ratio of sediment density to water density (i.e., specific gravity) is 2.65. The porosity of beach is $n = 0.4$. The K factor is assumed = 0.77, and κ is assumed = 1.

SOLUTION:

$a = 5,000/2 = 2,500$ m ; $Y_O = 50$ m

At fill ends $x/a = 1.0$

From Equation 2-26

$$\varepsilon = \frac{0.77 \ (3)^2 \ \sqrt{9.81 \cdot 3}}{8} \left(\frac{1}{2.65 \ - \ 1} \right) \left(\frac{1}{1 \ - \ 0.4} \right) \left(\frac{1}{3 \ + \ 7} \right)$$

$$\cong 0.47 \ \text{m}^2/\text{s}$$

After 3 days $(t = 259,200$ sec$)$

$$\frac{a}{2\sqrt{\varepsilon t}} = \frac{2,500}{2\sqrt{0.47 \ (259,200)}} = 3.58$$

From Figure III-2-36, for

$$\frac{x}{a} = 1.0 \qquad \frac{y}{Y_o} = 0.56 \quad \text{hence,} \ y = 0.56 \ (50) = 28 \ \text{m}$$

After 12 weeks $(t = 7,257,600$ sec$)$

$$\frac{a}{2\sqrt{\varepsilon t}} = \frac{2,500}{2\sqrt{0.47 \ (7,257,600)}} = 0.68$$

From Figure III-2-36, for

$$\frac{x}{a} = 1.0 \qquad \frac{y}{Y_0} = 0.46 \quad \text{hence,} \ y = 0.46 \ (50) = 23 \ \text{m}$$

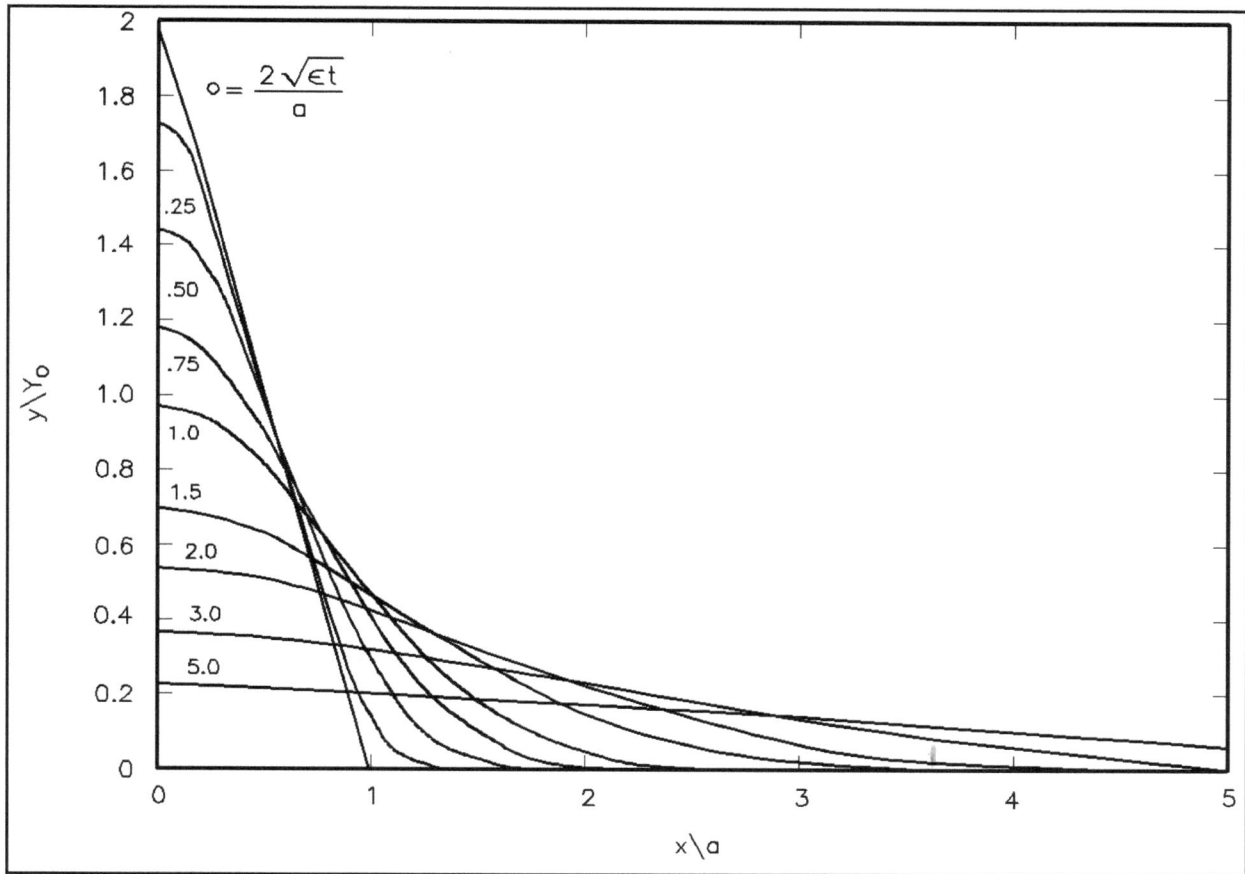

Figure III-2-39.   Nondimensional solution curves for triangular initial plan view fill area

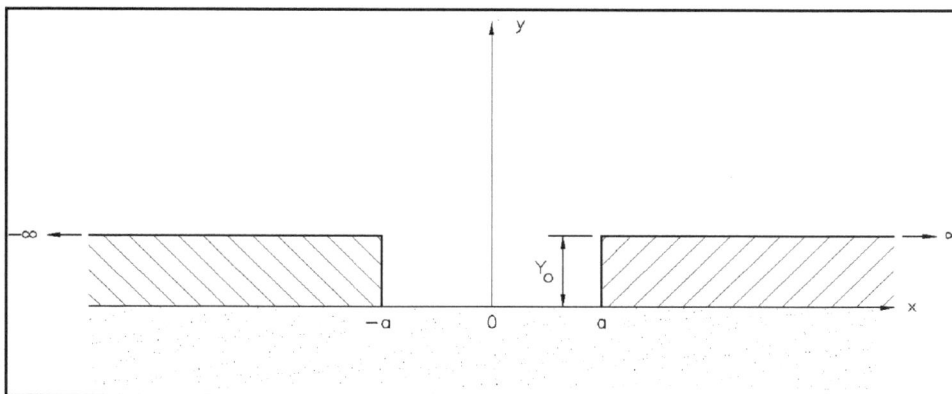

Figure III-2-40.   Long fill project with a gap (*t*=0)

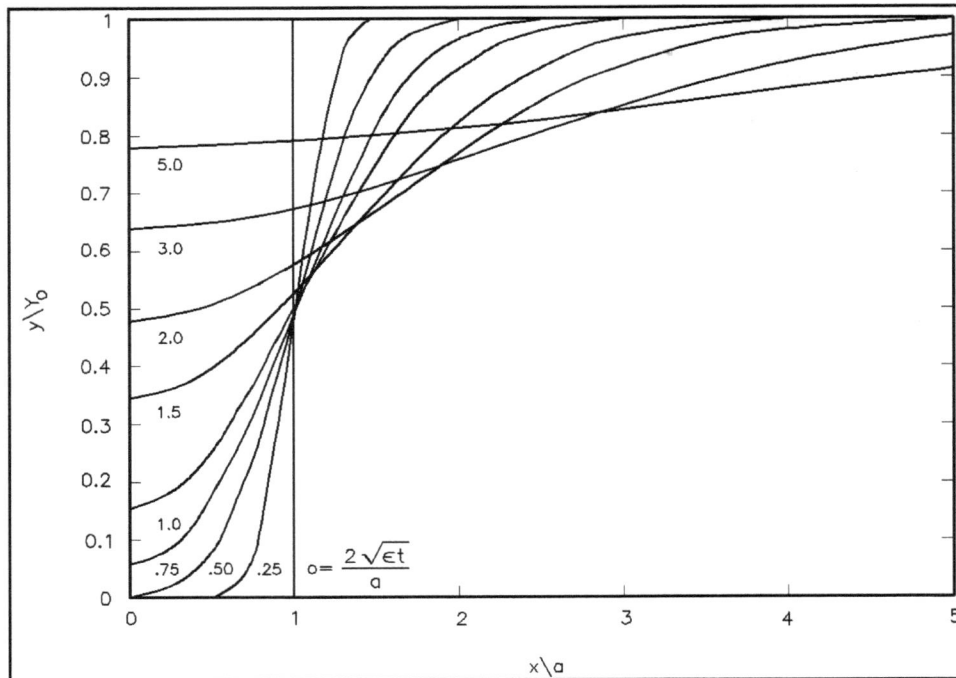

Figure III-2-41. Nondimensional solution curve for long fill with gap in plan view fill area

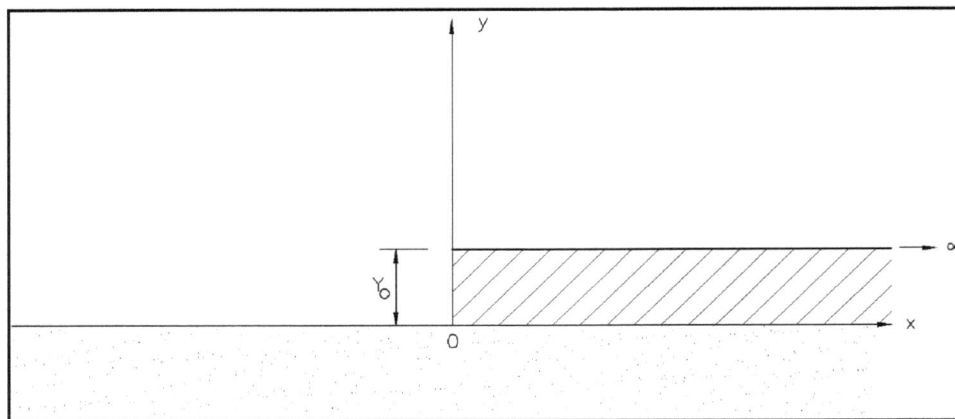

Figure III-2-42. Semi-infinite beach fill (*t*=0)

(20) For the situation in which the fill is in area $x < 0$, the graph can be utilized by flipping the solution presented around the y axis.

(21) A sixth case to be considered is the nourishment fill initially placed in the area $(x > 0)$ and maintained at the initial beach width Y within the project area. The solution for the planform beach adjacent to the fill $(x \leq 0)$ is given by

$$y = Y_o \left(1 + erf\left(\frac{x}{2\sqrt{\varepsilon t}} \right) \right)$$

(III-2-36)

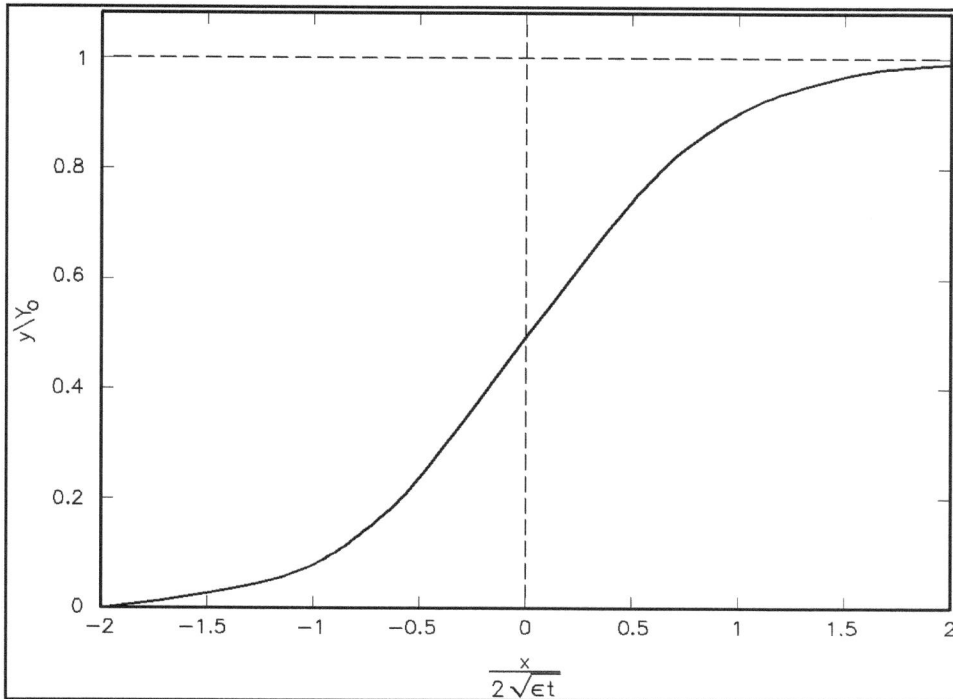

Figure III-2-43.   Nondimensional solution curve for semi-infinite plan view fill area

(22) For the analogous case to the above instance where the planform fill is placed in the area ($x < 0$) and maintained at the initial beach width Y in the project area, the solution (for $x \geq 0$) is:

$$y = Y_o \; erfc \left(\frac{x}{2\sqrt{\varepsilon t}} \right)$$

(III-2-37)

where the nondimensionalized solution graph has the same nondimensionalized solution as provided earlier in Figure III-2-34.

(23) A final case is presented for the situation in which groins and fill are implemented together as in Figure III-2-44. Where the initial fill is placed to the end of the groin, the solution is given by Dean (1984) as:

$$y = w - \ell \left(1 - \frac{x}{\ell} \right) \tan (\alpha_b) + \frac{2 \tan(\alpha_b)}{\ell} \sum_{n=0}^{\infty} \left[\frac{2\ell}{(2n + 1)\pi} \right]^2$$
$$\exp \left\{ - \varepsilon \frac{(2n + 1)^2 \pi^2 t}{4\ell^2} \right\} \cos \left[\frac{(2n + 1)\pi x}{2\ell} \right]$$

(III-2-38)

The nondimensionalized shoreline solution graph for this situation is provided in Figure III-2-45 for the parameters W/L = 0.25 and tan (α_b) = 0.1.

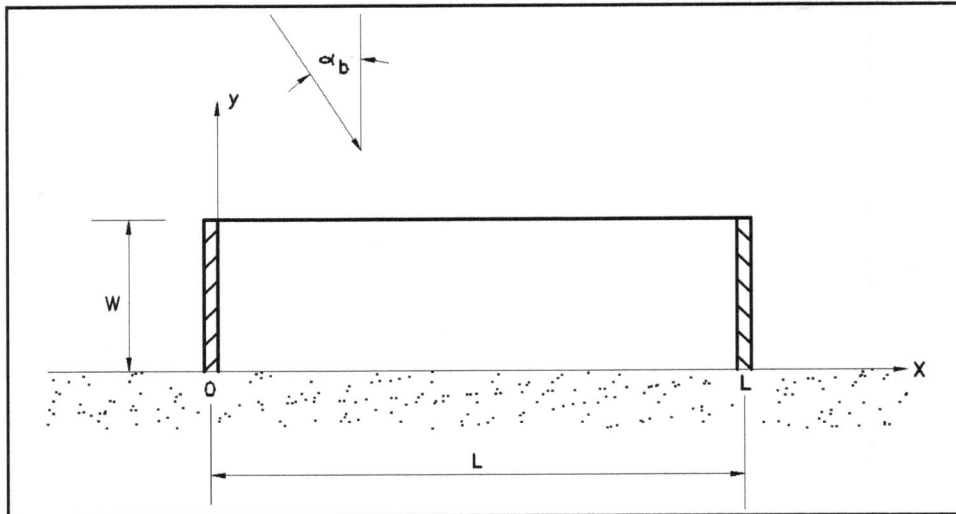

Figure III-2-44. Beach fill placed with groins ($t=0$)

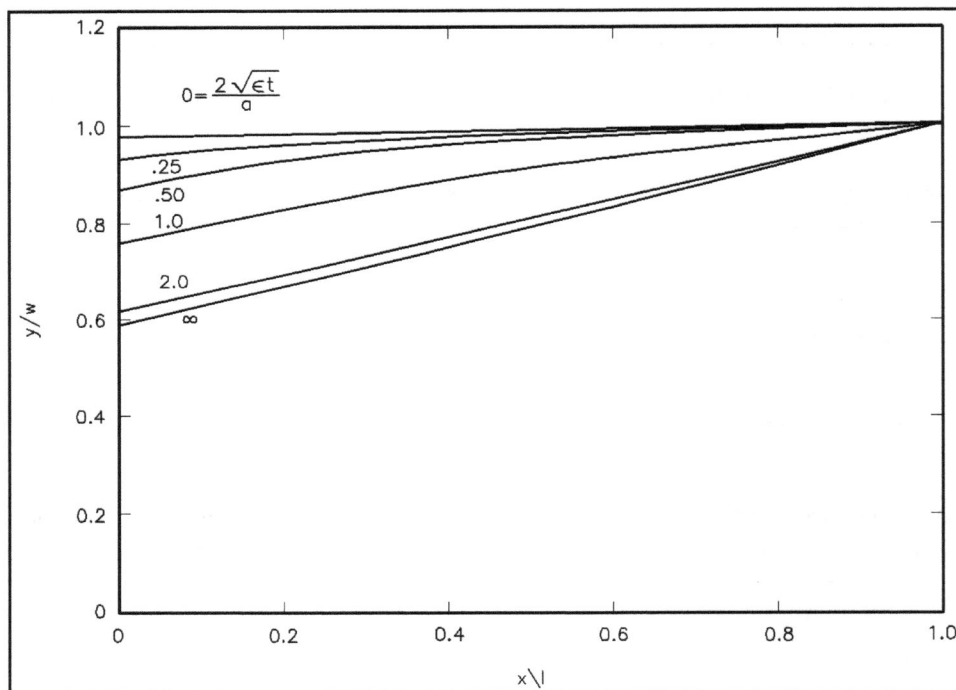

Figure III-2-45. Nondimensional solution curve for plan view of rectangular fill area between coastal structures (w/l = 0.25; tan α_b = 0.1)

(24) Since Equation 2-24 is linear, the above solutions can be combined to address more complex situations than those presented. As an example, consider the case in which staged construction of a beach nourishment project takes place over a long length of beach and the fill area of the beach is to be confined in the reach $x < 0$ and maintained at its fill width. Equation 2-37 could be utilized for solving this particular example in the case that the entire fill was placed at time $t = 0$. Instead, consider that the fill is placed in

stages with the plan view dimension of the fill as a function of time given in Figure III-2-46. For this particular scenario, the solution (at time t) analogous to Equation 2-37 only with the planform beach built in staged increments (as per Figure III-2-46) is:

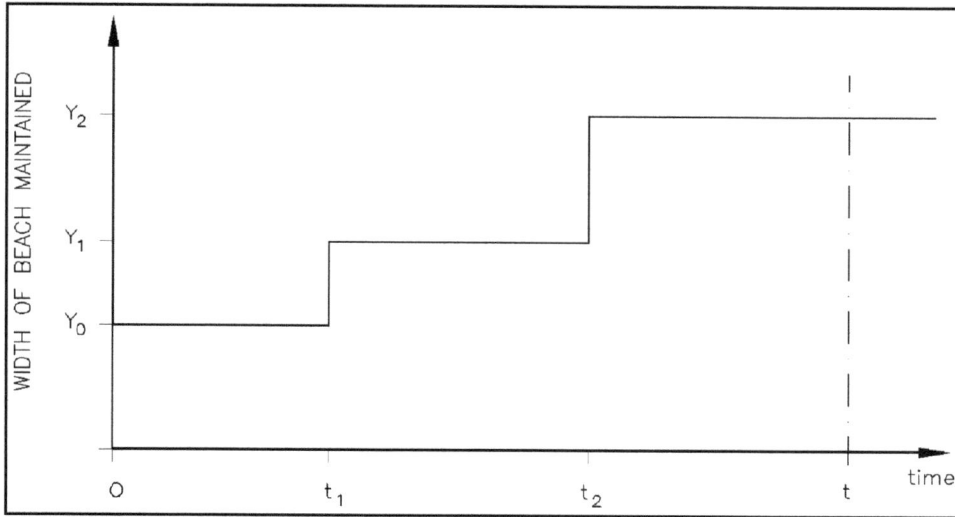

Figure III-2-46. Width of maintained beach ($x < 0$) as a function of time

$$y = Y_o \ erfc \left(\frac{x}{2\sqrt{\varepsilon t}} \right) + \left(Y_1 - Y_o \right) erfc \left(\frac{x}{2\sqrt{\varepsilon(t - t_1)}} \right) + \left(Y_2 - Y_1 \right) erfc \left(\frac{x}{2\sqrt{\varepsilon(t - t_2)}} \right) \qquad \text{(III-2-39)}$$

which can be nondimensionalized and solved utilizing the nondimensionalized solution graph provided by the simpler case of Equation 2-37 given in Figure III-2-34.

(25) It should be recognized that the linearized form of shoreline solution will produce a higher rate of shoreline change by overestimating the longshore transport rate because $2\alpha_b > \sin(2\alpha_b)$ in the linearized sand transport solution. Thus, under properly estimated parameters for idealized conditions, a higher rate of attenuation of beach fills will be obtained than is expected to occur in reality, thus providing a conservative answer to project losses.

(26) As a final point, it is noted that when wave angles are very large and the difference between the wave direction and shoreline orientation exceeds 45 deg, the true form of the shoreline diffusivity constant based on the nonlinear $\sin(2\alpha)$ term in the sand transport equation will be negative, which totally changes the characteristics of the shoreline change model equation. In these cases, the shoreline evolves in an unstable manner equivalent to running the previous stable solution forms backwards through time. In other words, a shoreline having a perturbation placed on it (such as a beach fill) would see a growth of the perturbation toward an elongated cuspate feature as time progresses rather than seeing the smoothing out of the perturbation as given in solutions of the preceding paragraphs. This type of shoreline instability may possibly explain certain shoreline features such as cuspate forelands which are ubiquitous on elongated bays where the dominant wave action is along the major axis of the bay and at large angles to the prevailing shoreline trends (Walton 1972).

EXAMPLE PROBLEM III-2-12

FIND:

The level of protection (i.e., width of beach) afforded to a historical lighthouse as a consequence of the nourishment project at the end of 3, 6, 8, 9, and 12 months after initial fill placement. The lighthouse is situated at the shoreline 1 mile ($\approx$ 1,609 m) south of the south end of the proposed beach at the initiation of the project fill.

GIVEN:

A long beach fill is placed along a north-south directed stretch of shoreline along the east coast of the United States (i.e., the azimuth of the offshore direction is 90 deg). The project is to be constructed and maintained continuously to design project width throughout the project fill area. The project will be constructed in two stages; Stage 1 will provide a fill width of 50 m, and Stage 2 (to be placed 3 months after initial placement) will add another 50 m of beach width, making the total finished project width = 100 m. Assume that: $H_b = 1$ m; $T = 10$ sec; $\alpha_b = 5$ deg (from the north) K is assumed = 0.77, and $\kappa = 1.0$. Sediment density to water density ratio (i.e., specific gravity) = $\rho_s/\rho = 2.65$; porosity $n = 0.4$; $d_B + d_c = 6$ m. Note that except for breaking wave height H_b, these are the same wave and sediment parameters as given in Example III-2-8.

SOLUTION:

For $H_b = 1$ m, with other parameters the same as the previous problem;

$$\varepsilon = \frac{0.77 \ (1)^2 \ \sqrt{9.81 \cdot 1}}{8} \left(\frac{1}{2.65 - 1}\right) \left(\frac{1}{1 - 0.4}\right) \left(\frac{1}{6}\right)$$

$$\cong 0.0508 \ \frac{m^2}{sec}$$

At the end of the first 3 months, the solution of Equation 2-37 (graphical solution provided by Figure III-2-34) can be utilized with $t = 3$ months $\approx 7.78 \times 10^6$ sec, and $x = 1$ mile $\approx 1,609$ m:

$$y \ (1 \ \text{mile}, \ 3 \ \text{months}) = 50 \cdot erfc \left(\frac{1,609}{2 \sqrt{0.0508 \cdot 7.78 \times 10^6}}\right) \ \text{meters}$$

$$\approx 50 \cdot erfc \ (1.28)$$

$$\approx 50 \ (0.07) \cong 3.5 \ m$$

(Continued)

Example Problem III-2-12 (Concluded)

For times beyond the 3-month initial fill placement and maintenance, a solution of the form of Equation 2-39 must be utilized. In this particular case, the solution would be of the form (where Y_o = 50 m, $Y_1 - Y_o$ = 50 m, and t_1 = 3 months):

$$y(x,\ t)\ =\ Y_o \cdot erfc\left(\frac{x}{2\sqrt{\varepsilon t}}\right)\ +\ (Y_1\ -\ Y_o) \cdot erfc\left(\frac{x}{2\sqrt{\varepsilon(t\ -\ t_1)}}\right)$$

which can be reformulated to be:

$$y(x,\ t)\ =\ 50 \cdot erfc\left(\frac{x}{2\sqrt{\varepsilon t}}\right)\ +\ 50\ erfc\left(\frac{x}{2\sqrt{\varepsilon t}} \cdot \frac{1}{\sqrt{1-\dfrac{t_1}{t}}}\right)$$

$$=\ term\ 1\ \ +\ \ term\ 2$$

The solution is provided in the table below:

(1)	(2)	(3)	(4)	(5)	(6)	(7)	(8)
t	$\dfrac{x}{2\sqrt{\varepsilon\ t}}$	$(2) \cdot \left(1-\dfrac{t_1}{t}\right)^{-1/2}$	$erfc((2))$	$erfc((3))$	term 1	term 2	y
6 mo. = 15.8×10^6 sec.	0.90	0.90 (1.414) = 1.27	0.20	0.07	10.0	3.5	13.5 m
9 mo. = 23.7×10^6 sec.	0.73	0.73 (1.225) = 0.89	0.30	0.21	15.0	10.5	25.5 m
12 mo. = 31.6×10^6 sec.	0.63	0.63 (1.155) = 0.73	0.37	0.30	18.5	15.0	33.5 m

Note that the solution form assumes that the fill is constantly maintained at its design width. This would probably not be achievable in practice, and solutions would have to be modified in accord with good engineering judgment.

III-2-4. Numerical Longshore Sand Transport Beach Change Models

As opposed to analytical solutions of shoreline change, which simplify the equations used to predict beach evolution, mathematical modeling facilitates generalization of these equations so that input parameters may vary in time and in the longshore, and possibly cross-shore, dimensions. Also, numerical models become necessary where difficult boundary conditions are encountered (say at groins or offshore breakwaters) because of shoreline morphology or wave transformation. Numerical models of beach change perform best when a perturbation is introduced to a system that is in equilibrium. The perturbation to the system might be an introduction or removal of littoral material (e.g., beach fill, sand mining, release of sediment due to a flooded river or landslide) or placement of a hardened structure (e.g., groins, detached breakwaters, seawalls, revetments). Historical trends of beach change and knowledge of the littoral budget are typically used to calibrate and verify the controlling equations, then forecasts may be simulated as a function of various engineering alternatives and/or wave climate scenarios. Therefore, beach response as a function of complex coastal processes may be readily examined in detail with mathematical models. However, the limitations inherent in the controlling equations, and assumptions implied in developing "representative" parameters require that model results be carefully interpreted, ideally within the context of other coastal engineering analyses.

a. Types of longshore transport models

(1) Fully three-dimensional models.

(a) In nature, nearshore beach change due to waves, circulation patterns, and longshore currents varies with time and location; therefore, equations to fully describe effects of these processes on beach evolution must be three-dimensional and time-dependent. Development of these equations is still an area of active research, and fully three-dimensional models are not available for routine engineering design.

(b) The intent of three-dimensional models is to describe bottom elevation changes which may vary in the cross-shore and longshore directions. These models provide insight into wave transformations and circulation for complicated bathymetry and in the vicinity of nearshore structures. However, they are less useful for making long-term shoreline evolution calculations because they are computationally intensive. These models also involve poorly known empirical coefficients such as those related to bottom friction, turbulent mixing, and sediment transport. Integrating the calculated local distributions of sediment transport over the cross-shore and for long time periods may lead to erroneous results because small local inaccuracies will be amplified over a long simulation. Because of their intent to predict local process parameters (e.g., waves, currents, sediment transport), they require a detailed data set for calibration, verification, and sensitivity testing, perhaps from a companion physical model study or field data collection.

(2) Schematic three-dimensional models. Schematic three-dimensional models simplify the controlling equations of fully three-dimensional models by, for example, restricting the shape of the profile or calculating global rather than point transport rates. Bakker (1968) has developed a two-line model which allows the evolution of two contours to be independently simulated. From this model, Perlin and Dean (1983) developed an n-line model that allows an arbitrary number of contour lines to represent the beach profile. Most schematic multi-line models developed to date are restricted to monotomic profile representations. For models that represent the profile by more than one contour, it is necessary to specify a relationship for cross-shore sediment transport. Schematic three-dimensional models have not yet reached the stage of wide application due to their complexity, requirement for considerable computational resources, and need for expertise in operational applications.

(3)   One-line (two-dimensional) models.

(a)   The Large Scale Sediment Processes Committee at the Nearshore Processes Workshop in St. Petersburg in 1989 concluded that long-term simulations of beach change are more reasonably formulated on the basis of total or bulk transport models such as Equation 2-7. These models have fewer coefficients than three-dimensional models and provide no details of the sediment transport profile. However, they may be calibrated and verified to include the integrated effect of all of the local processes on the total transport.

(b)   The shoreline change models developed from bulk transport models are often referred to as one-line models. One-line models assume that the beach profile is a constant shape; thus, the controlling equations may be solved for one contour line only (usually taken as the shoreline). Many shoreline change models have been developed and applied (e.g., Komar 1973a; LeMéhauté and Soldate 1977; Walton, Liu, and Hands 1988; and many others). Documented nonproprietary, one-line models that are also presently available include a model developed at the University of Florida (Dean and Grant 1989, Dean and Yoo 1992, 1994), and the computer model GENESIS (GENEralized model for SImulating Shoreline change) (Hanson 1987; Hanson and Kraus 1989; Gravens, Kraus, and Hanson 1991).

(c)   One-line models used to estimate longshore sand transport rates and long-term shoreline changes generally assume that the profile is displaced parallel to itself in the cross-shore direction. The profile may include bars and other features but is assumed to always maintain the same shape. This assumption is best satisfied if the profile is in equilibrium. The one-line model is formulated on the conservation equation of sediment in a control volume or shoreline reach, and a bulk longshore sand transport equation. It is assumed that there is an offshore closure depth d_C at which there are no significant changes in the profile, and the upper end of the active profile is at the berm crest elevation d_B. The constant profile shape moves in the cross-shore direction between these two limits. This implies that sediment transport is uniformly distributed over the active portion of the profile. The incremental volume of sediment in a reach is simply $(d_B + d_C)\Delta x\Delta y$, where Δx is the reach of shoreline segment, and Δy is the cross-shore displacement of the profile. Conservation of sediment volume may be written as

$$\frac{\Delta y}{\Delta t} + \frac{1}{d_B + d_C}\left(\frac{\Delta Q_\ell}{\Delta x} \pm q\right) = 0 \tag{III-2-40}$$

in which Q_ℓ is the longshore transport rate, q is a line source or sink of sediment along the reach, and t is time (Figure III-2-28). As examples, line sources of sediment may be rivers and coastal cliffs, and sinks may be produced by sand mining or dredging.

(d)   The longshore transport rate is evaluated using equations similar to Equation 2-7. These require measurement or calculation of the breaking wave angle relative to the beach. The local wave angle relative to the beach is the difference between the wave angle relative to a model baseline and the shoreline angle relative to the model baseline (Figure III-2-29),

$$\alpha_b = \alpha_{bg} - \alpha_{sg} = \alpha_{bg} - \tan^{-1}\left(\frac{dy}{dx}\right) \tag{III-2-41}$$

where x is the distance alongshore, and y is the distance offshore.

(e)   If the angle of the shoreline is small with respect to the x axis and simple relationships describe the waves, analytical solutions for shoreline change may be developed, as discussed in the previous section. For more complex conditions, such as time-varying wave conditions, large shoreline angles, variable longshore wave height (perhaps due to diffraction), multiple structures, etc., numerical models can be used in many instances.

   b.   *Shoreline Change Model GENESIS.*

   (1)   Overview.

   (a)   The numerical model GENESIS (Hanson 1987; Hanson and Kraus 1989; Gravens, Kraus, and Hanson 1991) is an example of a one-line shoreline change model that is supported for use both on personal computer and mainframe systems (see Cialone et al. (1992)) and has a companion system of support programs (Gravens 1992). GENESIS has been applied to numerous coastal engineering projects, and it calculates shoreline change due to spatial and temporal differences in longshore transport as produced by breaking waves. GENESIS is used in conjunction with grid-based wave transformation models that develop values of breaking wave height and angle for various representative wave periods and approach azimuths along the coast. Shoreline change at grid cells along the coastline is computed in the time domain as a function of these computed values of the breaking wave height and angle.

   (b)   As discussed by Hanson (1987) and Hanson and Kraus (1989), the empirical predictive formula for the longshore sand transport rate used in GENESIS is

$$Q_\ell = H^2_{b\ sig}\ C_{gb} \left(a_1\ sin2\alpha_b\ -\ a_2\ cos\alpha_b \frac{dH_{b\ sig}}{dx} \right) \qquad \text{(III-2-42)}$$

The nondimensional parameters a_1 and a_2 are given by

$$a_1 = \frac{K_1}{16 \left(\frac{\rho_s}{\rho} - 1 \right) (1 - n)\ (1.416)^{\frac{5}{2}}} \qquad \text{(III-2-43)}$$

and

$$a_2 = \frac{K_2}{8 \left(\frac{\rho_s}{\rho} - 1 \right) (1 - n)\ m\ (1.416)^{\frac{7}{2}}} \qquad \text{(III-2-44)}$$

where K_1 and K_2 are empirical coefficients, treated as calibration parameters, and m is the average bottom slope from the shoreline to the depth of active longshore sand transport. The factors involving 1.416 are used to convert the K_1 and K_2 coefficients from use with rms wave height to use with significant wave height (which is the statistical wave height required by GENESIS). That is, Equation 2-43 is presented such that K_1 is equivalent to K_{rms} (as opposed to K_{sig}) in Equation 2-7. Nonetheless, both longshore sand transport coefficients K_1 and K_2 should be viewed as calibration parameters that are to be adjusted to match measured positions of shoreline change (Hanson and Kraus 1989).

   (c)   The first term in Equation 2-42 corresponds to Equation 2-7, and accounts for longshore sand transport produced by obliquely incident breaking waves. The second term in Equation 2-42 is used to describe the effect of another generating mechanism for longshore sand transport, the longshore gradient in breaking wave height. The contribution arising from the longshore gradient in wave height is usually much smaller than that from oblique wave incidence in an open-coast situation. However, in the vicinity of structures, where diffraction produces a substantial change in breaking wave height over a considerable length of beach, inclusion of the second term provides an improved modeling result, accounting for diffraction effects.

(d)   The boundary conditions at the ends of a study area in a shoreline change modeling project must be specified. There are three common boundary conditions: no sand transport ($Q_\ell = 0$), free sand transport ($dQ_\ell/dx = 0$), and partial sand transport ($Q_\ell \neq 0$). The locations of the study area ends should be selected with these options in mind. Large headlands or jetties which completely block the longshore transport are good choices for model boundaries. At these locations $Q_\ell = 0$. Points where the position of the shoreline has not changed for many years are also good locations for boundaries. At these points, the gradient in longshore transport is small so that a free transport condition can be specified ($dQ_\ell/dx = 0$). At some locations, the longshore transport rate is known and can be used as a boundary condition (i.e., artificial sand bypassing at a jetty). If none of these "good" locations exist, engineering judgment must be used.

(e)   In all cases, results should be calibrated and verified using known shoreline positions and wave conditions for the longest period possible. The modeler also attempts to use wave data applicable to the period between the dates of the calibration shorelines. The GENESIS technical reference (Hanson and Kraus 1989) discusses in full the operation of shoreline change numerical simulation models.

(2)   Input data requirements and model output.

(a)   As discussed by Gravens (1991, 1992), preparation and analysis of the input and output data files occupy a substantial portion, perhaps the majority, of the time spent on a detailed shoreline change modeling project. Gravens (1991, 1992) stresses that the data gathering organization and analysis process cannot be overemphasized because (1) it forms the first necessary level in understanding coastal processes at the project site, and (2) the simulation results must be interpreted within the context of regional and local coastal processes, and the natural variability of the coastal system. Success in modeling shoreline change depends, to a large extent, on preparation and analysis of the input data.

(b)   General input information required by GENESIS includes the spatial and temporal ranges of the simulation, structure and beach fill configurations (if any), values of model calibration parameters, and simulated times when output is desired. Initial and measured (if available) shoreline positions as referenced to a baseline established for the simulation are also required. Offshore and nearshore (if available) wave information and associated reference depths are used to calculate longshore sand transport rates. Output information produced by GENESIS includes intermediate and final calculated shoreline positions, and net and gross longshore sand transport rates.

(3)   Capabilities and limitations.

(a)   GENESIS was designed to predict long-term trends of the beach plan shape in its evolution from one given initial condition. This change is usually caused by a notable perturbation; for example, by beach fill placement, sand mining, sand discharge from a river, construction of a detached breakwater, or jetties constructed at a harbor or inlet. In engineering applications and tests of GENESIS, modeled shoreline reaches have ranged from about 2 to 35 km with a grid resolution of 15 to 90 m, and simulation periods have spanned from approximately 6 months to 20 years, with wave data typically entered at simulated time intervals in the range of 30 min to 6 hr (Gravens, Kraus, and Hanson 1991).

(b)   Hanson and Kraus (1989) and Gravens, Kraus, and Hanson (1991) discuss the capabilities and limitations of GENESIS. The model allows an almost arbitrary number and combination of groins, jetties, detached breakwaters, beach fills, and seawalls, with a primary limitation in this regard being the size and speed of the computer and the maximum number of grid cells that can be accommodated by the program. Compound structures (such as T-shaped, Y-shaped, and spur groins) may be simulated with varying degrees of realism. At least five grid cells are required to reasonably model the shoreline behind and between structures. Sand bypassing around and transmission through groins and jetties may be simulated, as well as diffraction at detached breakwaters, jetties, and groins. Transmission through detached breakwaters may be simulated. GENESIS allows multiple wave trains to be input (such as from independent wave sources), and the wave information may include arbitrary values of wave height, period, and direction. As presented in Equation 2-42, sand transport is calculated due to oblique wave incidence and a longshore gradient in wave height.

(c) GENESIS is not applicable to simulating a randomly fluctuating beach system in which no trend in evolution of the shoreline is evident. In particular, GENESIS is not applicable to calculating shoreline change in the following situations which involve beach change that is not related to coastal structures, boundary conditions, or spatial differences in wave-induced longshore sand transport: beach change inside inlets or in areas dominated by tidal flow; beach change produced by wind-generated currents; storm-induced beach erosion in which cross-shore sediment transport processes are dominant; and scour at structures. GENESIS also does not include wave reflection from structures; cannot reliably simulate tombolo or salient development at a detached breakwater; and there is no direct provision for changing tide level. In addition, the basic assumptions in the development of shoreline change modeling theory apply.

(4) Example application - Bolsa Chica, California.

(a) As discussed by Hanson and Kraus (1989) and Gravens (1990b), GENESIS (either its predecessor, Version 1, or the current Version 2) has been applied at numerous project sites, including locations in Alaska (Chu et al. 1987); California (Gravens 1990a); Louisiana (Hanson, Kraus, and Nakashima 1989; Gravens 1994); New Jersey (Gravens, Scheffner, and Hubertz 1989); New York (Cialone et al. 1994); Florida (by the U.S. Army Engineers District, Jacksonville); and outside the United States (Hanson and Kraus 1986; Kraus 1988). Application of the model to assess a proposed structured inlet system at Bolsa Chica, California, is summarized below (Gravens 1990a) to illustrate model use in coastal project evaluation and refinement.

(b) GENESIS was applied to an approximately 10-mile-long shoreline reach from Anaheim Bay to the Santa Ana River, California (Figure III-2-47) as part of a comprehensive multi-tasked engineering investigation for the California State Lands Commission (SLC). The shoreline change modeling effort was directed towards quantifying the potential long-term impacts of the proposed entrance on adjacent shorelines, and to investigate mitigation of any adverse impacts induced by the entrance. Three major components were involved in the shoreline modeling effort: (1) a preliminary shoreline response study, in which available wave and shoreline data were used to provide preliminary estimates of the introduction of a littoral barrier in the local littoral cell; (2) a comprehensive wave hindcast of locally generated wind sea and North Pacific swell conditions from 1956 to 1975 at 3-hr intervals, and an 18-month hindcast of South Pacific swell; and (3) comprehensive shoreline response modeling using the hindcast wave data to predict response of the project area to various design alternatives. Discussion herein is focussed on this third component, including preparation of input data sets and analysis of model output, as they pertain to application of GENESIS. For a more complete description of the project, the reader is directed to Gravens (1990a).

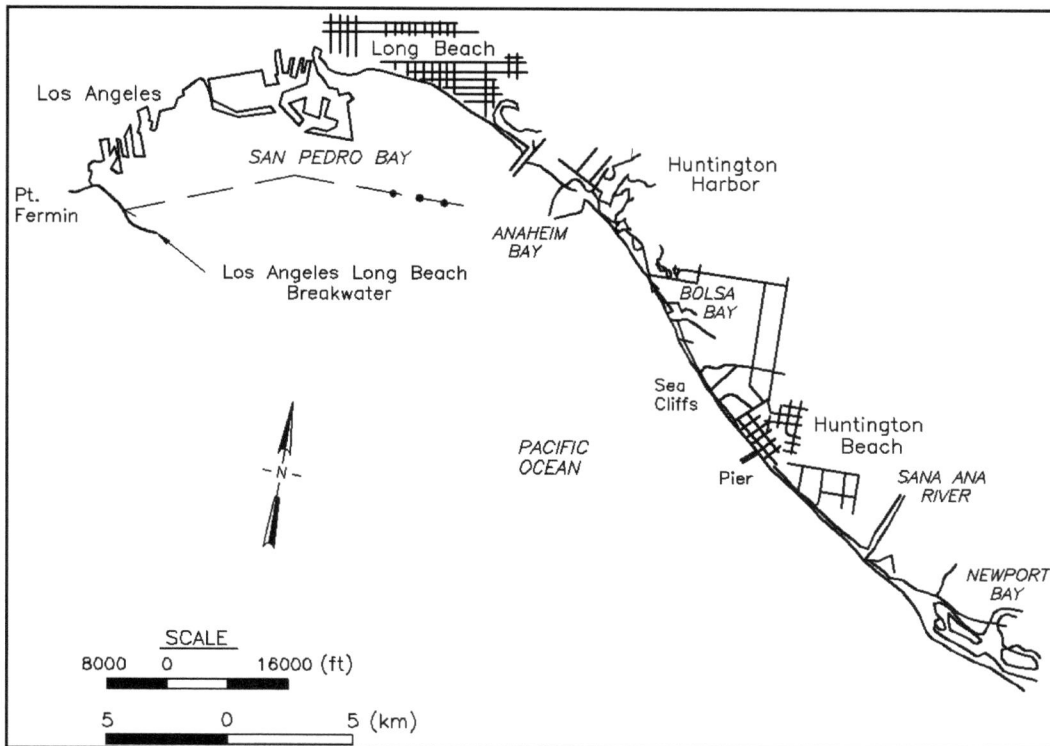

Figure III-2-47. Bolsa Chica, California, study area (Gravens 1990a)

(c) The modeled reach from Anaheim Bay to the Santa Ana River forms a littoral cell, including a complete barrier to littoral transport at Anaheim Bay, and a submarine canyon offshore of Newport Beach (Figure III-2-47). Coastal structures and features of importance within the model reach include the east Anaheim jetty, the sea cliffs at Huntington Beach, the Huntington Beach pier, and the north jetty at the mouth of the Santa Ana River. Each of these features influences the evolution of adjacent shoreline and was represented in the shoreline change model. The sea cliffs at Huntington Beach serve to pin the shoreline between the cliffs and Anaheim Bay to the northwest and the Santa Ana River to the southeast. The Huntington Pier and the east Anaheim Bay jetty modify the local breaking wave pattern and produce a local shoreline signature unique to these structures. Beaches between Anaheim Bay and the Santa Ana River have accreted an average of 1.3 m/year (4.4 ft/year) between 1934 and 1983.

(d) Ten shoreline position data sets dating from 1878 and 1983 were analyzed to determine historical and representative shoreline change trends. As a result of this analysis, the 1963, 1970, and 1983 shorelines were determined to be representative, and were summarized at 61-m (200-ft) intervals for model calibration (1963 to 1970) and verification (1970 to 1983).

(e) For the comprehensive shoreline response study, hindcast wave estimates at stations located near the lateral boundaries of the modeled shoreline reach were transformed from the hindcast stations to the offshore boundary of the GENESIS grid using the linear wave propagation model RCPWAVE (Regional Coastal Processes Wave Model; Ebersole 1985; Ebersole, Cialone, and Prater 1986) (Figure III-2-48). Use of two hindcast stations allowed systematic variations in the incident waves (alongshore wave height and angle) to be accounted for in RCPWAVE and ultimately in GENESIS.

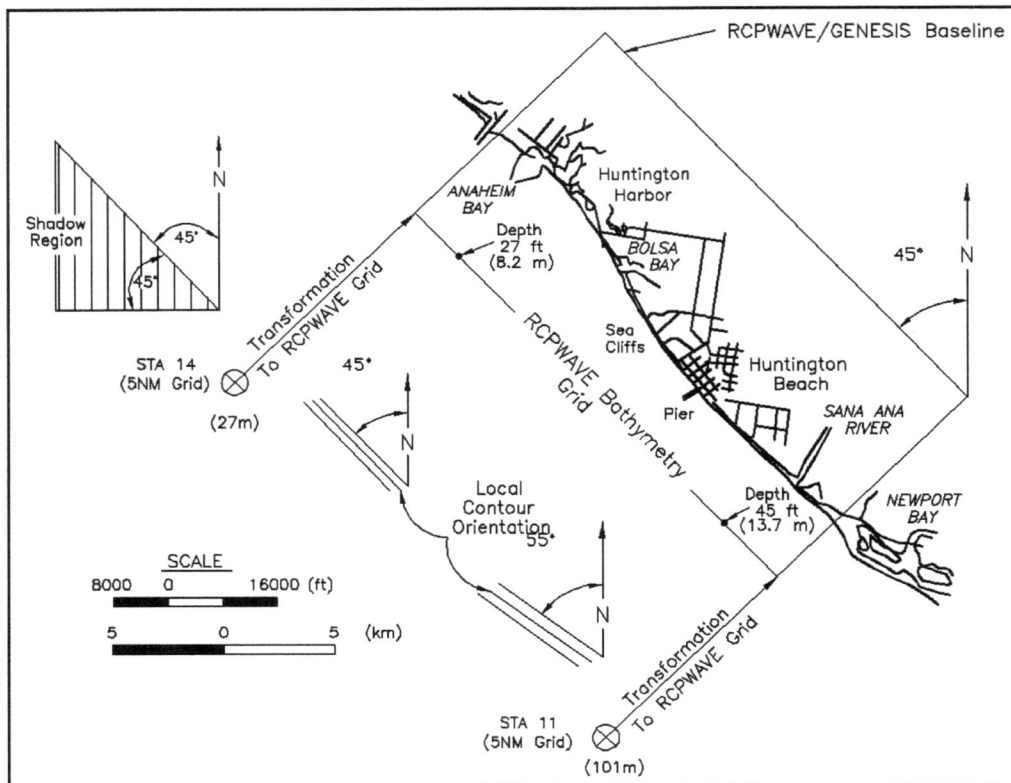

Figure III-2-48.   Wave transformation hindcast to RCPWAVE grid (Gravens 1990a)

(f)   Potential longshore sand transport rates were calculated using the transformed wave estimates and relationships similar to Equation 2-7 for the various shoreline orientations within the project reach. These potential transport rates are presented in the form of a total littoral drift rose (Walton 1972, Walton and Dean 1973) (Figure III-2-49).   The curve with the circular symbols in Figure III-2-49 represents the average downcoast littoral drift for the 20-year Northern Hemisphere hindcast of sea and swell wave conditions. The curves with "x" symbols and "∇" symbols in Figure III-2-49 represent the average upcoast littoral drift for the available 2 years of Southern Hemisphere swell wave estimates. It is interesting to note that there is a reversal in the direction of the average net longshore littoral drift and that this reversal occurs at different shoreline orientations depending on the time series of southern swell wave conditions used in the calculation.

(g)   The results of the final model calibration simulation, 1963 to 1970, are presented in Figure III-2-50. In model calibration, the calibration parameters K_1 and K_2 ranged between 0.8 and 0.2; values of these parameters that best estimated gross and net longshore sand transport rates and reproduced observed shoreline change were determined to be $K_1 = 0.45$ and $K_2 = 0.4$. Calibration results lead to three general observations. First, in the Anaheim Bay entrance area (between alongshore coordinates 220 and 260), there are significant differences between the calculated and measured shoreline positions. These differences are due in part to the reflection of waves from the east Anaheim Bay jetty (a process which was not modeled) and to a massive (4 million-cu-yd) renourishment of the Surfside-Sunset feeder beach in 1964. The percentage of fine material contained in the beach fill is unknown; consequently, the initial losses of fill material could not be estimated or accounted for in the model. Model results in this region should be viewed with caution.

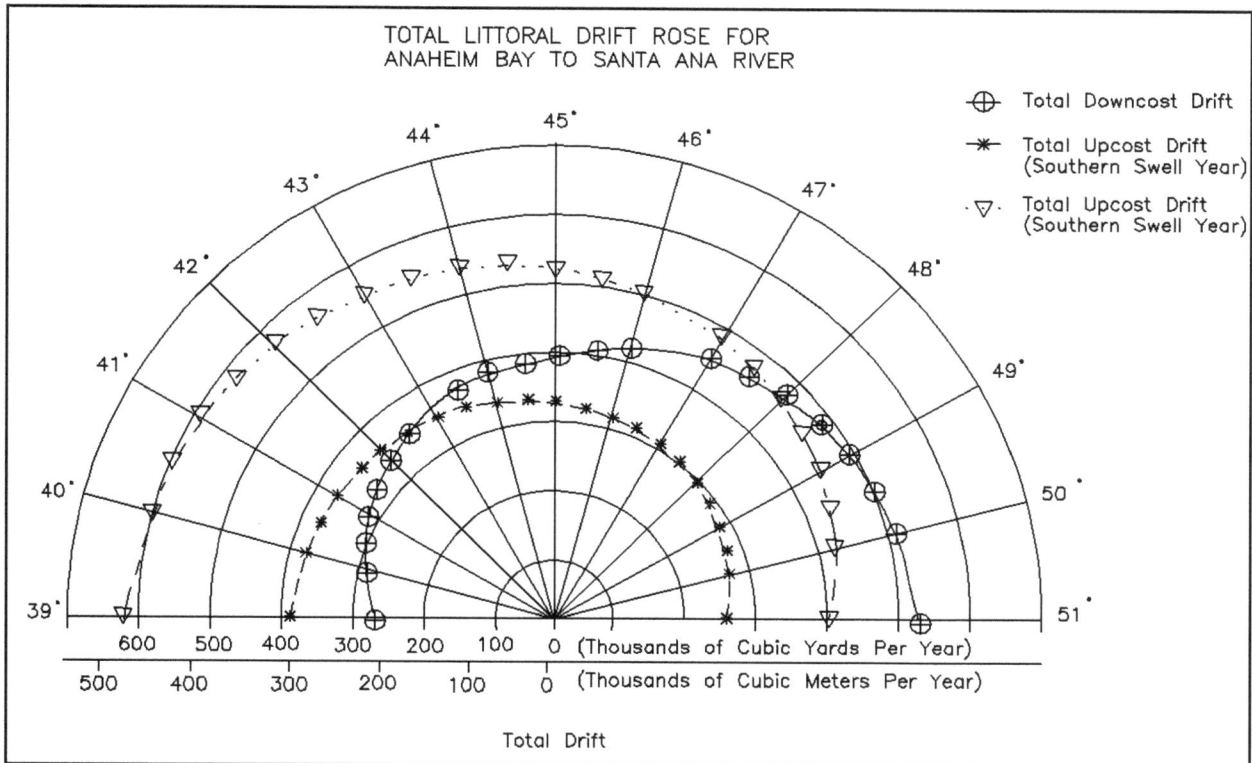

Figure III-2-49. Total littoral drift rose, Anaheim to Santa Ana River (Gravens 1990a)

(h)   Secondly, in the vicinity of the Huntington Pier (between alongshore coordinates 80 and 90), it is noted that the predicted shoreline positions do not agree well with the survey. The lack of agreement is due to limitations in the groin boundary condition used to simulate the effects of the pier. The imposed boundary condition at the pier was investigated in detail, and the conclusion was that the boundary condition imposed at the Huntington Pier had no significant effect on the model results northeast of the sea cliffs over the modeling interval.

(i)   Finally, in the vicinity of the proposed entrance system (between alongshore coordinates 155 and 220), the predicted and measured shoreline positions are in very good agreement. Model results for this region are considered to have high reliability.

(j)   The next step was to verify the model by performing a simulation using the same calibration parameters for a different time period. The verification time period (1970 to 1983) included two beach fill projects at the Surfside-Sunset feeder beach, the first in 1971 (2.3 million cu yd) and the second in 1979 (1.66 million cu yd). The results of the model verification are shown in Figure III-2-51. Although the agreement between the calculated and measured shoreline positions is not as close for the verification as it was for calibration, overall measured change in shoreline position was reproduced and was considered acceptable. The largest discrepancies between measured and predicted shorelines occur adjacent to the Anaheim Bay jetty, where initial losses of fine-grained beach fill to the offshore may have contributed to the differences.

BOLSA CHICA: MODEL CALIBRATION

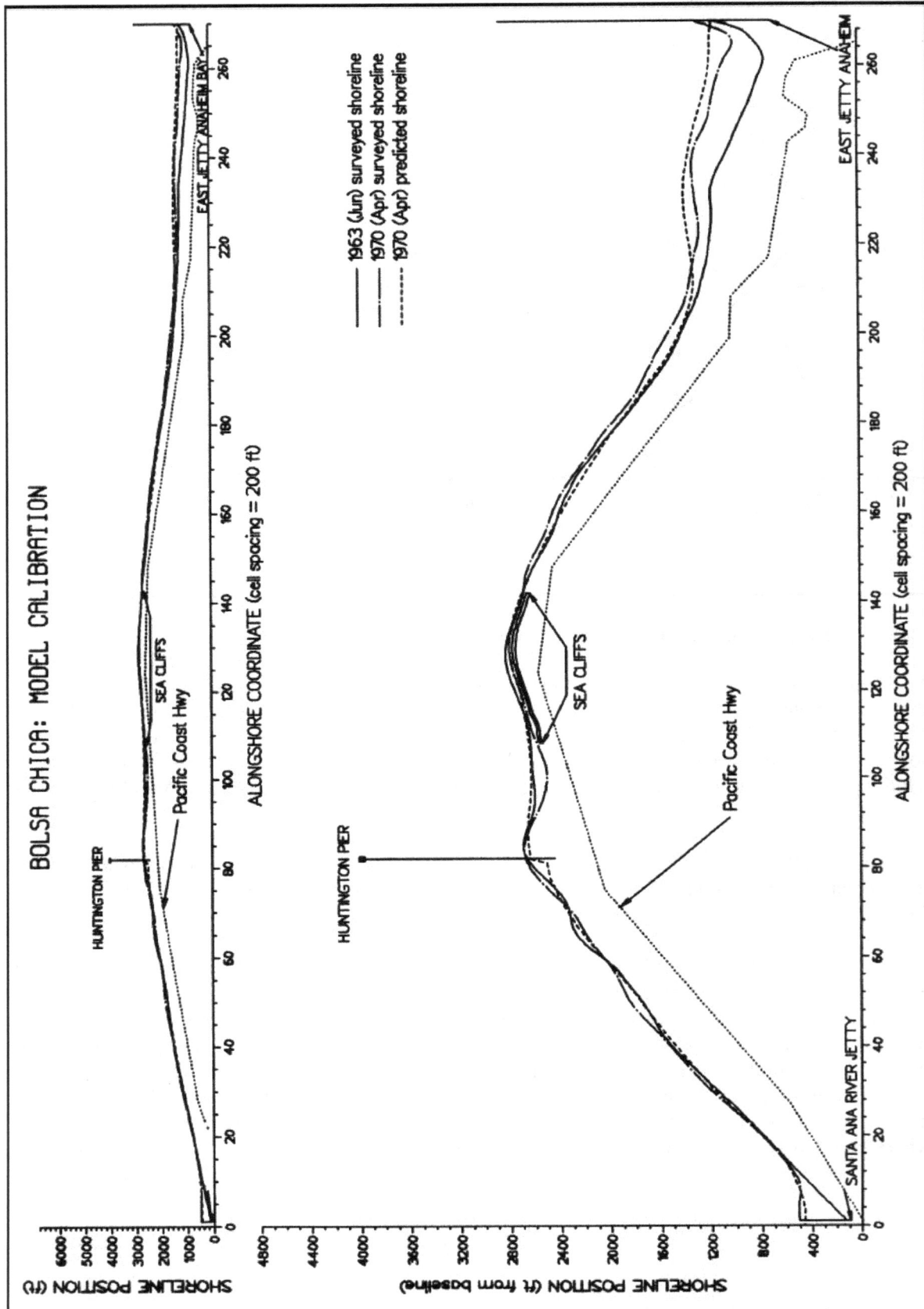

Figure III-2-50.   Model calibration results, Bolsa Chica (Gravens 1990a)

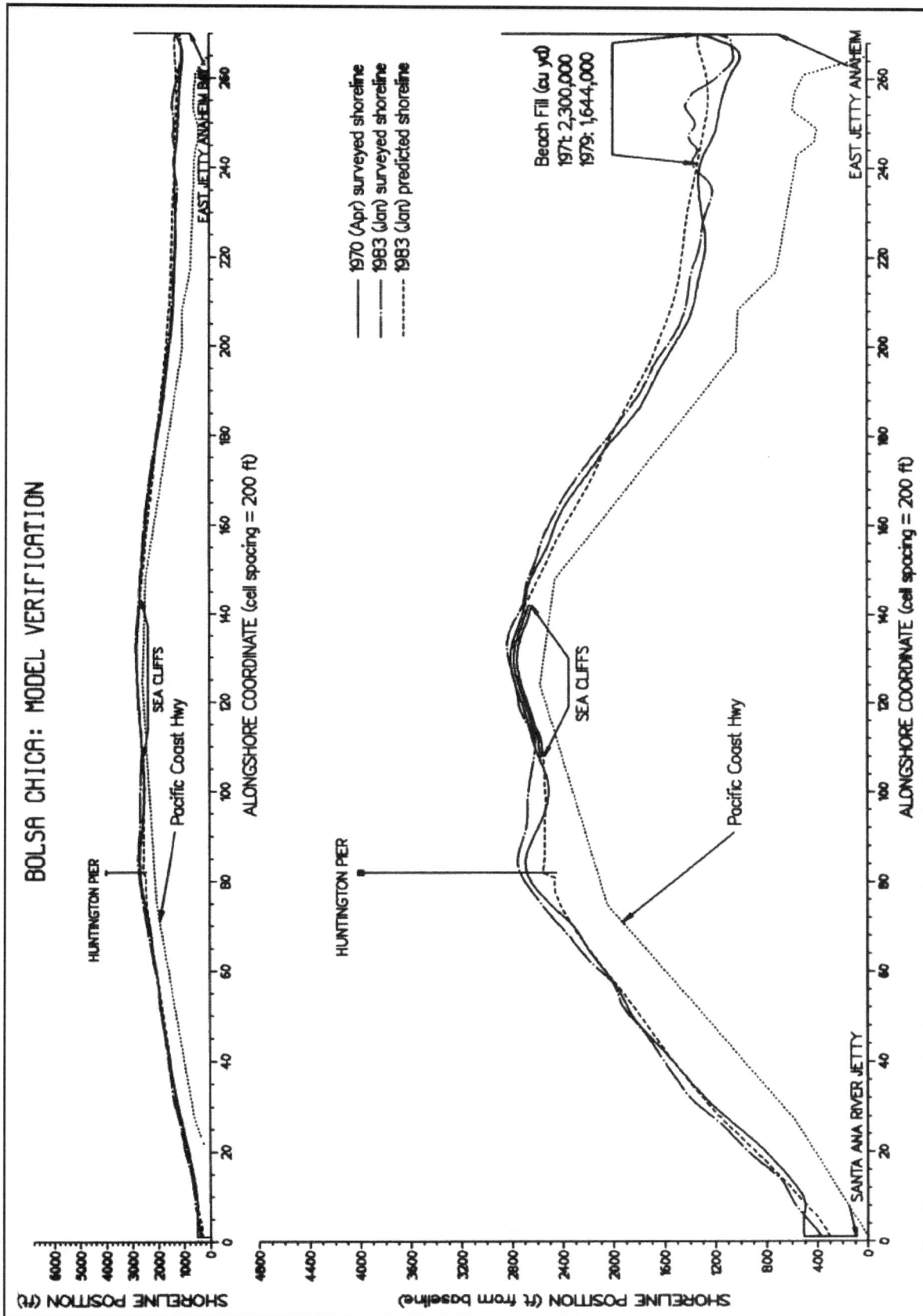

BOLSA CHICA: MODEL VERIFICATION

1970 (Apr) surveyed shoreline
1983 (Jan) surveyed shoreline
1983 (Jan) predicted shoreline

Beach Fill (cu yd)
1971: 2,300,000
1979: 1,644,000

Figure III-2-51. Model verification results, Bolsa Chica (Gravens 1990a)

(k)   After model calibration and verification, eight conceptual design alternatives were modeled, and several simulation variations were performed for each alternative. The intent of the simulations was to quantify the shoreline impacts of the proposed Bolsa Chica navigable ocean entrance system. In the simulation of Alternatives 1 and 3, no beach fill was included along the modeled reach. For Alternatives 2, 4, 5, 6, and 8, renourishment of the Surfside-Sunset feeder beach was specified at 1 million cu yd every 5 years. Alternatives 7 and 8 modeled impact mitigation sand management techniques. The 1983 shoreline was used as the initial shoreline, and all model tests were performed for 5- and 10-year simulation (prediction) periods using a randomly selected 10-year time history of Northern Hemisphere sea and swell conditions. The Southern Hemisphere swell component of the incident wave climate was varied from alternating available southern swell wave conditions, low-intensity southern swell, and high-intensity southern swell to predict a range of influence.

(l)   Model results and analysis from 24 production simulations are documented by Gravens (1990a); only results from one alternative are presented here for illustrative purposes. Predicted changes in 10-year post-project shoreline position with Alternative 8 are shown in Figure III-2-52. Alternative 8 includes two shore-perpendicular jetties spaced 245 m apart and extending approximately 425 m offshore, a detached breakwater composed of three sections located offshore of the entrance channel, a feeder beach at Surfside-Sunset, and impact mitigation sand management. The impacts of this entrance system with the specified sand management plan as compared to a without-project 10-year projection are shown in Figure III-2-53. This alternative satisfied the criteria established by the SLC for successful impact mitigation. The SLC specified that only sand accumulating within 460 m of the entrance jetties may be used by sand bypassing/backpassing, and that a successful sand management plan would predict more accretive, or equal or less erosive, conditions than would occur without the project in place.

(m)   Conclusions from shoreline change modeling of Bolsa Chica Bay were as follows:

- The proposed site of the new entrance system is located in a region of converging longshore sand transport.

- Locating the entrance system approximately 1.6 km up- or downcoast from the proposed site would not significantly change the predicted shoreline response.

- Implementation of a sand management plan would allow for the mitigation of adverse shoreline impacts.

- The Surfside-Sunset feeder beach nourishment program must be continued in order to maintain the shoreline within 3.2 km of the Anaheim Bay entrance. However, the proposed entrance system would neither aggravate nor improve the situation.

BOLSA CHICA: PREDICTED SHORELINE CHANGE

1983 (Jan) surveyed shoreline
5—yr (1988 Jan) predicted shoreline
10—yr (1993 Jan) predicted shoreline
seaward most shoreline position
landward most shoreline position

HUNTINGTON PIER

BREAKWATER

SEA CLIFFS

Pacific Coast Hwy

PROPOSED
ENTRANCE CHANNEL

ALONGSHORE COORDINATE (cell spacing = 200 ft)

SHORELINE POSITION (ft)

EAST JETTY ANAHEIM BAY

BREAKWATER

PROPOSED
ENTRANCE CHANNEL

BACKPASSING
BORROW AREA 1

BACKPASSING
BORROW AREA 2

BACKPASSING
FILL AREA 2

BACKPASSING
FILL AREA 1

SEA CLIFFS

Pacific Coast Hwy

Beach Fill (cu yd)
1983: 1,000,000
1988: 1,000,000

EAST JETTY ANAHEIM BAY

ALONGSHORE COORDINATE (cell spacing = 200 ft)

SHORELINE POSITION (ft from baseline)

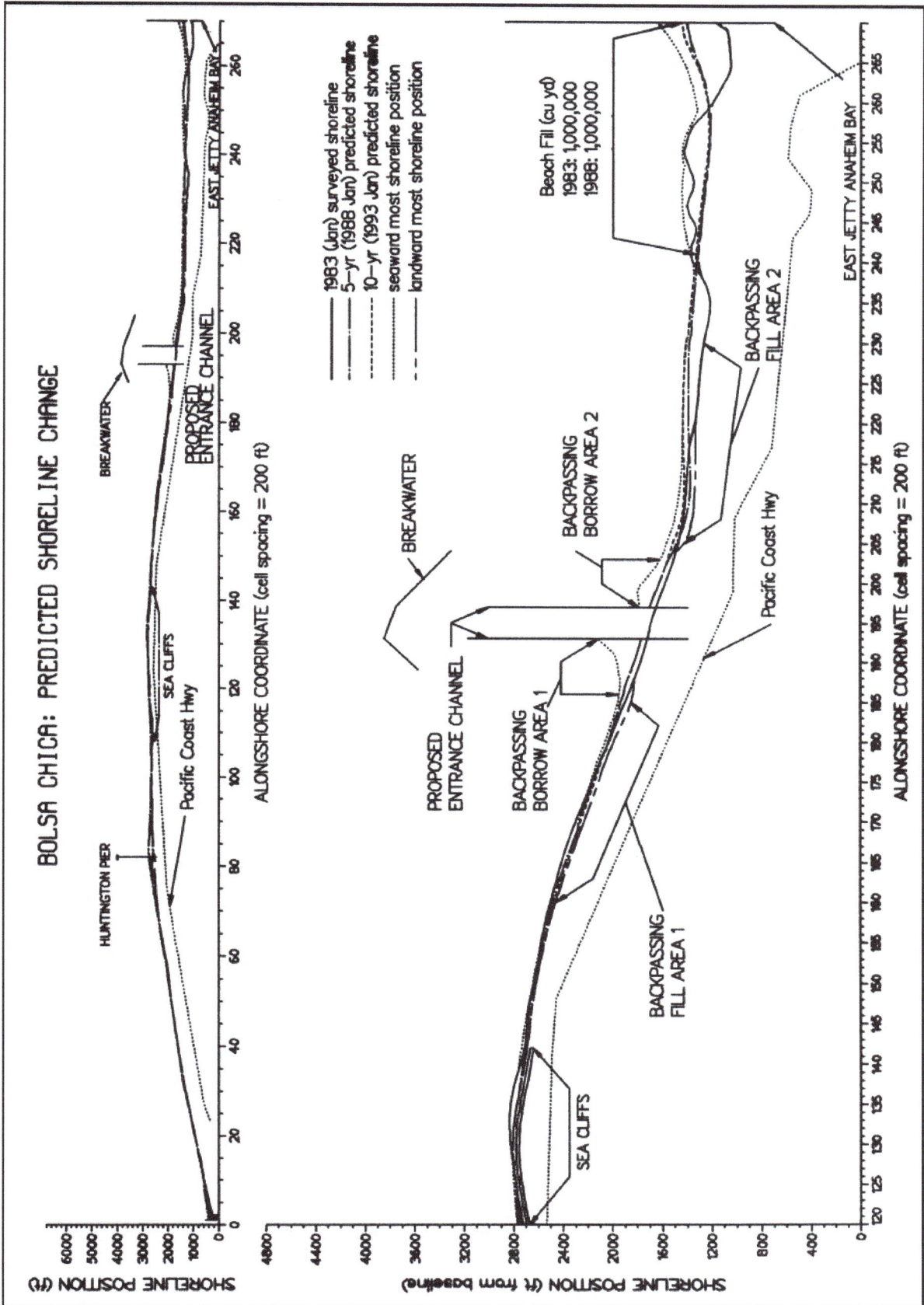

Figure III-2-52. Sand management alternative with feeder beach (Gravens 1990a)

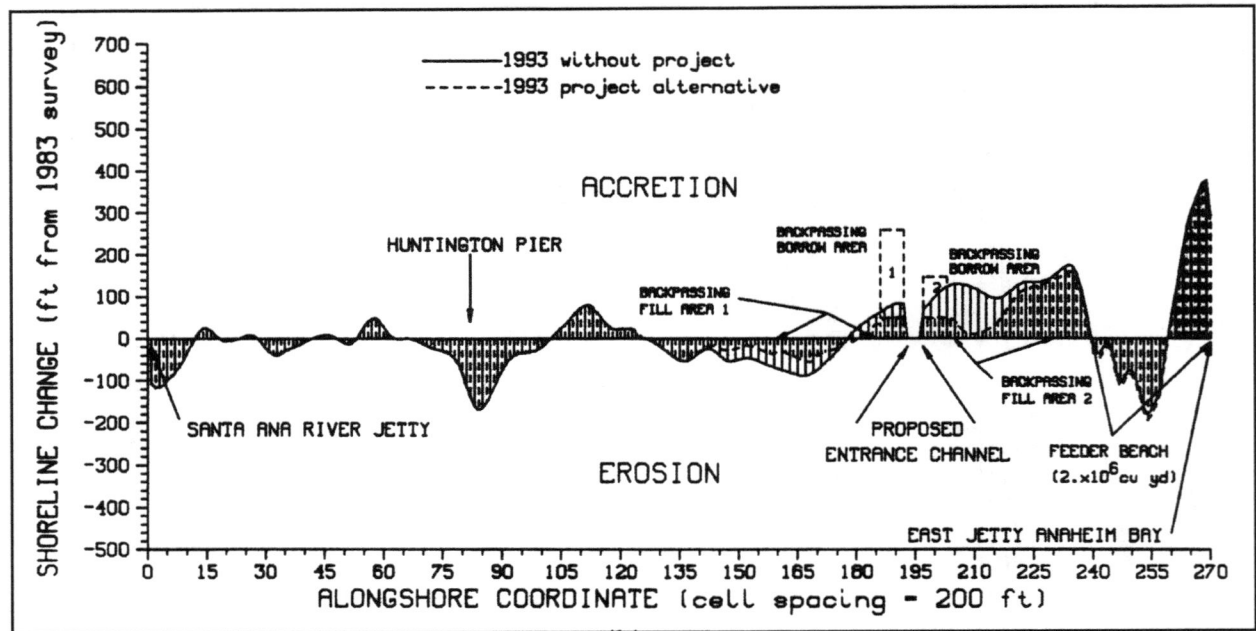

Figure III-2-53. Predicted shoreline change from 1983 shoreline position with sand management alternative and feeder beach (Gravens 1990a)

III-2-3. References

Aubrey 1980
Aubrey, D. G. 1980. "Our Dynamic Coastline," *Oceanus*, Vol 23, No. 4, pp 4-13.

Bagnold 1940
Bagnold, R. A. 1940. "Beach Formation by Waves: Some Model Experiments in a Wave Tank," *Journal Inst. Civil Engineering*, Vol 15, pp 27-52.

Bagnold 1963
Bagnold, R. A. 1963. "Beach and Nearshore Processes; Part I: Mechanics of Marine Sedimentation," *The Sea: Ideas and Observations, Vol 3,* M. N. Hill, ed., Interscience, New York, pp 507-528.

Bailard 1981
Bailard, J. A. 1981. "An Energetics Total Load Sediment Transport Model for a Plane Sloping Beach," *Journal of Geophysical Research*, Vol 86, No. C11, pp 10938-10954.

Bailard 1984
Bailard, J. A. 1984. "A Simplified Model for Longshore Sediment Transport," *Proceedings, 19th International Coastal Engineering Conference*, American Society of Civil Engineers, New York, pp 1454-1470.

Bailard and Inman 1981
Bailard, J. A., and Inman, D. L. 1981. "An Energetics Bedload Model for a Plane Sloping Beach: Local Transport," *Journal of Geophysical Research*, Vol 86, pp 2035-2043.

Bakker 1968
Bakker, W. T. 1968. "The Dynamics of a Coast with a Groin System," *Proceedings of the 11th Coastal Engineering Conference*, American Society of Civil Engineers, pp 492-517.

Bakker and Edelman 1965
Bakker, W. T., and Edelman, T. 1965. "The Coastline of River Deltas," *Proceedings of the 9th Coastal Engineering Conference*, American Society of Civil Engineers, pp 199-218.

Bakker, Klein-Breteler, and Roos 1971
Bakker, W. T., Klein-Breteler, E. H. J., and Roos, A. 1971. "The Dynamics of a Coast with a Groin System," *Proceedings of the 12th Coastal Engineering Conference*, American Society of Civil Engineers, pp 1001-1020.

Beach and Sternberg 1987
Beach, R. A., and Sternberg, R. W. 1987. "The Influence of Infragravity Motions on Suspended Sediment Transport in the Inner Surf Zone," *Proceedings of Coastal Sediments '87*, American Society of Civil Engineers, pp 913-928.

Bendat and Piersol 1971
Bendat, J. S., and Piersol, A. G. 1971. *Random Data: Analysis and Measurement Procedures*, Wiley-Interscience, New York.

Berek and Dean 1982
Berek, E. P., and Dean, R. G. 1982. "Field Investigation of Longshore Transport Distribution," *Proceedings, 18th International Conference on Coastal Engineering*, American Society of Civil Engineers, pp 1620-1639.

Bodge 1986
Bodge, K. R. 1986. "Short-Term Impoundment of Longshore Sediment Transport," Ph.D. diss., Department of Coastal and Oceanographic Engineering, University of Florida, Gainesville.

Bodge 1988
Bodge, K. R. 1988. "Longshore Current and Transport Across Non-Singular Equilibrium Beach Profiles," *Proceedings, 21st International Conf. on Coastal Engineering*, American Society of Civil Engineers, pp 1396-1410.

Bodge and Dean 1987a
Bodge, K. R., and Dean, R. G. 1987a. "Short-Term Impoundment of Longshore Transport," *Proceedings of Coastal Sediments '87*, American Society of Civil Engineers, pp 468-483.

Bodge and Dean 1987b
Bodge, K. R., and Dean, R. G. 1987b. "Short-Term Impoundment of Longshore Sediment Transport," Miscellaneous Paper CERC-87-7, U.S. Army Engineer Waterways Experiment Station, Vicksburg, MS.

Bodge and Kraus 1991
Bodge, K. R., and Kraus, N. C. 1991. "Critical Examination of Longshore Transport Rate Magnitude," *Proceedings, Coastal Sediments '91*, American Society of Civil Engineers, pp 139-155.

Bowen 1969
Bowen, A. J. 1969. "Rip Currents: 1. Theoretical Investigations," *Journal of Geophysical Research*, Vol 74, pp 5467-78.

Bowen 1973
Bowen, A. J. 1973. "Edge Waves and the Littoral Environment," *Proceedings 13th Conf. on Coast. Eng.*, American Society of Civil Engineers, pp 1313-20.

Bowen and Guza 1978
Bowen, A. J., and Guza, R. T. 1978. "Edge Waves and Surf Beat," *Journal Geophysical Research*, Vol 83, pp 1913-20.

Bowen and Inman 1966
Bowen, A. J., and Inman, D. L. 1966. "Budget of Littoral Sands in the Vicinity of Port Arguello, California," Technical Memorandum No. 19, Coastal Engineering Research Center, U.S. Army Corps Engineer Waterways Experiment Station, Vicksburg, MS.

Bowen and Inman 1969
Bowen, A. J., and Inman, D. L. 1969. "Rip Currents: 2. Laboratory and Field Observations," *Journal of Geophysical Research*, Vol 74, pp 5478-90.

Bowen and Inman 1971
Bowen, A. J., and Inman, D. L. 1971. "Edge Waves and Crescentic Bars," *Journal of Geophysical Research*, Vol 76, No. 76, pp 8662-71.

Box and Jenkins 1976
Box, G. E. P., and Jenkins, G. M. 1976. "Time Series Analysis, Forecasting, and Control," Holden-Day, San Francisco, CA, p 575.

Brenninkmeyer 1976
Brenninkmeyer, B. M. 1976. "In Situ Measurements of Rapidly Fluctuating, High Sediment Concentrations," *Marine Geology*, Vol 20, pp 117-128.

Brunn 1954
Brunn, P. 1954. "Migrating Sand Waves and Sand Humps, with Special Reference to Investigations Carried out on the Danish North Sea Coast," *Proceedings 5th Conf. on Coast. Eng.*, pp 269-95.

Bruno and Gable 1976
Bruno, R. O., and Gable, C. G. 1976. "Longshore Transport at a Total Littoral Barrier," *Proceedings of the 15th International Conference on Coastal Engineering*, American Society of Civil Engineers, pp 1203-1222.

Bruno, Dean and Gable 1980
Bruno, R. O., Dean, R. G., and Gable, C. G. 1980. "Littoral Transport Evaluations at a Detached Breakwater," *Proceedings of the 17th International Conference on Coastal Engineering*, American Society of Civil Engineers, pp 1453-1475.

Bruno, Dean, Gable, and Walton 1981
Bruno, R. O., Dean, R. G., Gable, C. G., and Walton, T. L. 1981. "Longshore Sand Transport Study at Channel Island Harbor, California," Technical Paper No. 81-2, Coastal Engineering Research Center, U.S. Army Engineer Waterways Experiment Station, Vicksburg, MS.

Bruun and Manohar 1963
Bruun, P., and Manohar, M. 1963. "Coastal Protection for Florida," Florida Engineering and Industrial Experiment Station Bulletin No. 113, University of Florida, Gainesville.

Caldwell 1956
Caldwell, J. M. 1956. "Wave Action and Sand Movement Near Anaheim Bay, California," U.S. Army Corps of Engineers, Beach Erosion Board, Technical Memorandum No. 68.

Chu, Gravens, Smith, Gorman and Chen 1987
Chu, Y., Gravens, M. B., Smith, J. M., Gorman, L. T., and Chen, H. S. 1987. "Beach Erosion Control Study, Homer Spit, Alaska," Miscellaneous Paper CERC-87-15, U.S. Army Engineer Waterways Experiment Station, Vicksburg, MS.

Cialone et al. 1992
Cialone, M. A., Mark, D. J., Chou, L. W., Leenknecht, D. A., Davis, J. E., Lillycrop, L. S., and Jensen, R. E. 1992. "Coastal Modeling System (CMS) User's Manual," Instruction Report CERC-91-1, U.S. Army Engineer Waterways Experiment Station, Vicksburg, MS.

Cialone, Neilans, Carson, and Smith (in preparation)
Cialone, M. A., Neilans, P. J., Carson, F. C., and Smith, J. M. "Long-term Shoreline Response Modeling at Westhampton Beach, New York," in preparation, U.S. Army Engineer Waterways Experiment Station, Vicksburg, MS.

Coakley and Skafel 1982
Coakley, J. P., and Skafel, M. G. 1982. "Suspended Sediment Discharge on a Non-Tidal Coast," *Proceedings of the 18th International Conference on Coastal Engineering*, American Society of Civil Engineers, pp 1288-1304.

Dalrymple 1975
Dalrymple, R. A. 1975. "A Mechanism for Rip Current Generation on an Open Coast," *Journal of Geophysical Research*, Vol 80, No. 24, pp 3485-3487.

Dalrymple and Lanan 1976
Dalrymple, R. A., and Lanan, G. A. 1976. "Beach Cusps Formed by Intersecting Waves," *Bull. Geo. Soc. Am.*, Vol 87, pp 57-60.

Darbyshire 1977
Darbyshire, J. 1977. "An Investigation of Beach Cusps in Hells Mouth Bay," *A Voyage of Discovery* (George Deacon 70th Anniversary Volume), M. Angel, ed., Pergamon, Oxford, pp 405-27.

Darbyshire and Pritchard 1978
Darbyshire, J., and Pritchard, E. 1978. *Riv. Ital. Geofisc. Sci. Affi*, Vol 5, pp 73-80.

Dean 1973
Dean, R. G. 1973. "Heuristic Models of Sand Transport in the Surf Zone," *Proceedings, Conference on Engineering Dynamics in the Surf Zone*, Sydney, Australia.

Dean 1984
Dean, R. G. 1984. "Principles of Beach Nourishment," Chapter 11, *CRC Handbook of Coastal Processes and Erosion*, P. D. Komar, ed., CRC Press, Inc., Boca Raton, FL, pp 217-232.

Dean 1987
Dean, R. G. 1987. "Measuring Longshore Transport with Traps," *Nearshore Sediment Transport*, Richard J. Seymour, ed., Plenum Press, New York.

Dean and Grant 1989
Dean, R. G., and Grant, J. 1989. "Development of Methodology for Thirty-Year Shoreline Projection in the Vicinity of Beach Nourishment Projects," University of Florida Report UFP/COEL-89/026, University of Florida, Gainesville.

Dean and Maurmeyer 1980
Dean, R. G., and Maurmeyer, E. M. 1980. "Beach Cusps at Point Reyes and Drakes Bay Beaches, CA," *Proceedings 17th Coastal Engineering Conference*, American Society of Civil Engineers, pp 863-884.

Dean and Yoo 1992
Dean, R. G., and Yoo, C. 1992. "Beach-Nourishment Performance Predictions," *Journal of Waterway, Port, Coastal, and Ocean Engineering*, Vol 118, No. 6, November/December, pp 567-586.

Dean and Yoo 1994
Dean, R. G., and Yoo, C. 1994. "Beach Nourishment in Presence of Seawall," *Journal of Waterway, Port, Coastal, and Ocean Engineering*, Vol 120, No. 3, May/June, pp 302-316.

Dean, Berek, Bodge, and Gable 1987
Dean, R. G., Berek, E. P., Bodge, K. R., and Gable, C. G. 1987. "NSTS Measurements of Total Longshore Transport," *Proceedings of Coastal Sediments '87*, American Society of Civil Engineers, pp 652-667.

Dean, Berek, Gable, and Seymour 1982
Dean, R. G., Berek, E. P., Gable, C. G., and Seymour, R. J. 1982. "Longshore Transport Determined by an Efficient Trap," *Proceedings of the 18th International Conference on Coastal Engineering*, American Society of Civil Engineers, pp 954-968.

del Valle, Medina, and Losada 1993
del Valle, R., Medina, R., and Losada, M. A. 1993. "Dependence of Coefficient K on Grain Size," Technical Note No. 3062, *Journal of Waterway, Port, Coastal, and Ocean Engineering*, Vol 119, No. 5, September/October, pp 568-574.

Dolan 1971
Dolan, R. 1971. "Coastal Landforms: Crescentic and Rhythmic," *Geol. Soc. Am. Bull.*, Vol 82, pp 177-80.

Dolan and Ferm 1968
Dolan, R., and Ferm, J. C. 1968. "Conentric Landforms along the Atlantic Coast of the United States," *Science*, Vol 1959, pp 627-69.

Dolan, Vincent, and Hayden 1974
Dolan, R., Vincent, L., and Hayden, B. 1974. "Crescentic Coastal Landforms," *Zeitschr. fur Geomorph.*, Vol 18, pp 1-12.a.

Douglass 1985
Douglass, S. L. 1985. "Longshore Sand Transport Statistics," M.S. thesis, Mississippi State University, Starkville, MS.

Downing 1984
Downing, J. P. 1984. "Suspended Sand Transport on a Dissipative Beach," *Proceedings of the 19th International Conference on Coastal Engineering*, American Society of Civil Engineers, pp 1765-1781.

Duane and James 1980
Duane, D. B., and James, W. R. 1980. "Littoral Transport in the Surf Zone Elucidated by an Eulerian Sediment Tracer Experiment," *Journal of Sedimentary Petrology*, Vol 50, pp 929-942.

Dubois 1978
Dubois, R. 1978. *Geol. Soc. Am. Bull.*, Vol 89, pp 1133-1139.

Eaton 1950
Eaton, R. O. 1950. "Littoral Processes on Sandy Coasts," *Proceedings, First Coastal Engineering Conference*, Long Beach, California, Council on Wave Research, Chapter 15, pp 140-154.

Ebersole 1985
Ebersole, B. A. 1985. "Refraction-Diffraction Model for Linear Water Waves," *Journal of Waterways, Port, Coastal, and Ocean Engineering*, American Society of Civil Engineers, Vol III (No. WW6) pp 939-953.

Ebersole, Cialone, and Prater 1986
Ebersole, B. A., Cialone, M. A., and Prater, M. D. 1986. "RCPWAVE - A Linear Wave Propagation Model for Engineering Use," Technical Report CERC-86-4, U.S. Army Engineer Waterways Experiment Station, Vicksburg, MS.

Escher 1937
Escher, B. G. 1937. "Experiments on the Formation of Beach Cusps," *Leidse Geologische Mededlingen*, Vol 9, pp 70-104.

Evans 1938
Evans, O. F. 1938. "Classification and Origin of Beach Cusps," *J. Geol.*, Vol 46, pp 615-627.

Fairchild 1972
Fairchild, J. C. 1972. "Longshore Transport of Suspended Sediment," *Proceedings of the 13th International Conference on Coastal Engineering*, American Society of Civil Engineers, pp 1069-1088.

Fairchild 1977
Fairchild, J. C. 1977. "Suspended Sediment in the Littoral Zone at Vetnor, New Jersey, and Nags Head, North Carolina," Technical Paper No. 77-5, U.S. Army Engineer Waterways Experiment Station, Vicksburg, MS.

Finkelstein 1982
Finkelstein, K. 1982. "Morphological Variations and Sediment Transport in Crenulate-Bay Beaches; Kodiak Island, Alaska," *Marine Geology*, Vol 47, pp 261-81.

Flemming 1964
Flemming, N. C. 1964. "Tank Experiments on the Sorting of Beach Material During Cusp Formation," *Journal Sediment. Petrol.*, Vol 34, pp 112-22.

Grant 1943
Grant, U. S. 1943. "Waves as a Transporting Agent," *American Journal of Science*, Vol 241, pp 117-123.

Gravens 1989
Gravens, M. B. 1989. "Estimating Potential Longshore Sand Transport Rates Using WIS Data," CETN-II-19, U.S. Army Engineer Waterways Experiment Station, Vicksburg, MS.

Gravens 1990a
Gravens, M. B. 1990a. "Bolsa Bay, California, Proposed Ocean Entrance System Study; Report 2, Comprehensive Shoreline Response Computer Simulation, Bolsa Bay, California," Miscellaneous Paper CERC-89-17, U.S. Army Engineer Waterways Experiment Station, Vicksburg, MS.

Gravens 1990b
Gravens, M. B. 1990b. "Computer Program: GENESIS Version 2," CETN-II-21, U.S. Army Engineer Waterways Experiment Station, Vicksburg, MS.

Gravens 1991
Gravens, M. B. 1991. "Development of an Input Data Set for Shoreline Change Modeling," *Proceedings, Coastal Sediments '91*, American Society of Civil Engineers, pp 1800-1813.

Gravens 1992
Gravens, M. B. 1992. "User's Guide to the Shoreline Modeling System (SMS)," Instruction Report CERC-92-1, U.S. Army Engineer Waterways Experiment Station, Vicksburg, MS.

Gravens (in preparation)
Gravens, M. B. "Shoreline Change Modeling At Grande Isle, Louisiana," in preparation, U.S. Army Engineer Waterways Experiment Station, Vicksburg, MS.

Gravens, Kraus, and Hanson 1991
Gravens, M. B., Kraus, N. C., and Hanson, H. 1991. "GENESIS: Generalized Model for Simulating Shoreline Change," Instruction Report CERC-89-19, U.S. Army Engineer Waterways Experiment Station, Vicksburg, MS.

Gravens, Scheffner, and Hubertz 1989
Gravens, M. B., Scheffner, N. W., and Hubertz, J. M. 1989. "Coastal Processes form Asbury Park to Manasquan, New Jersey," Miscellaneous Paper CERC-89-11, U.S. Army Engineer Waterways Experiment Station, Vicksburg, MS.

Guza and Bowen 1981
Guza, R. T., and Bowen, A. J. 1981. "On the Amplitude of Beach Cusps," *Journal Geophys Research*, Vol 86, pp 4125-4132.

Guza and Inman 1975
Guza, R. T., and Inman, D. L. 1975. "Edge Waves and Beach Cusps," *Journal Geophysical Research*, Vol 80, No. 21, pp 2997-3012.

Hanes, Vincent, Huntley, and Clarke 1988
Hanes, D. M., Vincent, C. E., Huntley, D. A., and Clarke, T. L. 1988. "Acoustic Measurements of Suspended Sand Concentration in the C^2S^2 Experiment at Stanhope Lane, Prince Edward Island," *Marine Geology*, Vol 81, pp 185-196.

Hanson 1987
Hanson, H. 1987. "GENESIS - A Generalized Shoreline Change Numerical Model for Engineering Use," Report No. 1007, Department of Water Resources Engineering, University of Lund, Lund, Sweden.

Hanson and Kraus 1986
Hanson, H., and Kraus, N. C. 1986. "Forecast of Shoreline Change Behind Multiple Coastal Structures," *Coastal Engineering in Japan*, Vol 29, pp 195-213.

Hanson and Kraus 1989
Hanson, H., and Kraus, N. C. 1989. "GENESIS: Generalized Numerical Modeling System for Simulating Shoreline Change; Report 1, Technical Reference Manual," Technical Report CERC-89-19, U.S. Army Engineer Waterways Experiment Station, Vicksburg, MS.

Hanson, Kraus and Nakashima 1989
Hanson, H., Kraus, N. C., and Nakashima, L. D. 1989. "Shoreline Change Behind Transmissive Detached Breakwaters," *Proceedings, Coastal Zone '89*, American Society of Civil Engineers, pp 568-582.

Hino 1974
Hino, M. 1974. "Theory on Formation of Rip-Current and Cuspidal Coast," *Proceedings of the 14th Coastal Engineering Conference*, American Society of Civil Engineers, pp 901-919.

Holman and Bowen 1979
Holman, R. A., and Bowen, A. J. 1979. "Edge Waves on Complex Beach Profiles," *Journal Geophysical Research*, Vol 84, pp 6339-46.

Holman and Bowen 1982
Holman, R. A., and Bowen, A. J. 1982. "Bars, Bumps, and Holes: Models for the Generation of Complex Beach Topography," *Journal Geophysical Research*, Vol 87, pp 457-468.

Hom-ma and Sonu 1963
Hom-ma, M., and Sonu, C. J. 1963. "Rhythmic Patterns of Longshore Bars Related to Sediment Characteristics," *Proceedings 8th Conf. on Coast. Eng.*, American Society of Civil Engineers, pp 248-78.

Hom-ma, Horikawa and Kajima 1965
Hom-ma, M., Horikawa, K., and Kajima, R. 1965. "A Study of Suspended Sediment Due to Wave Action," *Coastal Engineering in Japan*, Vol 3, pp 101-122.

Hsu and Evans 1989
Hsu, J. R. C., and Evans, C. 1989. "Parabolic Bay Shapes and Applications," *Proceedings, Instn. Civil Engrs.*, Vol 87, pp 557-70.

Hsu, Silvester and Xia 1987
Hsu, J. R. C., Silvester, R., and Xia, Y. M. 1987. "New Characteristics of Equilibrium Shaped Bays," *Proceedings, 8th Aust. Conference on Coastal and Ocean Eng.*, pp 140-44.

Hsu, Silvester and Xia 1989a
Hsu, J. R. C., Silvester, R., and Xia, Y. M. 1989. "Static Equilibrium Bays: New Relationships," *Journal Waterways, Port, Coastal and Ocean Engineering*, American Society of Civil Engineers, Vol 115, No. 3, pp 285-98.

Hsu, Silvester and Xia 1989b
Hsu, J. R. C., Silvester, R., and Xia, Y. M. 1989. "Generalities on Static Equilibrium Bays," *Journal Coastal Engineering*, Vol 12, pp 353-369.

Hubertz, Brooks, Brandon, and Tracy 1993
Hubertz, J. M., Brooks, R. M., Brandon, W. A., and Tracy, B. A. 1993. "Hindcast Wave Information for the U.S. Atlantic Coast," Wave Information Study Report 30, U.S. Army Engineer Waterways Experiment Station, Vicksburg, MS.

Huntley 1976
Huntley, D. A. 1976. "Long-Period Waves on a Natural Beach," *Journal of Geophysical Research*, Vol 81, No. 36, pp 6441-6449.

Huntley and Bowen 1973
Huntley, D. A., and Bowen, A. J. 1973. "Field Observations of Edge Waves," *Nature*, Vol 243, pp 160-1, Copyright© 1973 Macmillan Journals Ltd.

Huntley and Bowen 1975a
Huntley, D. A., and Bowen, A. J. (1975a). "Comparison of the Hydrodynamics of Steep and Shallow Beaches," *Nearshore Sediment Dynamics and Sedimentation*, J. R. Hails and A. Carr, ed., John Wiley, London, pp 69-109.

Huntley and Bowen 1975b
Huntley, D. A., and Bowen, A. J. (1975b). "Field Observations of Edge Waves and Their Effect on Beach Materials," *Quarterly Journal of the Geological Society (London)*, Vol 131, pp 68-81.

Huntley and Bowen 1979
Huntley, D. A., and Bowen, A. J. 1979. "Beach Cusps and Edge Waves," *Proceedings 16th Conf. Coastal Engineering.*, American Society of Civil Engineers, pp 1378-1393.

Ingle 1966
Ingle, J. C. 1966. "The Movement of Beach Sand; An Analysis using Fluorescent Grains," Department of Geology, University of Southern California, Los Angeles, California, Elsevier Publishing Company, New York.

Inman and Bagnold 1963
Inman, D. L., and Bagnold, R. A. 1963. "Beach and Nearshore Processes; Part II: Littoral Processes," contribution to *The Sea*, Vol 3, M. N. Hille, ed., John Wiley and Sons, pp 529-553.

Inman et al. 1980
Inman, D. L., Zampol, J. A., White, T. E., Hanes, D. M., Waldorf, B. W., and Kastens, K. A. 1980. "Field Measurements of Sand Motion in the Surf Zone," *Proceedings of the 17th International Conference on Coastal Engineering*, American Society of Civil Engineers, pp 1215-1234.

Inman, Komar, and Bowen 1968
Inman, D. L., Komar, P. D., and Bowen. 1968. "Longshore Transport of Sand," *Proceedings of the 11th International Conference on Coastal Engineering*, American Society of Civil Engineers, pp 248-306.

Inman, Tait, and Nordstrom 1971
Inman, D. L., Tait, and Nordstrom, 1971. "Mixing in the Surf Zone," *Journal of Geophysical Research*, Vol 76, pp 3493-3514.

Johnson 1910
Johnson, D. W. 1910. "Beach Cusps," *Geol. Soc. Am. Bull.*, Vol 21, pp 604-24.

Johnson 1919
Johnson, D. W. 1919. *Shore Processes and Shoreline Development*, 1965 facsimile, Hafner, New York.

Johnson 1956
Johnson, J. W. 1956. "Dynamics of Nearshore Sediment Movement," *Bulletin of the American Society of Petroleum Geologists*, Vol 40, pp 2211-2232.

Johnson 1957
Johnson, J. W. 1957. "The Littoral Drift Problem at Shoreline Harbors," *Journal of the Waterways and Harbors Division*, American Society of Civil Engineers, Vol 83, pp 1-37.

Johnson 1992
Johnson, C. N. 1992. "Mitigation of Harbor Caused Shore Erosion with Beach Nourishment, St. Joseph Harbor, MI," *Proceedings, Coastal Engineering Practice '92,* American Society of Civil Engineers, pp 137-153.

Kamphuis 1991
Kamphuis, J. W. 1991. "Alongshore Sediment Transport Rate," *Journal of Waterway, Port, Coastal, and Ocean Engineering*, Vol 117, No. 6, pp 624-640.

Kamphuis and Readshaw 1978
Kamphuis, J. W., and Readshaw, J. S. 1978. "A Model Study of Alongshore Sediment Transport Rate," *Proceedings, 16th International Coastal Engineering Conference*, American Society of Civil Engineers, pp 1656-1674.

Kamphuis and Sayao 1982
Kamphuis, J. W., and Sayao, O. 1982. "Model Tests on Littoral Sand Transport Rate," *Proceedings, 16th International Conference on Coastal Engineering,* American Society of Civil Engineers, pp 1305-1325.

Kamphuis, Davies, Nairn and Sayao 1986
Kamphuis, J. E., Davies, M. H., Nairn, R. B., and Sayao, O. J. 1986. "Calculation of Littoral Sand Transport Rate," *Journal Coastal Engineering Conference*, Vol 10, No. 1, pp 1-21.

Kana 1977
Kana, T. W. 1977. "Suspended Sediment Transport at Price Inlet, S.C.," *Proceedings of Coastal Sediments '77*, American Society of Civil Engineers, pp 366-382.

Kana 1978
Kana, T. W. 1978. "Surf Zone Measurements of Suspended Sediment," *Proceedings of the 16th International Conference on Coastal Engineering*, American Society of Civil Engineers, pp 1725-1743.

Kana and Ward 1980
Kana, T. W., and Ward, L. G. 1980. "Nearshore Suspended Sediment Load During Storms and Post-Storm Conditions," *Proceedings, 17th International Conference on Coastal Engineering*, pp 1158-1171.

Katoh, Tanaka and Irie 1984
Katoh, K., Tanaka, N., and Irie, I. 1984. "Field Observation on Suspended-Load in the Surf Zone," *Proceedings of the 19th International Conference on Coastal Engineering,* American Society of Civil Engineers, pp 1846-1862.

Knoth and Nummedal 1977
Knoth, J. S., and Nummedal, D. 1977. "Longshore Sediment Transport Using Fluorescent Tracer," *Proceedings of Coastal Sediments '77*, American Society of Civil Engineers, pp 383-398.

Komar 1971
Komar, P. D. 1971. "Nearshore Cell Circulation and the Formation of Giant Cusps," *Geol. Soc. Am. Bull.*, Vol 81, pp 2643-50.

Komar 1973a
Komar, P. D. 1973a. "Computer Models of Delta Growth Due to Sediment Input from Waves and Longshore Currents," *Geological Society of America Bulletin*, Vol 84, pp 2217-2226.

Komar 1973b
Komar, P. D. 1973b. "Observations of Beach Cusps at Mono Lake, California," *Geol. Soc. Am. Bull.*, Vol 84, pp 3593-600.

Komar 1976
Komar, P. D. 1976. *Beach Processes and Sedimentation*, Prentice-Hall, Inc., Englewood Cliffs, NJ.

Komar 1977
Komar, P. D. 1977. "Beach Sand Transport: Distribution and Total Drift," *Journal of the Waterway, Port, Coastal and Ocean Engineering Division*, American Society of Civil Engineers, Vol 103, pp 225-239.

Komar 1978
Komar, P. D. 1978. "Wave Conditions on the Oregon Coast During the Winter of 1977-78 and the Resulting Erosion of Nestucca Spit," *Shore and Beach*, Vol 46, pp 3-8.

Komar 1983
Komar, P. D. 1983. "Rhythmic Shoreline Features and their Origins," *Mega-Geomorphology*, R. Garnder and H.G. Scoging, eds., Claredon Press, Oxford, U.K., pp 92-112.

Komar 1988
Komar, P. D. 1988. "Environmental Controls on Littoral Sand Transport," *21st International Coastal Engineering Conference*, American Society of Civil Engineers, pp 1238-1252.

Komar and Inman 1970
Komar, P. D., and Inman, D. L. 1970. "Longshore Sand Transport on Beaches," *Journal of Geophysical Research*, Vol 75, No. 30, pp 5914-5927.

Komar and McDougal 1988
Komar, P.D., and McDougal, W. G. 1988. "Coastal Erosion and Engineering Structures: The Oregon Experience," *Journal Coastal Research*, Vol 4, pp 77-92.

Komar and Rea 1976
Komar, P. D., and Rea, C. C. 1976. "Erosion of Siletz Spit, Oregon," *Shore and Beach*, Vol 44, pp 9-15.

Kraus 1988
Kraus, N. C. 1988. "Part IV: Prediction Models of Shoreline Change," Chapter 6, "Case Studies of Application of the Shoreline Change Model," *Nearshore Dynamics and Coastal Processes: Theory, Measurement, and Predictive Models*, K. Horikawa, ed., University of Tokyo Press, Tokyo, Japan, pp 355-366.

Kraus and Dean 1987
Kraus, N. C., and Dean, J. L. 1987. "Longshore Sand Transport Rate Distributions Measured by Trap," *Proceedings, Coastal Sediments '87*, N. C. Kraus, ed., American Society of Civil Engineers, pp 881-896.

Kraus and Harikai 1983
Kraus, N. C., and Harikai, S. 1983. "Numerical Model of the Shoreline Change at Oarai Beach," *Coastal Engineering*, Vol 7, pp 1-28.

Kraus and Larson 1991
Kraus, N. C., and Larson, M. 1991. "NMLONG: Numerical Model for Simulating the Longshore Current; Report 1: Model Development and Tests," Technical Report DRP-91-1, U.S. Army Engineer Waterways Experiment Station, Vicksburg, MS.

Kraus et al. 1982
Kraus, N. C., Isobe, M., Igarashi, H., Sasaki, T. O., and Horikawa, K. 1982. "Field Experiments on Longshore Sand Transport in the Surf Zone," *Proceedings, 18th International Conference on Coastal Engineering*, American Society of Civil Engineers, pp 969-988.

Kraus, Gingerich, and Rosati 1988
Kraus, N. C., Gingerich, K. J., and Rosati, J. D. 1988. "Toward an Improved Empirical Formula for Longshore Sand Transport," *Proceedings of the 21st International Conference on Coastal Engineering*, American Society of Civil Engineers, pp 1182-1196.

Kraus, Gingerich and Rosati 1989
Kraus, N. C., Gingerich, K. J., and Rosati, J. D. 1989. "DUCK85 Surf Zone Sand Transport Experiment," Technical Report CERC-89-5, U.S. Army Engineer Waterways Experiment Station, Vicksburg, MS.

Krumbein 1944a
Krumbein, W. C. 1944a. "Shore Currents and Sand Movement on a Model Beach," Beach Erosion Board Tech. Memo. No. 7, U.S. Army Engineer Waterways Experiment Station, Vicksburg, MS.

Krumbein 1944b
Krumbein, W. C. 1944b. "Shore Processes and Beach Characteristics," Beach Erosion Board Tech. Memo. No. 37, U.S. Army Engineer Waterways Experiment Station, Vicksburg, MS.

Kuenen 1948
Kuenen, Ph.H. 1948. "The Formation of Beach Cusps," *Journal Geol.*, Vol 56, pp 34-40.

Larson, Hanson and Kraus 1987
Larson, M., Hanson, H., and Kraus, N. C. 1987. "Analytical Solutions of the One-Line Model of Shoreline Change," Technical Report CERC-87-15, U.S. Army Engineer Waterways Experiment Station, Vicksburg, MS.

LeBlond 1972
LeBlond, P. H. 1972. "On the Formation of Spiral beaches," *Proceedings 13th Int. Conf. on Coastal Engineering*, American Society of Civil Engineers, pp 1331-45.

LeBlond 1979
LeBlond, P. H. 1979. "An Explanation of the Logarithmic Spiral Plan Shape of Headland Bay Beaches," *Journal Sedimentary Petrology*, Vol 49, No. 4, pp 1093-1100.

Lee 1975
Lee, K. K. 1975. "Longshore Currents and Sediment Transport in West Shore of Lake Michigan," *Water Resources Research*, Vol 11, pp 1029-1032.

Leenknecht, Szuwalski, and Sherlock 1992
Leenknecht, D. A., Szuwalski, A., and Sherlock, A. R. 1992. "Automated Coastal Engineering System, User Guide and Technical Reference, Version 1.07," U.S. Army Engineer Waterways Experiment Station, Vicksburg, MS.

LeMéhauté and Brebner 1961
LeMéhauté, B., and Brebner, A. 1961. "An Introduction to Coastal Morphology and Littoral Processes," Report No. 14, Civil Engineering Department, Queens University at Kingston, Ontario, Canada.

LeMéhauté and Soldate 1977
LeMéhauté, B., and Soldate, M. 1977. "Mathematical Modeling of Shoreline Evolution," Miscellaneous Report 77-10, U.S. Army Engineer Waterways Experiment Station, Vicksburg, MS.

Lippman and Holman 1990
Lippman, T. C., and Holman, R. A. 1990. "Spatial and Temporal Variability of Sand Bar Morphology," *Journal of Geophysical Research,* Vol 95, No. C7, pp 11575-11590.

Longuet-Higgins 1970
Longuet-Higgins, M. S. 1970. "Longshore Currents Generated by Obliquely Incident Sea Waves, Parts 1 and 2," *Journal of Geophysical Research*, Vol 75, No. 33, November, pp 6778-6801.

Longuet-Higgins and Parkin 1962
Longuet-Higgins, M. S., and Parkin, D. W. 1962. "Sea Waves and Beach Cusps," *Geog. J.*, Vol 128, pp 194-201.

Mann and Dalrymple 1986
Mann, D. W., and Dalrymple, R. A. 1986. "A Quantitative Approach to Delaware's Nodal Point," *Shore and Beach,* Vol 54, No. 2, pp 13-16.

McDougal and Hudspeth 1981
McDougal, W. G., and Hudspeth, R. T. 1981. "Wave Induced Setup/Setdown and Longshore Current; Non-Planar Beaches: Sediment Transport," *Proceedings, Oceans*, IEEE, pp 834-846.

McDougal and Hudspeth 1983a
McDougal, W. G., and Hudspeth, R. T. 1983a. "Wave Setup/Setdown and Longshore Current on Non-Planar Beaches," *Journal Coastal Engineering*, Vol 7, pp 103-117.

McDougal and Hudspeth 1983b
McDougal, W. G., and Hudspeth, R. T. 1983b. "Longshore Sediment Transport on Non-Planar Beaches," *Journal Coastal Engineering,* Vol 7, pp 119-131.

McDougal and Hudspeth 1989
McDougal, W. G., and Hudspeth, R. T. 1989. "Longshore Current and Sediment Transport on Composite Beach Profiles," *Journal Coastal Engineering*, Vol 12, pp 315-338.

McDougal, Sturtevant, and Komar 1987
McDougal, W. G., Sturtevant, M. A., and Komar, P. D. 1987. "Laboratory and Field Investigations of the Impact of Shoreline Stabilization Structures on Adjacent Properties," *Proceedings, Coastal Sediments '87,* American Society of Civil Engineers, pp. 961-73.

Meisburger 1989
Meisburger, E. P. 1989. "Oolites as a Natural Tracer in Beaches of Southeastern Florida," Miscellaneous Paper CERC-89-10, U.S. Army Engineer Waterways Experiment Station, Vicksburg, MS.

Mogi 1960
Mogi, A. 1960. "On the Topographical Change of the Beach at Tokai, Japan," *Japan Geographical Review*, 398-411.

Moore and Cole 1960
Moore, G. W., and Cole, J. Y. 1960. "Coastal Processes in the Vicinity of Cape Thompson, Alaska; Geologic Investigations in Support of Project Chariot in the Vicinity of Cape Thompson, Northwestern Alaska - Preliminary Report," U.S. Geological Survey Trace Elements Investigation Report 753.

Munch-Peterson 1938
Munch-Peterson, J. 1938. "Littoral Drift Formula," *Beach Erosion Board Bulletin,* U.S. Army Engineer Waterways Experiment Station, Vicksburg, MS, Vol 4, No. 4, pp 1-31.

Niedoroda and Tanner 1970
Niedoroda, A. W., and Tanner, W. F. 1970. "Preliminary Study of Transverse Bars," *Mar. Geol.*, Vol 9, pp 41-62.

Otvos 1964
Otvos, E. G. 1964. "Observations of Beach Cusps and Pebble Ridge Formation on the Long Island Sound," *Journal Sediment. Petrol.*, Vol 34, pp 554-60.

Ozhan 1982
Ozhan, E. 1982. "Laboratory Study of Breaker Type Effect on Longshore Sand Transport," *Proceedings, Euromech 156, Mechanics of Sediment Transport,* Istanbul, A. Balkema Publishers.

Parker and Quigley 1980
Parker, G. F., and Quigley, R. M. 1980. "Shoreline Embayment Growth Between Two Headlands at Port Stanley, Ontario," *Proceedings, Canadian Coastal Conference*, pp 380-93.

Pelnard-Considere 1956
Pelnard-Considere, R. 1956. "Essai de theorie de l'Evolution des Forms de Rivages en Plage de Sable et de Galets," *4th Journees de l'Hydralique*, les energies de la Mer, Question III, Repport No. 1, pp 289-298.

Perlin and Dean 1983
Perlin, M., and Dean, R. G. 1983. "A Numerical Model to Simulate Sediment Transport in the Vicinity of Coastal Structures," Miscellaneous Report No. 83-10, Coastal Engineering Research Center, U.S. Army Engineer Waterways Experiment Station, Vicksburg, MS.

Rea and Komar 1975
Rea, C. C., and Komar, P. D. 1975. "Computer Simulation Models of a Hooked Beach Shoreline Configuration," *Journal Sedimentary Petrology*, Vol 45, pp 866-72.

Rosati and Kraus 1989
Rosati, J. D., and Kraus, N. C. 1989. "Development of a Portable Sand Trap for Use in the Nearshore," Technical Report CERC-89-11, U.S. Army Engineer Waterways Experiment Station, Vicksburg, MS.

Russell and McIntire 1965
Russell, R. J., and McIntire, W. G. 1965. "Beach Cusps," *Geol. Soc. Am. Bull.*, Vol 76, pp 307-20.

Sallanger 1979
Sallanger, A. H., Jr. 1979. "Beach Cusp Formation," *Marine Geology*, Vol 29, pp 23-37.

Savage 1962
Savage, R. P. 1962. "Laboratory Determination of Littoral Transport Rates," *Journal of the Waterway, Port, Coastal, and Ocean Division*, American Society of Civil Engineers, No. WW2, pp 69-92.

Sawaragi and Deguchi 1978
Sawaragi, T., and Deguchi, I., 1978. "Distribution of Sand Transport Rate Acros a Surf Zone," *Proceedings, 16th International Conference on Coastal Engineering*, American Society of Civil Engineers, pp 1596-1613.

Sayao 1982
Sayao, O. 1982. "Beach Profiles and Littoral Sand Transport," Ph.D. diss., Department of Civil Engineering, Queen's University, Kingston, Ontario, Canada.

Scripps Institute of Oceanography 1947
Scripps Institute of Oceanography. 1947. "A Statistical Study of Wave Conditions at Five Locations along the California Coast," Wave Report No. 68, University of California, San Diego.

Seymour and Aubrey 1985
Seymour, R. J., and Aubrey, D. G. 1985. *Marine Geology*, Vol 65, pp 289-304.

Shepard 1952
Shepard, F. P. 1952. "Revised Nomenclature for Depositional Ccoastal Features," *Bull. Am. Assoc. Petrol Geol.*, Vol 36, pp 1902-12.

Shepard 1973
Shepard, F. P. 1973. *Submarine Geology,* 3rd ed., Harper & Row, New York.

Shore Protection Manual 1977
Shore Protection Manual. 1977. 3rd ed., 2 Vol, U. S. Army Engineer Waterways Experiment Station, U. S. Government Printing Office, Washington, DC.

Shore Protection Manual 1984
Shore Protection Manual. 1984. 4th ed., 2 Vol, U. S. Army Engineer Waterways Experiment Station, U. S. Government Printing Office, Washington, DC.

Silvester 1970
Silvester, R. 1970. "Development of Crenulate Shaped Bays to Equilibrium," *Journal Waterways and Harbors Div.*, American Society of Civil Engineers, Vol 96(WW2), pp 275-87.

Silvester and Ho 1972
Silvester, R., and Ho, S. K. 1972. "Use of Crenulate-Shaped Bays to Stabilize Coasts," *Proceedings, 13th Int. Conf. on Coastal Engineering*, American Society of Civil Engineers, pp 1347-65.

Silvester and Hsu 1993
Silvester, R., and Hsu, J. R. C. 1993. *Coastal Stabilization: Innovative Concepts*, Prentice Hall, Inc., Englewood Cliffs, NJ.

Silvester, Tsuchiya, and Shibano 1980
Silvester, R., Tsuchiya, T., and Shibano, T. 1980. "Zeta Bays, Pocket Beaches and Headland Control," *Proceedings, 17th Int. Conf. on Coastal Eng.*, American Society of Civil Engineers, pp 1306-19.

Sonu 1969
Sonu, C. J. 1969. "Collective Movement of Sediment in Littoral Environment," *Proceedings 11th Conf. on Coast. Eng.*, American Society of Civil Engineers, pp 373-400.

Sonu 1972
Sonu, C. J. 1972. "Edge Wave and Crescentic Bars," Comments on paper by A. J. Bowen and D. L. Inman, *Journal Geophysical Research*, No. 33, pp 6629-31.

Sonu 1973
Sonu, C. J. 1973. "Three-dimensional Beach Changes," *Journal Geol.*, Vol 81, pp 42-64.

Sonu and Russell 1967
Sonu, C. J., and Russell, R. J. 1967. "Topographic Changes in the Surf Zone Profile," *Proceedings 10th Conf. on Coast Eng.*, American Society of Civil Engineers, pp 502-24.

Sonu and van Beek 1971
Sonu, C. J., and van Beek, J. L. 1971. "Systematic Beach Changes on the Outer Banks, North Carolina," *Journal Geol.* Vol 79, pp 416-25.

Sonu, McCloy, and McArthur 1967
Sonu, C. J., McCloy, J. M., and McArthur, D. S. 1967. "Longshore Currents and Nearshore Topographies," *Proc. 10th Conf. on Coast Eng.*, American Society of Civil Engineers, pp 5255-49.

Sternberg, Shi, and Downing 1984
Sternberg, R. W., Shi, N. C., and Downing, J. P. 1984. "Field Investigation of Suspended Sediment Transport in the Nearshore Zone," *Proceedings of the 19th International Conference on Coastal Engineering*, American Society of Civil Engineers, pp 1782-1798.

Thieke and Harris 1993
Thieke, R. J., and Harris, P. S. 1993. "Application of Longshore Transport Statistics to the Evaluation of Sand Transfer Alternatives at Inlets," *Journal of Coastal Research,* Vol 18, pp 125-145.

Thornton 1972
Thornton, E. B. 1972. "Distribution of Sediment Transport Across the Surf Zone," *Proceedings of the 13th International Conference on Coastal Engineering*, American Society of Civil Engineers, pp 1049-1068.

Thornton and Guza 1981
Thornton, E. B., and Guza, R. T. 1981. "Longshore Currents and Bed Shear Stress," *Proceedings of the Conference on Directional Wave Spectra Applications*, R. L. Weigel, ed., University of California, Berkeley.

Thornton and Morris 1977
Thornton, E. B., and Morris, W. D. 1977. "Suspended Sediments Measured Within the Surf Zone," *Proceedings of Coastal Sediments '77*, American Society of Civil Engineers, pp 655-668.

Trask 1952
Trask, P. D. 1952. "Source of Beach Sand at Santa Barbara, California, as Indicated by Mineral Grain Studies," Beach Erosion Board Tech. Memo. No. 28, U.S. Army Engineer Waterways Experiment Station, Vicksburg, MS.

Trask 1955
Trask, P. D. 1955. "Movement of Sand Around Southern California Promontories," Beach Erosion Board Tech. Memo. No. 76, U.S. Army Engineer Waterways Experiment Station, Vicksburg, MS.

U.S. Army Corps of Engineers 1966
U.S. Army Corps of Engineers. 1966. *Shore Protection, Planning, and Design*, Technical Report No. 4, 3rd ed., Coastal Engineering Research Center, U.S. Army Engineer Waterways Experiment Station, Vicksburg, MS.

van Bendegom 1949
van Bendegom, L. 1949. "Consideration of the Fundamentals of Coastal Protection," Ph.D. diss., Technical University, Delft, The Netherlands [in Dutch].

Verhagen 1989
Verhagen, H. J. 1989. "Sand Waves Along the Dutch Coast," *Coastal Engineering,* Vol 13, pp 129-147.

Vitale 1981
Vitale, P. 1981. "Moveable-Bed Laboratory Experiments Comparing Radiation Stress and Energy Flux Factor as Predictors of Longshore Transport Rate," Miscellaneous Report No. 81-4, U.S. Army Engineer Waterways Experiment Station, Coastal Engineering Research Center, Vicksburg, MS.

Walton 1972
Walton, T. L., Jr. 1972. "Littoral Drift Computations Along the Coast of Florida by Means of Ship Wave Observations," Technical Report No. 15, Coastal and Oceanographic Engineering Laboratory, University of Florida, Gainesville.

Walton 1976
Walton, T. L., Jr. 1976. "Uses of Outer Bars of Inlets as Sources of Beach Nourishment Material," *Shore and Beach,* Vol 44, No. 2.

Walton 1977
Walton, T. L., Jr. 1977. "Equilibrium Shores and Coastal Design," *Proceedings, Coastal Sediments '77*, American Society of Civil Engineers.

Walton 1979
Walton, T. L. 1979. "Littoral Sand Transport on Beaches," Ph.D. diss., University of Florida, Gainesville.

Walton 1980
Walton, T. L. 1980. "Littoral Sand Transport from Longshore Currents," Technical Note, *Journal of the Waterway, Port, Coastal, and Ocean Division*, American Society of Civil Engineers, Vol 106, No. WW4, November, pp 483-487.

Walton 1982
Walton, T. L. 1982. "Hand-held Calculator Algorithms for Coastal Engineering; Second Series," CETA 82-4, Coastal Engineering Research Center, U.S. Army Engineer Waterways Experiment Station, Vicksburg, MS.

Walton 1989
Walton, T. L., Jr. 1989. "Discussion of Sand Bypassing Simulation Using Synthetic Longshore Transport Data," *Journal Waterway, Port, Coastal and Ocean Division*, American Society of Civil Engineers, Vol 115, No. 4, pp 576-557.

Walton 1994
Walton, T. L., Jr. 1994. "Shoreline Solution for Tapered Beach Fill," *Journal Waterway, Port, Coastal and Ocean Engineering*, Vol 120, No. 6, pp 1-5.

Walton and Borgman 1991
Walton, T. L., Jr., and Borgman, L. B. 1991. "Simulation of Nonstationary, Non-Gaussion Water Levels on Great Lakes," *Journal of Waterway, Port, Coastal, and Ocean Engineering*, Vol 116, No. 6, American Society of Civil Engineers, New York.

Walton and Bruno 1989
Walton, T. L., and Bruno, R. O. 1989. "Longshore Transport at a Detached Breakwater, Phase II," *Journal of Coastal Research*, Vol 5, No. 4, pp 679-691.

Walton and Chiu 1977
Walton, T., and Chiu, T. 1977. "Sheltered Shorelines in Florida," *Shore and Beach*, Vol 45, No. 4.

Walton and Chiu 1979
Walton, T., and Chiu, T. 1979. "A Review of Analytical Techniques to Solve the Sand Transport Equation and Some Simplified Solutions," *Proceedings of Coastal Structures '79*, American Society of Civil Engineers, pp 809-837.

Walton and Dean 1973
Walton, T. L., Jr., and Dean, R. G. 1973. "Application of Littoral Drift Roses to Coastal Engineering Problems," *Proceedings, First Australian Conference on Coastal Engineering*, National Committee on Coastal and Ocean Engineering Institution of Engineers, Australia.

Walton and Dean 1976
Walton, T. L., Jr., and Dean, R. G. 1976. "Use of Outer Bars of Inlets as Sources of Beach Nourishment Material," *Shore and Beach*, Vol 44, No. 2.

Walton and Douglass 1985
Walton, T. L., Jr., and Douglass, S. L. 1985. "Stochastic Sand Transport Using ARIMA Modeling," *Proceedings 21st Congress, International Association for Hydraulic Research*, Melbourne, Australia, Vol 4, pp 166-172.

Walton and Sensabaugh 1979
Walton, T. L., and Sensabaugh, W. 1979. "Seawall Design on the Open Coast," Sea Grant Report No. 29, University of Florida, Gainesville.

Walton, Liu and Hands 1988
Walton, T. L., Liu, P. L-F., and Hands, E. B. 1988. "Shoreline at a Jetty Due to Cyclic and Random Waves," *Proceedings, 21st International Conference on Coastal Engineering*, American Society of Civil Engineers, pp 1911-1921.

Walton, Thomas and Dickey 1985
Walton, T. L., Thomas, J. L., and Dickey, M. D. 1985. "Cross-shore Distribution of Sediment Transport at a Weir Jetty," *Australasian Conference on Coastal and Ocean Engineering,* Christchurch, New Zealand, 2-6 December, pp 525-535.

Watts 1953a
Watts, G. M. 1953a. "A Study of Sand Movement at South Lake Worth Inlet, Florida," Beach Erosion Board Tech. Memo. No. 42, U.S. Army Engineer Waterways Experiment Station, Vicksburg, MS.

Watts 1953b
Watts, G. M. 1953b. "Development and Field Test of a Sampler for Suspended Sediment in Wave Action," Beach Erosion Board Tech. Memo. No. 34, U.S. Army Engineer Waterways Experiment Station, Vicksburg, MS.

Weggel 1972
Weggel, J. R. 1972. "Maximum Breaker Height," *Journal of the Waterways, Ports, and Coastal Engineering Division*, Vol 98, ww 4, pp 529-548.

Weggel and Perlin 1988
Weggel, J. R., and Perlin, M. 1988. "Statistical Description of Longshore Transport Environment." *Journal Waterway, Port, Coastal and Ocean Engineering*, American Society of Civil Engineers, Vol 114, No. 2, pp 125-145.

White 1987
White, T. E. 1987. "Nearshore Sand Transport," Ph.D. diss., University of California, San Diego.

Williams 1973
Williams, A. T. 1973. "The Problem of Beach Cusp Development," *Journal Sediment Petrol.*, Vol 43, pp 857-66.

Wright et al. 1979
Wright, L. D., Chappell, J., Thorn, B. G., Bradshaw, M. P., and Cowell, P. 1979. "Morphodynamics of Reflective and Dissipative Beach and Inshore Systems: Southeastern Australia," *Marine Geology*, Vol 32, pp 105-40.

Yasso 1965
Yasso, W. E. 1965. "Plan Geometry of Headland Bay Beaches," *Journal Geology*, Vol 73, pp 702-14.

Zenkovitch 1960
Zenkovitch, V. P. 1960. "Fluorescent Substances as Tracers for Studying the Movement of Sand on the Sea Bed, Experiments Conducted in the U.S.S.R.," *Dock and Harbor Authority*, Vol 40, pp 280-283.

Zenkovitch 1964
Zenkovitch, V. P. 1964. "Cyclic Cuspate Sand Spits and Sediment Transport Efficiency," discussion, *Journal Geol.*, Vol 72, pp 879-80.

III-2-6. Definition of Symbols

α	Wave crest angle relative to bottom contours [deg]
α_b	Wave breaker angle relative to the shoreline [deg]
β	Geometric parameter in the expression to approximate the shoreline in the lee of headland-type features (Equation III-2-24 III-2-24 and Figure III-2-27) [deg]
γ	Energy flux dissipation rate [dimensionless]
Γ_{stb}	Stable wave height to water depth ratio for wave re-formation
ε	Shoreline diffusivity parameter (Equation III-2-26) [length^2/time]
θ	Geometric parameter in the expression to approximate the shoreline in the lee of headland-type features (Equation III-2-24 and Figure III-2-27) [deg]
θ_n	Azimuth angle of the outward normal to the shoreline (Equation III-2-17) [deg]
κ	Breaker index H_b/d_b [dimensionless]
Λ_{mix}	Lateral mixing coefficient [dimensionless]
ξ_b	Surf similarity parameter [dimensionless]
ρ	Mass density of water (salt water = 1,025 kg/m^3 or 2.0 slugs/ft^3; fresh water = 1,000kg/m^3 or 1.94 slugs/ft^3) [force-time^2/length^4]
ρ_s	Mass density of sediment grains (2,650 kg/m^3 or 5.14 slugs/ft^3 for quartz-density sand) [force-time^2/length^4]
a_1, a_2	Dimensionless parameters used in the GENESIS empirical predictive formula for longshore sand transport rate (Equation III-2-42)
C_0, C_1, C_2	Coefficients in expression to approximate the shoreline in the lee of headland-type features (Equation III-2-24)
C_b	Wave group speed at the breaker line [length/time]
C_f	Friction coefficient
C_g	Wave group speed [length/time]
C_{g0}	Wave group speed in deep water [length/time]
C_{gb}	Wave group speed at the breaker line [length/time]
d	Water depth in the surf zone, including wave-induced setup [length]
d_b	Water depth at the breaker line [length]
d_B	Berm crest elevation or beach berm height above still-water [length]
d_c	Depth of appreciable sand transport as measured from still-water level [length]

d_C	Offshore closure depth [length]
E	Total wave energy in one wavelength per unit crest width [length-force/length^2]
E_b	Wave energy at the breaker line per unit crest width [length-force/length]
$erf(\ )$	Error function
$erfc(\ )$	Complementary error function
g	Gravitational acceleration (32.17 ft/sec^2, 9.807m/sec^2) [length/time^2]
H	Wave height in the surf zone [length]
H_b	Wave height at breaking [length]
H_{mo}	Average wave height [length]
H_{sig}	Significant wave height [length]
I_ℓ	Immersed weight transport rate [force/sec]
K	Empirical proportionality coefficient [dimensionless]
K_1, K_2	Empirical coefficients, treated as calibration parameters, used in the GENESIS empirical predictive formula for longshore sand transport rate (Equation III-2-42)
$K_{a,b}$	Constants [dimensionless]
k_q	Dimensional normalizing constant [time^3/force]
m	Average bottom slope from the shoreline to the depth of active longshore sand transport [length-rise/length-run]
n	In-place sediment porosity (≈ 0.4) [dimensionless]
$p(t)$	Proportion of fill left within project boundaries at a given time after project initiation (Equation III-2-32)
P_ℓ	Potential longshore sediment transport rate [force/time]
q	Line source or sink of sediment
Q	Potential volumetric longshore transport rate [length^3/time]
$q_x(y)$	Local longshore transport per unit width offshore [length^3/time/length]
$Q_{\ell R}$	Annual longshore transport to the right (looking seaward) [length^3/time]
$Q_{\ell NET}$	Net annual longshore transport [length^3/time]
$Q_{\ell L}$	Annual longshore transport to the left (looking seaward) [length^3/time]
$Q_{\ell GROSS}$	Gross annual longshore transport [length^3/time]
Q_ℓ	Volume longshore transport rate [length^3/time]

R	Geometric parameter in the expression to approximate the shoreline in the lee of headland-type features (Equation III-2-24 and Figure III-2-27) [length]
R_0	Distance between the ends of headlands that define a given section of shoreline (Equation III-2-24 and Figure III-2-27) [length]
T	Wave period [time]
T_0	Length of a wane record [time]
t_f	Time required for structure to fill to capacity (Equation III-2-29)
T_p	Average peak spectral wave period [time]
u_{mb}	Maximum oscillatory velocity magnitude (Equation III-2-9) [length/time]
V_0	Theoretical longshore velocity at breaking point for the no-lateral-mixing case [length/time]
V_ℓ	Longshore current speed [length/time]
W	Width of the surf zone [length]
w_f	Sediment fall velocity [length/time]
y	Distance seaward the shoreline will extend at a structure [length]
Y	Distance to the measured current from the shoreline (Equation III-2-12) [length]
Y	Length of structure (Equation III-2-29) [length]

III-2-7. Acknowledgments

Authors of Chapter III-2, "Longshore Sediment Transport:"

Julie D. Rosati, Coastal and Hydraulics Laboratory (CHL), Engineer Research and Development Center (ERDC), Vicksburg, Mississippi.
Todd L. Walton, Ph.D., CHL
Kevin Bodge, Ph.D., Olsen Associates, Jacksonville, Florida.

Reviewers:

James Bailard, Ph.D., Private consultant
James R. Houston, Ph.D., ERDC
David L. Kriebel, Ph.D., Department of Ocean Engineering, United States Naval Academy, Annapolis, Maryland.
Paul A. Work, Ph.D., School of Civil and Environmental Engineering, Georgia Institute of Technology, Atlanta, Georgia.

Chapter 3
CROSS-SHORE SEDIMENT
TRANSPORT PROCESSES

EM 1110-2-1100
(Part III)
1 August 2008 (Change 2)

Table of Contents

Page

Cross-Shore Sediment Transport Processes

List of Tables

List of Figures

Chapter III-3
Cross-Shore Sediment Transport Processes

III-3-1. Introduction

a. Overview/purpose.

(1) Sediment transport at a point in the nearshore zone is a vector with both longshore and cross-shore components (see Figure III-3-1). It appears that under a number of coastal engineering scenarios of interest, transport is dominated by either the longshore or cross-shore component and this, in part, has led to a history of separate investigative efforts for each of these two components. The subject of total longshore sediment transport has been studied for approximately five decades. There is still considerable uncertainty regarding certain aspects of this transport component including the effects of grain size, barred topography, and the cross-shore distribution of longshore transport. A focus on cross-shore sediment transport is relatively recent, having commenced approximately one decade ago and uncertainty in prediction capability (including the effects of all variables) may be considerably greater. In some cases the limitations on prediction accuracy of both components may be due as much to a lack of good wave data as to an inadequate understanding of transport processes.

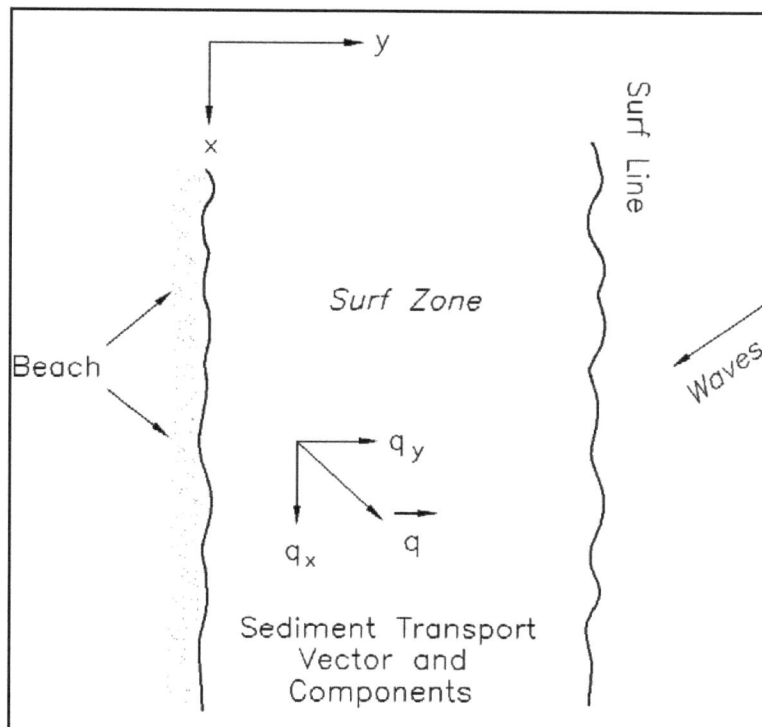

Figure III-3-1. Longshore (q_x) and cross-shore (q_y) sediment transport components

(2) Cross-shore sediment transport encompasses both offshore transport, such as occurs during storms, and onshore transport, which dominates during mild wave activity. Transport in these two directions appears to occur in significantly distinct modes and with markedly disparate time scales; as a result, the difficulties in predictive capabilities differ substantially. Offshore transport is the simpler of the two and tends to occur

with greater rapidity and as a more regular process with transport more or less in phase over the entire active profile. This is fortunate since there is considerably greater engineering relevance and interest in offshore transport due to the potential for damage to structures and loss of land. Onshore sediment transport within the region delineated by the offshore bar often occurs in "wave-like" motions referred to as "ridge-and-runnel" systems in which individual packets of sand move toward, merge onto, and widen the dry beach. A complete understanding of cross-shore sediment transport is complicated by the contributions of both bed and suspended load transport. Partitioning between the two components depends in an unknown way on grain size, local wave energy, and other variables.

(3) Cross-shore sediment transport is relevant to a number of coastal engineering problems, including: (a) beach and dune response to storms, (b) the equilibration of a beach nourishment project that is placed at slopes steeper than equilibrium, (c) so-called "profile nourishment" in which the sand is placed in the nearshore with the expectation that it will move landward nourishing the beach (this involves the more difficult problem of onshore transport), (d) shoreline response to sea level rise, (e) seasonal changes of shoreline positions, which can amount to 30 to 40 m, (f) overwash, the process of landward transport due to overtopping of the normal land mass due to high tides and waves, (g) scour immediately seaward of shore-parallel structures, and (h) the three-dimensional flow of sand around coastal structures in which the steeper and milder slopes on the updrift and downdrift sides of the structure induce seaward and landward components, respectively. These problems are schematized in Figure III-3-2.

b. *Scope of chapter.*

(1) This chapter consists of two additional sections. The first section describes the general characteristics of equilibrium beach profiles and cross-shore sediment transport. This section commences with a qualitative description of the forces acting within the nearshore zone, the characteristics of an equilibrium beach profile, and a discussion of conditions of equilibrium when the forces are balanced, as well as the ensuing sediment transport when conditions change, causing an imbalance. The general profile characteristics across the continental shelf are reviewed with special emphasis on the more active nearshore zone. Bar morphology and short- and long-term changes of beach profiles due to storms and sea level rise are examined, along with effects of various parameters on the profile characteristics, including wave climate and sediment characteristics. Survey capabilities to quantify the profiles are reviewed.

(2) The second section deals with quantitative aspects of cross-shore sediment transport with special emphasis on engineering applications and the prediction of beach profile change. First, the general shape of the equilibrium beach profile is quantified in terms of sediment grain size and basic wave parameters. Equilibrium profile methods are then used to develop analytical solutions to several problems of interest in beach nourishment design. Similar analytical solutions are developed for the steady-state beach profile response to elevated water levels, including both the long-term response to sea level rise and the short-term response to storm surge. For simplified cases, analytical methods are then presented for estimating the dynamic profile response during storms. For more general applications, numerical modelling approaches are required and these are briefly reviewed.

III-3-2. General Characteristics of Natural and Altered Profiles

a. *Forces acting in the nearshore.*

(1) There are several identifiable forces that occur within the nearshore active zone that affect sediment motion and beach profile response. The magnitudes of these forces can be markedly different inside and

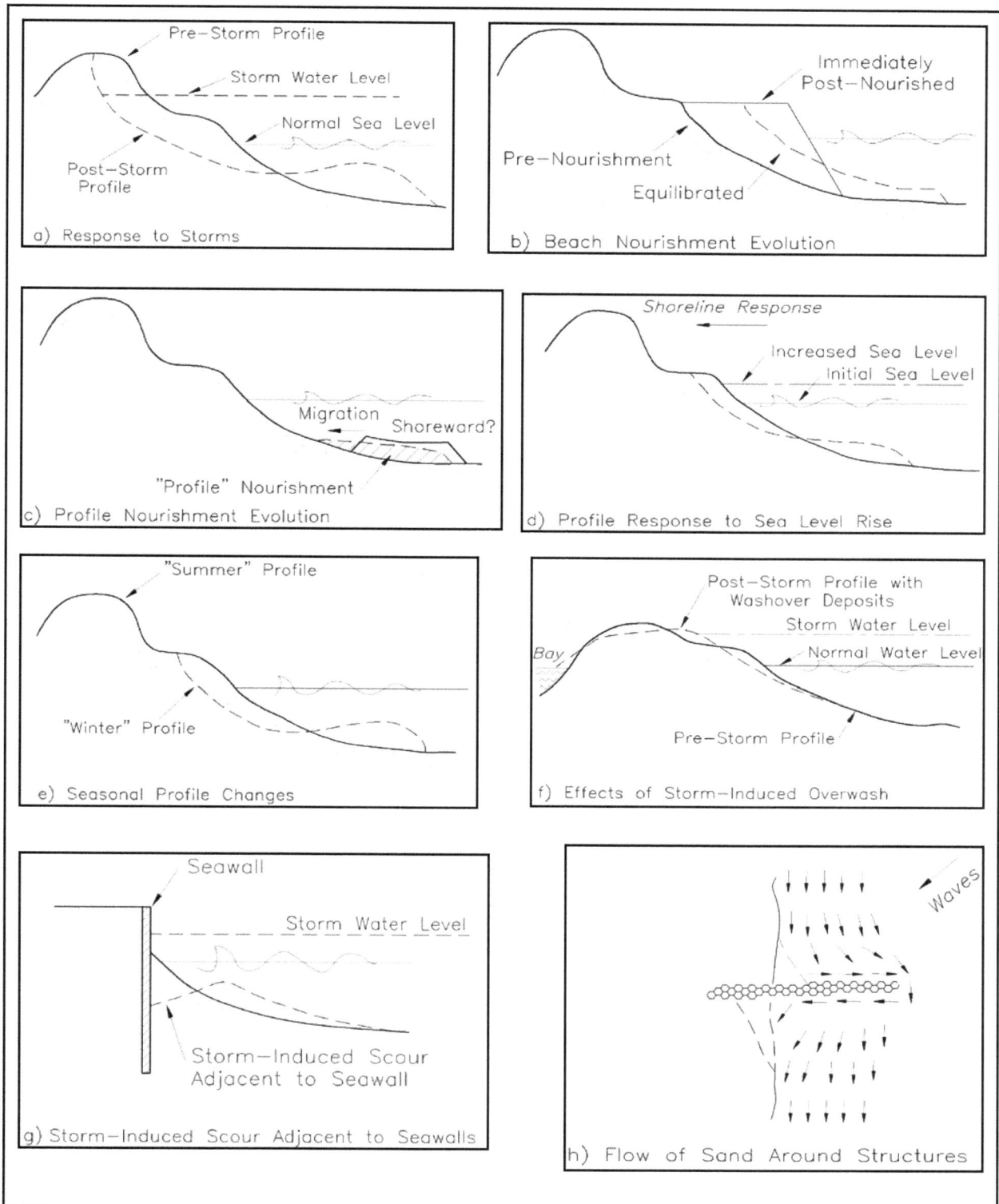

Figure III-3-2.   Problems and processes in which cross-shore sediment transport is relevant

outside the surf zone. Under equilibrium conditions, these forces are in balance and although there is motion of the individual sand grains under even low wave activity, the profile remains more or less static. Cross-shore sediment transport occurs when hydrodynamic conditions within the nearshore zone change, thereby modifying one or more of the forces resulting in an imbalance and thus causing transport gradients and profile change. Established terminology is that onshore- and offshore-directed forces are referred to as "constructive" and "destructive," respectively. These two types of forces are briefly reviewed below; however, as will be noted, the term "forces" is used in the generic sense. Moreover it will be evident that some forces could behave as constructive under certain conditions and destructive under others.

(2) As noted, constructive forces are those that tend to cause onshore sediment transport. For classic nonlinear wave theories (Stokes, Cnoidal, Solitary, Stream Function, etc.), the wave crests are higher and of shorter duration than are the troughs. This feature is most pronounced just outside the breaking point and also applies to the water particle velocities. For oscillatory water particle velocities expressed as a sum of phase-locked sinusoids such as for the Stokes or Stream Function wave theories, even though the time mean of the water particle velocity is zero, the average of the bottom shear stress τ_b expressed as

$$\overline{\tau_b} = \rho \, \frac{f}{8} \, \overline{|v_b| v_b} \qquad \text{(III-3-1)}$$

can be shown to be directed onshore. In the above, ρ is the mass density of water, f is the Darcy-Weisbach friction coefficient which, for purposes here is considered constant over a wave period, and v_b is the instantaneous wave-induced water particle velocity at the bottom. A definition sketch is provided in Figure III-3-3. An example of the time-varying shear stress due to a nonlinear (Stream Function) wave is shown in Figure III-3-4. Dean (1987a) has developed the average bottom shear stress based on the Stream Function wave theory and presented the results in the nondimensional form shown in Figure III-3-5.

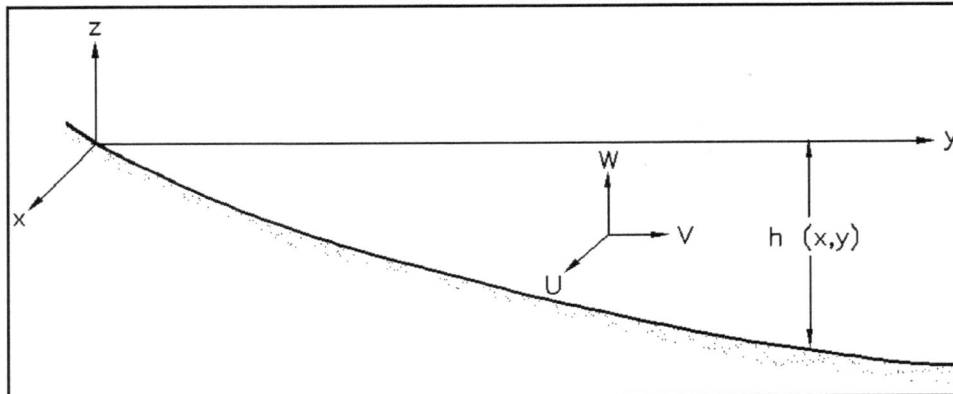

Figure III-3-3.   Definition sketch

(3) A second constructive force originates within the bottom boundary layer, causing a net mean velocity in the direction of propagating water waves. This streaming motion was first observed in the laboratory by Bagnold (1940) and has been quantified by Longuet-Higgins (1953) as due to the local transfer of momentum associated with energy losses by friction. For the case of laminar flows, the maximum (over depth) value of this steady velocity v_s is surprisingly independent of the value of the viscosity and is given by

$$v_s = -\frac{3 \sigma k H^2}{16 \, \sinh^2 kh} \qquad \text{(III-3-2)}$$

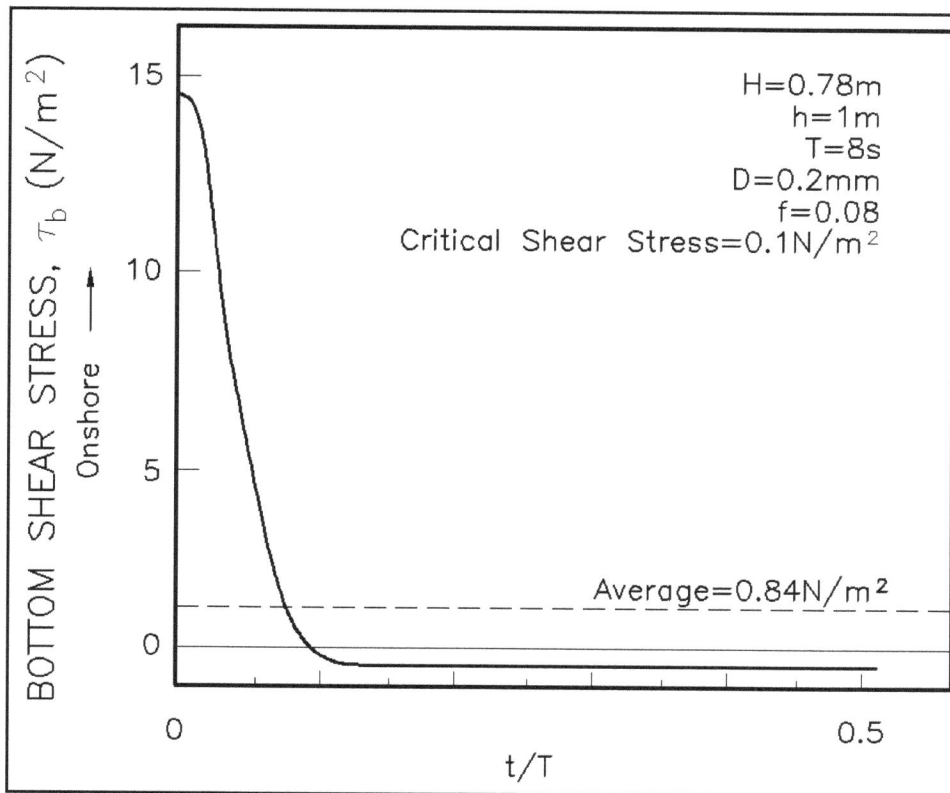

Figure III-3-4. Variation with time of the bottom shear stress under a breaking nonlinear wave. H = 0.78 m, h= 1.0 m, T = 8.0 s, and D = 0.2 mm

which, for the case of shallow water and a wave height proportional to the breaking depth, will be shown to be 1.5 times the <u>average</u> of the return flow due to the mass transport. In Equation 3-2, σ is the wave angular frequency, k is the wave number, and H the wave height. Although the maximum velocity is independent of the viscosity, the bottom shear stress τ_{bs} induced by the streaming velocity is not and is given by

$$\tau_{bs} = - \frac{\rho\ \epsilon^{\frac{1}{2}}\ \sigma^{\frac{3}{2}}\ H^2\ k}{8\ \sqrt{2}\ \sinh^2 kh} \tag{III-3-3}$$

in which ϵ is the eddy viscosity.

(4) Within the surf zone, cross-shore transport may be predominantly due to sediment in suspension. If the suspension is intermittent, occurring each wave period, the average water particle velocity during the period that the particle is suspended determines the direction of cross-shore transport. Although this cause of sediment transport is not a true force, it does represent a contributing mechanism. Turbulence, although also not a true force, can be effective in mobilizing sediment and dependent on whether the net forces are shoreward or seaward at the time of mobilization, can be constructive or destructive, respectively. Dean (1973) noted that suspended sediment can move either onshore (constructive) or offshore (destructive), depending on how high a sand grain is suspended off the bottom. Under the wave crest, if the sediment particle is suspended a distance above the bottom proportional to the wave height H, and if the particle has a fall velocity w, then the time required for the grain to fall back to the bottom would be proportional to H/w. If this fall time is less than one-half of the wave period, then the particle should experience net onshore

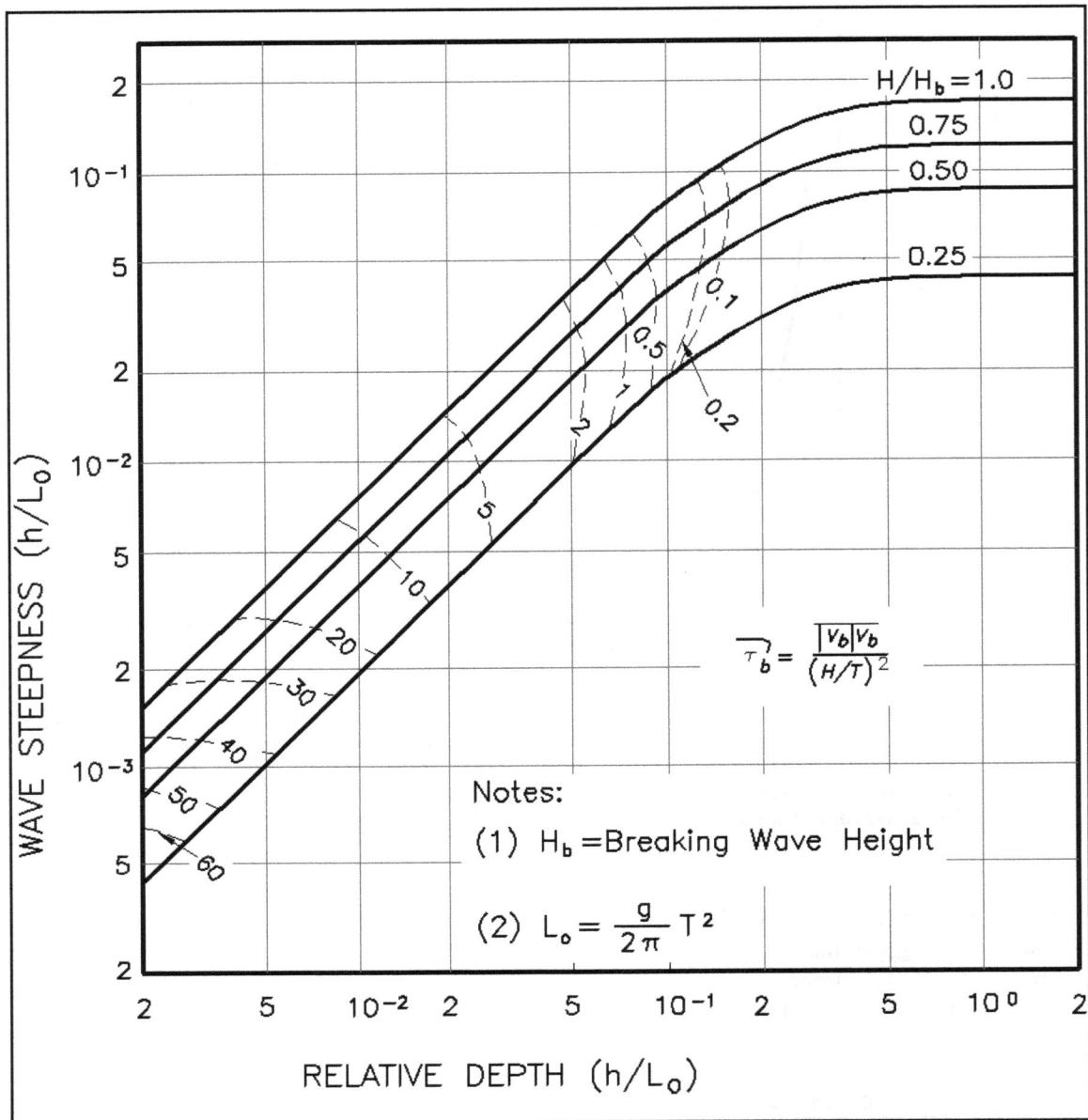

Figure III-3-5.   Isolines of nondimensional average bottom shear stress T_b versus relative depth and wave steepness (Dean 1987a). Note that bottom shear stresses are directed landward

motion, whereas the particle should move offshore if the fall time is greater than one-half the wave period. While such an approach is overly simplistic, and does not include the effects of mean cross-shore currents, it has been shown that net onshore or offshore sediment transport can be correlated to the so-called fall time parameter H/wT, which will be discussed later in this chapter.

(5) Gravity is the most obvious destructive force, acting downslope and in a generally seaward direction for a monotonic profile. However, for the case of a barred profile, gravity can act in the shoreward direction over portions of the profile. Gravity tends to "smooth" any irregularities that occur in the profile. If gravity were the only force acting, the only possible equilibrium profile would be horizontal and sandy beaches as we know them would not exist. It should be recognized, however, that gravity may also serve as a stabilizing force, since sediment particles cannot be mobilized from the bed unless: (a) upward-directed forces

associated with fluid turbulence can exceed the submerged weight of the sediment, and/or (b) slope-parallel fluid shear forces can exceed the frictional resistance of sediment. Also, as noted, gravity causes suspended sediment to settle out of the water column, with fall velocity w, which may cause suspended sediment to move shoreward if not suspended too high in the water column.

(6) Other destructive forces are generally related to the vertical structure of the cross-shore currents. The undertow, the seaward return flow of wave mass transport, induces a seaward stress on the bottom sediment particles. For linear waves, the time-averaged seaward discharge due to the return flow of shoreward mass transport Q is (Dean and Dalrymple 1991)

$$Q = \frac{E}{\rho C} \tag{III-3-4}$$

where E is the wave energy density and C is the wave celerity. If the return flow due to mass transport were distributed uniformly over the water depth, it can be shown from linear shallow-water wave theory that the mean velocity would be

$$\overline{V} = \frac{\sqrt{g}\, H^2}{8\, h^{3/2}} \tag{III-3-5}$$

which, as noted for shallow water, is two-thirds of the maximum streaming velocity. Within the surf zone, the wave height can be considered to be proportional to the local depth, as $H = \kappa h$, so that the mean velocity further simplifies to $0.08\,(gh)^{1/2}$ for $\kappa \approx 0.78$ where $(gh)^{1/2}$ is the wave celerity in shallow water. In storm events where there is overtopping of the barrier island, a portion or all of the potential return flow due to mass transport can be relieved through strong landward flows, thereby eliminating this destructive force and resulting instead in constructive forces.

(7) It is well-known that associated with wave propagation toward shore is a shoreward flux of linear momentum (Longuet-Higgins and Stewart 1964). When waves break, the momentum is transferred to the water column, resulting in a shoreward-directed thrust and thus a wave-induced setup within the surf zone, the gradient of which is proportional to the local bottom slope. This momentum is distributed over depth, as shown in Figure III-3-6. In shallow water, linear water wave theory predicts that one-third of the momentum flux originates between the trough and crest levels and has its centroid at the mean water level. The remaining two-thirds originates between the bottom and the mean water level, is uniformly distributed over this dimension, and thus has its centroid at the mid-depth of the water column. Because of the contribution at the free surface, breaking waves induce an equivalent shear force on the water surface which will be quantified later. This causes a seaward bottom shear stress within the breaking zone. The bottom shear stress is dependent on the rate of energy dissipation. This effective shear force due to momentum transfer must be balanced by the bottom shear stress and the pressure forces due to the slope of the water surface.

(8) Often during major storm events, strong onshore winds will be present in the vicinity of the shoreline. These winds cause a shoreward-directed surface flow and a seaward-directed bottom flow, as shown in Figure III-3-7. Of course, seaward-directed winds would cause shoreward-directed bottom velocities and thus constructive forces. Thus, landward- and seaward-directed winds result in destructive and constructive forces, respectively.

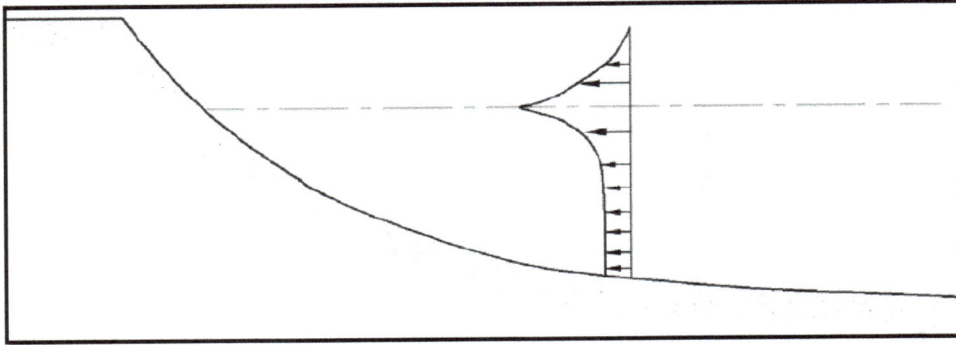

Figure III-3-6. Distribution over depth of the flux of the onshore component of momentum

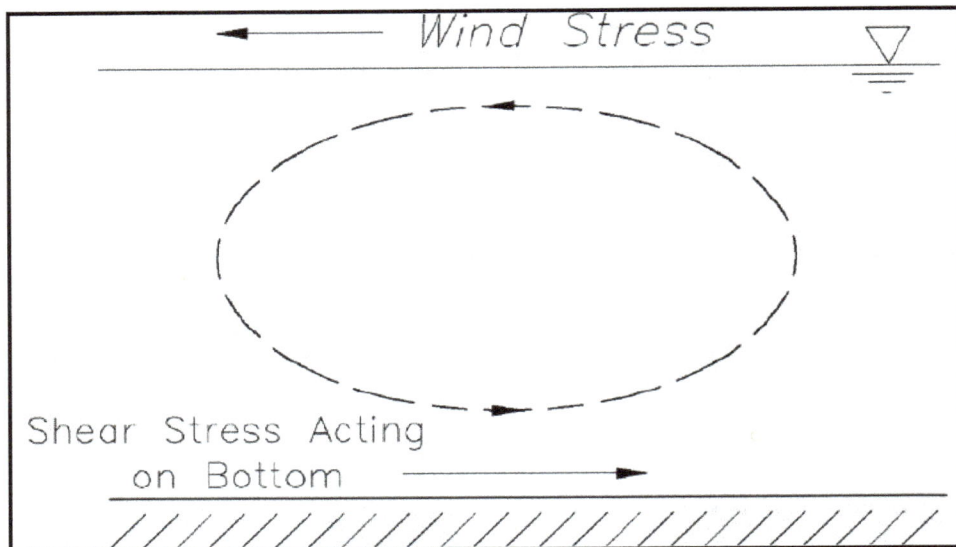

Figure III-3-7. Bottom stresses caused by surface winds

(9) Considering linear wave theory, and a linear shear stress relationship with eddy viscosity ϵ, the distribution of the mean velocity over depth for the case of return of mass transport (no overtopping) and without including the contribution of bottom streaming can be shown to be (Dean and Dalrymple 2000)

$$v(z) = \frac{h}{\rho \epsilon} \left[2\tau_\eta - \frac{\partial E}{\partial y} \right] \left[\frac{3}{8} \left(\frac{z}{h} \right)^2 + \frac{1}{2} \left(\frac{z}{h} \right) + \frac{1}{8} \right] + \frac{3}{2} \frac{Q}{h} \left[1 - \left(\frac{z}{h} \right)^2 \right] \tag{III-3-6}$$

In this expression, the first term is associated with the surface wind stress τ_η. The second term is associated with the vertical gradient of momentum flux and is expressed as a function of the cross-shore gradient in wave energy $\partial E/\partial y$. It is noted that this term is zero outside the breakpoint and contributes only inside the surf zone where energy is dissipated. The third term is associated with the seaward return flow of mass transport, where Q represents the net seaward discharge over the water column as given by Equation 3-4.

(10) For the three effects considered in Equation 3-6, the shear stress associated with the vertical velocity distribution may be computed for any elevation z as

$$\tau = \rho \epsilon \frac{\partial v}{\partial z} \qquad\qquad (III\text{-}3\text{-}7)$$

where ϵ is the turbulent eddy viscosity. The resulting seaward-directed shear stress at the bottom ($z = -h$) is then given by

$$\tau_b = \frac{1}{4}\frac{\partial E}{\partial y} - \frac{\tau_\eta}{2} + 3\frac{\epsilon E}{Ch^2} \qquad\qquad (III\text{-}3\text{-}8)$$

(11) Velocity distributions and shear stress distributions inside and outside the surf zone based on Equations III-3-6 and III-3-7 are shown in Figure III-3-8 for a condition of no surface wind stress and for cases of no overtopping and full overtopping both inside and outside the breakpoint. Profile conditions assumed for this example are shown in the figure caption and assume an equilibrium beach profile where wave breaking is assumed to occur at a depth of 1 m. For the case with no overtopping, most of the seaward velocity shown in Figure III-3-8 is due to the return flow required to balance the shoreward flows near the surface. This is further illustrated in the cases with overtopping where it is assumed that the shoreward flows overtop the profile so that there is no net return flow due to mass transport.

(12) Table III-3-1 summarizes the mechanisms identified as contributing to constructive and/or destructive forces and, where possible, provides an estimate of their magnitudes. For purposes of these calculations, the following conditions have been considered: an equilibrium beach profile with a grain size of $D = 0.2$ mm, $h = 1$ m, $H = 0.78$ m, $T = 8$ s, $\epsilon = 0.04$ m^2/s, wind speed $= 20$ m/s. It is seen that of the bottom stresses that can be quantified, those associated with undertow due to mass transport and momentum flux transfer are dominant.

b.   Equilibrium beach profile characteristics.

(1)   In considerations of cross-shore sediment transport, it is useful to first examine the case of equilibrium in which there is no net cross-shore sediment transport. The competing forces elucidated in the previous section can be fairly substantial, exerting tendencies for both onshore and offshore transport. A change will bring about a disequilibrium that causes cross-shore sediment transport. The concept of an equilibrium beach profile has been criticized, since in nature the forces affecting equilibrium are always changing with the varying tides, waves, currents, and winds. Although this is true, the concept of an equilibrium profile is one of the coastal engineer's most valuable tools in providing a framework to consider disequilibrium and thus cross-shore sediment transport. Also, many useful and powerful conceptual and design relationships are based on profiles of equilibrium.

(2)   When applying equilibrium profile concepts to problems requiring an estimate of profile retreat or advance, a related concept of importance is the principle of conservation of sand across the profile. Under conditions where no longshore gradients exist in the longshore transport, onshore-offshore transport causes a redistribution of sand across the profile but does not lead to net gain or loss of sediment. Most engineering methods applied to the prediction of profile change ensure that the total sand volume is conserved in the active profile, so that erosion of the exposed beach face requires a compensating deposition offshore, while deposition on the exposed beach face must be accompanied by erosion of sediment in the surf zone. For cases where longshore gradients in longshore transport do exist, it is then common to assume that the profile

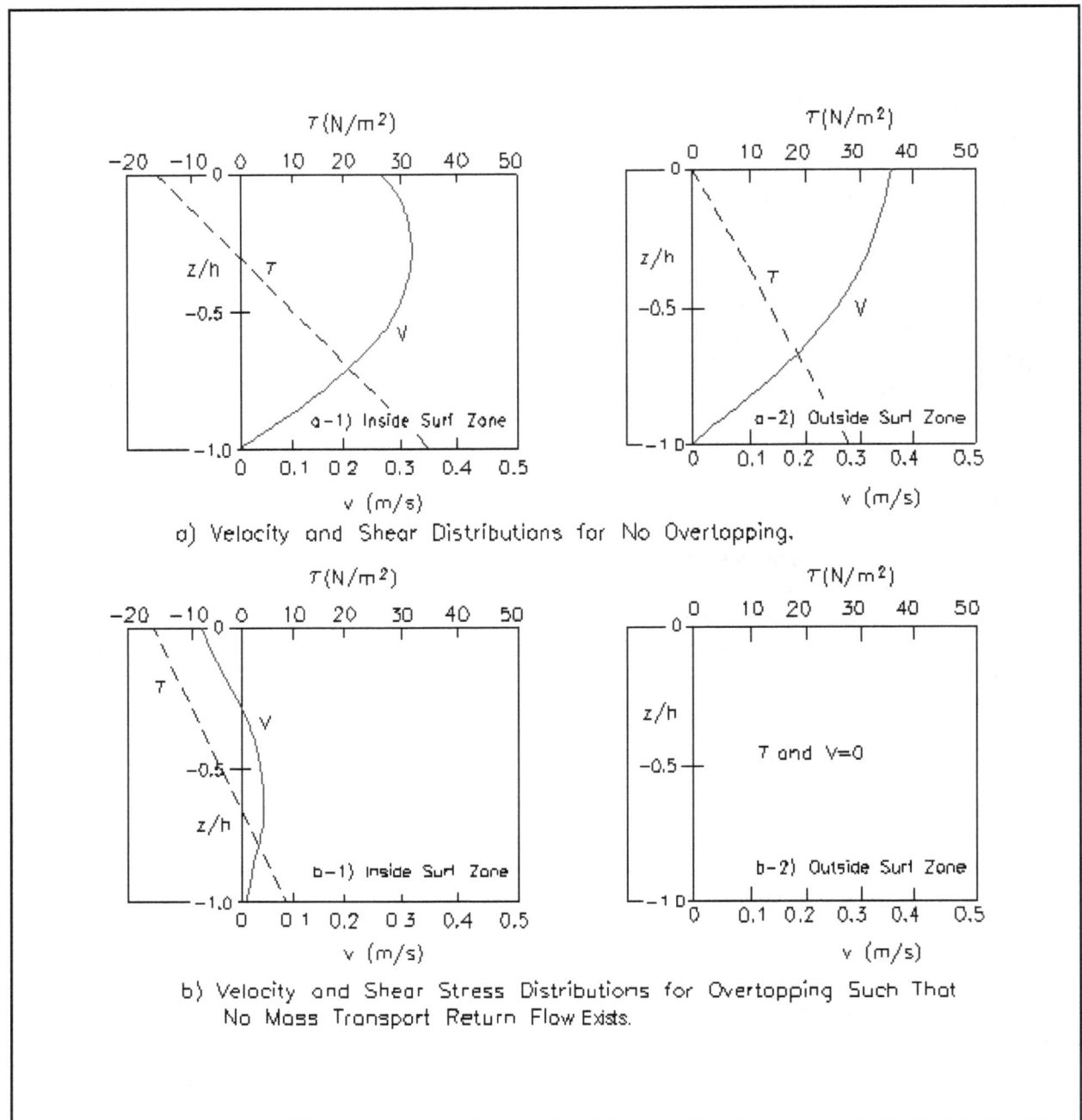

Figure III-3-8. Velocity distributions inside and outside the surf zone for no surface wind stress and cases of no overtopping and full overtopping both inside and outside the surf zone

advances or retreats uniformly at all active elevations while maintaining its shape across the profile. In this way, sediment volume can be added or removed from the profile without changing the shape of the active profile. As a result, most methods for predicting beach profile change treat the longshore and cross-shore components separately so that the final profile form and location are determined by superposition.

Cross-Shore Sediment Transport Processes

Table III-3-1
Constructive and Destructive Cross-shore "Forces" in Terms of Induced Bottom Shear Stresses

Constructive or Destructive	Description of Force	Magnitude of Force (N/m²)	
		Breaking Waves	Nonbreaking Waves
Constructive	Average Bottom Shear Stress Due to Nonlinear Waves[1]	0.84	0.84
	Streaming Velocities[2]	28.9	28.9
	Overtopping	28.6	28.6
	Gravity[3]	0.046	0.046
Destructive	Undertow Due to Mass Transport	28.6	28.6
	Undertow Due to Momentum Flux Transfer	7.9	0
Constructive or Destructive	Intermittent Suspension	?	?
	Turbulence	Relatively Large	Relatively Small
	Wind Effects[4]	0.95	0.95

Notes:
For the calculations resulting in the values in this table: H = 0.78 m, h = 1.0 m, T = 8 s.
[1] f = 0.08
[2] ϵ = 0.04 m²/s
[3] Equil brium profile with D = 0.2 mm
[4] Wind speed = 20 m/s.

(3) Generally observed properties of equilibrium profiles are as follows: (a) they tend to be concave upward, (b) the slopes are milder when composed of finer sediments, (c) the slopes tend to be flatter for steeper waves, and (d) the sediments tend to be sorted with the coarser and finer sediments residing in the shallower and deeper waters, respectively. The effects of changes that induce cross-shore sediment transport can be deduced from these known general characteristics. For example, an increase in water level will cause a disequilibrium, as can be seen by noting that due to the concave upward nature of the profile, the depth at a particular reference distance from the new shoreline is now greater than it was before the increased water level. If the equilibrium profile had been planar, then the increase in water level would not change the depth at a distance from the new shoreline and there would be no disequilibrium. It will be shown that without the introduction of additional sediment into the system, the only way in which the profile can reattain equilibrium is to recede, thus providing sediment to fill the bottom to a depth consistent with the equilibrium profile and the new (elevated) water level.

(4) Since profiles are generally flatter for steeper waves, an increase or decrease in wave steepness will also induce seaward or landward sediment flows, respectively. Naturally, onshore and offshore winds will cause seaward and landward sediment transport, respectively. As an example of the shoreline response to storms, Figure III-3-9 presents results from Katoh and Yanagishima (1988) in which the offshore waves, shoreline position, and beach face slope were measured over a period of approximately 7 months. It is seen that the shoreline retreats abruptly during the higher wave events and advances more gradually during periods of milder wave activity. The beach slope and shoreline changes, of course, correlate with the slope becoming milder during periods of shoreline retreat. The authors also found it of interest that the rate of shoreline advancement during the recovery phase was almost constant at 0.68 m/day.

(5) Many beaches in nature have one or more longshore bars present. At some locations, these bars are seasonal and at some they are more or less permanent. Figure III-3-10a presents a profile from

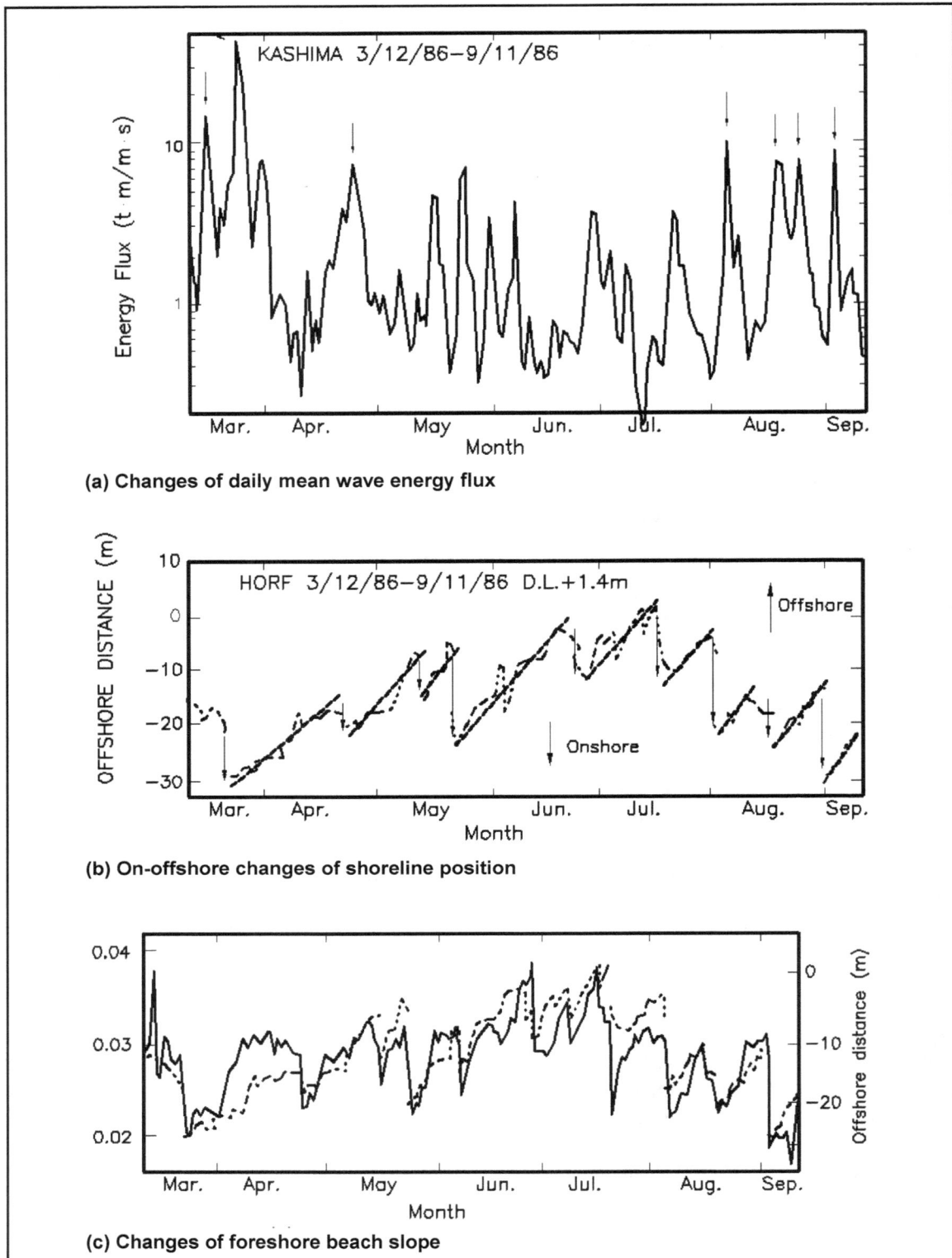

(a) Changes of daily mean wave energy flux

(b) On-offshore changes of shoreline position

(c) Changes of foreshore beach slope

Figure III-3-9. Effects of varying wave energy flux (a) on: (b) shoreline position, and (c) foreshore beach slope (dots are shoreline position in (b) and (c), solid curve is trend line in (b), foreshore slope in (c)) (Katoh and Yanagishima 1988)

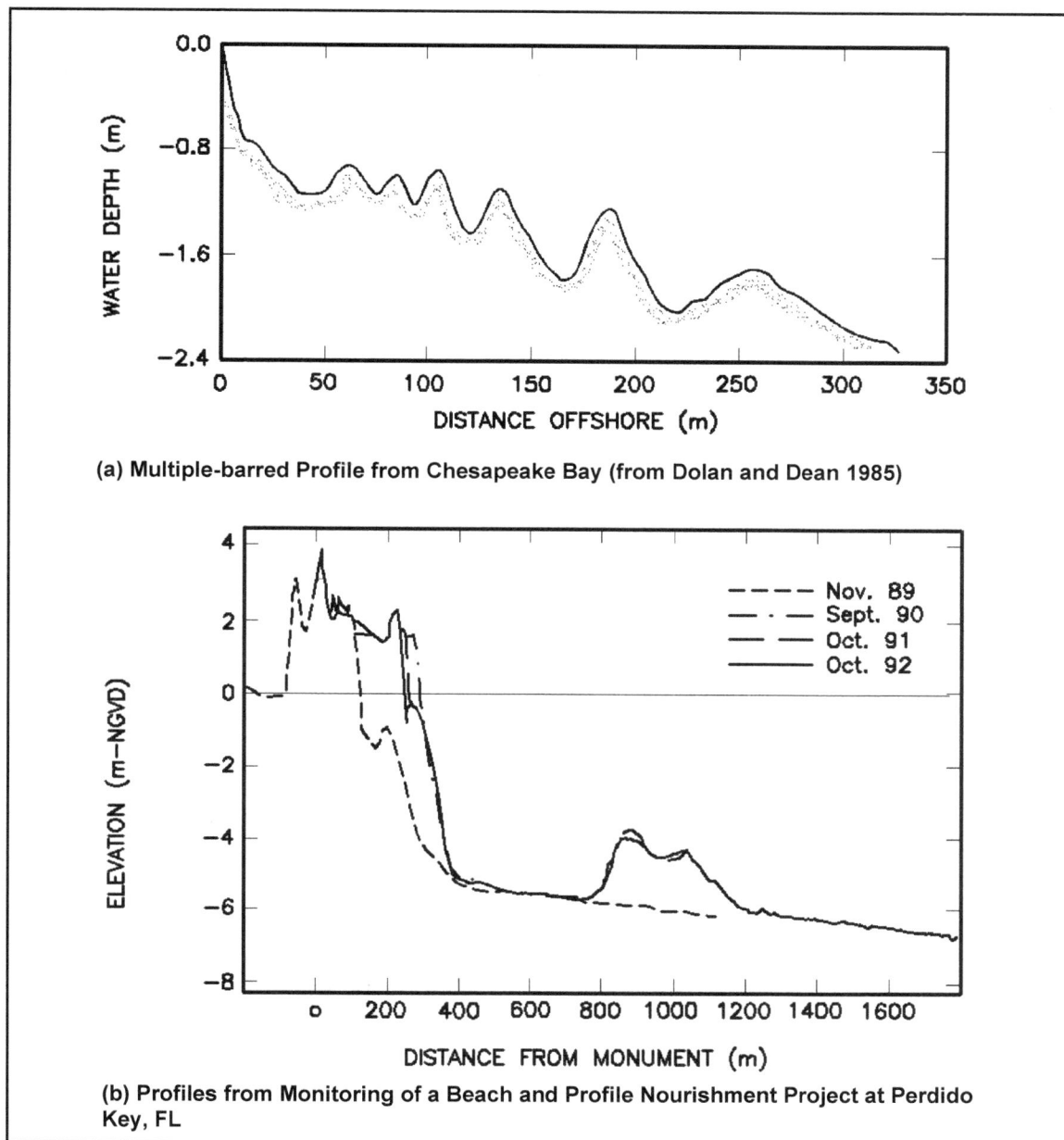

(a) Multiple-barred Profile from Chesapeake Bay (from Dolan and Dean 1985)

(b) Profiles from Monitoring of a Beach and Profile Nourishment Project at Perdido Key, FL

Figure III-3-10.   Examples of two offshore bar profiles

Chesapeake Bay in which at least six bars are evident and Figure III-3-10b shows profiles measured in a monitoring program to document the evolution of a beach nourishment project at Perdido Key, FL. This project included both beach nourishment in the form of a large seaward buildup of the berm and foreshore and profile nourishment in the form of a large offshore mound. As seen from Figure III-3-10b, a bar was present before nourishment and gradually re-formed in depths of less than 1 m as the profile equilibrated during the 2-year period shown in Figure III-3-10b.

(6)   It will be shown later that the presence of bars depends on wave and sediment conditions and at a particular beach, bars may form or move farther seaward during storms. It appears that the outer bars on some profiles are relict and may have been caused by a past large storm which deposited the sand in water too deep

for fair- weather conditions to return the sand to shore. At some beaches with more than one bar, the inner bar will exhibit more rapid response to changing wave conditions than those farther offshore. Figure III-3-11 presents results from Duck, NC, in which profile surveys were conducted over a period of approximately 11 years. It is seen in this case that both the outer and inner bars undergo significant changes in position whereas the shoreline remains relatively fixed, possibly due to coarser sediment in shallow water and at the shoreline. As an example of the potential rates of change of bar position, Birkemeier (1984) shows examples of offshore migration of the outer bar at Duck, NC, during three successive storms in the fall of 1981 averaging almost 4 m/day while onshore migration of the outer bar following the storm season averaged almost 0.5 m/day.

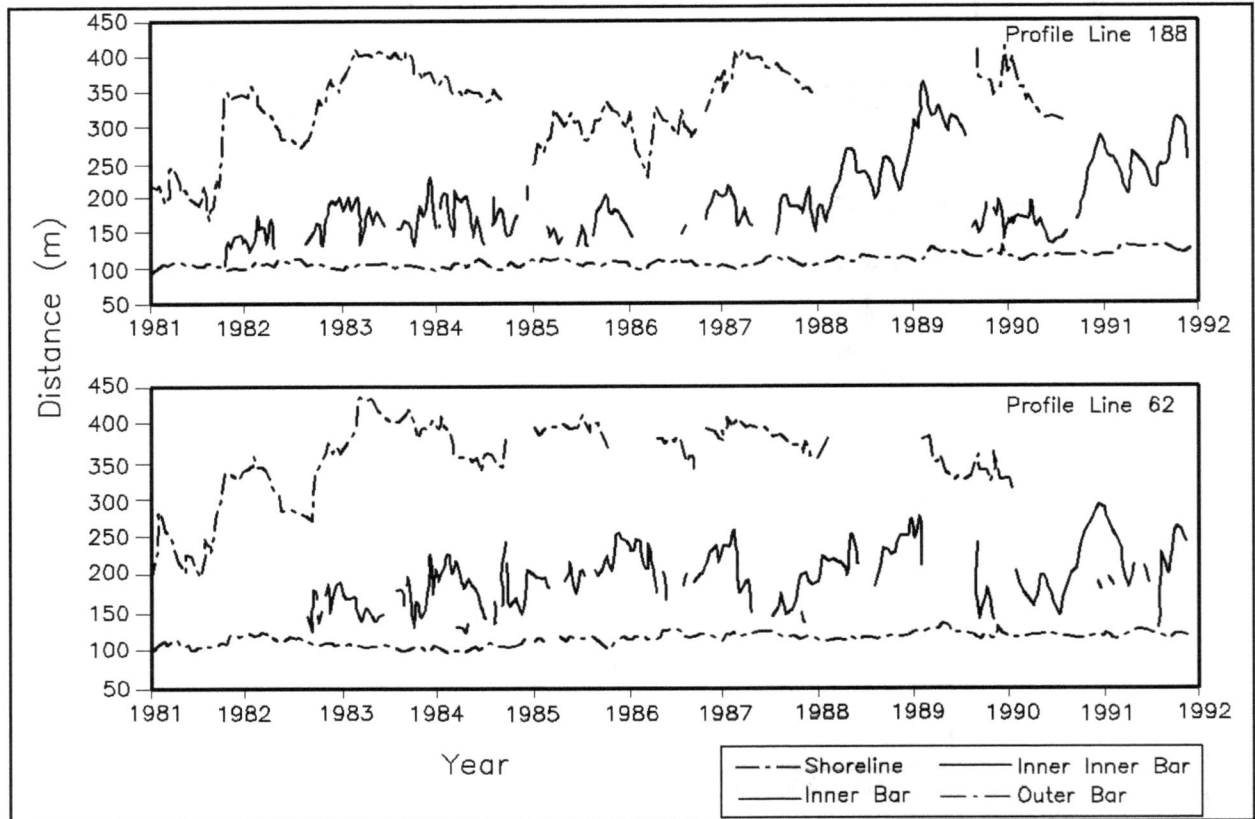

Figure III-3-11. Variation in shoreline and bar crest positions, Duck, NC (Lee and Birkemeier 1993)

(7) Keulegan (1945, 1948) reported on studies which included both laboratory and field data to determine relationships for bar formation. A focus of these studies was the geometric characteristics of the longshore bars. In examining bars from nature, an attempt was made to select sites with small tidal effects. The bar geometry was defined in terms of the depth over the bar crest h_{CR} the depth of the bar trough h_T and the depth to the bar base h_D at the position of the bar crest. These definitions are shown in Figure III-3-12. Keulegan found that the ratio of depths of bar crest to bar base h_{CR}/h_D was approximately 0.58 for both the laboratory and field cases. The ratio of depths of trough to crest h_T/h_{CR} ranged from 1.6 to 1.8. It was also found that bars in nature are considerably broader than those produced in the laboratory. This is probably due to varying wave heights in nature and, to a lesser extent, to varying water levels. Figure III-3-13 compares laboratory and field bar geometries. The field bar is approximately twice as wide as the bar produced in the laboratory.

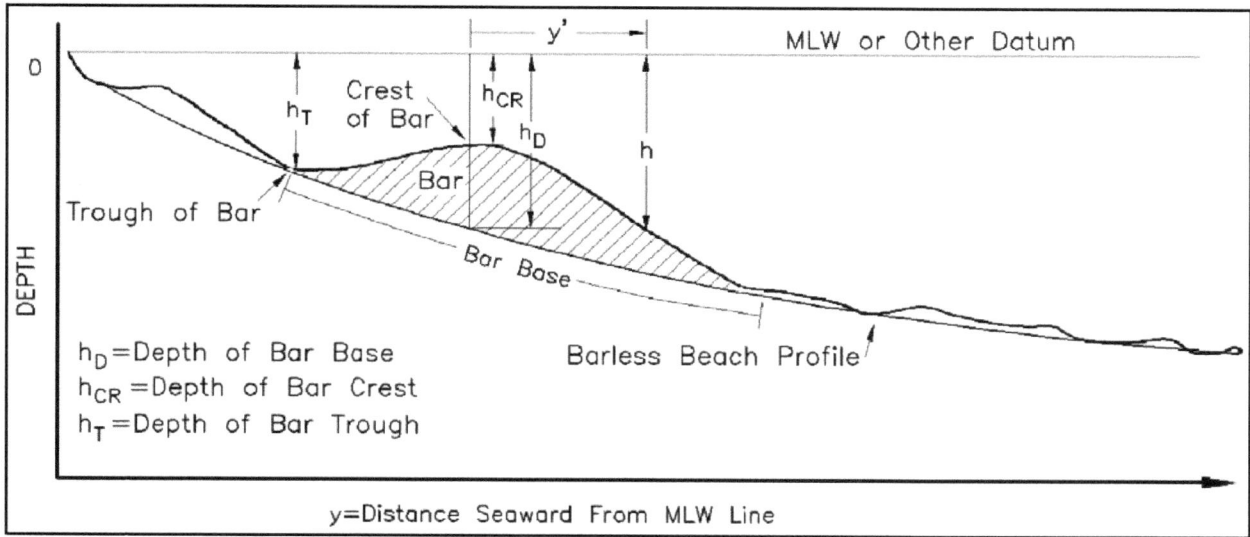

Figure III-3-12.   Definition of offshore bar characteristics (Keulegan 1945)

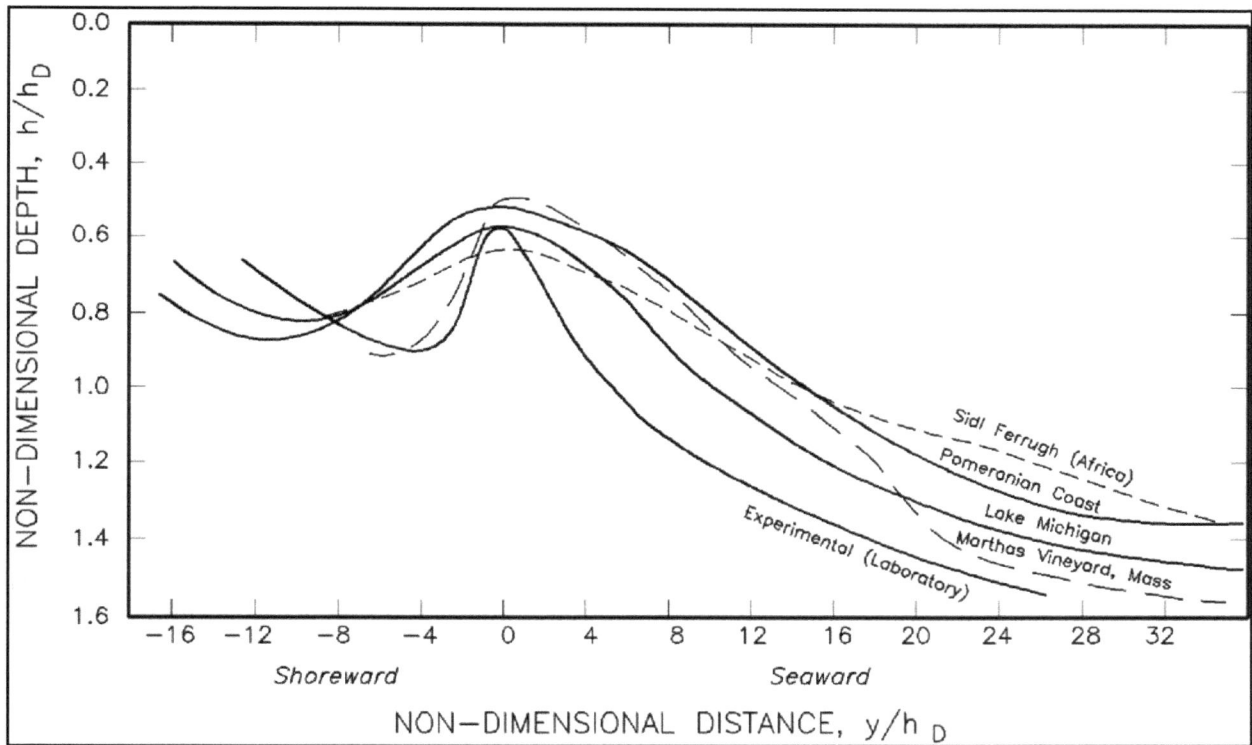

Figure III-3-13.   Nondimensional geometries of natural bars compared with those produced in the laboratory (Keulegan 1948)

(8)   The shapes of profiles across the continental shelf are less predictable than those within the more active zone of generally greater engineering interest. This may be due to the presence of bottom material different than sand (including rock and peat outcrops), the much greater time constants required for equilibration in these greater depths, the greater role of currents in shaping the profile and the effects of past sea level variations. In general, the slopes seaward of the more active zone are quite small if the bottom is composed of sand or smaller-sized materials. Figure III-3-14 presents three examples of profiles extending off the East and Gulf coasts of Florida. It is seen that at this scale, the profile may be approximated by a nearshore slope that extends to 5 to 18 m and milder seaward slopes, which are on the order of 1/2,000 to 1/10,000.

   c.   Interaction of structures with cross-shore sediment transport.

The structure that interacts most frequently with cross-shore sediment transport is a shore-parallel structure such as a seawall or revetment. During storm events, a characteristic profile fronting a shore-parallel structure is one with a trough at its base, as shown in Figure III-3-15, from Kriebel (1987) for a profile affected by Hurricane Elena in Pinellas County, Florida, in September 1985. This trough is due to large transport gradients immediately seaward of the structure. Although the <u>hydrodynamic</u> cause of this scour is not well-known, it has been suggested that it is due to a standing wave system with an antinode at the structure. A second, more heuristic explanation is that sand removal is prevented behind the seawall and the transport system removes sand from as near as possible to where removal would normally occur. Barnett and Wang (1988) have reported on a model study to evaluate the interaction of a seawall with the profile and have found that the additional volume represented by the scour trough is approximately 62 percent of what would have been removed landward of the seawall if it had not been present. During mild wave activity, it appears that the profile recovers nearly as it would have if the seawall had not been present. The reader is referred to the comprehensive review by Kraus (1988) for additional information on shore-parallel structures and their effects on the shoreline.

   d.   Methods of measuring beach profiles.

(1)   Introduction. Changes in beach and nearshore profiles are a result of cross-shore and longshore sediment transport. If the longshore gradients in the longshore component can be considered small, it is possible, through the continuity equation, to infer the volumetric cross-shore transport from two successive profile surveys.

(2)   Clausner, Birkemeier, and Clark (1986) have carried out a comprehensive field test of four nearshore survey systems, including: (a) the standard fathometer system, (b) the CRAB, which is a self-propelled platform on which a survey prism is mounted, (c) a sea sled, which also carries a prism but is towed by a boat or a cable from shore, and (d) a hydrostatic profiler, which utilizes a cable for towing and an oil-filled tube to sense the elevation difference between the shore and the location of the point being surveyed. Each of these systems is reviewed briefly below and their performance characteristics are described and summarized in Table III-3-2.

(a)   Fathometer. This method of measuring nearshore profiles requires knowledge of the water level as a reference datum. To provide a complete description of the active profile, fathometer surveys must be complemented with surveys of the shallow-water and above-water portions of the profile. In the field tests, the fathometer was mounted on a 47-ft vessel and the surveys were conducted under favorable wave conditions, which should result in a lower estimate of the error. Characteristics of this system and results obtained from the field measurements are presented in Table III-3-2.

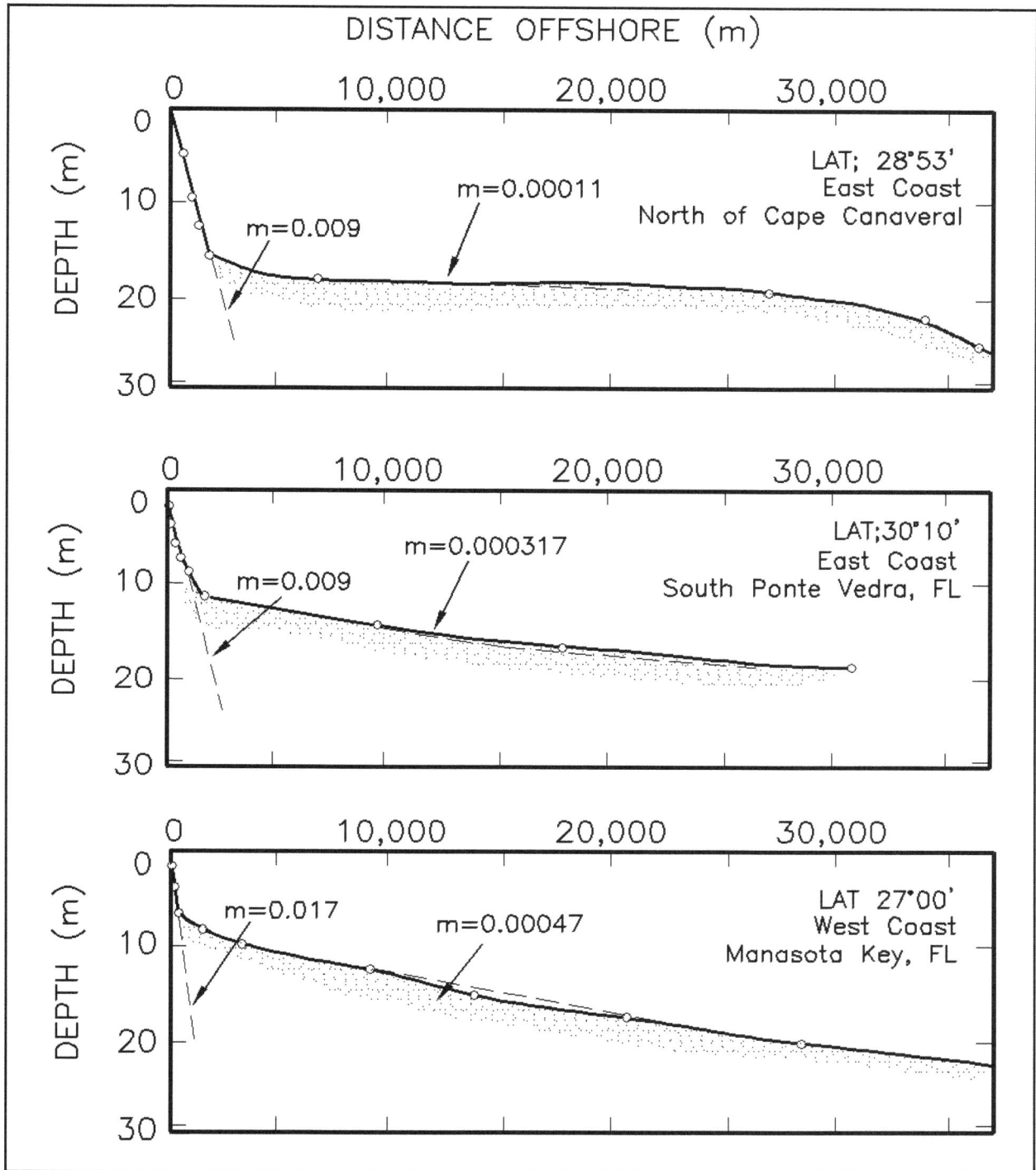

Figure III-3-14.   Profiles extending across the continental shelf for three locations along the East and Gulf coastlines of the United States (Dean 1987a)

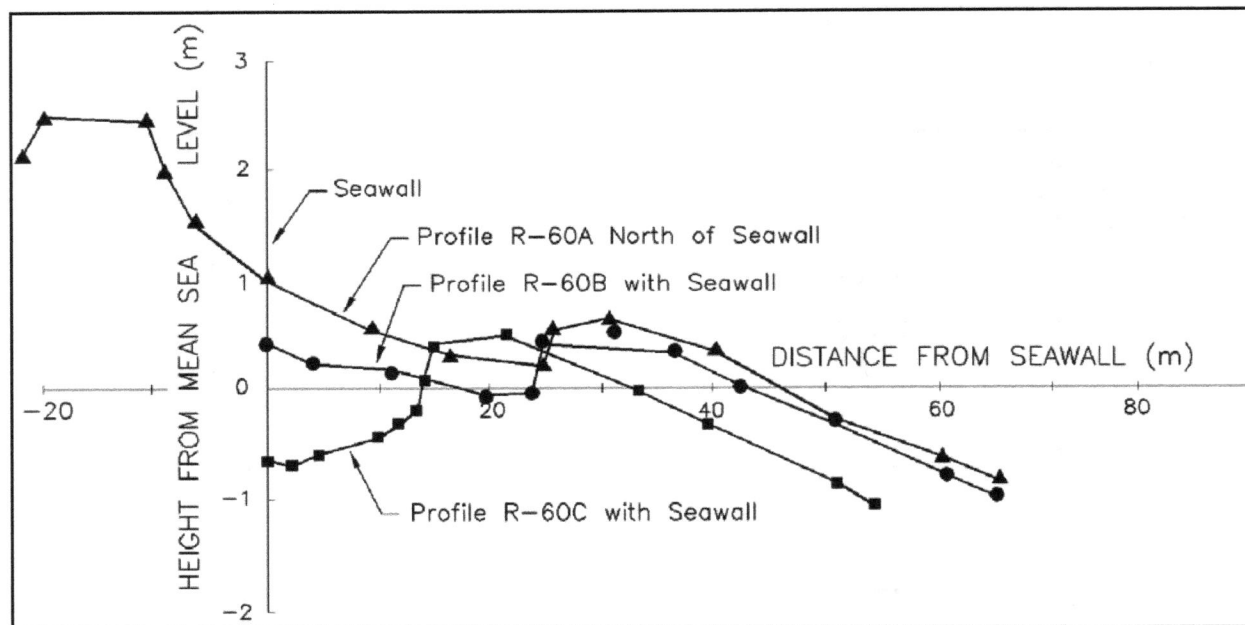

Figure III-3-15. Comparison of response of natural and seawalled profiles to Hurricane Elena, September 1985 (after Kriebel (1987))

Table III-3-2
Summary of Field Evaluation of Various Nearshore Survey Systems (Based on Clausner, Birkemeier, and Clark (1986))

| | Operating Requirements | | | Field Performance Characteristics | | | |
| | | | | | Vertical Accuracy | | Horizontal Accuracy |
System	Personnel Required	Wave Heights (m)	Number of Profiles Measured	Profiles per Day	Average Difference From Mean (cm)	Average Vertical Envelope (cm)	Average Distance Off Line (m)
Fathometer	4	< 1	6 (5)[1]	16	9 (6)[1]	31 (20)[1]	1.3
CRAB	2	< 2	5	7	2	5	0.4
Sea Sled	3 - 4	< 1	5	9	1	3	1.6
Hydrostatic Profiler	2 - 3	< 1	4	3	3	7	3.6

[1] Based on the smoothed analog records. All other fathometer data based on digital records.

(b) CRAB. The CRAB (Coastal Research Amphibious Buggy) is a self-propelled vehicle that has a survey prism mounted on it and is used in conjunction with a laser survey system. At the time of the Clausner report (Clausner, Birkemeier, and Clark 1986), it was necessary to stop the CRAB to take a reading. More recently, the system has been upgraded to an automatic self-tracking mode such that the CRAB can be moved continuously and readings taken at predetermined time increments. Since the CRAB avoids the need for a water level datum, the vertical accuracy is inherently superior to that of fathometer measurements.

(c) Sea sled. The sea sled incorporates many of the inherent survey advantages of the CRAB, since dependency on the water level is avoided. The major difference is that the CRAB is self-propelled whereas the sea sled is towed by either a boat or a truck on the shore. Since the sea sled is dependent on some vehicle to transport it through the surf zone and this vehicle is usually a boat, it will be more limited by wave conditions than the CRAB, which can operate in sea states of 2 m.

(d) Hydrostatic profiler. The hydrostatic profiler was developed by Seymour and Boothman (1984) and consists of a long (about 600 m) oil-filled tube extending from the shoreline to a small weighted sled at the measurement location. A pressure sensor at the sled "weighs" the vertical column of oil from the shore to the sled location which can be interpreted as the associated elevation difference. In general, the hydrostatic profiler has not been widely used due to inherent limitations in its performance, related to sensitivity to pressure surges and to longshore currents.

(2) Summary. In summary, referring to Table III-3-2, the CRAB emerges as the overall best system. The sea sled provides slightly better overall vertical accuracy; however, as noted, the CRAB now utilizes an automatic tracking mode, which should reduce possibility of human-induced error. The main disadvantages of the CRAB system are the limited availability of such systems and the difficulties of transporting from one site to another. The reader is referred to the report by Clausner, Birkemeier, and Clark (1986) for additional details of the four systems and the results of the field tests.

III-3-3. Engineering Aspects of Beach Profiles and Cross-shore Sediment Transport

a. Introduction. Previous sections have discussed the natural characteristics of beach profiles in equilibrium, the effects that cause disequilibrium, and the associated profile changes. Also shown in Figure III-3-2 were the numerous possible engineering applications of equilibrium beach profiles. This section presents some of the applications, illustrates these with examples, and investigates approaches to calculation of cross-shore sediment transport and the associated profile changes.

b. Limits of cross-shore sand transport in the onshore and offshore directions.

(1) The long-term and short-term limits of cross-shore sediment transport are important in engineering considerations of profile response. During short-term erosional events, elevated water levels and high waves are usually present and the seaward limit of interest is that to which significant quantities of sand-sized sediments are transported and deposited. It is important to note that sediment particles are in motion to considerably greater depths than those to which significant profile readjustment occurs. This readjustment occurs most rapidly in the shallow portions of the profile and, during erosion, transport and deposition from these areas cause the leading edge of the deposition to advance into deeper water. This is illustrated in Figure III-3-16 from Vellinga (1983), in which it is seen that with progressively increasing time, the evolving profile advances into deeper and deeper water. It is also evident from this figure that the rate of profile evolution is decreasing consistent with an approach to equilibrium. For predicting cross-shore profile change, the depth of limiting motion is not that to which the sediment particles are disturbed but rather these award limit to which the depositional front has advanced. Vellinga recommends that this depth be 0.75 H_s in which H_s is the deepwater significant wave height computed from the breaking wave height using linear water wave theory. In general, the limit of effective transport for short-term (storm) events is commonly taken as the breaking depth h_b based on the significant wave height.

(2) The onshore limit of profile response is also of interest as it represents the maximum elevation and landward limit of sediment transport. During normal erosion/accretion cycles, the upper limit of significant beach profile change coincides with the wave runup limit. Under constructive conditions, as the beach face builds seaward, this upper limit of sediment deposition is usually well-defined in the form of a depositional

beach berm. During erosion conditions, the berm may retreat more or less uniformly. In some cases, the berm may be so high that runup never reaches its crest, in which case an erosion scarp will form above the runup limit. This is also evident in the case of eroding dunes, which are not overtopped by wave runup. In these cases, the slope of the eroding scarp may be quite steep, approaching vertical in some cases. A common assumption is that the eroding scarp will form at more or less the angle of repose of the sediment. Vellinga (1983), based on results shown in Figure III-3-16, suggests adopting a 1:1 slope for this erosion scarp. In other cases, the berm may be significantly overtopped by either the water level (storm surge) or by the wave runup. If overwash occurs, the landward limit may be controlled by the extent to which the individual uprush and overwash events are competent to transport sediment. Often this distance is determined by loss of transporting power due to percolation into the beach or by water impounded by the overwash event itself. In the latter case, the landward depositional front will advance at more or less the angle of repose into the impounded water.

(3) The seaward limit of effective profile fluctuation over long-term (seasonal or multi-year) time scales is a useful engineering concept and is referred to as the "closure depth," denoted by h_c. Based on laboratory and field data, Hallermeier (1978, 1981) developed the first rational approach to the determination of closure depth. He defined two depths, the shallowest of which delineates the limit of intense bed activity and the deepest seaward of which there is expected to be little sand transport due to waves. The shallower of the two appears to be of the greatest engineering relevance and will be discussed here. Based on correlations with the Shields parameter, Hallermeier defined a condition for sediment motion resulting from wave conditions that are relatively rare. Effective significant wave height H_e and effective wave period T_e were based on conditions exceeded only 12 hr per year; i.e., 0.14 percent of the time. The resulting approximate equation for the depth of closure was determined to be

$$h_c = 2.28 H_e - 68.5 \left(\frac{H_e^2}{g T_e^2} \right)$$ (III-3-9)

in which H_e can be determined from the annual mean significant wave height $\bar{H}$ and the standard deviation of significant wave height σ_H as

$$H_e = \bar{H} + 5.6 \sigma_H$$ (III-3-10)

(4) Based on this relationship, Hallermeier also proposed a form of Equation 3-9 that did not depend on the effective wave period in the form

$$h_c = 2\bar{H} + 11 \sigma_H$$ (III-3-11)

(5) Birkemeier (1985) evaluated Hallermeier's relationship using high-quality field measurements from Duck, NC, and found that the following simplified approximation to the effective depth of closure provided nearly as good a fit to the data

$$h_c = 1.57 H_e$$ (III-3-12)

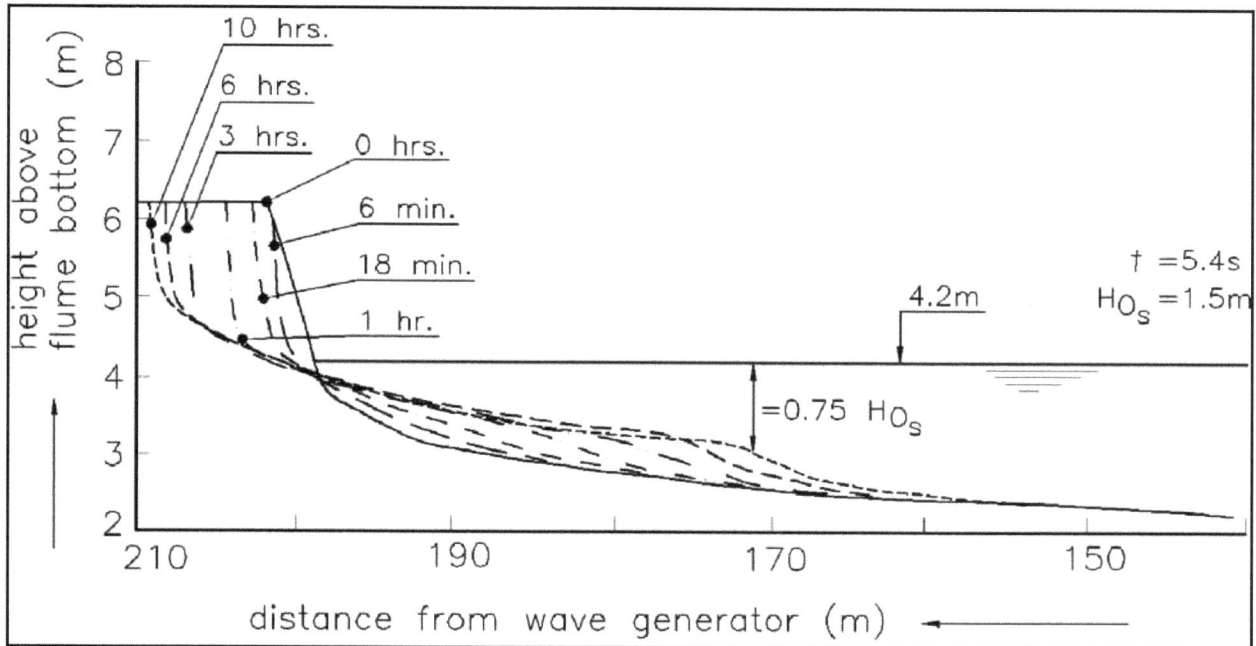

Figure III-3-16. Erosional profile evolution, large wave tank results (Vellinga 1983)

(6)   In the applications to follow, it will be assumed that h_c is an appropriate representation of the closure depth for profile equilibration and for significant beach profile change over long time scales. This quantity will be denoted as h_* in most of the examples presented when applied to beach nourishment problems. For short-term profile changes such as those that occur during a storm, the breaking depth h_b will be assumed to delineate the active profile. It should be noted that other approaches to "channel depth" are discussed in the literature (Hands 1983).

c.   *Quantitative description of equilibrium beach profiles.*

(1)   Various models have been proposed for representing equilibrium beach profiles (EBP). Some of these models are based on examination of the geometric characteristics of profiles in nature and some attempt to represent in a gross manner the forces active in shaping the profile. One approach that has been utilized is to recognize the presence of the constructive forces and to hypothesize the dominance of various destructive forces. This approach can lead to simple algebraic forms for the profiles for testing against profile data.

(2)   Dean (1977) has examined the forms of the EBPs that would result if the dominant destructive forces were one of the following:

(a)   Wave energy dissipation per unit water volume.

(b)   Wave energy dissipation per unit surface area.

(c)   Uniform average longshore shear stress across the surf zone. It was found that for all three of these destructive forces, by using linear wave theory and a simple wave breaking model, the EBP could be represented by the following simple algebraic form

$$h = A y^n \tag{III-3-13}$$

in which A, representing a sediment scale parameter, depends on the sediment size D. This form with an exponent n equal to 2/3 had been found earlier by Bruun (1954) based on an examination of beach profiles in Denmark and in Monterey Bay, CA. Dean (1977) found the theoretical value of the exponent n to be 2/3 for the case of wave energy dissipation per unit volume as the dominant force and 0.4 for the other two cases. Comparison of Equation 3-13 with approximately 500 profiles from the east coast and Gulf shorelines of the United States showed that, although there was a reasonably wide spread of the exponents n for the individual profiles, a value of 2/3 provided the best overall fit to the data. As a result, the following expression is recommended for use in describing equilibrium beach profiles

$$h = A y^{\frac{2}{3}} \tag{III-3-14}$$

This allows the appealing interpretation that the wave energy dissipation per unit water volume causes destabilization of the sediment particles through the turbulence associated with the breaking waves. Thus dynamic equilibrium results when the level of destabilizing and constructive forces are balanced.

(3) The sediment scale parameter A and the equilibrium wave energy dissipation per unit volume D_* are related by (Dean 1991)

$$A = \left[\frac{24}{5} \frac{D_*}{\rho g \sqrt{g} \kappa^2} \right]^{\frac{2}{3}} \tag{III-3-15}$$

(4) Moore (1982) and Dean (1987b) have provided empirical correlations between the sediment scale parameter A as a function of sediment size D and fall velocity w_f as shown in Figure III-3-17. These results are based on a least-squares fit of Equation 3-14 to measured beach profiles. Figure III-3-18 presents an expanded version of the A versus D relationship for grain sizes more typical of beach sands and Table III-3-3 provides a tabulation of A values over the size range $D = 0.10$ mm to $D = 1.09$ mm. Although Table III-3-3 provides A values to four decimal places at diameter increments of 0.01 mm, this should not be interpreted as signifying that understanding of EBP justifies this level of quantification. Rather the values are presented for consistency by different users and possibly for use in sensitivity tests.

(5) The equilibrium profile parameter A may also be correlated to sediment fall velocity. In Figure III-3-17, a relationship is suggested between A and w_f that is valid over the entire range of sediment sizes shown. Kriebel, Kraus, and Larson (1991) developed a similar correlation over a range of typical sand grain sizes from $D = 0.1$ mm to $D = 0.4$ mm and found the following relationship

$$A = 2.25 \left(\frac{w_f^2}{g} \right)^{\frac{1}{3}} \tag{III-3-16}$$

(6) This dependence of A on fall velocity to the two-thirds power has also been suggested by Hughes (1994) based on dimensional analysis.

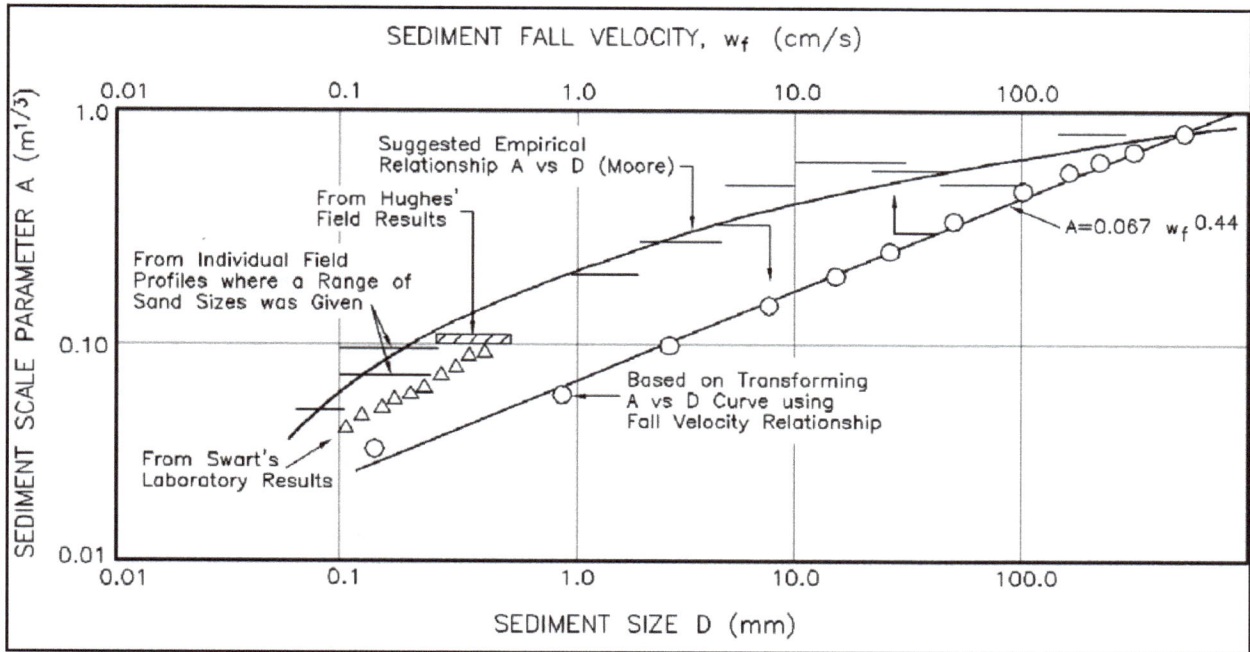

Figure III-3-17. Variation of sediment scale parameter *A* with sediment size *D* and fall velocity *w*f (Dean 1987b)

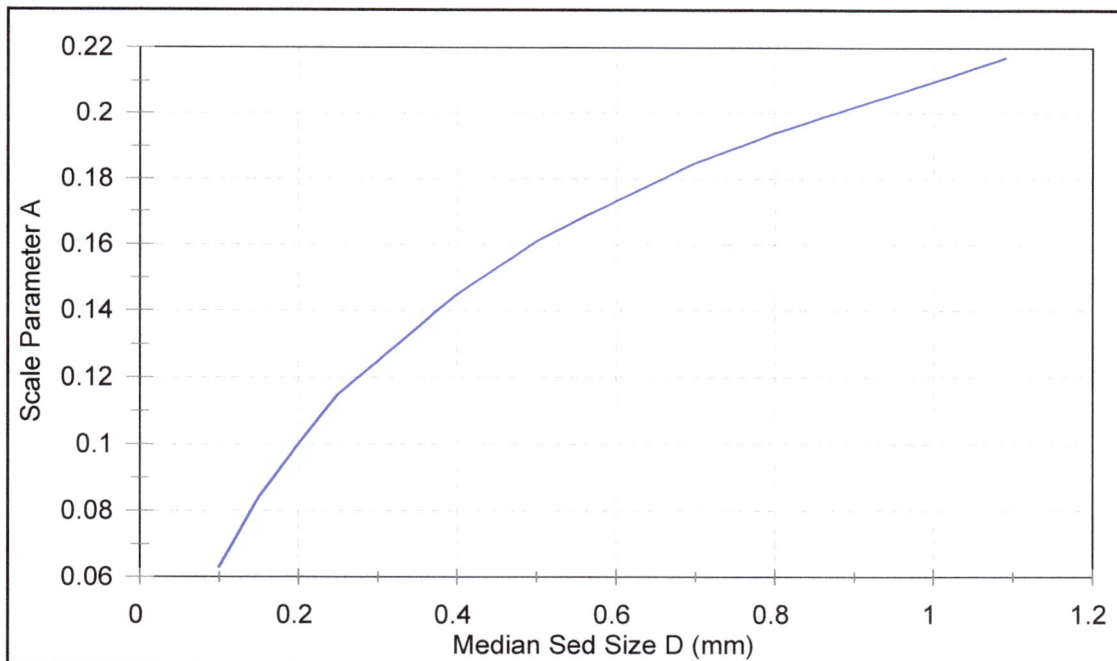

Figure III-3-18. Variation of sediment scale parameter *A*(*D*) with sediment size *D* for beach sand sizes (based on Dean 1978b, values recomputed by Dean, June 2001)

Table III-3-3
Summary of Recommended *A* Values (Units of *A* Parameter are m$^{1/3}$)

D(mm)	0.00	0.01	0.02	0.03	0.04	0.05	0.06	0.07	0.08	0.09
0.1	0.063	0.0672	0.0714	0.0756	0.0798	0.084	0.0872	0.0904	0.0936	0.0968
0.2	0.100	0.103	0.106	0.109	0.112	0.115	0.117	0.119	0.121	0.123
0.3	0.125	0.127	0.129	0.131	0.133	0.135	0.137	0.139	0.141	0.143
0.4	0.145	0.1466	0.1482	0.1498	0.1514	0.153	0.1546	0.1562	0.1578	0.1594
0.5	0.161	0.1622	0.1634	0.1646	0.1658	0.167	0.1682	0.1694	0.1706	0.1718
0.6	0.173	0.1742	0.1754	0.1766	0.1778	0.179	0.1802	0.1814	0.1826	0.1838
0.7	0.185	0.1859	0.1868	0.1877	0.1886	0.1895	0.1904	0.1913	0.1922	0.1931
0.8	0.194	0.1948	0.1956	0.1964	0.1972	0.198	0.1988	0.1996	0.2004	0.2012
0.9	0.202	0.2028	0.2036	0.2044	0.2052	0.206	0.2068	0.2076	0.2084	0.2092
1.0	0.210	0.2108	0.2116	0.2124	0.2132	0.2140	0.2148	0.2156	0.2164	0.2172

Notes:
(1) The *A* values above, some to four places, are not intended to suggest that they are <u>known</u> to that accuracy, but rather are presented for consistency and sensitivity tests of the effects of variation in grain size.
(2) As an example of use of the values in the table, the *A* value for a median sand size of 0.24 mm is: $A = 0.112$ m$^{1/3}$. To convert *A* values to feet$^{1/3}$ units, multiply by $(3.28)^{1/3} = 1.49$.

(7) There are two inherent limitations of Equation 3-14. First, the slope of the beach profile at the water line (y=0) is infinite. Second, the beach profile form is monotonic; i.e., it cannot represent bars. It has been shown that the first limitation can be overcome by recognizing gravity as a significant destabilizing force when the profile becomes steep. In this case with the beach face slope denoted as m_o, the form is

$$y = \frac{h}{m_o} + \left(\frac{h}{A}\right)^{\frac{3}{2}}$$
(III-3-17)

which, unfortunately, is significantly more cumbersome to apply. Larson (1988) and Larson and Kraus (1989) have shown that an EBP of the form of Equation 3-17 results by replacing the simple breaking wave model leading to Equation 3-14 by the more complex breaking model of Dally et al. (1985). Bodge (1992) and Komar and McDougal (1994) have each proposed slightly different forms of an <u>exponential</u> beach profile. The form proposed by Bodge is

$$h(y) = h_o(1 - e^{-ky})$$
(III-3-18)

in which h_o is the asymptotic depth at a great offshore distance, and k is a decay constant. The form suggested by Komar and McDougal is quite similar with $h_o = m_o/k$ in which m_o is the beach face slope. Bodge fit his profile to the averages of the ten data sets provided by Dean (1977) and found that the majority (about 80 percent) fit the exponential form better than the $Ay^{2/3}$ expression. The exponential forms have two free constants which are determined to provide the best fit and thus should agree better than for the case in which n is constrained to the 2/3 value. Since the exponential profile form requires determination of the two free parameters from the individual profile being represented, it can be applied in a diagnostic manner but not prognostically. In another approach Inman, Elwany, and Jenkins (1993) discuss the fitting of compound beach profile to a number of beaches. The curve-fitting approach requires up to seven free parameters and appears to require subjectivity in parameter choice. This method cannot be applied in a prognostic manner.

d. Computation of equilibrium beach profiles. The most simple application is the calculation of equilibrium beach profiles for various grain sizes, assumed uniform across the profile. This application is illustrated by the following example.

The extension of the equilibrium profile form to cases where the grain size varies across the profile is discussed in Dean (1991).

e. Application of equilibrium profile methods to nourished beaches.

(1) In the design of beach nourishment projects, it is important to estimate the dry beach width after profile equilibration. Most profiles are placed at slopes considerably steeper than equilibrium and the equilibration process, consisting of a redistribution of the fill sand across the active profile out to the depth of closure, occurs over a period of several years. In general, the performance of a beach fill, in terms of the resulting gain in dry beach width relative to the volume of sand placed on the beach, is a function of the compatibility of the fill sand with the native sand. Based on equilibrium beach profile concepts, it should be evident that since profiles composed of coarser sediments assume steeper profiles, beach fills using coarser sand will require less sediment to provide the same equilibrium dry beach width Δy than fills using sediment that is finer than the native sand.

(2) It can be shown that three types of nourished profiles are possible, depending on the volumes added and on whether the nourishment is coarser or finer than that originally present on the beach. These profiles are termed "intersecting," "nonintersecting," and "submerged," respectively, and are shown in Figure III-3-20. It can be shown that an intersecting profile requires the added sand to be coarser than the native sand, although this condition does not guarantee intersecting profiles, since the intersection may be at a depth in excess of the depth of closure. Nonintersecting or submerged profiles always occur if the sediment is of the same diameter or finer than the native sand.

(3) Several more general examples will assist in understanding the significance of the sand and volume characteristics. Denoting the sediment scale parameters for the <u>native</u> and <u>fill</u> sediments as A_N and A_F, respectively, Figure III-3-21 presents the variation in dry beach width for a native sand size of 0.20 mm and various fill diameters ranging from 0.15 mm to 0.40 mm. These results are illustrated for a closure depth h_* of 6 m, a berm height B of 2 m, and a volumetric addition per unit beach length of 340 m^3/m. In the upper panel, the fill sediment is coarser than the native sand and the profiles are intersecting, resulting in an equilibrium additional dry beach width of 92.4 m. In the second panel, the fill sand is of the same size as the native (nonintersecting profiles) and the added beach width is 45.3 m. The third and fourth panels illustrate the effects of further decreases in sediment sizes with an incipient submerged profile in the last panel. These examples have considered the effects only of cross-shore equilibration. In design of beach nourishment projects, the additional effects of more rapid spreading out of the nourishment project due to longshore sediment transport due to fine sediments should also be considered. The next generic example, presented in Figure III-3-22, illustrates the effects of adding greater amounts of sediment that are finer than the native. For small amounts, the profile is totally submerged. However, as greater and greater amounts are added, the landward extremity of the nourished profile advances toward land, and ultimately the profile becomes emergent.

EXAMPLE PROBLEM III-3-1

FIND:
The equilibrium beach profiles.

GIVEN:
Consider grain sizes of 0.3 mm and 0.66 mm.

SOLUTION:
From Figure III-3-18 and/or Table III-3-3, the associated A values are 0.125 m$^{1/3}$ and 0.18 m$^{1/3}$, respectively. Applying Equation 3-14, the two profiles are computed and are presented in Figure III-3-19. The profile composed of the coarser sand is considerably steeper than that for the finer material.

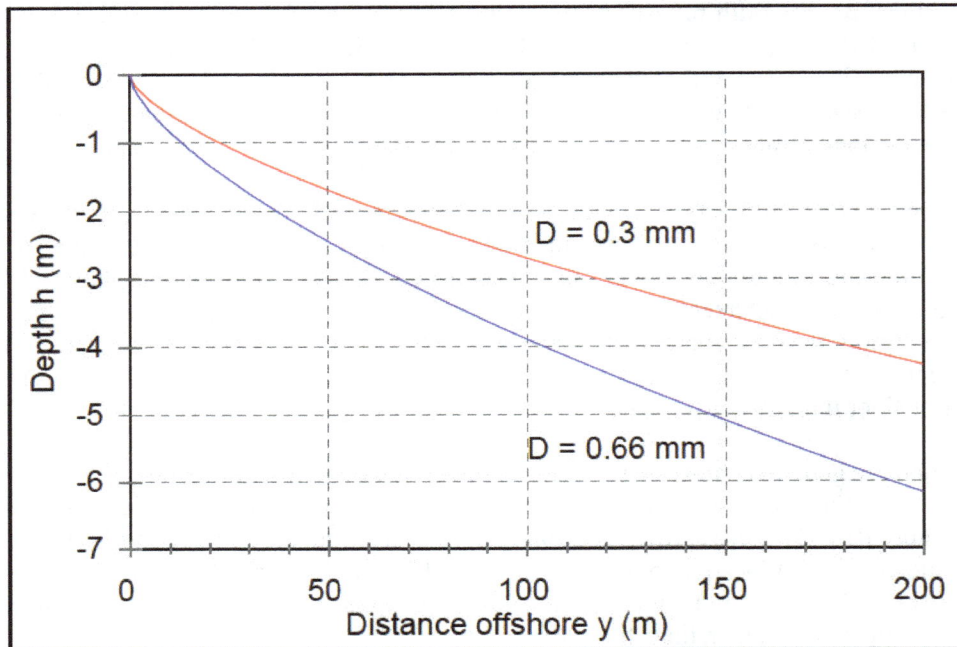

Figure III-3-19. Equilibrium beach profiles for sand sizes of 0.3 mm and 0.66 mm
A(D = 0.3 mm) = 0.125 m$^{1/3}$, A(D = 0.66 mm) = 0.18 m$^{1/3}$

f.   Quantitative relationships for nourished profiles.

(1)   In order to investigate the conditions of profile type occurrence and additional quantitative aspects, it is useful to define the following nondimensional quantities: $A' = A_F/A_N$, $\Delta y' = \Delta y/W_*$, $B' = B/h_*$, and $V' = V/(B\,W_*)$, where the symbol V denotes added volume per unit beach length, B is the berm height, and h_* is the depth to which the nourished profile will equilibrate as shown in Figure III-3-21. In general, this will be considered to be the closure depth. It is important to note that the width W_* is based on the native sediment scale parameter A_N as given by

$$W_* = \left(\frac{h_*}{A_N}\right)^{\frac{3}{2}}$$

(III-3-19)

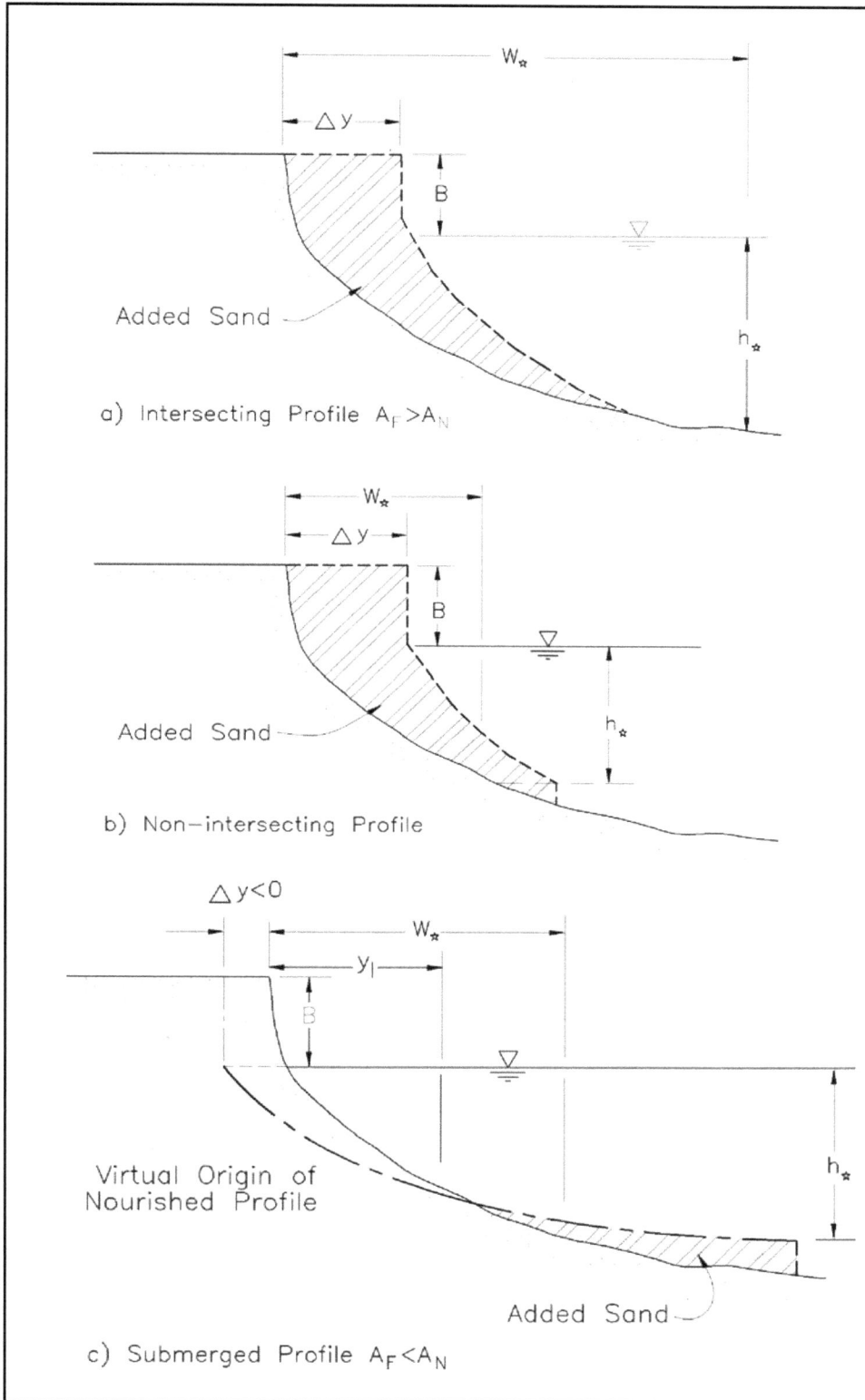

Figure III-3-20.   Three generic types of nourished profiles. (a) intersecting, (b) nonintersecting, and (c) submerged profiles (Dean 1991)

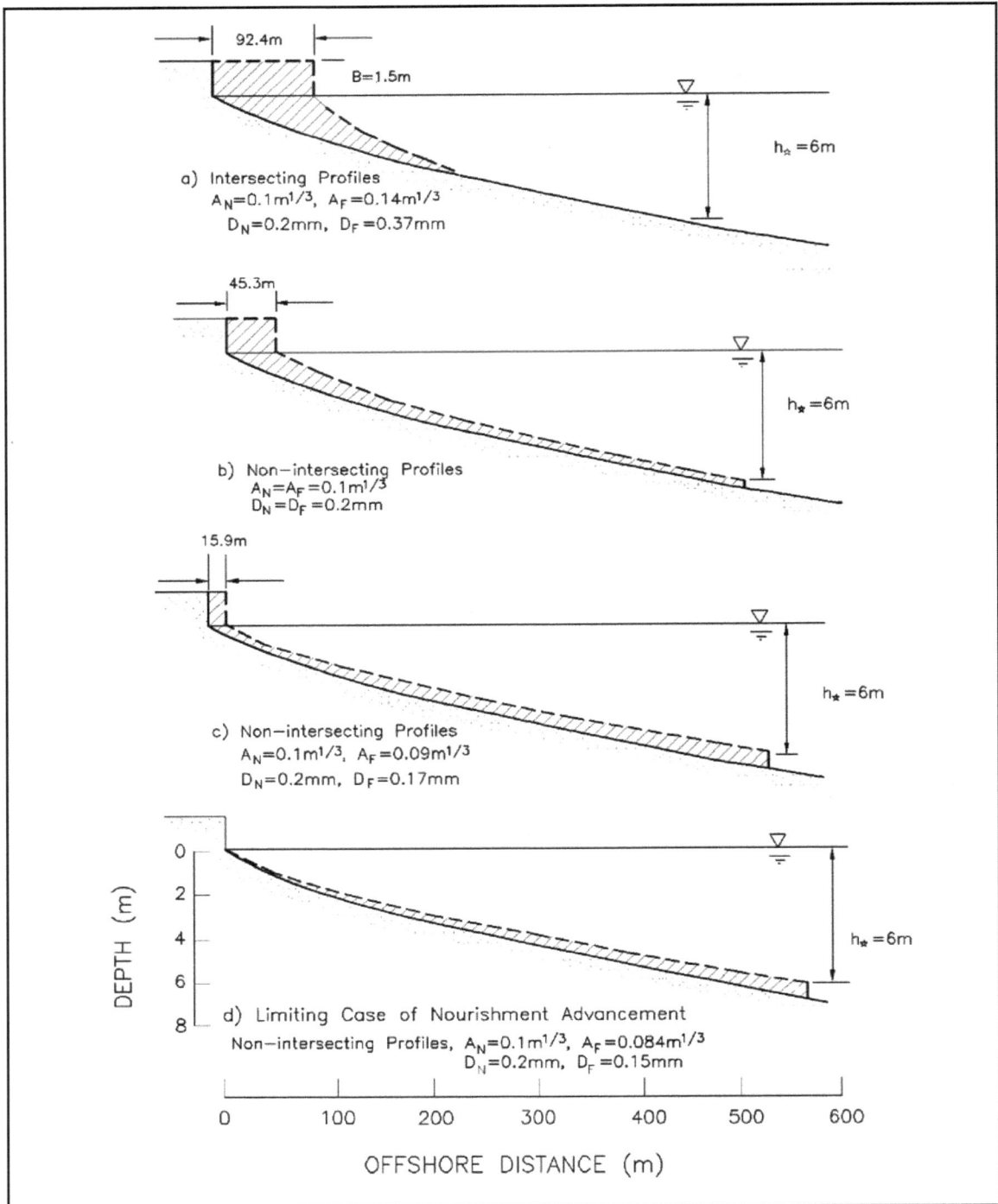

Figure III-3-21. Effect of nourishment material scale parameter A_F on width of resulting dry beach. Four examples of decreasing A_F with same added volume per unit beach length (Dean 1991)

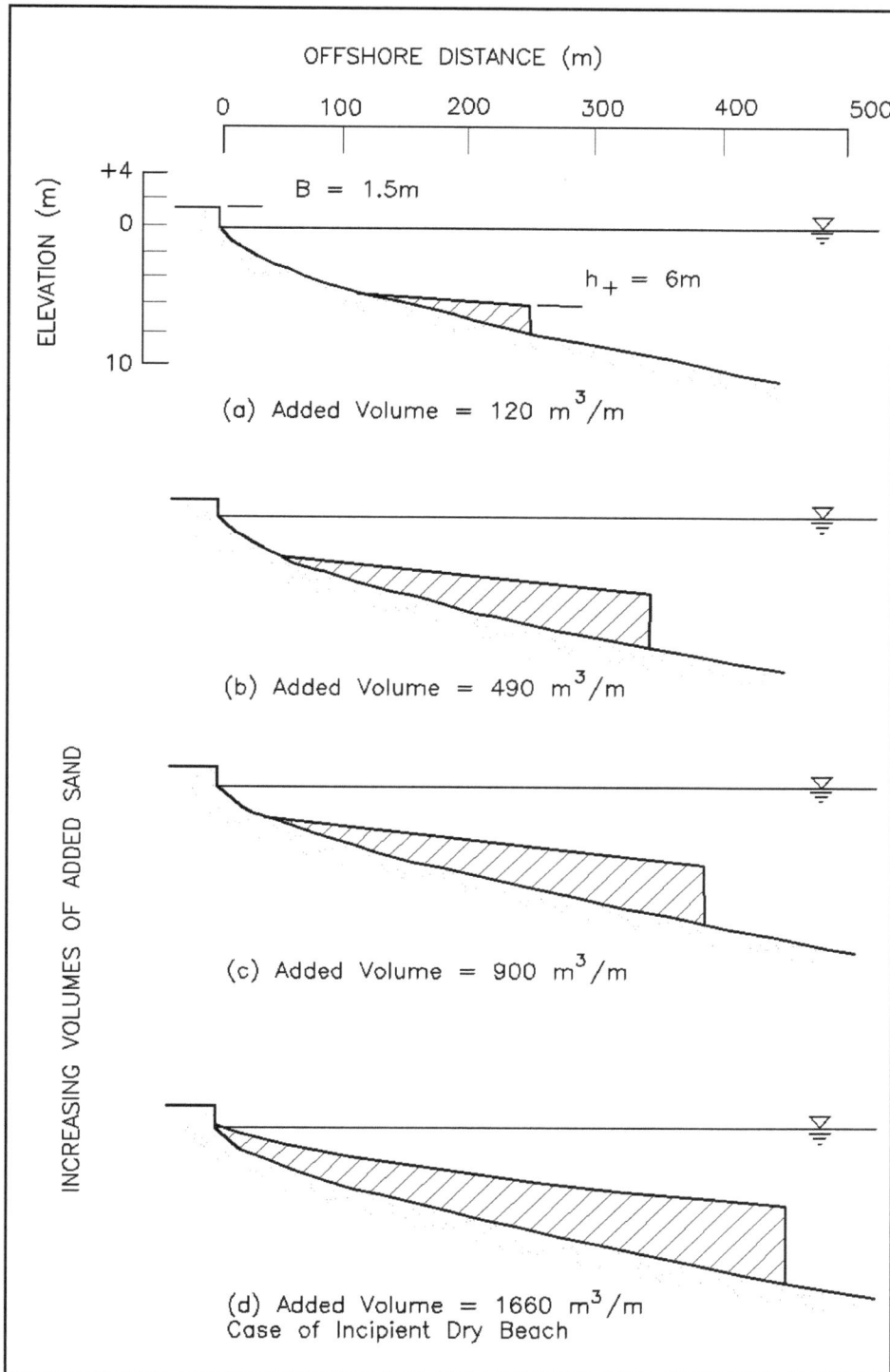

Figure III-3-22.   Effect of increasing volume of sand added on resulting beach profile. A_F = 0.1 m$^{1/3}$, A_N = 0.2 m$^{1/3}$, h_* = 6.0 m, B = 1.5 m (Dean 1991)

It is possible to show that the nondimensional equilibrium dry beach width $\Delta y'$ can be presented in terms of three nondimensional quantities

$$\Delta y' = f(B', V', A') \tag{III-3-20}$$

(2) The relationships governing the conditions for intersecting/nonintersecting profiles are

$$\Delta y' + \left(\frac{1}{A'}\right)^{\frac{3}{2}} - 1 \begin{cases} < 0, \text{ intersecting profiles} \\ > 0, \text{ non-intersecting profiles} \end{cases} \tag{III-3-21}$$

given that the fill sediment scale parameter is greater than or equal to the native sediment scale parameter.

The critical volume of sand delineating intersecting and nonintersecting profiles is

$$(V')_{cl} = \left(1 + \frac{3}{5B'}\right) \left[1 - \left(\frac{1}{A'}\right)^{\frac{3}{2}}\right] \tag{III-3-22}$$

which applies only for $A'>1$, since for $A'<1$, the profiles will always be nonintersecting although it should be recognized that nonintersecting profiles can also exist for $A'>1$. If $A'>1$, but $V' > V_{cl}$, then the profile will be nonintersecting. Also of interest is the critical volume of sand V_{c2} that will just yield a finite shoreline displacement for the case of sand that is finer than the native ($A'<1$)

$$(V')_{c2} = \frac{3}{5B'} \left(\frac{1}{A'}\right)^{\frac{3}{2}} \left(\frac{1}{A'} - 1\right) \tag{III-3-23}$$

(3) Figure III-3-23 presents these two critical volumes versus the scale parameter A' for the special case $B'=0.25$.

(4) For intersecting profiles, the nondimensional volume required to yield an advancement $\Delta y'$ is

$$V_1' = \Delta y' + \frac{3}{5B'} (\Delta y')^{\frac{5}{3}} \frac{1}{\left[1 - \left(\frac{1}{A'}\right)^{\frac{3}{2}}\right]^{\frac{2}{3}}} \tag{III-3-24}$$

This equation would apply for the example in Figure III-3-20a.

(5) For nonintersecting but emergent profiles, the corresponding volume V_2' is

$$V_2' = \Delta y' + \frac{3}{5B'} \left\{\left[\Delta y' + \left(\frac{1}{A'}\right)^{\frac{3}{2}}\right]^{\frac{5}{3}} - \left(\frac{1}{A'}\right)^{\frac{3}{2}}\right\} \tag{III-3-25}$$

This equation would apply for Figure III-3-20b.

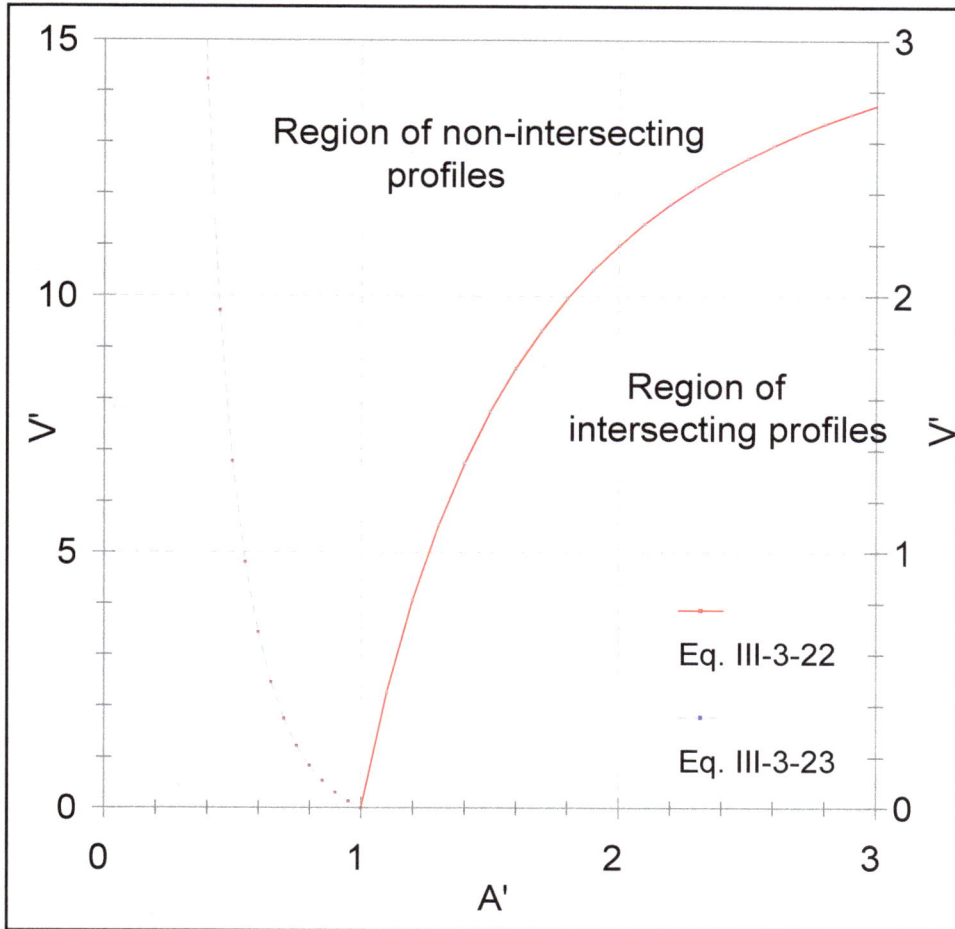

Figure III-3-23. (1) Volumetric requirement for finite shoreline advancement (Equation 3-23); (2) Volumetric requirement for intersecting profiles (Equation 3-22). Results presented for special case $B' = 0.25$

(6) For submerged profiles, referring to Figure III-3-20c, it can be shown that

$$\frac{\Delta y}{y_I} = 1 - \left(\frac{1}{A'}\right)^{\frac{3}{2}} \tag{III-3-26}$$

where $\Delta y' < 0$, $A' < 1$, and the nondimensional volume of sediment can be expressed as

$$V_4' = \frac{3}{5B'}\left\{\left[\Delta y' + \left(\frac{1}{A'}\right)^{\frac{3}{2}}\right]^{\frac{5}{3}} + \frac{(-\Delta y')^{\frac{5}{3}}}{\left[\left(\frac{1}{A'}\right)^{\frac{3}{2}}-1\right]^{\frac{2}{3}}} - \left(\frac{1}{A'}\right)^{\frac{3}{2}}\right\} \tag{III-3-27}$$

where $(\frac{1}{A'})^{3/2} \geq |\Delta y'|$

This equation would apply for Figure III-3-20c but is of limited value since no beach width would be added.

(7)   Equations 3-24 and 3-25 can be displayed in a useful form for calculating the volume required for a particular equilibrium additional dry beach width. However, as is evident from Equation 3-20, there are three independent variables: B', A', and V'. Thus, since only two independent variables can be displayed on a single plot, it is necessary to have a series of plots. Three are presented here, one each for $B' = 0.5$ (Figure III-3-24) $B' = 0.333$ (Figure III-325) and $B' = 0.25$ (Figure III-3-26). The information contained in these plots will be discussed by reference to Figure III-3-24.

(8)   The vertical axis is the nondimensional added beach width $\Delta y'$, the horizontal axis is the nondimensional sediment scale parameter A', and the isolines are the nondimensional volumes V'. For a given A' and V', the value of $\Delta y'$ is readily determined. It is seen that $\Delta y'$ increases with increasing V' and A'. The heavy dashed line delineates the regions of intersecting and nonintersecting profiles (Equation 3-23). With decreasing A' and constant V', the value of $\Delta y'$ decreases to the asymptotes for a submerged profile. Several examples will be presented illustrating the application of Figures III-3-24, III-3-25, and III-3-26.

   g.   Longshore bar formation and seasonal shoreline changes.

(1)   Longshore bars were discussed briefly in Part III-3-2. They are elongated mounds more or less parallel to the shoreline and are known to be more prevalent for storm conditions and for finer sediments. Bars may be present as single features or may occur as a series (Figure III-3-10). Additionally, bars can be seasonal or perennial. In most locations where bars are seasonal, their formation is associated with a seaward transport of sediment and a retreat of the shoreline. At a particular location, the amount of seasonal fluctuation depends on the number and intensity of storms during a particular year. Figure III-3-30 shows results of measurements by Dewall and Richter (1977) from Jupiter Island, Florida, where the seasonal fluctuations appear to be on the order of 15 m. Figure III-3-31, from Dewall (1979) shows shoreline and volume changes (above mean sea level) from Westhampton, Long Island, New York, where the seasonal changes may be on the order of 20 to 40 m.

(2)   Although the prediction of bar geometry and the associated shoreline changes have not advanced to a reliable stage, parameters have been proposed and correlated successfully with conditions for which bars form. Based on field observations, Dean (1973) hypothesized that sediment was suspended during the crest phase position and that if the fall time were less or greater than one-half the wave period, the net transport would be landward or seaward, respectively, resulting in bar formation in the latter case. This mechanism would be consistent with the wave-breaking cause. Further rationalizing that the suspension height would be proportional to the wave height resulted in identification of the so-called fall velocity parameter H_b/w_fT. Although there is no agreement on the cause of longshore bar formation, it appears to result from wave breaking, with edge waves and other phenomena proposed as possible causes.

(3)   Examination of small-scale laboratory data for which the deep water reference wave height H_o values were available led to the following relationship (Dean 1973) for offshore sediment transport leading to bar formation

$$\frac{H_o}{w_fT} \geq 0.85 \qquad\qquad\qquad \text{(III-3-28)}$$

(4)   Later, Kriebel, Dally, and Dean (1986) examined only prototype and large-scale laboratory data and found a constant of approximately 2.8 rather than 0.85 as in Equation 3-28. Kraus, Larson, and Kriebel (1991) examined only large wave tank data and proposed the following two relationships for bar formation

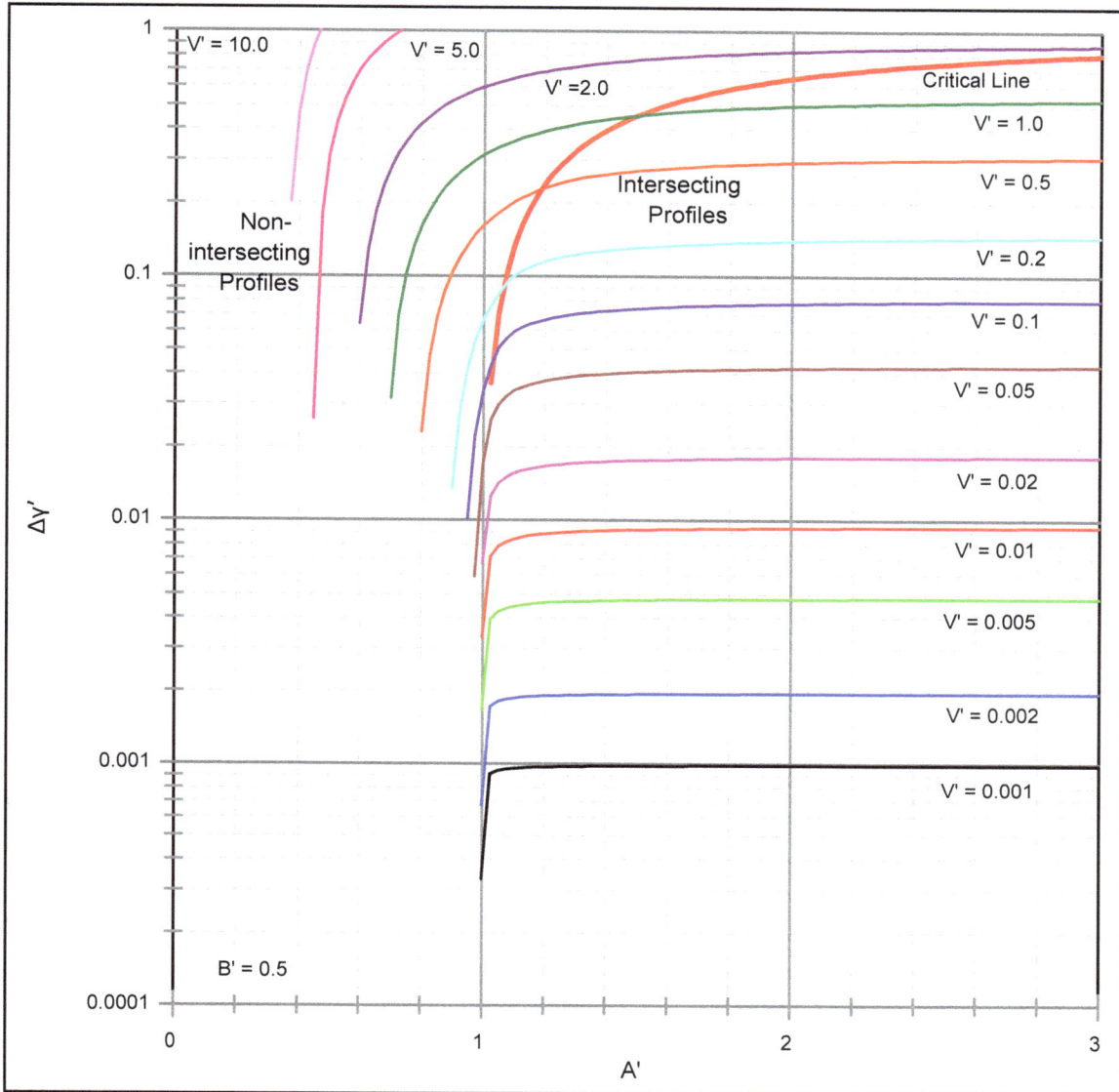

Figure III-3-24. Variation of nondimensional shoreline advancement $\Delta y/W_*$, with A' and V. Results shown for $H_o/B = 2.0$ ($B' = 0.5$) (based on Dean (1991), values recomputed by Dean, May 2001). Intersecting and non-intersecting profiles divided by critical line; definition sketches shown in Figure III-3-20.

$$\frac{H_o}{L_o} \geq 115 \left(\frac{\pi w_f}{gT} \right)^{\frac{3}{2}} \qquad\qquad\qquad \text{(III-3-29)}$$

and

$$\frac{H_o}{L_o} \leq 0.0007 \left(\frac{H_o}{w_f \, T} \right)^3 \qquad\qquad\qquad \text{(III-3-30)}$$

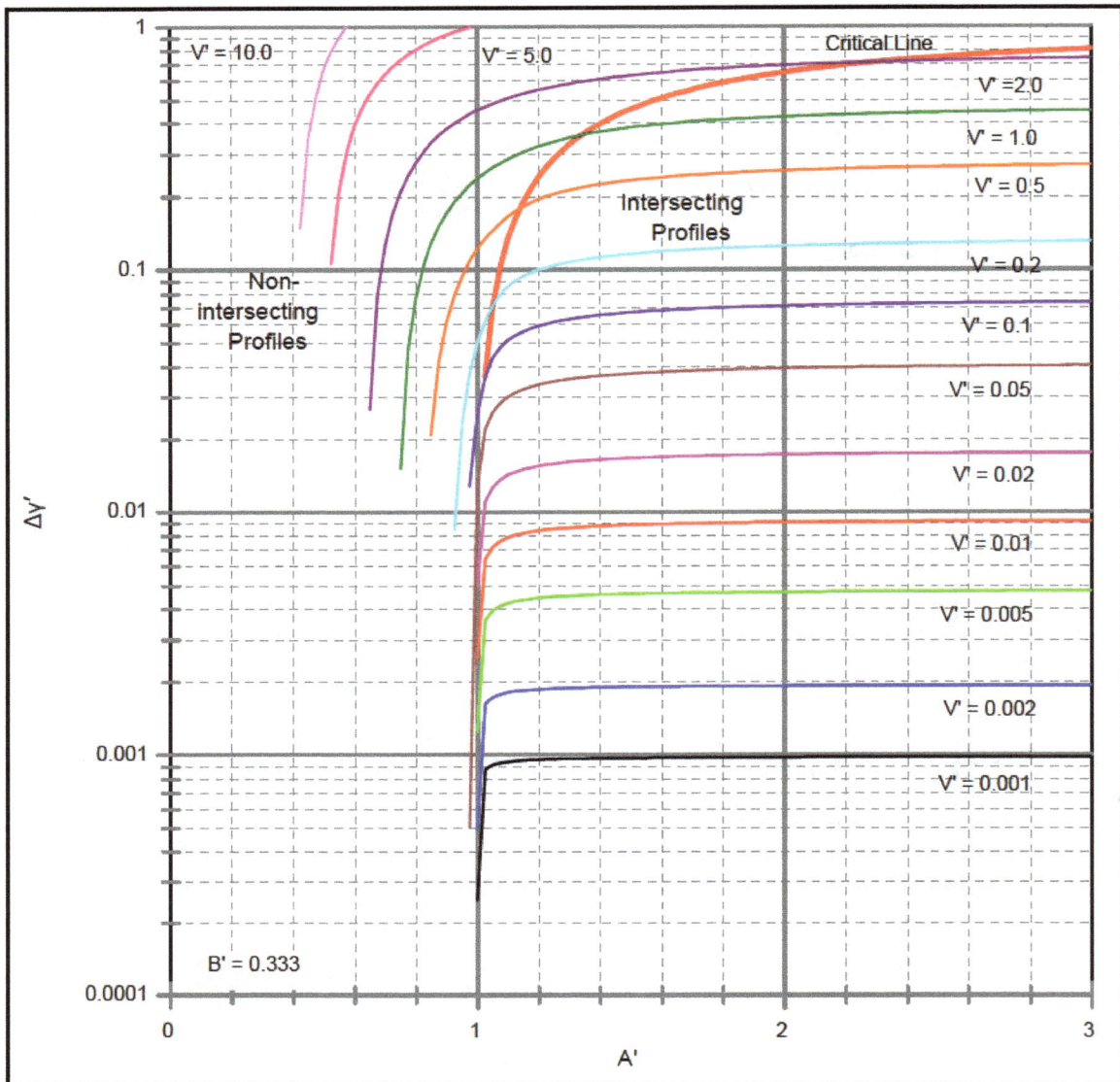

Figure III-3-25. Variation of nondimensional shoreline advancement Δy/W., with A′ and V. Results shown for H./B = 3.0 (B′ = 0.333) (based on Dean (1991), values recomputed by Dean, May 2001). Intersecting and non-intersecting profiles divided by critical line; definition sketches shown in Figure III-3-20.

in which H_o is the average deepwater wave height. For field data in which the significant deepwater wave height was used, the constant in Equation 3-30 was modified to

$$\frac{H_{o_s}}{L_o} \le 0.00027 \left(\frac{H_{o_s}}{w_f T} \right)^3 \tag{III-3-31}$$

(5) It is interesting that Equation 3-30 provides a better fit to the laboratory data than a fixed value of the fall velocity parameter; however, for field data, a fixed value of the fall velocity parameter provides a

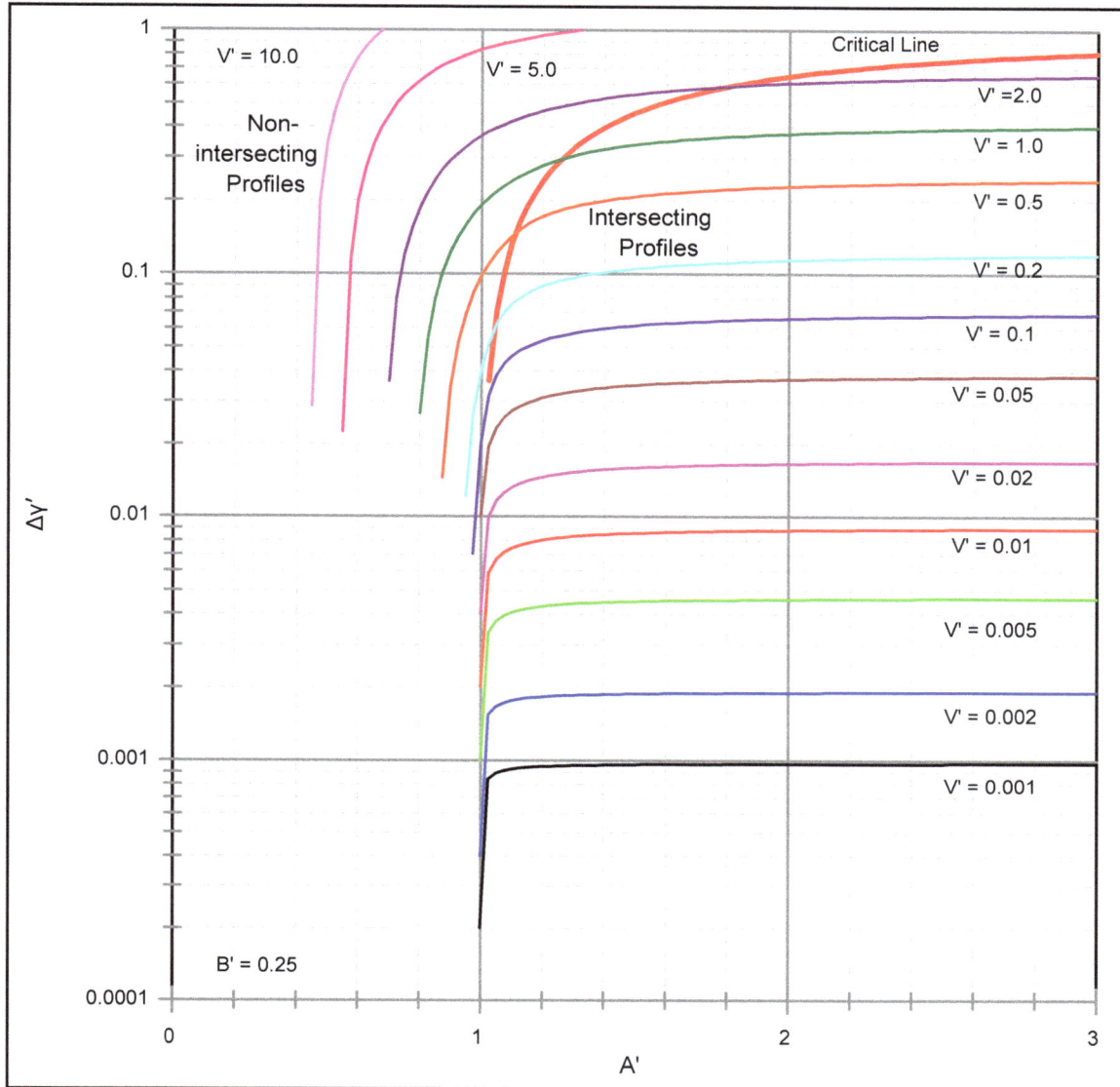

Figure III-3-26. Variation of nondimensional shoreline advancement $\Delta y/W$., with A′ and V. Results shown for H./B = 4.0 (B′ = 0.25), based on Dean (1991), values recomputed by Dean, May 2001). Intersecting and non-intersecting profiles divided by critical line; definition sketches shown in Figure III-3-20.

better fit than Equation 3-30 (Kraus, Larson, and Kriebel 1991). Dalrymple (1992) has shown that Equations 3-29 and 3-30 can be represented in terms of a single profile parameter P where

$$P = \frac{g H_o^2}{w_f^3 T} \qquad (III-3-32)$$

and that the criterion for bar formation is that P exceeds about 10,000.

EXAMPLE PROBLEM III-3-2

FIND:
   The equilibrated additional dry beach width Δy due to cross-shore transport.

GIVEN:
   $D_N = 0.2$ mm, $D_F = 0.24$ mm, $B = 2$ m, $V = 400$ m^3/m.
$\bar{H} = 1.5$ m, $\sigma_H = 0.64$ m as provided in WES and CEDRS databases discussed in Part II-8.

SOLUTION:
   Based on the above, the sediment scale parameters are determined from Figure III-3-18 and/or Table III-3-3 to be: $A_N = 0.1$ m$^{1/3}$, $A_F = 0.11$ m$^{1/3}$. The effective wave height H_e is determined from Equation 3-10 as:

$$H_e = \bar{H} + 5.6\,\sigma_H = 1.5 + (5.6)(0.64) \cong 5.1 \text{ m}$$

The closure depth h_* is determined from Equation 3-12 as:

$$h_* = h_C = 1.57\,H_e = (1.57)\,(5.11) \cong 8 \text{ m}$$

The reference width of effective motion W_* is based on Equation 3-19

$$\boldsymbol{W_* = \left(\frac{h_*}{A_N}\right)^{\frac{3}{2}} = \left(\frac{8}{0.1}\right)^{\frac{3}{2}} \approx 716\,m}$$

With this information, it is possible to determine the following nondimensional quantities:

$$B' = B/h_* = 2/8 = 0.25, \quad A' = A_F/A_N = 0.11/0.1 = 1.1$$

$$V' = V/(BW_*) = 400/(2)(716) \cong 0.28$$

Since $B' = 0.25$, Figure III-3-26 is applicable and for $A' = 1.1$ and $V' = 0.28$, it is found that $\Delta y' = 0.092$. Thus $\Delta y = (0.092)(716) \cong 65.9$ m. Also, it is seen from Figure III-3-26 that this solution is near the boundary of the intersecting/nonintersecting profiles. The native and nourished profiles are shown in Figure III-3-27. The solution is next carried out with the appropriate equations for comparison with the graphical procedure. Applying Equation 3-22 to determine whether the profiles will be intersecting or nonintersecting

$$\boldsymbol{(V')_{cl} = \left(1 + \frac{3}{5B'}\right)\left[1 - \left(\frac{1}{A'}\right)^{\frac{3}{2}}\right]}$$

$$\boldsymbol{= \left(1 + \frac{3}{(5)(0.25)}\right)\left[1 - \left(\frac{1}{1.1}\right)^{\frac{3}{2}}\right] \approx 0.45}$$

compared with the applied V' of 0.28. Thus since $V' < (V')_{cl}$ the solution is an intersecting profile. Applying Equation 3-24 requires an iterative solution for $\Delta y'$. This equation can be reduced to

$$0.28 = \Delta y' + 9.20(\Delta y')^{5/3}$$

the solution to which yields $\Delta y' = 0.0955$ or $\Delta y = 68.4$ m

(Continued)

Example Problem III-3-2 (Concluded)

Thus, for this example, the graphical solution is reasonable. The calculated intersection distance for the two profiles y_I is determined from

$$h_I = A_N y_I^{\frac{2}{3}} = A_F(y_I - \Delta y)^{\frac{2}{3}}$$

or

$$y_I = \frac{\Delta y \left(\frac{A_F}{A_N}\right)^{\frac{3}{2}}}{\left(\frac{A_F}{A_N}\right)^{\frac{3}{2}} - 1} = \frac{68.4(1.1)^{\frac{3}{2}}}{(1.1)^{\frac{3}{2}} - 1} = 513 \text{ m}$$

and

$$h_I = A_N y_I^{\frac{2}{3}} \approx 6.4 \text{ m}$$

By comparison, the corresponding values from the graphical solution are $y_I = 495$ m and $h_I = 6.25$ m. Figure III-3-27 presents the results of the graphical solution.

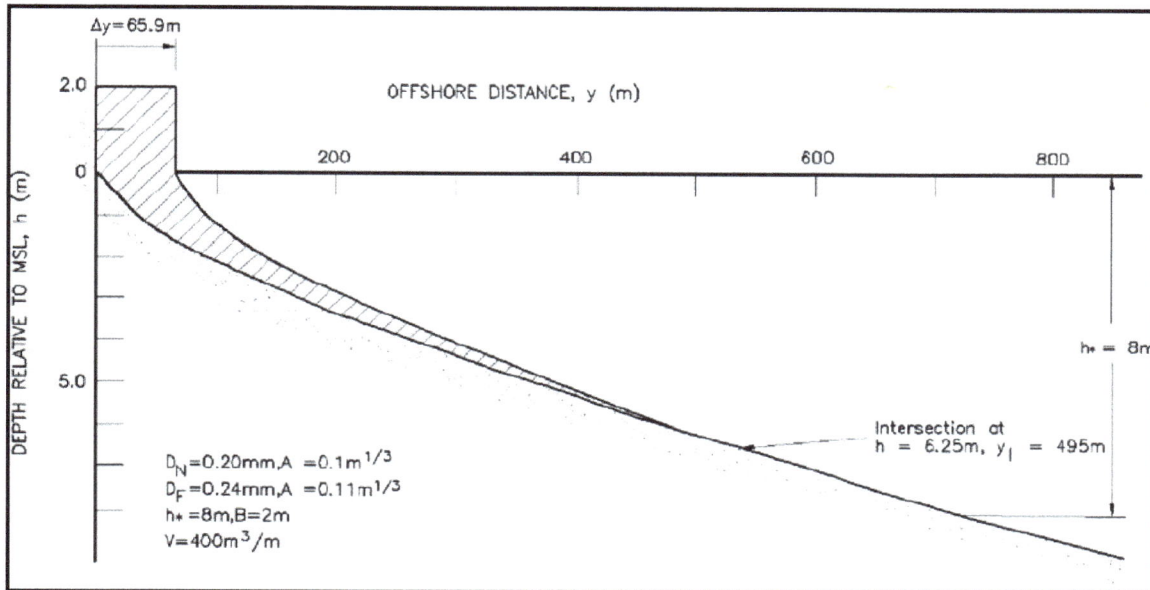

Figure III-3-27.   Nourishment with coarser sand than native (intersecting profiles)

EXAMPLE PROBLEM III-3-3

FIND:
   V, the volume of sand necessary to achieve the additional dry beach width Δy of 50 m.

GIVEN:
   $B=2$ m, $D_N = D_F = 0.25$ mm, $\Delta y = 50$ m, where $\bar{H}=1.5$ m, $\sigma_H = 0.4$ m.

SOLUTION:

The value of h_* is computed as:

$$h_* = 1.57\,(\bar{H} + 5.6\,\sigma_H) = 1.57\,[1.5 + (5.6)(0.4)] \cong 5.9 \text{ m}$$

For these values, the associated A values are determined from Figure III-3-18 and/or Table III-3-3 to be: $A_N = A_F = 0.115$ m$^{1/3}$. The reference width of active motion W_* is

$$W_* = \left(\frac{h_*}{A_N}\right)^{\frac{3}{2}} \approx 367 \text{ m}$$

and the required nondimensional quantities are:

$$\Delta y/W_* \cong 50/367 \cong 0.136,\; B' = B/h_* \cong 0.34,\; A' = A_F/A_N = 0.115/0.115 = 1.0$$

Since the value of B' lies between the two values represented in Figures III-3-24 and III-3-26, it is necessary to interpolate. The values from these two figures are: $V'(B' = 0.25) = 0.75$, $V'(B' = 0.5) = 0.35$. Interpolating linearly, V' is found to be 0.606 for the desired B' value of 0.34, from which the volume V is determined as $V = V'B\,W_* = (0.606)\,(2)\,(367) \cong 445$ m^3/m. Since the fill and native sediments are of the same size, it is clear that the two profiles will be nonintersecting and that Equation 3-25 can be used to compute the nondimensional volume directly

$$V' = \Delta y' + \frac{3}{5B'}\left\{\left[\Delta y' + \left(\frac{1}{A'}\right)^{\frac{3}{2}}\right]^{\frac{5}{3}} - \left(\frac{1}{A'}\right)^{\frac{3}{2}}\right\}$$

$$= 0.136 + \frac{3}{(5)(0.34)}\left\{[0.136 + 1]^{\frac{5}{3}} - 1\right\} \approx 0.554$$

(Continued)

Example Problem III-3-3 (Concluded)

which yields a volume of sand, $V = V'B W_* = (0.554)(2)(367) = 407 \text{ m}^3/\text{m}$, which is about 9 percent less than the value determined by interpolating between the values from the two figures. Since in this particular case, the two sand sizes are the same and thus <u>every</u> contour must be displaced by the same amount, an approximate equation for the required volume density is the product of the contour displacement (50 m) and the full depth of active motion: $V = (\Delta y)(h_* + B) = (50)(5.9 + 2) = 395 \text{ m}^3/\text{m}$. This result differs slightly from the values determined above because a small wedge-shaped sand volume has been neglected near the depth of closure. The native and nourished profiles are plotted in Figure III-3-28.

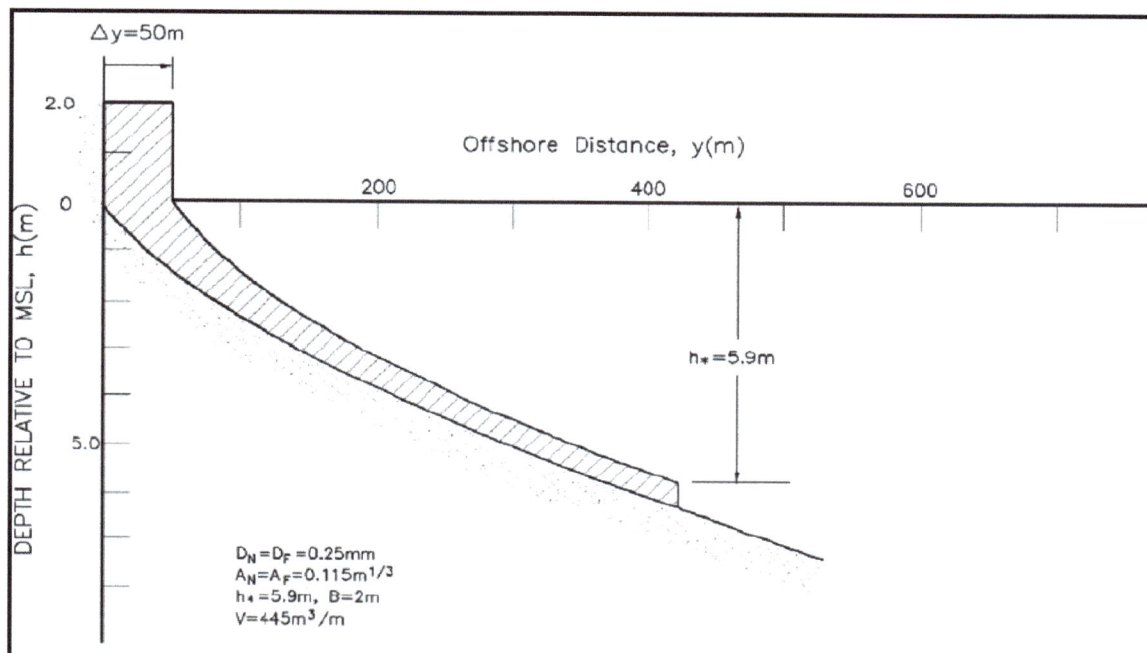

Figure III-3-28.   Example III-3-3. Nourishment with same-sized sand as native (nonintersecting profiles)

EXAMPLE PROBLEM III-3-4

FIND:
    The variation in equilibrated dry beach width with volume added for three candidate fill sand sizes.

GIVEN:
    Three candidate borrow sites with representative sand sizes: $D_{F_1} = 0.15$ mm, $D_{F_2} = 0.2$ mm and D_{F_3} = 0.25 mm. The native sand size, $D_N = 0.2$ mm. The berm height $B = 1.5$ m and $h_* = 6$ m.

SOLUTION:
    The associated values of the sediment scale parameters are determined from Figure III-3-18 and/or Table III-3-3: $A_N = 0.1$ m$^{1/3}$, $A_{F_1} = 0.084$ m$^{1/3}$, $A_{F_2} = 0.1$ m$^{1/3}$, and $A_{F_3} = 0.115$ m$^{1/3}$, respectively.

The procedures illustrated in Example Problems III-3-2 and III-3-3 produce the results shown in Figure III-3-29. It is seen that there is a nearly linear relationship for the sand that is of the same size as the native in accordance with: $V = (\Delta y)(h_* + B)$. For the fill sand, which is coarser than the native, for volumes less than approximately 450 m^3/m, the increase in dry beach width for each volume added is greater than for the same-sized sand. For this region, the profiles are intersecting. For greater volumes, the profiles are nonintersecting and the slope of the relationship is nearly the same as for $A_F/A_N = 1.0$. For the sand smaller than the native, the profiles are submerged for the smaller volumes and later become emergent in accordance with Equation 3-23. For larger volumes, the relationship has approximately the same slope as that for sand the same size as the native. The explanation for this is that once the profiles become emergent and nonintersecting, additional volumes of sand added simply displace the profile with all contours moving the same distance over the active depth. Thus the slope in Figure III-3-29 for this situation is nearly independent of the grain size.

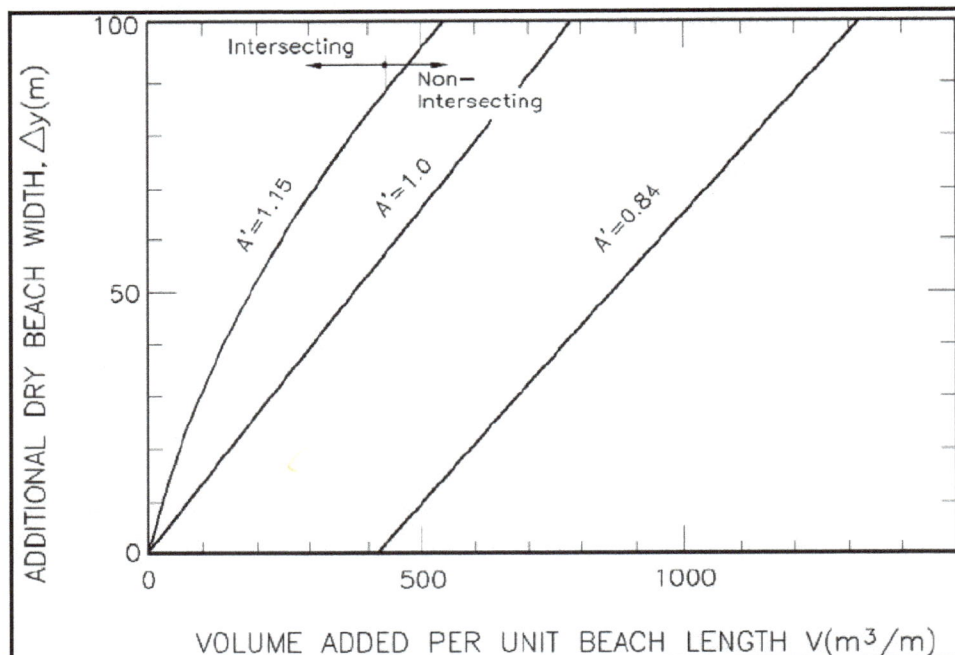

Figure III-2-29. Illustration of effect of volume added V and fill sediment scale parameter A_F on additional dry beach width Δy. Example conditions: $B = 1.5$ m, $h_* = 6$ m, $A_N = 0.1$ m$^{1/3}$

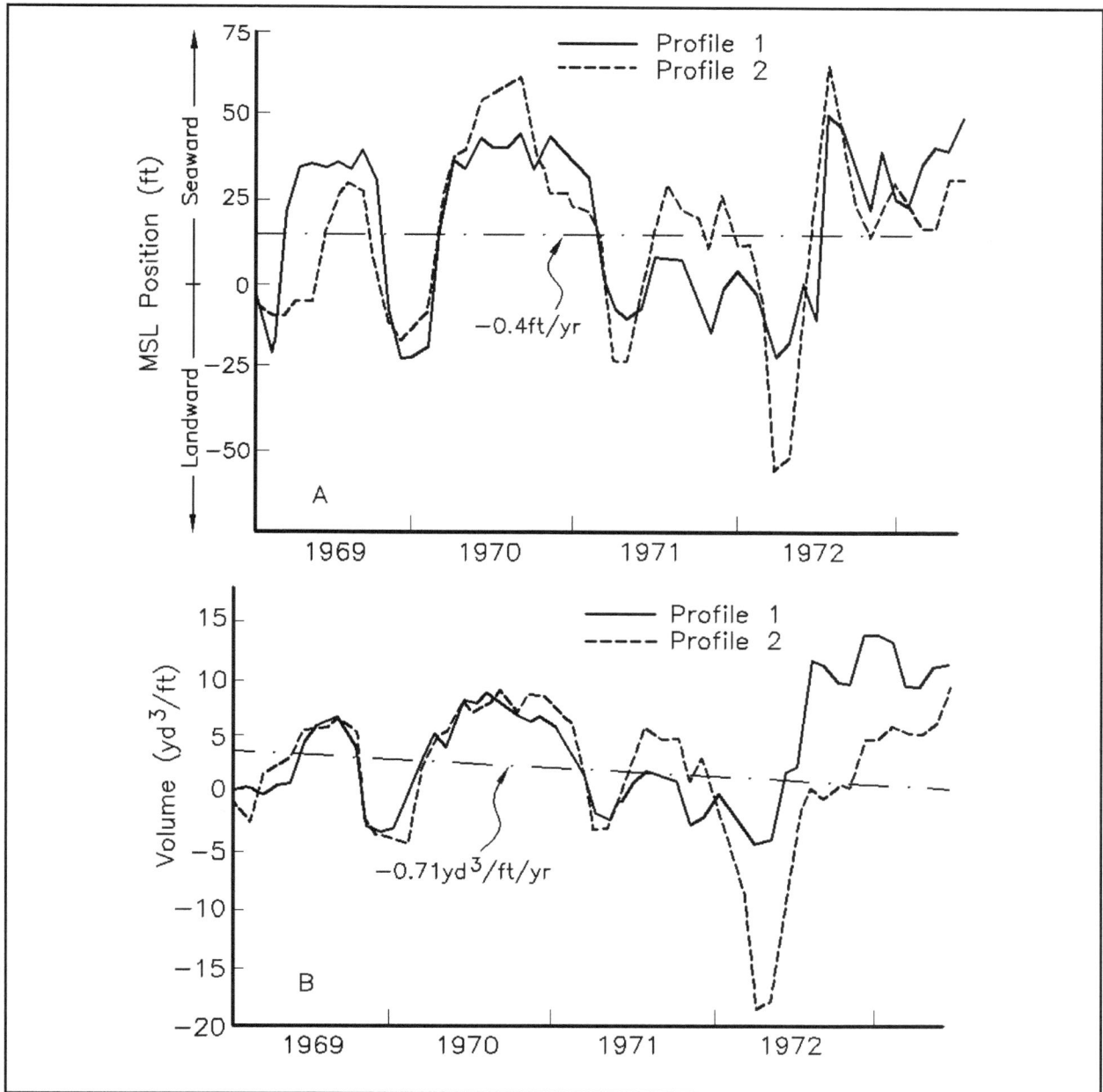

Figure III-3-30. Mean monthly shoreline position (A) and unit volume (B) at Jupiter Island, FL, referenced to first survey (Dewall and Richter 1977)

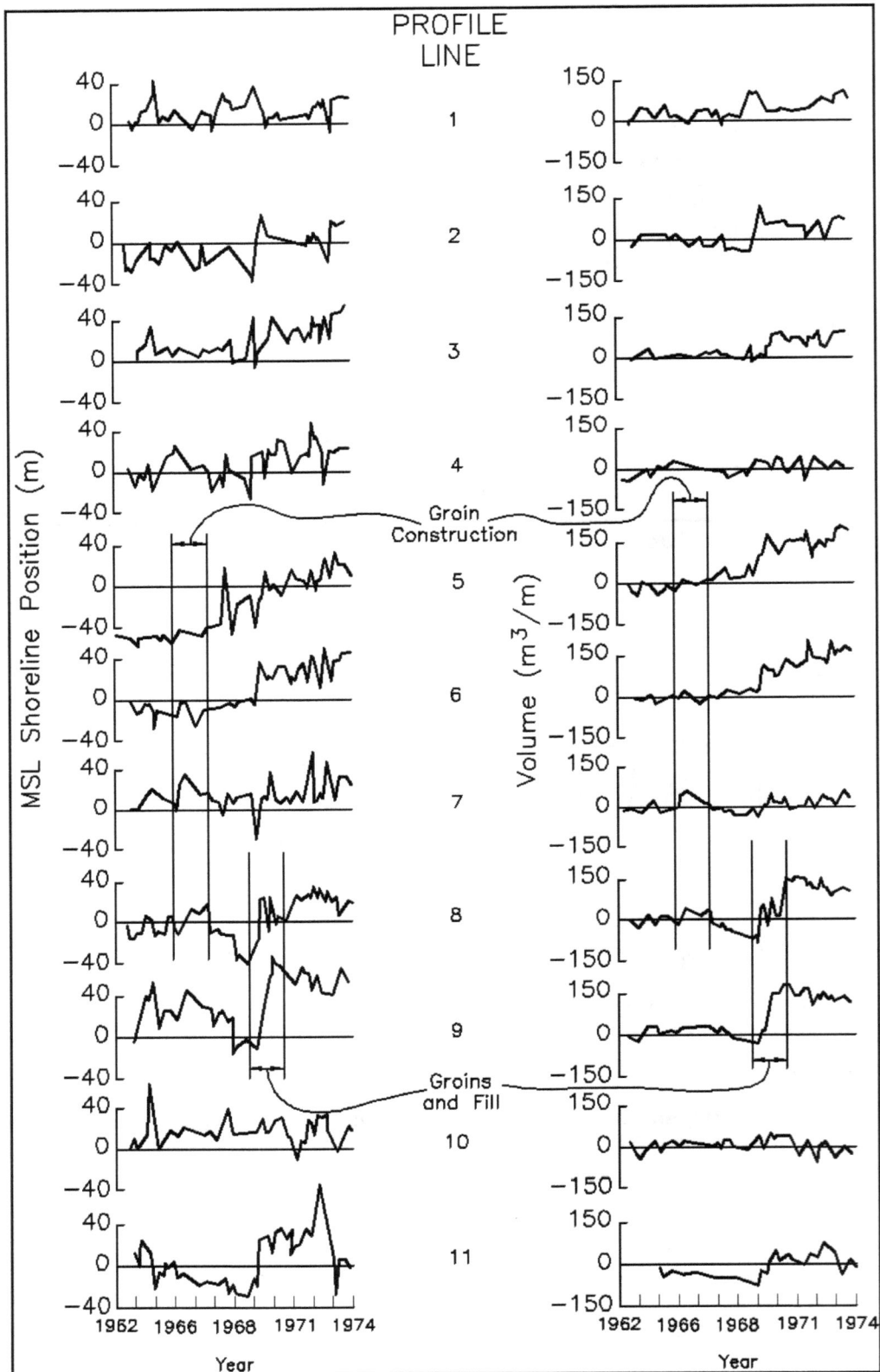

Figure III-3-31.   Changes in shoreline position and unit volume at Westhampton Beach, New York (Dewall 1979)

h. Static models for shoreline response to sea level rise and/or storm effects.

(1)   As water level and/or wave conditions change, the profile will respond toward a new equilibrium. If the conditions change very slowly, the profile changes will nearly maintain pace with the changed conditions and static models are applicable. However, rapidly changing conditions require dynamic models that account for the time scales of response of the profile. This section presents several useful static models for profile response.

(2)   First, consider the long-term profile response to sea level rise. On a worldwide basis, the average sea level has risen approximately 12 cm during the last century. However, the <u>relative sea level</u> changes (difference between absolute sea level rise and vertical land movements) at a particular location can differ substantially from the average, ranging from locations at which the relative sea level (RSL) change is a rise of four times the average (Louisiana (Penland, Suter, and McBride 1987)) and locations where the RSL change is decreasing at a rate of almost 1 m per century (Alaska (Hicks, Debaugh, and Hickman 1983)). Human-induced activities can cause considerable subsidence, primarily from extraction of ground fluids and the consequent reduction of pore pressures (National Research Council 1984), and can be a reason for relative sea level changes higher than average. Uplifting due to tectonic activity is the typical reason for relative sea level changes lower than average. As noted previously, any rise in mean water level on a beach profile that is otherwise in equilibrium must result in a redistribution of sand with erosion of the foreshore and with deposition of sand offshore near the depth of closure to maintain the profile shape relative to the rising water level. In the following discussion, equilibrium profile methods are applied to determine analytical solutions for the shoreline recession, here denoted by the symbol R, which will be more convenient notation than negative values of y.

(3)   Bruun (1962) proposed the following relationship for equilibrium shoreline response R_∞ to sea level rise S

$$R_\infty = S \frac{L_*}{h_* + B}$$

(III-3-33)

in which L_* and $(h_* + B)$ are the width and vertical extent of the active profile and the subscript "∞" indicates a static response. The basis for this equation is seen in Figure III-3-32 in which the two components of the response are: (1) a retreat of the shoreline R_∞, which produces a sediment "yield" $R_\infty(h_* + B)$, and (2) an increase in elevation of the equilibrium profile by an amount of the sea level rise S, which causes a sediment "demand" equal to SL_*. Equating the demand and the yield results in Equation 3-33, which is known as the "Bruun Rule." It is noted that the Bruun Rule does not depend on the particular profile shape.

(4)   The Bruun Rule has been modified to account for several features of natural beaches that were not accounted for in the original development. Bruun (1988), for example, shows that Equation 3-33 may be modified to account for loss or "winnowing" of fine sediment out of the profile or for loss of sediment to deepwater canyons or other "sinks" in the offshore. Similar corrections may be made to account for unbalanced sediment flux into or out of the beach profile due to gradients in the net longshore sediment transport, as shown, for example, by Everts (1985).

(5)   Despite these modifications, several aspects of the Bruun Rule have remained problematic. For example, the upper limit of the active profile is not clearly defined in Figure III-3-32 so that it is difficult to establish a realistic profile width L_* in Equation 3-33. Dean and Maurmeyer (1983) later extended Bruun's result to apply to the case of a barrier island in the form

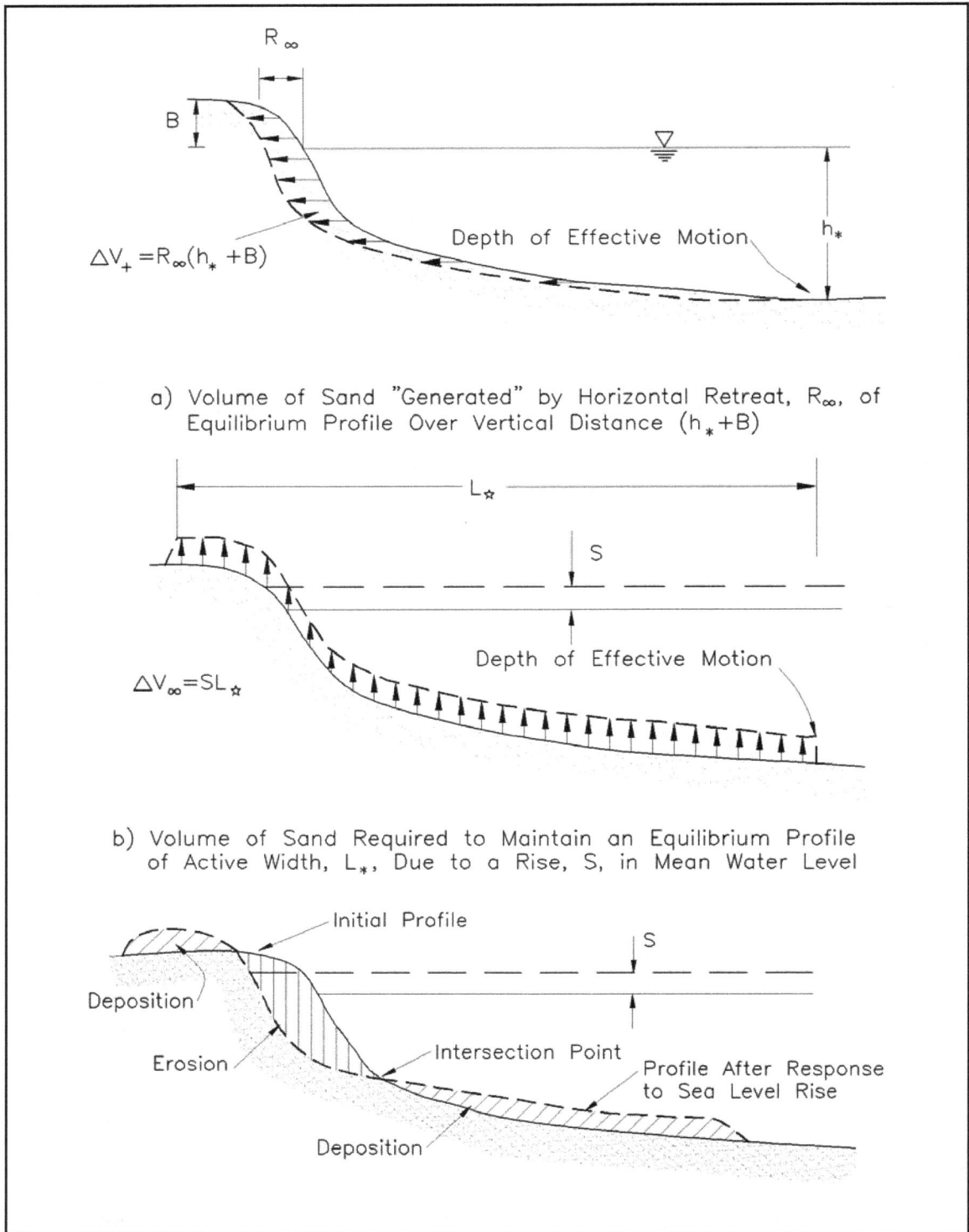

R_∞

B

∇

Depth of Effective Motion

h_*

$\Delta V_+ = R_\infty (h_* + B)$

a) Volume of Sand "Generated" by Horizontal Retreat, R_∞, of Equilibrium Profile Over Vertical Distance $(h_* + B)$

$L_\star$

S

Depth of Effective Motion

$\Delta V_\infty = SL_\star$

b) Volume of Sand Required to Maintain an Equilibrium Profile of Active Width, L_*, Due to a Rise, S, in Mean Water Level

Initial Profile

S

Deposition

Erosion

Intersection Point

Profile After Response to Sea Level Rise

Deposition

Figure III-3-32. Components of sand volume balance due to sea level rise and associated profile retreat according to the Bruun Rule

$$R_\infty = S \frac{L_* + L_W + L_L}{h_* - h_L}$$

(III-3-34)

and the various terms are explained in Figure III-3-33. In general, the shoreline retreat is some 50 to 200 times the sea level rise with the greater factors associated with the milder beach slopes and more energetic wave conditions (i.e., greater h_*). These factors for the case of a barrier island would be considerably greater due to the difference term in the denominator (Equation 3-34). Situations where the region behind the island is deep may explain the geological evidence for some barrier islands which are believed to have "drowned in place" rather than to have migrated landward.

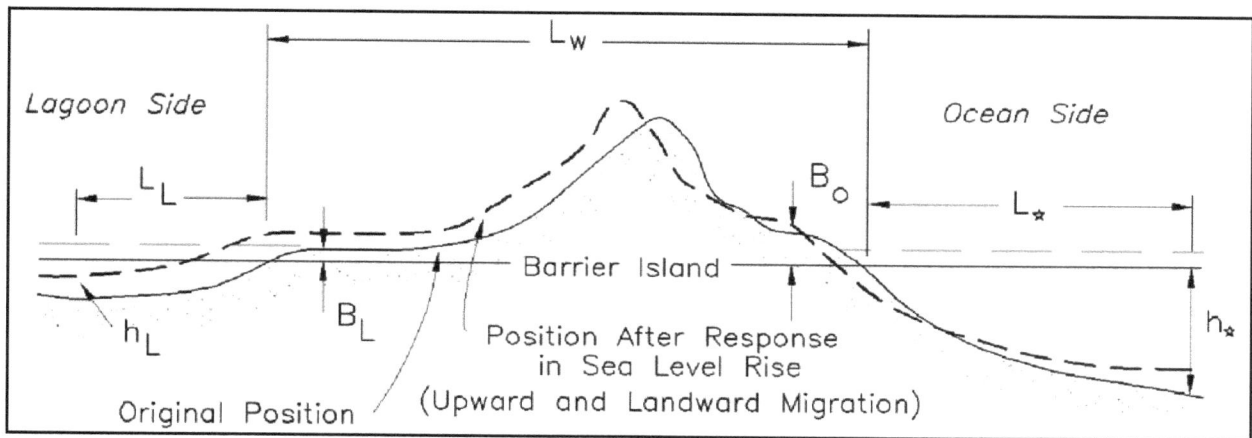

Figure III-3-33.   The Bruun Rule generalized for the case of a barrier island that maintains its form relative to the adjacent ocean and lagoon (Dean and Maurmeyer 1983)

(6)   The Bruun Rule has been subjected to verifications both in the laboratory and in the field. Hands (1983), evaluated the Bruun Rule in Lake Michigan using 25 beach profiles over a 50-km length of shoreline subjected to a 0.2-m water level rise over a 7-year period from 1969 to 1976. The on- and offshore limits of profile response were determined directly from measured beach profiles. Thus, the depth of closure was identified as the maximum depth of significant profile change observed from beach surveys. Likewise, the upper limit of profile change was selected as the natural vegetation line in the foredunes. With these empirical input parameters, calculated profile retreat over the 7-year period then agreed to within 10 percent of measured values. Over shorter periods of time for example, over a 3-year period, Hands found that the Bruun Rule initially over-estimated the profile response due to the time lag between elevated water levels and the profile response. Storm processes were then identified as being responsible for causing rapid equilibration of the profiles and for causing the profiles to "catch up" or equilibrate relative to the water level.

(7)   Edelman (1972) modified the Bruun Rule to make it more appropriate for larger values of increased water levels and for time-varying storm surges. It was assumed that the profile maintained pace with the rising sea level and thus, at each time, the following equation is valid

$$\frac{dR}{dt} = \frac{dS}{dt} \left[\frac{W_b}{h_b + B(t)} \right]$$

(III-3-35)

Cross-Shore Sediment Transport Processes

where now $B(t)$ represents the instantaneous total height of the active profile <u>above</u> the current water level. As shown in Figure III-3-34, $B(t)=B_o-S(t)$ where B_o is the original berm height. For application to storm events, Edelman also adopted the breaking depth h_b (and surf zone width W_b) rather than the offshore depth of closure h_* (and corresponding W_*) as would be appropriate for long-term sea level rise. Substituting Equation 3-35 and integrating gives

$$R(t) = W_b \ \ln \left[\frac{h_b + B_o}{h_b + B_o - S(t)} \right] \tag{III-3-36}$$

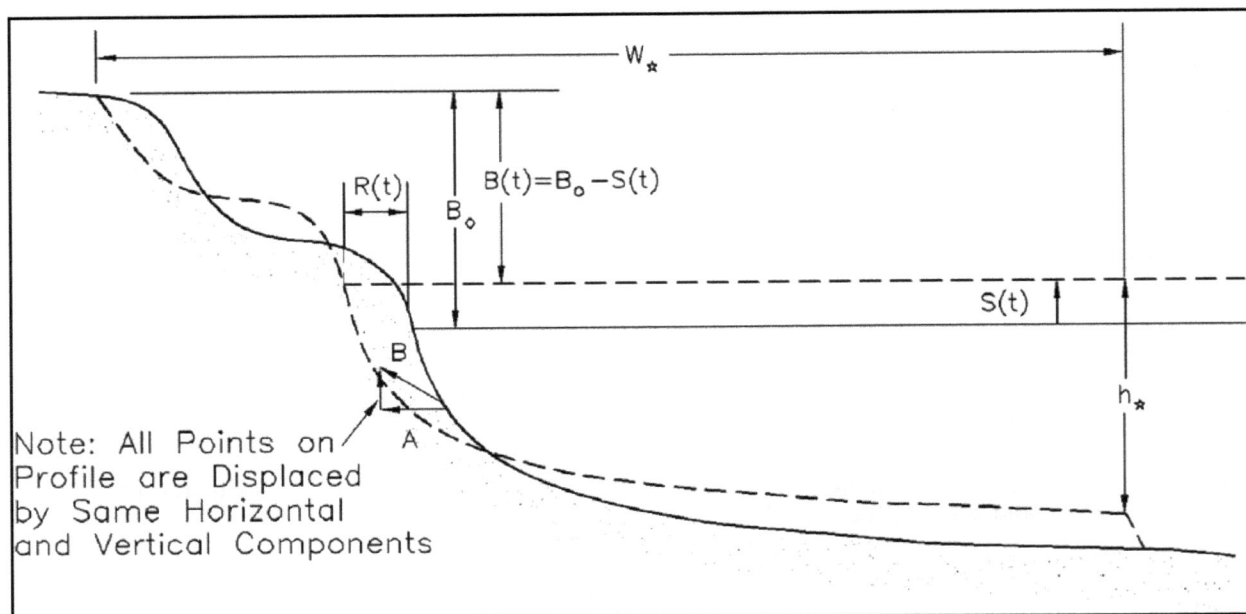

Figure III-3-34. Elements of the Edelman model

(8) Using the small argument approximation for the natural logarithm, it is readily shown that to the first approximation, Edelman's equation is equivalent to the Bruun Rule.

(9) Dean (1991) derived similar solutions for storm-induced berm retreat based on theoretical pre- and post-storm profile forms given by the equilibrium profile in Equation 3-14. Because pre- and post-storm profiles were defined by an analytical form, the equilibrium beach response could be obtained by integrating the areas (volume per unit length) between the initial and final equilibrium profiles and by equating the resulting eroded and deposited areas. Solutions were obtained for both the case of a uniform water level rise and for the case where breaking waves create a distribution of wave setup across the surf zone. One interesting result is the case where water levels are elevated by both a storm surge and by breaking-induced wave setup. For this case, an approximate solution for the steady-state erosion is given as

$$R_\infty = (S + 0.068 \, H_b) \, \frac{W_b}{B + h_b} \tag{III-3-37}$$

where W_b is the width of the surf zone, defined for the equilibrium profile as $W_b = (h_b/A)^{3/2}$. The solution for erosion due to combined storm surge and wave setup is similar in form to the Bruun Rule in Equation 3-33. In this case, wave setup causes a general rise in water level in the surf zone so that it functions much like storm surge. It is noted, however, that storm surge has a much larger effect than wave setup. For this reason, the other analytical solutions that follow do not include wave setup effects and, instead, assume that beach response is driven primarily by storm surge.

(10) Kriebel and Dean (1993) considered both profiles with a vertical face at the water line as shown in Figure III-3-35a and profiles with a sloping beach face as shown in Figure III-3-35b. They showed that by accounting for the small wedge-shaped sand volume offshore of the breaking depth, somewhat improved expressions could be developed for the potential beach recession due to elevated water levels. The beach profile with a vertical face is a special limiting case of the profile with a sloping beach face, thus only the results for the sloping beach face are given here. As shown by Kriebel and Dean (1993), the general result for equilibrium berm recession due to a storm surge level S is given as

$$R_\infty = S \frac{W_b - \dfrac{h_b}{m_o}}{B + h_b - \dfrac{S}{2}} \tag{III-3-38}$$

where m_o is the slope of the beach profile at the waterline. This slope is joined to the concave equilibrium profile at a depth where the slope of the equilibrium profile is equal to m_o. As a result, the surf zone width can be shown to be equal to

$$W_b = y_o + \left(\frac{h_b}{A}\right)^{\frac{3}{2}} \tag{III-3-39}$$

where y_o is a small offset of the shoreline between the sloping beach face and the imaginary or virtual origin of the equilibrium profile, given by $y_o = (4A^3)/(27m_o^3)$. For most conditions, this offset is negligible and can be neglected when estimating the surf zone width, as will be illustrated in Example Problem III-3-6. For engineering application, it is also of interest to compute the volume of sand eroded between the initial and final profiles per unit length of beach. For the case with a sloping beach face, the volume eroded from the berm above the initial still-water level due to a storm surge level S is given by

$$V_\infty = R_\infty B + \frac{S^2}{2m_o} - \frac{2}{5} \frac{S^{\frac{5}{2}}}{A^{\frac{3}{2}}} \tag{III-3-40}$$

(11) When the beach face slope becomes infinite, the solutions from Equation 3-38 for the vertical beach face depicted in Figure III-3-35a are similar to those obtained by the Bruun Rule in Equation 3-33 or by Dean (1991) in Equation 3-37. The major difference is the term $S/2$ in the denominator of Equation 3-38, which is the result of considering the small wedge-shaped volume of sand near the breakpoint. For more realistic beach face slopes, the results in Equation 3-38 will yield smaller estimates of the potential berm recession than the Bruun or Dean solutions, since a portion of the rise in water level is accommodated by shifting the shoreline higher on the sloping beach face and less berm retreat is then required. Kriebel and Dean (1993) also provide analytical solutions for cases where the beach profile has a distinct dune on top of the berm.

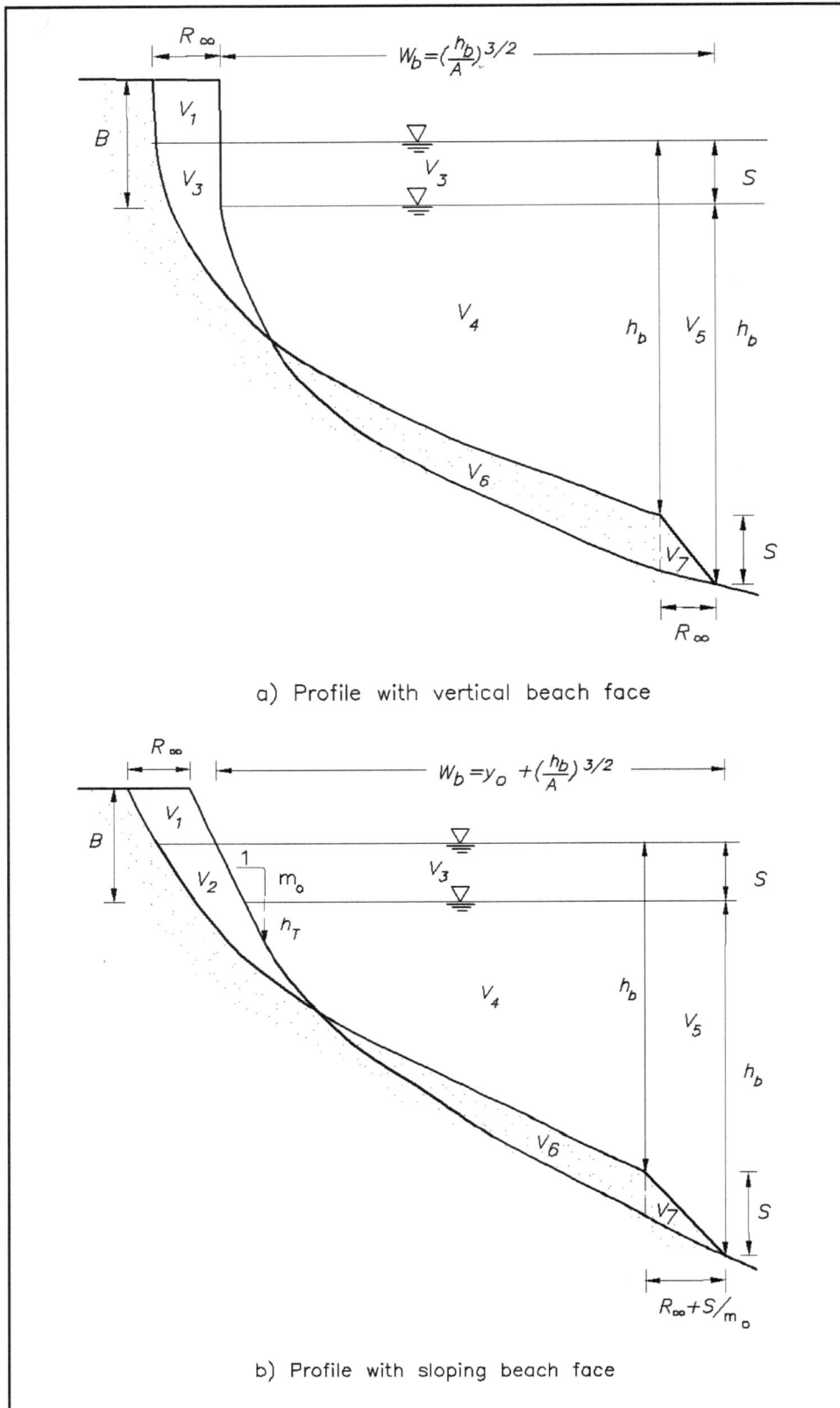

a) Profile with vertical beach face

b) Profile with sloping beach face

Figure III-3-35. Profile forms considered by Kriebel and Dean (1993)

Cross-Shore Sediment Transport Processes

EXAMPLE PROBLEM III-3-5

FIND:

The rate of shoreline retreat, according to the Bruun Rule, then find the rate at which beach nourishment would be required to maintain the shoreline position.

GIVEN:

$h_* = 6$ m, $B = 2$ m, $D = 0.2$ mm, and a rate of sea level rise of $\dfrac{dS}{dt} = 0.003$ m/year.

SOLUTION:

The sediment scale parameter is determined from Figure III-3-18 and/or Table III-3-3 as $A = 0.1$ m$^{1/3}$. The width W_* of the active nearshore zone is determined from Equation 3-19 as

$$W_* = \left(\frac{h_*}{A}\right)^{\frac{3}{2}} \approx 465 \text{ m}$$

and from Equation 3-33, assuming the berm stays at a constant elevation, the rate of shoreline retreat is given by

$$\frac{dR_\infty}{dt} = \frac{dS}{dt}\frac{W_*}{h_* + B} \approx 0.17 \text{ m/yr}$$

so that the ratio of shoreline retreat to sea level rise is about 58.

When considering sea level rise and beach nourishment, from Equation 3-42, the rate at which sand must be added to offset the erosion due to sea level rise is

$$\frac{dV}{dt} = W_* \frac{dS}{dt} = (465\text{m})(0.003\text{m/year}) \approx 1.4\text{m}^3/\text{m/year}$$

EXAMPLE PROBLEM III-3-6

FIND:

The potential berm retreat, along with the volume of sand eroded from above the mean sea level, for the storm conditions given below.

GIVEN:

Equilibrium beach profile with sand grain size $D = 0.25$ mm, beach face slope $m_o = 0.05$, and berm height $B = 2$ m. Storm conditions with a peak surge level $S = 2$ m and a breaking depth $h_b = 3.0$ m.

SOLUTION:

The equilibrium berm retreat may be determined from Equation 3-38, which first requires knowledge of the breaking depth h_b and the surf zone width W_b, which, from Equation 3-39, are given by $W_b = y_o + (h_b/A)^{3/2}$. From Table III-3-3, the so-called A parameter for the equilibrium beach profile is found to be $A = 0.115$ m$^{1/3}$ based on the grain size of 0.25 mm.

In calculating the surf zone width, the term $(h_b/A)^{3/2}$ is equal to $(3.0m/0.115m^{1/3})^{3/2} = 133.2$ m while the small shoreline offset is given by $y_o = (4A^3)/(27m_o^3) = 1.8$ m. The total surf zone width is then $W_b = 1.8$ m $+ 133.2$ m $= 135.0$ m. As noted, however, the offset is negligible and could be neglected for simplicity.

From Equation 3-38, the potential berm retreat is now determined as

$$R_\infty = \frac{S\left(W_b - \dfrac{h_b}{m_o}\right)}{B + h_b - \dfrac{S}{2}} = \frac{2\left(135.0 - \dfrac{3.0}{0.05}\right)}{2 + 3.0 - \dfrac{2}{2}} = 37.5 \text{ m}$$

From Equation 3-40, the potential volume eroded from above the mean sea level datum is given by

$$V_\infty = R_\infty B + \frac{S^2}{2m_o} - \frac{2\,S^{\frac{5}{2}}}{5\,A^{\frac{3}{2}}}$$

$$= (37.5)(2) + \frac{(2)^2}{2\,(0.05)} - \frac{2\,(2)^{\frac{5}{2}}}{5\,(0.115)^{\frac{3}{2}}} = 57.0 \text{ m}^3/\text{m}$$

These solutions are known to generally overestimate erosion associated with severe storms. For example, Chiu (1977) compared both the Edelman and Dean methods to erosion measured after Hurricane Eloise on the Florida coast and found that both methods gave erosion estimates that were as much as a factor of 5 higher than observed. The reason for this overprediction is that these methods assume the profile responds instantly to changes in water level while they neglect the transient or time-dependent approach to equilibrium. Such equilibrium solutions are still useful from an engineering perspective since they place a conservative upper bound on the actual beach response.

(12) While these idealized analytical solutions are useful, such computations of the maximum potential response can also be performed numerically for a measured (surveyed) beach and dune profile. In this case,

it may be assumed that the existing profile is in a stable equilibrium configuration, possibly with an offshore bar. The solution for the profile response to a water level rise would then be carried out by shifting the measured profile form upward and landward until a mass balance is achieved between the sand eroded from the berm and dune and the sand deposited offshore near the breakpoint.

(13) The problem of shoreline stabilization through beach nourishment with compatible sands in an era of sea level rise may be treated by combining the two effects. The rate of shoreline retreat is given as

$$\frac{dR_\infty}{dt} = \frac{dS}{dt}\frac{W_*}{h_*+B} - \frac{1}{(h_*+B)}\frac{dV}{dt}$$ (III-3-41)

in which dV/dt is the rate at which sand is added per unit length of beach. In order for the shoreline retreat due to sea level rise to be offset by the advancement due to nourishment, $(dR_\infty/dt = 0)$

$$\frac{dV}{dt} = W_*\frac{dS}{dt}$$ (III-3-42)

which can be interpreted as adding sufficient sand to just fill the active profile of width W_* at the rate of sea level rise. This result could have been foreseen by referring to Figure III-3-32.

i. Computational models for dynamic response to storm effects.

(1) Introduction. Dynamic computational models are distinguished from the static models discussed earlier by accounting for the transient nature of the profile adjustment. As an illustration, for Example III-3-2, if the initial placement of nourished sand was different (usually steeper) than the equilibrium profile in Figure III-3-27, it is possible to determine from the equation of continuity (conservation of sand) the total cross-shore volumetric transport; however, it is not possible to determine the _rate_ at which the sand was transported to reach equilibrium. The equilibration process could have required 1 year or a decade. In many problems of coastal engineering interest, the _rates_ are extremely important. As examples, it will be shown that a rapidly moving storm may cause only a fraction of its erosion potential due to the relatively long time scales of the sediment transport processes and to the relatively short duration of the more energetic conditions caused by the storm. A second problem in which the time scales are of interest is that of profile equilibration of a beach nourishment project. Although it is accepted that equilibration occurs within 1 to 5 years, and certainly depends on the frequency of energetic storms, the economic value of that portion of added dry beach width associated with disequilibrium during evolution can be substantial.

(2) Numerical and analytical models. More than a dozen numerical models and at least two analytical models have been developed to represent dynamic cross-shore sediment transport processes. These models require a continuity equation and a transport (or dynamic) equation (or equivalent) that governs the rate at which the processes occur. The conservation equation balances the differences between inflows and outflows from a region as predicted by the transport equation. In addition, boundary conditions must be employed at the landward and seaward ends of the active region. These boundary conditions can usually be expressed in terms of a maximum limiting slope such that if the slope is exceeded, adjustment of the profile will occur, a condition sometimes referred to as "avalanching." The locations of the seaward and landward boundaries are usually taken at the limits of wave breaking and wave runup, respectively. Examples of numerical models for which computer codes are available include those of Kriebel and Dean (1985), Kriebel (1986), Larson (1988), and Larson and Kraus (1989, 1990). Analytical models have been published by Kobayashi (1987), and Kriebel and Dean (1993).

(3) General description of numerical models. In numerical modelling, two representations of the physical domain have been considered as shown in Figure III-3-36. In the first type, shown in

Figure III-3-36a, the cells are finite increments of the distance variable y. Thus the distance is the independent variable and the dependent variable is depth, which varies with time. In the second type shown in Figure III-3-36b, the computational cells are formed by finite increments of the depth h. In this case, the independent variable is h and y varies with time for each h value. There is an inherent advantage of the first type since the presence of bars can be represented with no difficulties. All dynamic models require a continuity equation and a transport (dynamic) equation.

(a) Conservation equation. The conservation equation is very straightforward and for the computational cell type in Figure III-3-36a is given by

$$\frac{\partial h}{\partial t} = \frac{\partial q_y}{\partial y} \tag{III-3-43}$$

in which y and t are the independent variables. If h and t are regarded as the independent variables as in Figure III-3-36b, the conservation equation is

$$\frac{\partial y(h)}{\partial t} = - \frac{\partial q_y}{\partial h} \tag{III-3-44}$$

in which it is noted that for each depth value (h), there is an associated distance value (y).

(b) Transport relationships. Sediment transport relationships fall within the categories of "closed loop," which converge to a target profile and "open loop," which are not a priori constrained to the final (equilibrium) profile. Transport relationships of the "closed loop" type will be reviewed first.

(c) Closed loop transport relationships. One of the first closed loop transport relationships was that proposed by Kriebel and Dean (1985). Recalling that the equilibrium beach profile (EBP) represented by Equation 3-14 is consistent with uniform wave energy dissipation per unit water volume D_*, Kriebel and Dean adopted a simple transport relationship in the form

$$q_y = K'(D - D_*) \tag{III-3-45}$$

such that at equilibrium $D = D_*$, and an equilibrium profile results. The parameter K' is then used to calibrate the model by correlating the sediment transport rate to the excess energy dissipation. The transport relationship above may be modified to be consistent with a profile with beach face slope m_o

$$q_y = K''(D - D_*) + \epsilon \frac{\partial h}{\partial y} \tag{III-3-46}$$

where ϵ is an additional model parameter as suggested by Larson and Kraus (1989). In this expression, the calibration parameter K'' differs from that used in Equation 3-45, since the additional gravitational effects, represented by the slope term, assist in moving sediment offshore.

(d) Open loop transport relationships. These transport relationships depend on the detailed hydrodynamics and attempt to incorporate the actual processes more faithfully than the closed loop variety. Usually both bed load and suspended load transport components are represented based on the hydrodynamic properties averaged over a wave period. Open loop models will not be considered further in this chapter; however, an excellent review of open loop models is given by Roelvink and Broker (1993). In general, these models can be grouped according to the physical processes that are assumed to be dominant for cross-shore

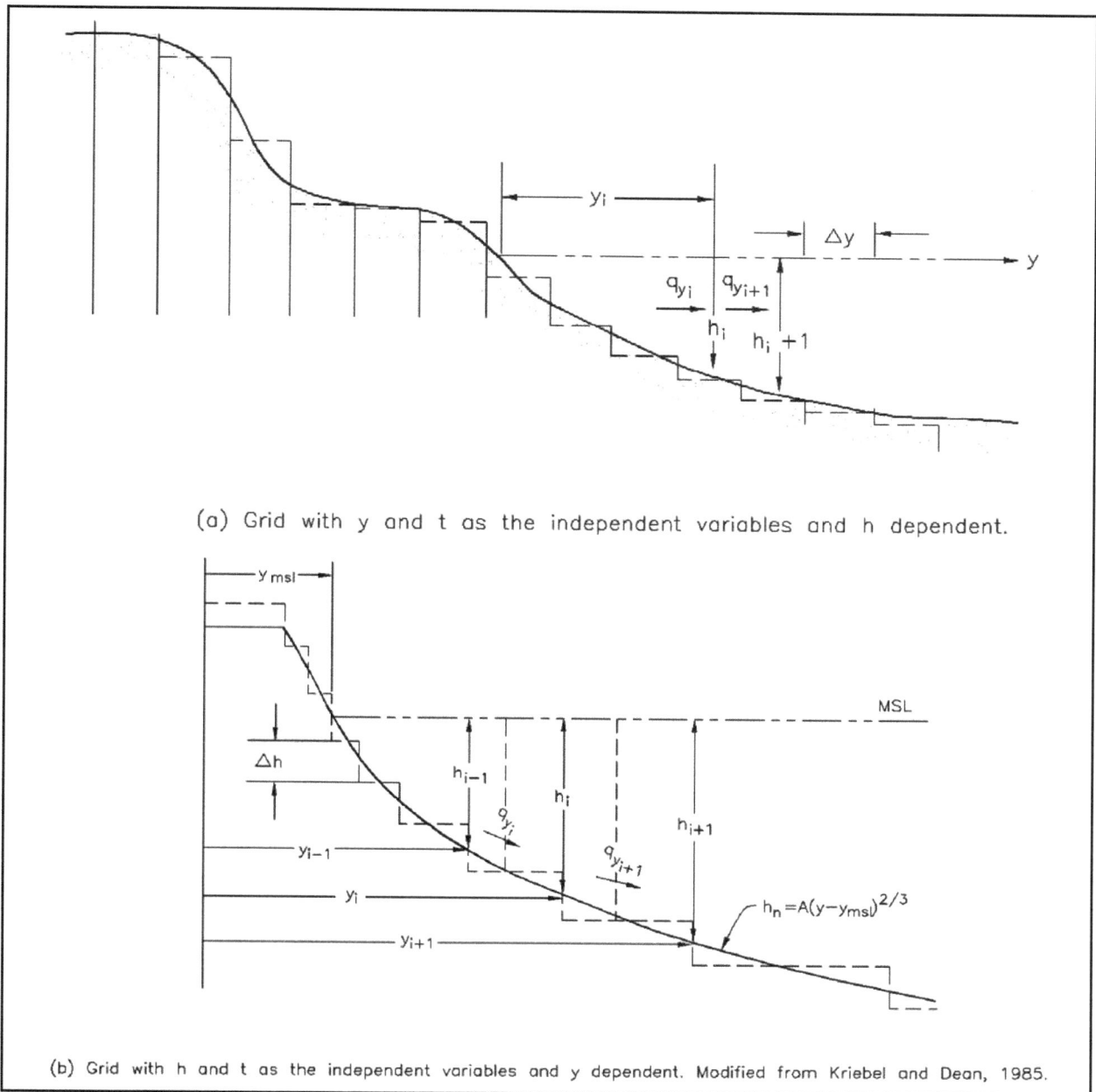

(a) Grid with y and t as the independent variables and h dependent.

(b) Grid with h and t as the independent variables and y dependent. Modified from Kriebel and Dean, 1985.

Figure III-3-36. Two types of grids employed in numerical modelling of cross-shore sediment transport and profile evolution

sediment transport. Several models compute transport rates by vertically integrating the distributions of suspended sediment concentration and cross-shore currents. Examples of such models include those of Dally and Dean (1984), Stive and Battjes (1984), and Broker-Hedegaard, Deigaard, and Fredsoe (1991). Other models consider instead that bed shear stress is the dominant forcing mechanism. An example is the model by Watanabe et al. (1980). A third widely used approach has been to compute the combined bed and suspended load transport first, based on the energetics approach proposed by Bagnold (1966). Examples of models based on this approach include those of Roelvink and Stive (1989) and Nairn and Southgate (1993).

(4)   General description of analytical models.

(a)   The analytical model of profile evolution by Kobayashi (1987) incorporates the conservation and transport relationships (Equations 3-41 and 3-43). The transport equation is linearized such that a diffusion equation results. The landward and seaward receding and advancing limits of the evolving profile are treated as moving boundary conditions. Kobayashi (1987) presents an analytical solution for the case of a constant elevated water level in the form of fairly complex error functions. For more complete water level scenarios, a numerical scheme must be used. As presented, the model does not lend itself readily to engineering applications.

(b)   A simpler analytical model for predicting dynamic profile response during storms is the so-called Convolution Method of Kriebel and Dean (1993). This method is based on the observation that beaches tend to respond toward a new equilibrium exponentially over time. For laboratory conditions, where a beach is suddenly subjected to steady wave action, the time-dependent shoreline response $R(t)$ may be approximated by the form

$$R(t) = R_\infty \left(1 - e^{-\frac{t}{T_S}} \right) \qquad \text{(III-3-47)}$$

where R_∞ is the equilibrium beach response and T_S is the characteristic time-scale of the system. An exponential response of this kind has been observed in wave tank experiments by Swart (1974), Dette and Uliczka (1987), and Larson and Kraus (1989).

(c)   A more general result for the dynamic erosion response may be obtained by noting that Equation 3-47 suggests that the rate of profile response is proportional to the difference between the instantaneous profile form and the ultimate equilibrium form. An approximate differential equation governing the profile response to time-dependent variations in water level may be assumed in the form

$$\frac{dR(t)}{dt} = \frac{1}{T_S} [R_\infty f(t) - R(t)] \qquad \text{(III-3-48)}$$

where $f(t)$ represents a unit-amplitude function of time that describes the storm surge hydrograph, while R_∞ represents the equilibrium beach response for the peak water level. The general solution to this system may be expressed as a convolution integral as

$$R(t) = \frac{R_\infty}{T_S} \int_0^t f(\tau)\, e^{-\frac{(t-\tau)}{T_S}}\, d\tau \qquad \text{(III-3-49)}$$

(d)   As a result, several important characteristics of dynamic beach profile response are evident. First, a beach has a certain "memory," so that the beach response at any one time is dependent on the forcing conditions applied over some preceding time period. As a result, the beach response will *lag* behind the erosion forcing. In addition, because of the exponential response characteristics of the beach system, the beach response will be *damped* so that the actual maximum response will be less than the erosion potential of the system.

j. Example application of an analytical model.

(1) Of the two analytical models, the Kriebel and Dean (1993) analytical model is simpler to apply and thus will be discussed and illustrated by several examples herein.

(2) A useful application of the convolution method is to analyze the erosion associated with an idealized storm surge hydrograph. Consider the case where the storm surge is approximated by the function

$$S(t) = S \sin^2(\sigma t) = S f(t) \qquad \text{(III-3-50)}$$

with $\sigma = \pi/T_D$ and where T_D is the total storm surge duration. The maximum storm surge level S would be used to determine the maximum potential erosion R_∞ according to Equation 3-38 or one of the other expressions for static profile response developed in the preceding section. As shown by Kriebel and Dean (1993), solution of the convolution integral in Equation 3-49, with the unit-amplitude forcing term $f(t)$ equal to the sine-squared function, gives the following time-dependent erosion response

$$\frac{R(t)}{R_\infty} = \frac{1}{2}\left\{1 - \frac{\beta^2}{1+\beta^2}\exp\left(-\frac{2\sigma t}{\beta}\right) - \frac{1}{1+\beta^2}[\cos(2\sigma t) + \beta\sin(2\sigma t)]\right\} \qquad \text{(III-3-51)}$$

where β is the ratio of the erosion time scale to the storm duration, which is given as $\beta = 2\pi T_S/T_D$. The predicted beach response is shown in Figure III-3-37 for two different values of β, corresponding approximately to a short-duration hurricane ($\beta = 10.6$) and to a long-duration northeaster ($\beta = 0.76$). The examples illustrate the role of storm duration in determining the maximum erosion response such that short-duration storms may only achieve a small percentage of their potential equilibrium response.

(3) The magnitude of the beach response from the sine-squared storm surge can be summarized in terms of the expected maximum dynamic erosion relative to the potential static or equilibrium response. This may be shown to be a function of β, as illustrated in Figure III-3-38. In general, short-duration storms fall to the right of this curve such that the predicted maximum erosion may be only 20 to 40 percent of the maximum erosion potential. For long-duration storms, the maximum erosion may be from 40 to 90 percent of the maximum erosion potential. When the storm duration is equal to the erosion time scale, ($\beta = 2\pi$), the dynamic erosion response is only 36 percent of the static response.

(4) The time scale of dynamic profile response T_S has not been as widely considered in coastal engineering as the equilibrium erosion R_∞ and, thus far, the time scale has not been derived analytically. As a result, empirical descriptions of the time scale are required. These have been developed from results of the numerical erosion model of Kriebel (1986, 1990) for various combinations of profile geometry and breaking wave conditions. From these numerical tests, it was found that the time scale was approximately independent of the storm surge level, but varied strongly with sediment size (through the A parameter) and breaking wave height, and varied less significantly as a function of beach profile geometry. Numerical results were analyzed by dimensional analysis to arrive at the following empirical relationship

$$T_S = 320\,\frac{H_b^{\frac{3}{2}}}{g^{1/2}A^3}\left(1 + \frac{h_b}{B} + \frac{m_o W_b}{h_b}\right)^{-1} \qquad \text{(III-3-52)}$$

(5) In Figure III-3-39, the numerically generated values of the erosion time scale are plotted as a function of the expression on the right-hand side of the equation (above). In general, beaches composed of

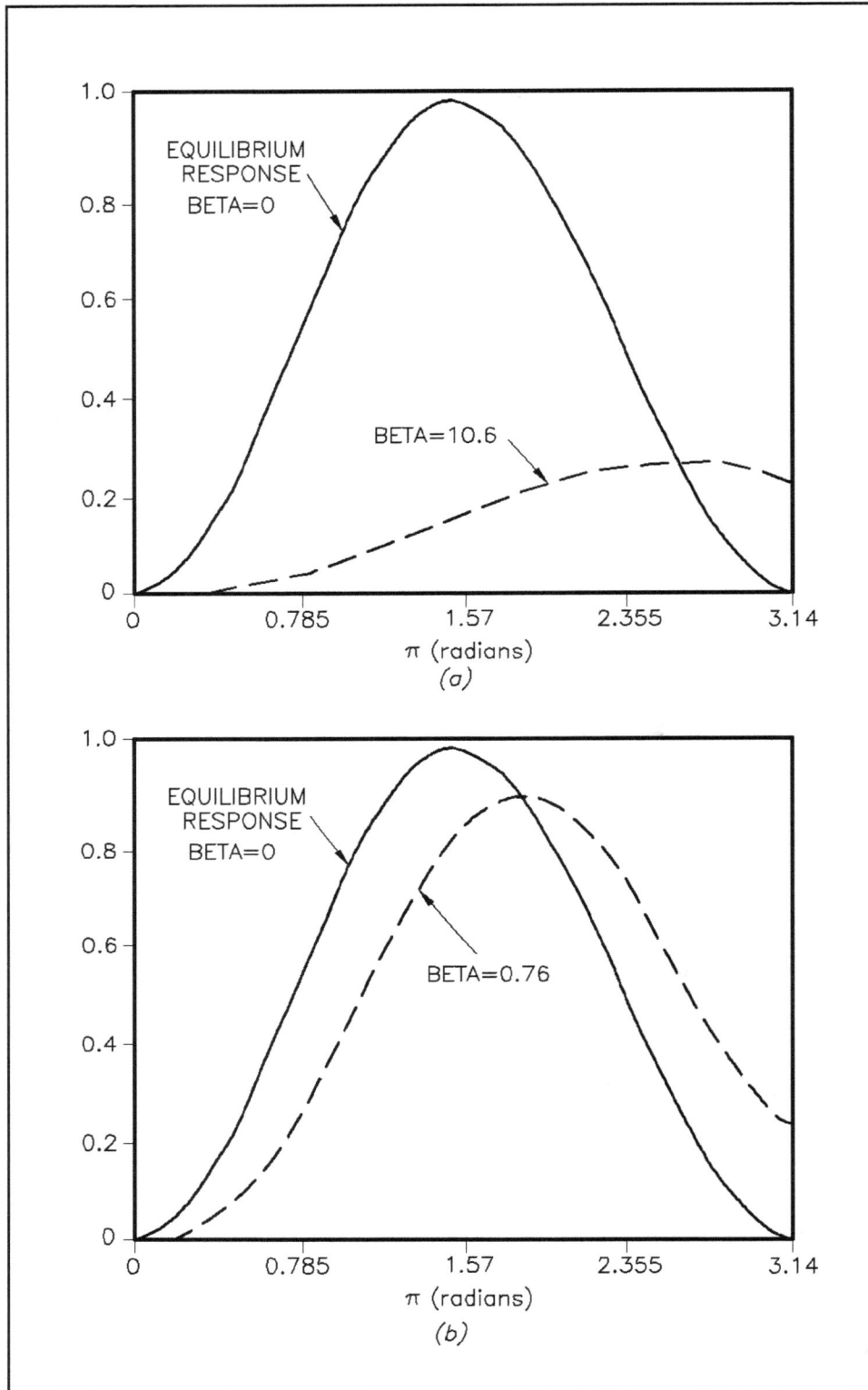

Figure III-3-37.   Examples of profile response to idealized sine-squared storm surge: (a) Short-duration hurricane, and (b) Long-duration northeaster (Kriebel and Dean 1993)

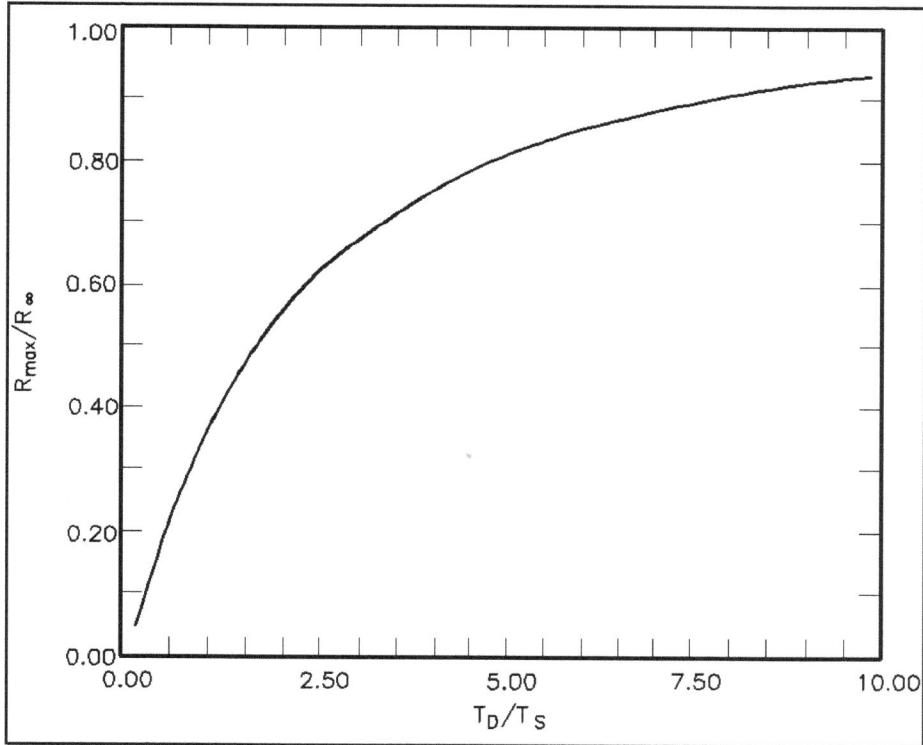

Figure III-3-38. Maximum relative erosion versus ratio of storm duration to profile time scale, T_D/T_s

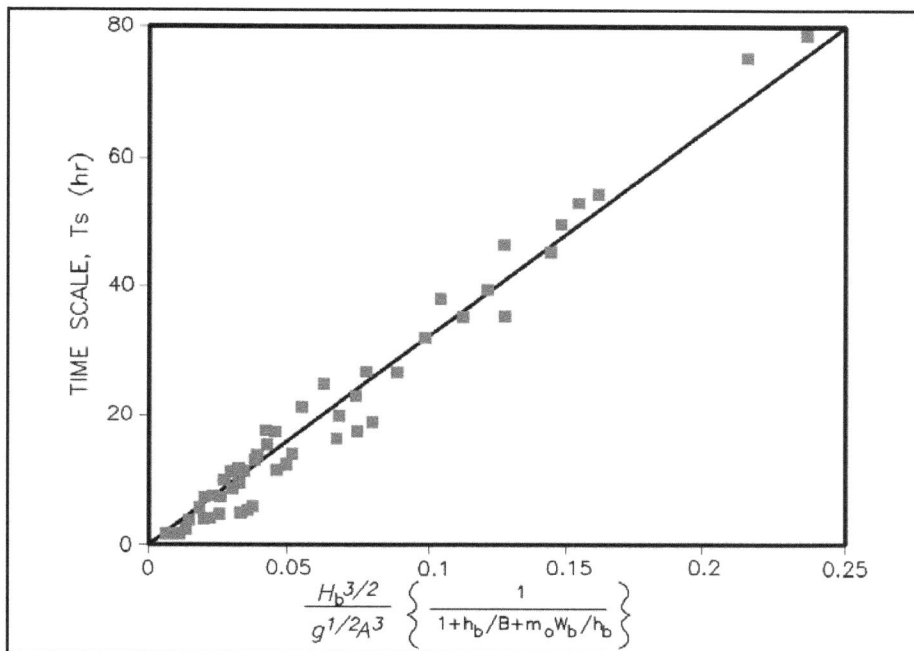

Figure III-3-39. Empirical relationship for determination of erosion time scale, T_s

EXAMPLE PROBLEM III-3-7

FIND:
The shoreline recession, $R(t)$, as a function of time and quantify the maximum erosion, R_{max}.

GIVEN:
$H_b = 3$ m, $h_b = 3.85$m, $D = 0.2$ mm, $B = 2$ m, $m_o = 1:10$, $T_D = 10$ hr, and $S_{max} = 1.5$ m.

SOLUTION:

As in previous examples, the value of the profile scale parameter is determined from Figure III-3-18 and/or Table III-3-3 as 0.1 $m^{1/3}$. The active width of the surf zone W_b is calculated from Equation 3-19 as

$$W_b = \left(\frac{h_b}{A}\right)^{3/2} = \left(\frac{3.85}{0.1}\right)^{3/2} = 238.9 \text{ m}$$

The equilibrium value of the shoreline response based on the maximum water level S_{max} is determined from Equation 3-38 as

$$R_\infty = \frac{S\left(W_b - \dfrac{h_b}{m_o}\right)}{B + h_b - \dfrac{S}{2}}$$

$$= \frac{1.5\left(238.9 - \dfrac{3.85}{0.1}\right)}{2 + 3.85 - \dfrac{1.5}{2}} = 58.9 \text{ m}$$

The morphological time scale T_s is determined from Figure III-3-39 by calculating the value of the abscissa in Figure III-3-39 as

$$\frac{H_b^{3/2}}{g^{1/2}A^3}\left(\frac{1}{1 + \dfrac{h_b}{B} + \dfrac{m_o W_b}{h_b}}\right)$$

$$\frac{3^{3/2}}{(9.8)^{1/2}(0.1)^3}\left(\frac{1}{1 + \dfrac{3.85}{2} + \dfrac{(0.1)(238.9)}{3.85}}\right)$$

$$= 181.8 \text{ sec} \approx 0.053 \text{ hr}$$

(Continued)

Example Problem III-3-7 (Concluded)

The morphological time scale is determined from Figure III-3-39 to be approximately 17.0 hr. The time-varying shoreline recession is determined from Equation 3-49 and is presented in Figure III-3-40. The maximum erosion can be determined from Equation 3-49 or directly from Figure III-3-40 as $R_{max}/R_\infty =$ 0.236 or $R_{max} = 13.9$ m.

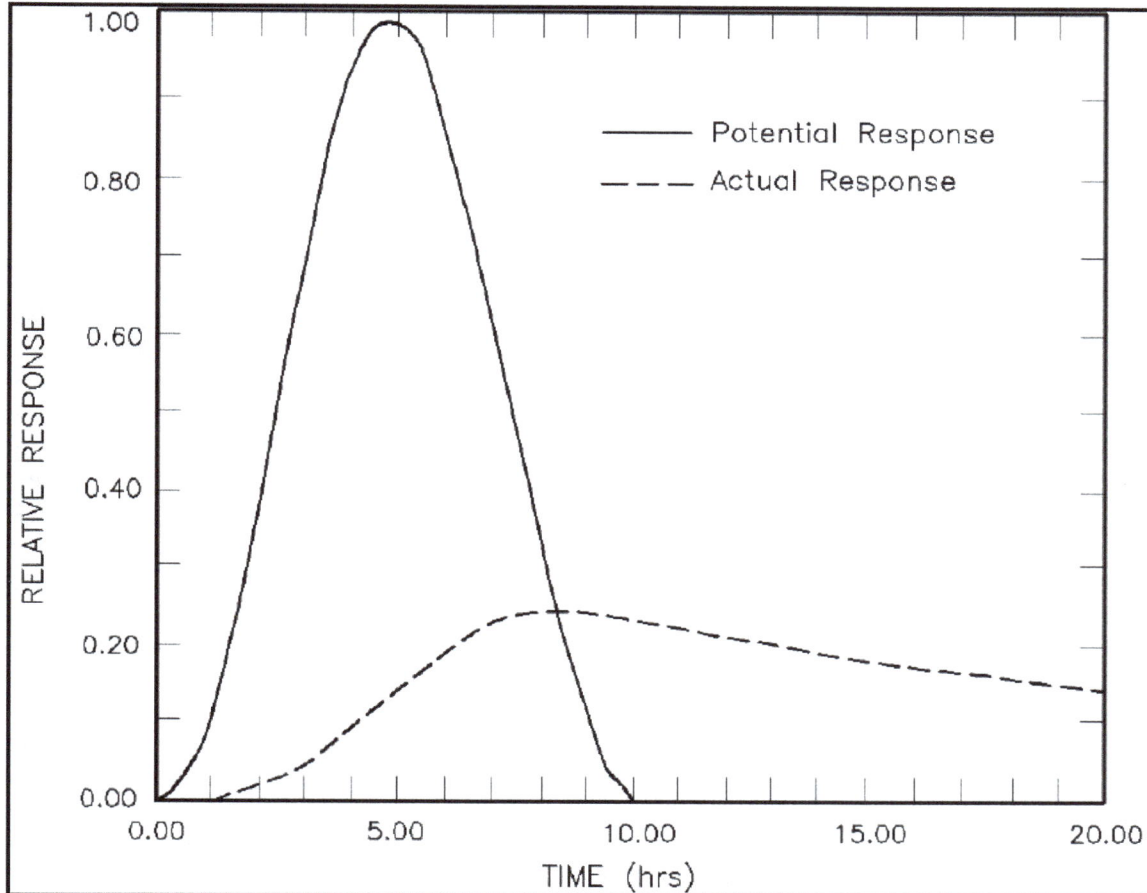

Figure III-3-40. Potential and actual shoreline response based on Kriebel and Dean (1993) model

very fine sand, subjected to very large breaking wave heights, have extremely long time scales such that they will experience only a small percentage of their equilibrium erosion potential during a typical storm.

k. Examples of numerical models.

(1) The numerical model described by Kriebel and Dean (1985) and later Kriebel (1986, 1990), was the first widely used numerical model developed to simulate storm-induced erosion based on equilibrium beach profile concepts. This model assumes that a beach profile will evolve toward an equilibrium form in response

to changing water levels and wave conditions and that the sediment transport rate is proportional to the "disequilibrium" existing between the profile at any instant in time and the equilibrium profile form.

This is quantified in terms of the "excess" energy dissipation per unit volume in the surf zone, as given by Equation 3-45. The energy dissipation per unit volume at any location in the surf zone is given by

$$D_* = \frac{1}{h}\frac{\partial(EC_g)}{\partial y} \approx \frac{5}{16}\rho g^{\frac{3}{2}}\kappa^2 h^{\frac{1}{2}}\frac{\partial h}{\partial y} \qquad \text{(III-3-53)}$$

where the last form is based on the assumption of shallow-water breaking waves. Based on the equilibrium beach profile given by Equation 3-14, the equilibrium energy dissipation per unit volume is given by

$$D_* = \frac{5}{24}\rho g^{\frac{3}{2}}\kappa^2 A^{\frac{3}{2}} \qquad \text{(III-3-54)}$$

where the equilibrium profile parameter A is determined either from the sediment grain size, as suggested in Table III-3-3 or from a best-fit of the equilibrium profile equation $h = Ay^{2/3}$ to the measured beach profile.

   (2)   The numerical solution for profile response is based on a finite difference solution to the sediment conservation equation given in Equation 3-44. In this case, the profile is gridded in a "stair-step" form as shown in Figure III-3-36(b) so that erosion or accretion (retreat or advance) of each elevation contour is determined by vertical gradients in the sediment transport rate. At each time-step, the local depth h in Equation 3-53 is the depth to the sand bed below the time-varying storm surge. As a result, beach profile change is driven primarily by changes in water level associated with storm surge. Breaking waves are treated very simply by shallow-water, spilling-breaker assumptions and, as a result, have a secondary effect on profile response. In effect, the breakpoint serves to separate the two main computational domains in the model: the offshore region, where the sediment transport rate is assumed to equal zero, and the surf zone, where the transport rate is given by Equation III-3-45.

   (3)   In this model, an increase in water level due to storm surge allows waves to break closer to shore, thus temporarily decreasing the width of the surf zone and increasing the energy dissipation per unit volume above the equilibrium level. According to Equation III-3-45, this leads to offshore directed sediment transport, the gradients of which cause erosion of the foreshore, deposition near the breakpoint, and an overall widening of the profile toward a new equilibrium form. At each time-step in the finite-difference solution, the profile responds toward equilibrium, but this is rarely, if ever, achieved due to the limited storm durations.

   (4)   The transport relationship used in the Kriebel and Dean model, given by Equation 3-45, requires calibration of a single empirical parameter K'. Kriebel (1986) first determined this parameter from numerical simulations of both large wave tank tests and hurricane-induced beach profile change as observed in Hurricane Eloise on the Gulf Coast of Florida. Figure III-3-41 shows numerical profile development along with results of large wave tank tests of Saville (1957). This illustrates the time-dependent profile development. It is noted that in the offshore region near the breakpoint, numerical results appear reasonable; however, the model does not simulate bar and trough features. For the Hurricane Eloise data, one profile was calibrated and the calibrated model was then applied to 20 additional profiles. Overall results showed that the model was capable of predicting the volume eroded from the dune to within about 25 to 40 percent, with little bias toward either under- or overestimating the volume eroded. Figure III-3-42 shows numerical model simulation of a long-duration extra-tropical storm from Point Pleasant, New Jersey. In these cases, measured

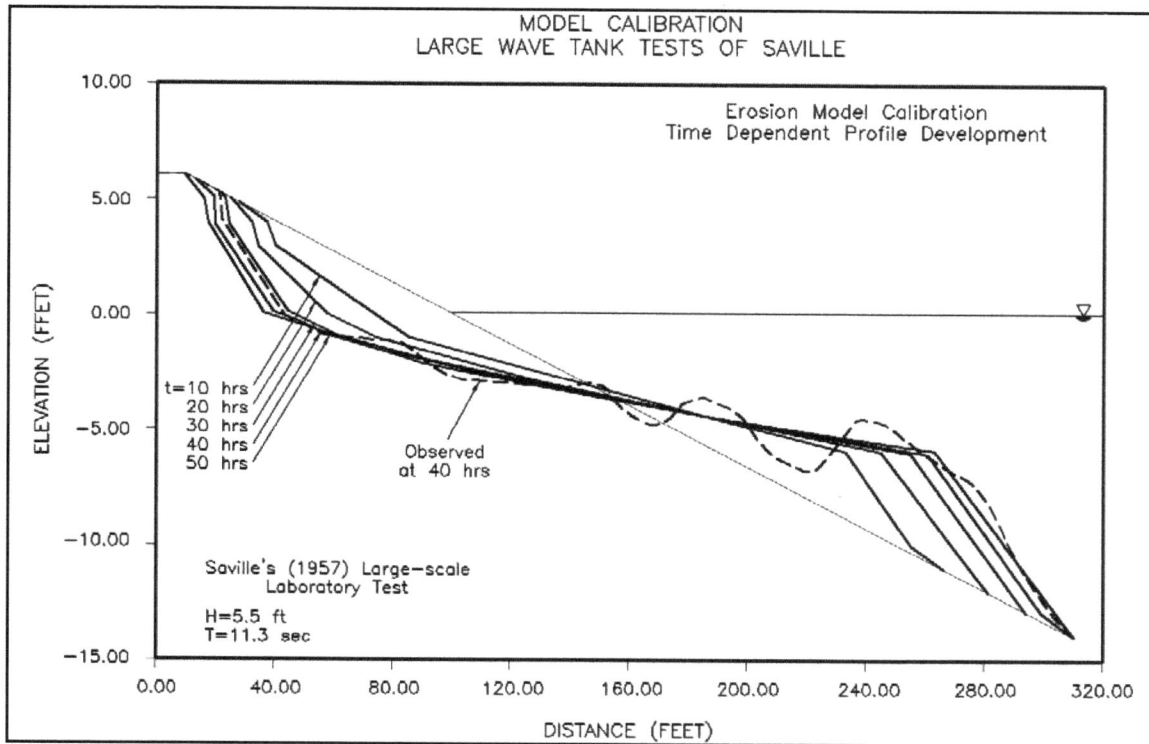

Figure III-3-41. Example of Kriebel and Dean erosion model calibration using large-wave tank data of Saville (1957) (from Kriebel 1990)

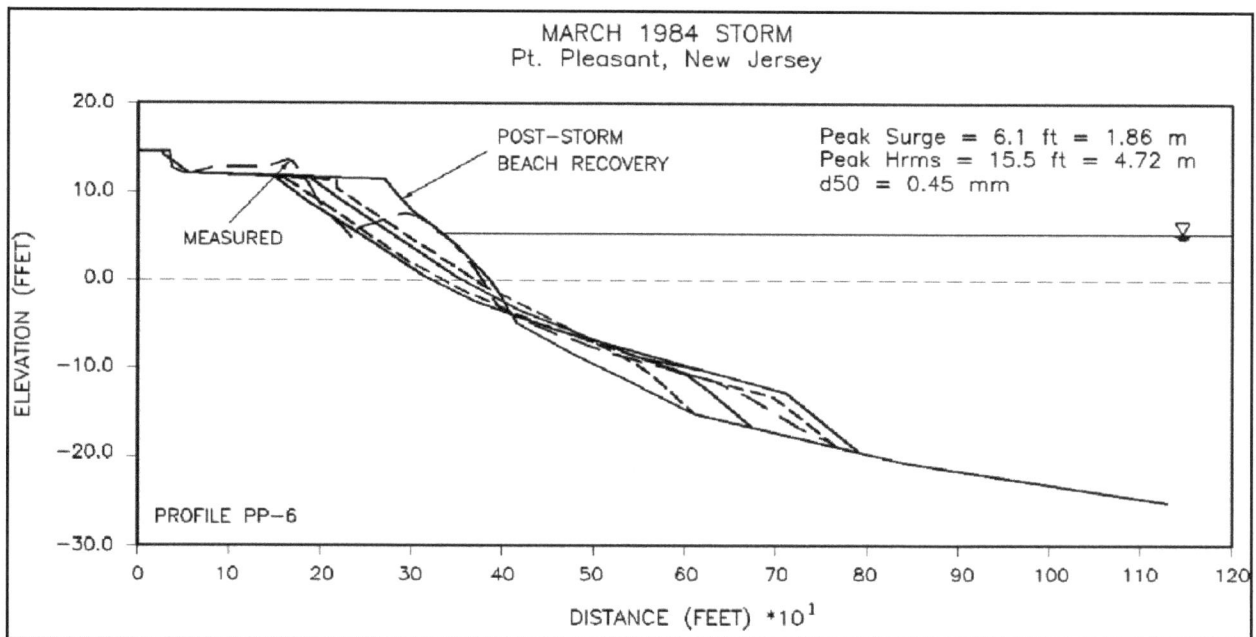

Figure III-3-42. Comparison of Kriebel and Dean erosion model to measured profiles from northeast storm at Point Pleasant, NJ (from Kriebel 1990)

post-storm profiles show evidence of post-storm beach recovery, which is not predicted in the numerical model.

An earlier version of the Kriebel and Dean model is provided in the ACES software package (Leenknecht, Szuwalski, and Sherlock 1992).

(5) The numerical model by Larson (1988) and Larson and Kraus (1989, 1990), SBEACH, is conceptually similar to the model of Kriebel and Dean (1985) but contains a more detailed description of breaking wave transformation and sediment transport across the beach profile, especially near the breakpoint. This model approximates the equation for conservation of sand in Equation 3-43 in finite difference form based on the profile gridding depicted in Figure III-3-36(a). Thus, vertical changes in water depth are determined by horizontal gradients in sediment transport rate. In contrast to the Kriebel and Dean model, this allows simulation of breakpoint bar formation and evolution.

(6) In the Larson and Kraus model, sediment transport rates in the surf zone are generally determined by Equation 3-46 in terms of excess energy dissipation, but with an additional effect of the local bottom slope. Because of this additional term, Equation 3-46 requires calibration through adjustment of two parameters, K'' and ϵ. The breaking wave model employed in the Larson and Kraus model is more sophisticated than that used by Kriebel and Dean, and is based on the breaking wave model of Dally, Dean, and Dalrymple (1985). This breaking wave model introduces gradients in the breaking wave height and energy dissipation that, in turn, lead naturally to gradients in sediment transport that produce bar/trough formations. Because of this improved breaking wave model, beach profile changes can be driven by changes in wave conditions, in addition to changes in water level.

(7) The computational domain used in the Larson and Kraus model is divided into four regions across the beach profile and the exact sediment transport relationship is adjusted somewhat for each of the four regions. In the surf zone, which is the major region for cross-shore sediment transport in the model, transport directions are first determined from the following critical value of wave steepness

$$\frac{H_o}{L_o} > \text{ or } < 0.0007 \left(\frac{H_o}{w_f T} \right)^3 \tag{III-3-55}$$

which is recognized as Equation 3-30 presented earlier. At each time-step in the solution, if the actual value of wave steepness exceeds the critical value given above, then transport is directed offshore over the entire active profile. Transport is then onshore if wave steepness is smaller than this critical value. Transport magnitudes in the surf zone are then determined by Equation 3-44, although the transport is "turned off" if the wave energy dissipation is below a critical value since this might cause a reversal in transport direction in conflict with the transport direction determined from wave steepness. It is noted, however, that the relationship determining transport direction (Equation 3-53) does not include any of the profile characteristics. In some cases where the profile is very steep, as may occur after beach nourishment, for example, the transport direction utilized in the model may be onshore based on wave and sediment characteristics whereas the actual transport would likely be offshore, even under mild wave conditions, due to the artificially high profile slope.

(8) The Larson and Kraus model has been compared with laboratory and field data as shown in Figures III-3-43 and III-3-44. Figure III-3-43 presents results for the two sets of available large-scale wave tank data. The upper panel shows a comparison from the Saville (1957) data and the comparisons include calculations at various times and the measured profile at 40 hr. The lower panel is for data from the Japan large tank and presents calculations at various times and the measured profile at 30-1/2 hr. The calculations

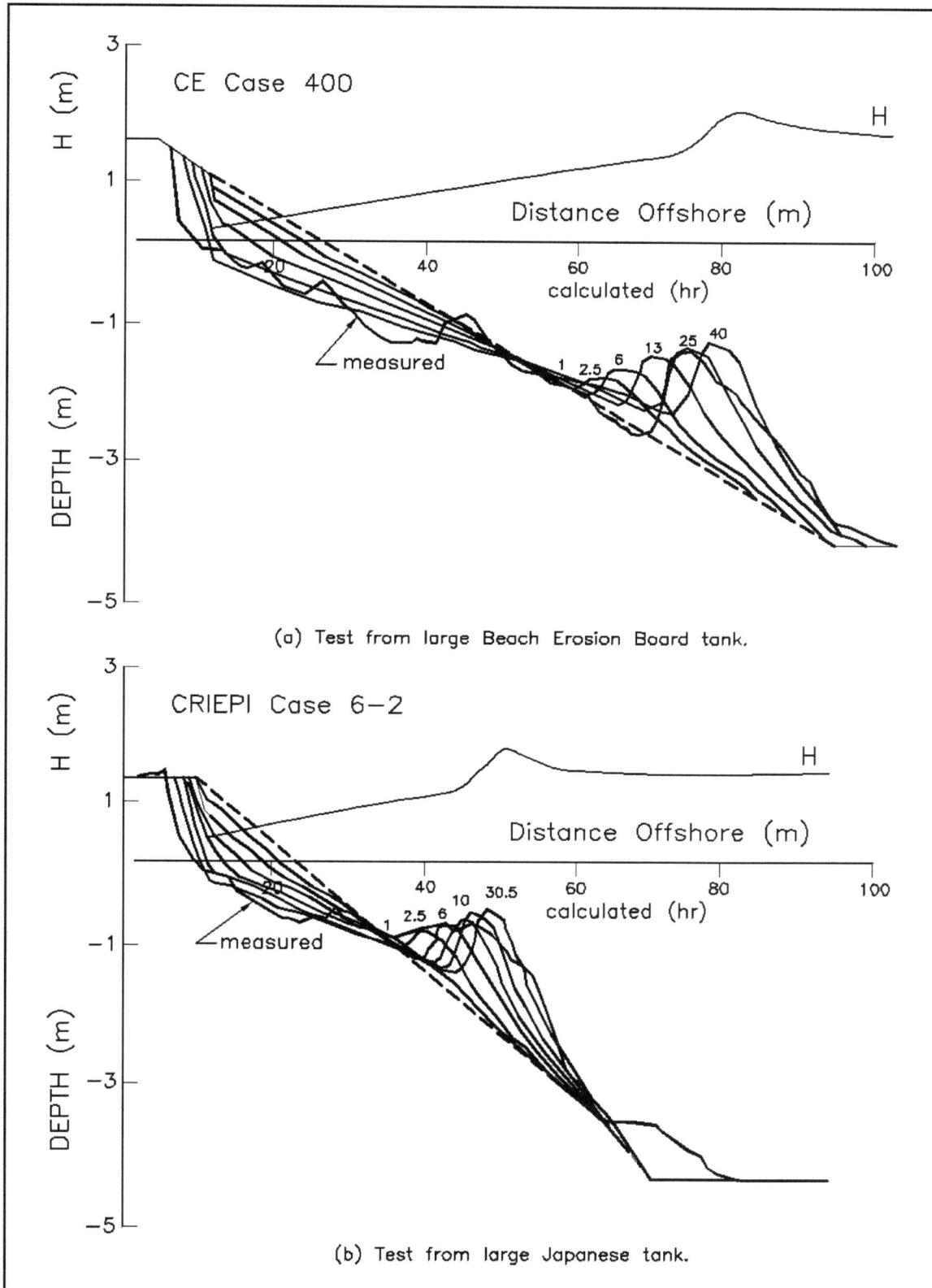

(a) Test from large Beach Erosion Board tank.

(b) Test from large Japanese tank.

Figure III-3-43. SBEACH compared to two tests from large-scale wave tanks (Larson and Kraus 1989)

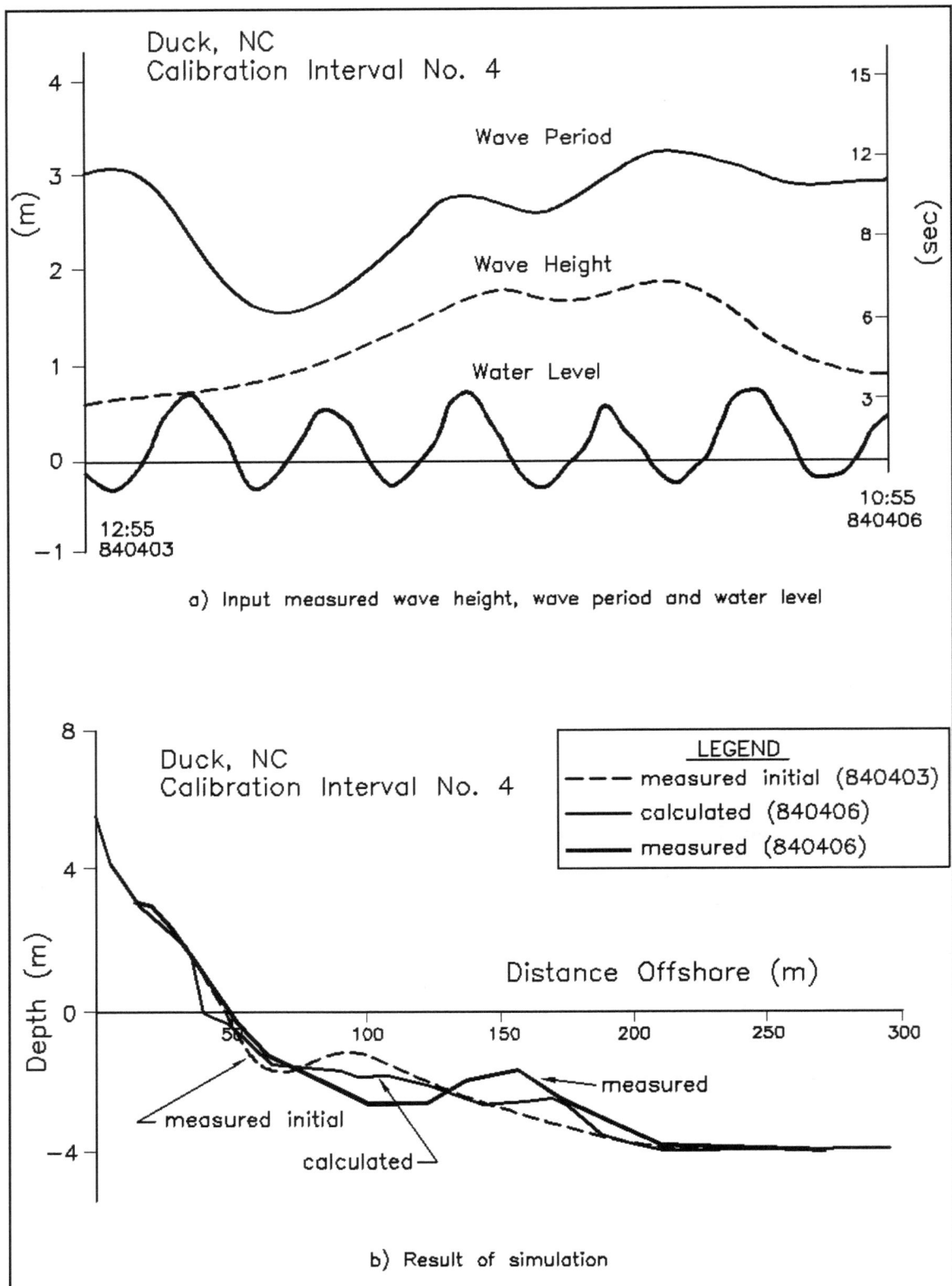

a) Input measured wave height, wave period and water level

b) Result of simulation

Figure III-3-44.   SBEACH tested against profile evolution data from Duck, NC (Larson and Kraus 1989)

are in quite good agreement with the data. The regular waves used in these tests were monochromatic, which tend to favor the well-developed and concentrated bars that are simulated quite well by the numerical model. Figure III-3-44, from Duck, NC, compares the evolution of a profile over a 3-day period in which the wave heights were on the order of 1.5 m. The initial profile included a bar located approximately 40 m from shore. During the period of interest, the bar migrated seaward approximately 65 m. The simulations provide a reasonable qualitative representation of the evolution. The bar becomes more subdued at its initial location, but in contrast to measurements, is still present at the final time. Also, the bar had started to emerge slightly seaward of the measured location. The calculations showed substantially greater erosion at the shoreline than measured. This example demonstrates the extreme difficulty in simulating an actual event in nature. In the SBEACH model, sediment transport rates from Equation 3-46 require calibration through adjustment of two parameters, K'' and ϵ.

l. Physical modeling of beach profile response.

(1) Physical modeling of beach profile response is carried out with the model being a scaled version of the prototype. In recent years, physical modelling of profiles has been employed predominantly as a research method to understand transport processes rather than as a means to investigate profile response to a particular scenario of water level and wave conditions. The ratio of quantities in the model to those in the prototype is termed the "scale ratio" and will be designated here by a subscript "*r*." For example, the length and wave period ratios would be L_r and T_r, respectively. In some models, it is appropriate to utilize a distorted model in which the vertical scale ratio is different from that of the horizontal scale. Modelling of cross-shore sediment transport requires the determination of the appropriate scaling relationships for both the waves and sediments. Hughes (1994) presents a complete discussion of scaling laws as applied to predicting cross-shore sediment transport.

(2) Noda (1972) carried out a study of profile modelling and has found that distorted models were appropriate. The horizontal and vertical scale ratios were recommended as

$$D_r\,(S_r)^{1.85}\;=\;h_r^{0.55} \tag{III-3-56}$$

$$l_r\;=\;h_r^{1.32}\,S^{-0.386} \tag{III-3-57}$$

in which D_r is the grain size ratio, s_r is the submerged specific weight ratio, $s_r = ((\rho_s - \rho)/\rho)_r$ in which ρ_s and ρ are the mass density of sediment and water, respectively, l_r is the horizontal length scale, and h_r is the vertical length ratio. Figure III-3-45 presents Noda's recommended scaling relationships. Usually sand is the common material in both the model and prototype, $\rho_s = 2650$ kg/m^3, and Equation III-3-56 and III-3-57 become

$$D_r\;=\;h_r^{0.55} \tag{III-3-58}$$

and

$$l_r\;=\;h_r^{1.32} \tag{III-3-59}$$

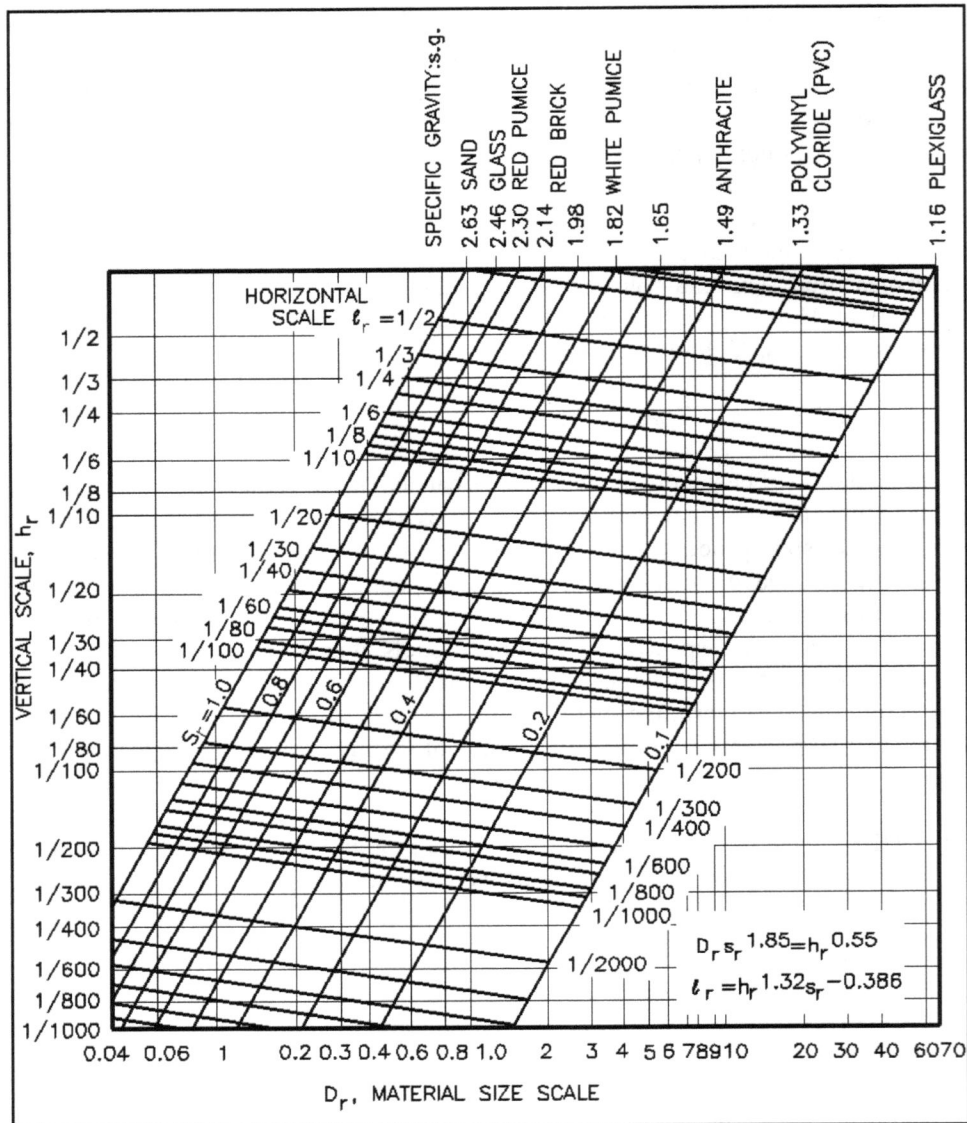

Figure III-3-45.   Noda's recommendation for profile modeling (Noda 1972)

(3)   Thus, according to Noda's relationships, the diameter would be scaled in accordance with the depth ratio; however, the scaling factor for diameter would be closer to unity than that for the depth ratio and the length ratio would be smaller than the depth ratio. This type of distortion is common for hydraulic models in which it is necessary to represent a large horizontal extent.

(4)   Dean (1973) carried out a study of conditions that would lead to bar formation and suggested that the fall time of a sediment particle suspended at the wave crest phase position relative to the wave period would be relevant in modelling applications. This led Dean (1973, 1985) to identify the following combination of parameters which should be maintained the same in model and prototype

$$\frac{H}{w_f T}$$

(III-3-60)

(5)   This combination of terms will be termed herein as the "Fall Velocity Parameter" (FVP). It was further suggested that an appropriate model for cross-shore sediment transport is one based on undistorted Froude modelling for the wave characteristics and one that maintains the FVP the same in the model and prototype. This simple approach leads to the following requirement for scaling the sediment fall velocity w_f

$$(w_f)_r = \sqrt{L_r} \qquad\qquad\qquad\qquad \text{(III-3-61)}$$

which is the standard relationship for velocity scaling for a Froude model.

(6)   Evaluation of cross-shore modelling according to the FVP has been carried out by Vellinga (1983), Kriebel, Dally, and Dean (1986) and Hughes and Fowler (1990). Each of these studies concluded that the FVP was effective in scaling the erosion process. Figures III-3-46 and III-3-47, from Kriebel, Dally, and Dean (1986) and Hughes and Fowler (1990), compare different scales while maintaining the same FVP in model and prototype. Considering the FVP as valid leads to the following valuable transport relationship for numerical models

$$q_r = L_r^{\frac{3}{2}} \qquad\qquad\qquad\qquad \text{(III-3-62)}$$

(7)   One limitation for scaling by the FVP is that the length ratio L_r and the prototype diameter can result in designated model sediments so small that cohesive forces would be significant. Although there are no strict guidelines for a minimum sediment size, values smaller than approximately 0.08 to 0.09 mm should be avoided.

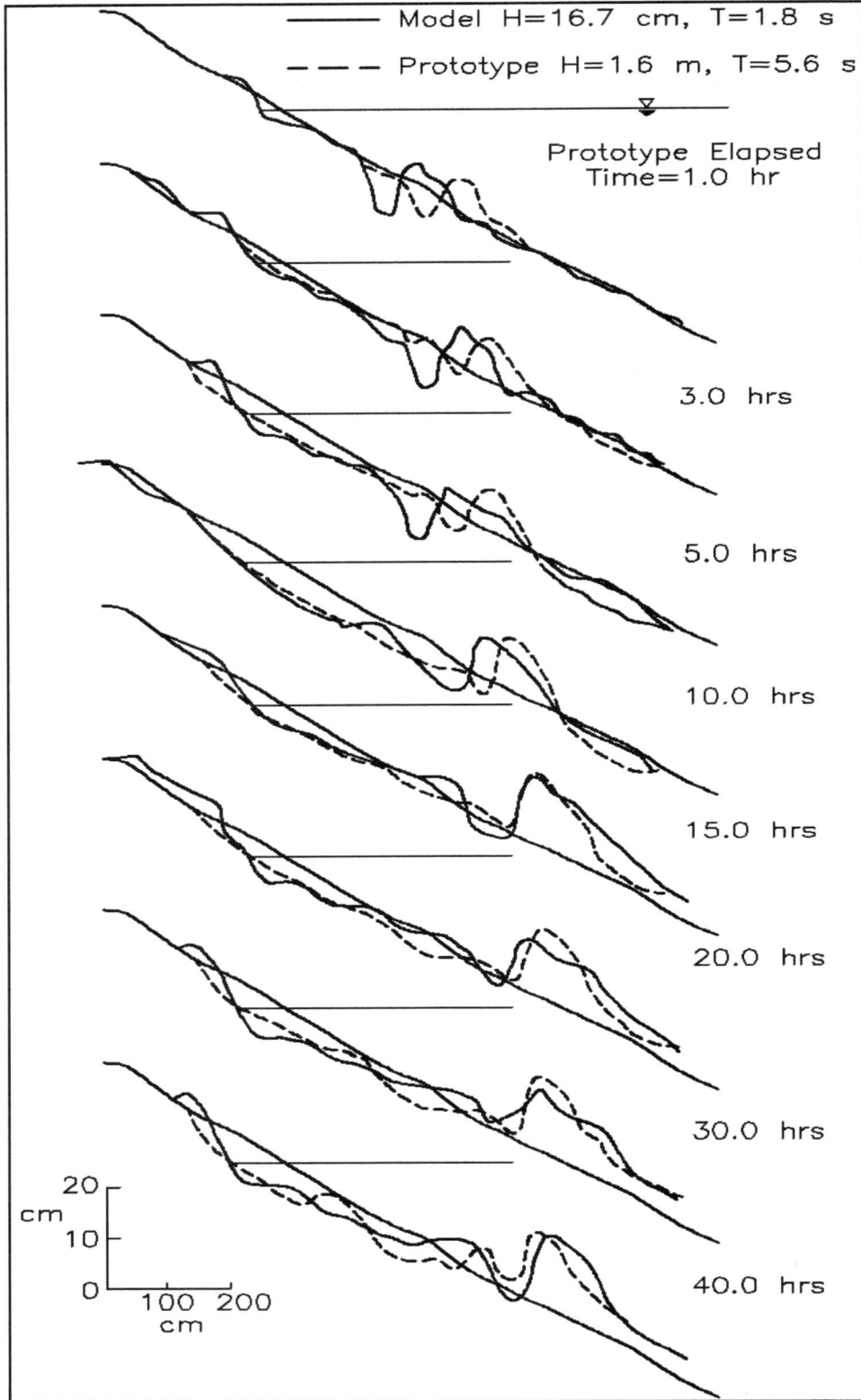

Figure III-3-46.   Profile evolution by small- and large-scale wave tank tests. Based on maintaining the same fall velocity parameter. Length ratio = 1:9.6 (Kriebel, Dally, and Dean 1986)

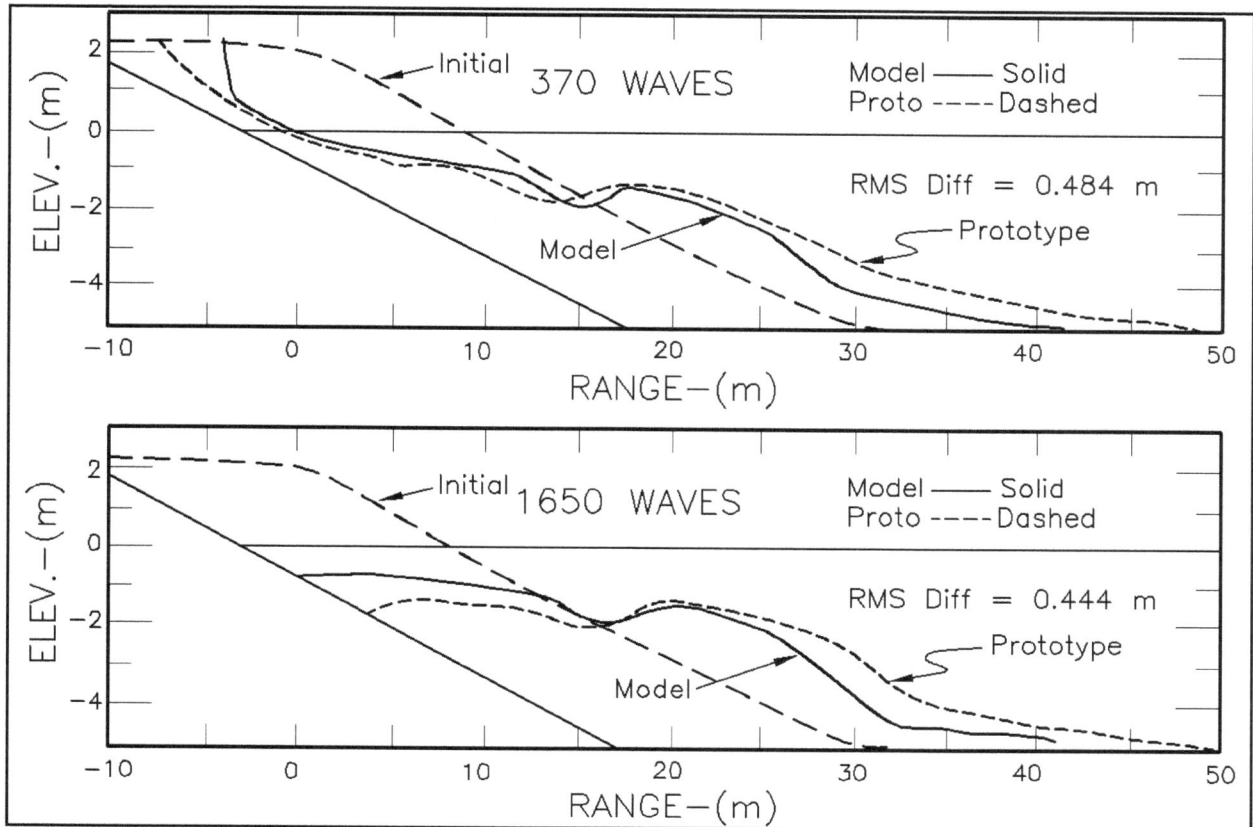

Figure III-3-47.   Profile evolution by small- and large-scale wave tank tests. Case of sloping seawall. Based on maintaining the same fall velocity parameter. Length ratio = 1:7.5 (Hughes and Fowler 1990)

III-3-4. References

Bagnold 1940
Bagnold, R. A. 1940. "Beach Formation by Waves: Some Model Experiments in a Wave Tank," *Journal of the Institution of Civil Engineers,* Vol 15, pp 27-52.

Bagnold 1966
Bagnold, R. A. 1966. "An Approach to the Sediment Transport Problem from General Physics," U.S. Geological Survey Professional Paper 422-I, U. S. Dept of Interior.

Bailard 1981
Bailard, J. A. 1981. "An Energetics Total Load Sediment Transport Model for a Plane Sloping Beach," *Journal of Geophysical Research*, Vol 86, No. C11, pp 10938-10954.

Barnett and Wang 1988
Barnett, M., and Wang, H. 1988. "Effects of a Vertical Seawall on Profile Response," American Society of Civil Engineers, *Proceedings of the Twenty-first International Conference on Coastal Engineering,* Chapter 111, pp 1493-1507.

Birkemeier 1984
Birkemeier, W. A. 1984. "Time Scales of Nearshore Profile Change," *Proceedings, 19th International Conference on Coastal Engineering,* Chapter 102, pp 1501-1521.

Birkemeier 1985
Birkemeier, W. A. 1985. "Field Data on Seaward Limit of Profile Change," *Journal of the Waterways, Port Coastal and Ocean Engineering,* American Society of Civil Engineers, Vol 111, No. 3, pp 598-602.

Bodge 1992
Bodge, K. R. 1992. "Representing Equilibrium Beach Profiles with an Exponential Expression," *Journal of Coastal Research,* Vol 8, No. 1, pp 47-55.

Broker-Hedegaard, Deigaard, and Fredsoe 1991
Broker-Hedegaard, I., Deigaard, R., and Fredsoe, J. 1991. "Onshore/Offshore Sediment Transport and Morphological Modelling of Coastal Profiles,"*Proceedings Coastal Sediments '91 Conference*, pp 643-657.

Bruun 1954
Bruun, P. 1954. "Coast Erosion and the Development of Beach Profiles," Beach Erosion Board Technical Memorandum No. 44, U.S. Army Engineer Waterways Experiment Station, Vicksburg, MS.

Bruun 1962
Brunn, P. 1962. "Sea-Level Rise as a Cause of Shore Erosion," *Journal of Waterways and Harbor Division,* American Society of Civil Engineers, Vol 88, pp 117-130.

Bruun 1988
Bruun, P. 1988. "The Bruun Rule of Erosion by Sea Level Rise: A Discussion on Large-Scale Two- and Three-Dimensional Usages," *Journal of Coastal Research*, Vol 4, No. 4, pp 627-648.

Chiu 1977
Chiu, T. Y. 1977. "Beach and Dune Response to Hurricane Eloise of September 1975. *Proceedings of Coastal Sediments '77,* pp 116-134.

Clausner, Birkemeier, and Clark 1986
Clausner, J. E., Birkemeier, W. A., and Clark, G. R. 1986. "Field Comparison of Four Nearshore Survey Systems," Miscellaneous Paper CERC-86-6, U.S. Army Engineer Waterways Experiment Station, Vicksburg, MS.

Dally and Dean 1984
Dally, W. R., and Dean, R. G. 1984. "Suspended Sediment Transport and Beach Profile Evolution," *Journal of Waterway, Port, Coastal, and Ocean Engineering,* Vol 110, No. 1, pp 15-33.

Dally, Dean, and Dalrymple 1985
Dally, W. R., Dean, R. G., and Dalrymple, R. G. 1985. "Wave Height Variation Across Beaches of Arbitrary Profile," *Journal of Geophysical Research*, Vol 90, No. C6, pp 11917-11927.

Dalrymple 1992
Dalrymple, R. A. 1992. "Prediction of Storm/Normal Beach Profiles," *Journal of Waterways, Port, Coastal and Ocean Engineering,* American Society of Civil Engineers, Vol 118, No. 2, pp 193-200.

Dean 1973
Dean, R. G. 1973. "Heuristic Models of Sand Transport in the Surf Zone," *Proceedings, Conference on Engineering Dynamics in the Surf Zone,* Sydney, Australia.

Dean 1977
Dean, R. G. 1977. "Equilibrium Beach Profiles: U.S. Atlantic and Gulf Coasts," Department of Civil Engineering, Ocean Engineering Report No. 12, University of Delaware, Newark, DE.

Dean 1985
Dean, R. G. 1985. "Physical Modelling of Littoral Processes," *Physical Modelling in Coastal Engineering,* R. A. Dalrymple, ed., A. A. Balkema, pp 119-139.

Dean 1987a
Dean, R. G. 1987a. "Additional Sediment Input into the Nearshore Region," *Shore and Beach,* Vol 55, Nos. 3-4, pp 76-81.

Dean 1987b
Dean, R. G. 1987b. "Coastal Sediment Processes: Toward Engineering Solutions," *Coastal Sediments '87,* American Society of Civil Engineers, New Orleans, LA, Vol 1, pp 1-24.

Dean 1991
Dean, R. G. 1991. "Equilibrium Beach Profiles: Characteristics and Applications," *Journal of Coastal Research,* Vol 7, No. 1, pp 53-84.

Dean and Dalrymple 1991
Dean, R. G., and Dalrymple, R. A. 1991. *Water Wave Mechanics for Engineers and Scientists,* World Scientific Pub. Co., Teaneck, NJ.

Dean and Dalrymple 2001
Dean, R. G., and Dalrymple, R. A. 2001. *Coastal Processes with Engineering Applications,* Cambridge University Press, New York.

Dean and Maurmeyer 1983
Dean, R. G., and Maurmeyer, E. M. 1983. "Models for Beach Profile Response," *CRC Handbook on Beach Erosion and Coastal Processes,* P. D. Komar, ed., Chapter 7, pp 151-166.

Dette and Uliczka 1987
Dette, H., and Uliczka, K. 1987. "Prototype Investigation on Time-Dependent Dune Recession and Beach Erosion," *Proceedings Coastal Sediments '87,* New Orleans, pp 1430-1444.

Dewall 1979
Dewall, A. E. 1979. "Beach Changes at Westhampton Beach, New York," MR 79-5, Coastal Engineering Research Center, U.S. Army Engineer Waterways Experiment Station, Vicksburg, MS.

Dewall and Richter 1977
Dewall, A. E., and Richter, J. J. 1977. "Beach and Nearshore Processes in Southeastern Florida," *Coastal Sediments '77,* pp 425-443.

Dolan and Dean 1985
Dolan, T., and Dean, R. G. 1985. "Multiple Longshore Sand Bars in the Upper Chesapeake Bay," *Estuarine Coastal & Shelf Science*, Vol 21, pp 727-743.

Edelman 1972
Edelman, T. 1972. "Dune Erosion During Storm Conditions," *Proceedings of the Thirteenth International Conference on Coastal Engineering,* pp 1305-1312.

Everts 1985
Everts, C. H. 1985. "Sea Level Rise Effects on Shoreline Position," *Journal of Waterway, Port, and Coastal Engineering*, Vol 111, No. 6, pp 985-999.

Hallermeier 1978
Hallermeier, R. J. 1978. "Uses for a Calculated Limit Depth to Beach Erosion," *Proceedings of the 16th International Conference on Coastal Engineering,* American Society of Civil Engineers, Hamburg, pp 1493-1512.

Hallermeier 1981
Hallermeier, R. J. 1981. "A Profile Zonation for Seasonal Sand Beaches from Wave Climate," *Coastal Engineering,* Vol 4, pp 253-277.

Hands 1983
Hands, E. 1983. "Erosion of the Great Lakes Due to Changes in the Water Level," *CRC Handbook of Coastal Processes and Erosion*, P. D. Komar, ed., CRC Press, pp 167-189.

Hicks, Debaugh, and Hickman 1983
Hicks, S. D., Debaugh, H. A., Jr., and Hickman, E. 1983. "Sea Level Variations for the United States, 1955-1980," National Ocean Service, National Oceanic and Atmospheric Administration.

Hughes 1994
Hughes, S. A. 1994. *Physical Models and Laboratory Techniques in Coastal Engineering.* World Scientific, River Edge, NJ.

Hughes and Fowler 1990
Hughes, S. A., and Fowler, J. E. 1990. "Validation of Movable-Bed Modeling Guidance," *Proceedings of the 22nd International Conference on Coastal Engineering,* American Society of Civil Engineers, Chapter 186, pp 2457-2470.

Inman, Elwany, and Jenkkins 1993
Inman, D. L., Elwany, M. H. S., and Jenkkins, S. A. 1993. "Shoreline and Bar-Berm Profiles on Ocean Beaches," *Journal of Geophysical Research*, Vol 98, No. C10, pp 18,181 - 18,199.

Katoh and Yanagishima 1988
Katoh, K., and Yanagishima, S. 1988. "Predictive Model for Daily Changes of Shoreline," *Proceedings of the Twenty-First International Conference on Coastal Engineering,* American Society of Civil Engineers, Hamburg, Chapter 93, pp 1253-1264.

Keulegan 1945
Keulegan, G. H. 1945. "Depths of Offshore Bars," Engineering Notes No. 8, Beach Erosion Board, U.S. Army Engineer Waterways Experiment Station, Vicksburg, MS.

Keulegan 1948
Keulegan, G. H. 1948. "An Experimental Study of Submarine Sand Bars," Technical Report No. 8, Beach Erosion Board, U.S. Army Engineer Waterways Experiment Station, Vicksburg, MS.

Kobayashi 1987
Kobayashi, N. 1987. "Analytical Solutions for Dune Erosion by Storms," *Journal of the Waterway, Coastal and Ocean Engineering,* American Society of Civil Engineers, Vol 113, No. 4, pp 401-418.

Komar and McDougal 1994
Komar, P. D., and McDougal, W. G. 1994. "The Analysis of Exponential Beach Profiles," *Journal of Coastal Research,* Vol 10, pp 59-69.

Kraus 1988
Kraus, N. C. 1988. "The Effects of Seawalls on the Beach: An Extended Literature Review," *Journal of Coastal Research*, Special Issue No. 4, pp 1-28.

Kraus, Larson, and Kriebel 1991
Kraus, N. C., Larson, M., and Kriebel, D. L. 1991. "Evaluation of Beach Erosion and Accretion Predictors," *Proceedings of Conference on Coastal Sediments '91,* American Society of Civil Engineers, pp 572-587.

Kriebel 1986
Kriebel, D. L. 1986. "Verification Study of a Dune Erosion Model," *Shore and Beach,* Vol 54, No. 3, pp 13-20.

Kriebel 1987
Kriebel, D. L. 1987. "Beach Recovery Following Hurricane Elena," *Proceedings of Conference on Coastal Sediments '87,* American Society of Civil Engineers, pp 990-1005.

Kriebel 1990
Kriebel, D. L. 1990. "Advances in Numerical Modeling of Dune Erosion," *22nd Intl. Conf. on Coastal Engineering,* American Society of Civil Engineers, pp 2304-2317.

Kriebel and Dean 1985
Kriebel, D. L., and Dean, R. G. 1985. "Numerical Simulation of Time-Dependent Beach and Dune Erosion," *Coastal Engineering,* Vol 9, pp 221-245.

Kriebel and Dean 1993
Kriebel, D. L., and Dean, R. G. 1993. "Convolution Method for Time-Dependent Beach-Profile Response," *Journal of Waterway, Port, Coastal and Ocean Engineering,* American Society of Civil Engineers, Vol 119, No. 2, pp 204-227.

Kriebel, Dally, and Dean 1986
Kriebel, D. L., Dally, W. R., and Dean, R. G. 1986. "Undistorted Froude Model for Surf Zone Sediment Transport," *Proceedings of the Twentieth International Conference on Coastal Engineering,* American Society of Civil Engineers, pp 1296-1310.

Kriebel, Kraus, and Larson 1991
Kriebel, D. L., Kraus, N. C., and Larson, M. 1991. "Engineering Methods for Predicting Beach Profile Response," *Proceedings of Conference on Coastal Sediments '91,* American Society of Civil Engineers, pp 557-571.

Larson 1988
Larson, M. 1988. "Quantification of Beach Profile Change," *Report No. 1008,* Department of Water Resources and Engineering, University of Lund, Lund, Sweden.

Larson and Kraus 1989
Larson, M., and Kraus, N. C. 1989. "SBEACH: Numerical Model for Simulating Storm-induced Beach Change; Report 1: Empirical Foundation and Model Development," Technical Report CERC-89-9, U.S. Army Engineer Waterways Experiment Station, Vicksburg, MS.

Larson and Kraus 1990
Larson, M., and Kraus, N. C. 1990. "SBEACH: Numerical Model for Simulating Storm-induced Beach Change; Report 2: Numerical Formulation and Model Tests," Technical Report CERC-89-9, U.S. Army Engineer Waterways Experiment Station, Vicksburg, MS.

Lee and Birkemeier 1993
Lee, G. H., and Birkemeier, W. A. 1993. "Beach and Nearshore Survey Data: 1985-1991 CERC Field Research Facility," Technical Report CERC-93-3, U.S. Army Engineer Waterways Experiment Station, Vicksburg, MS.

Leenknecht, Szuwalski, and Sherlock 1992
Leenknecht, D. A., Szuwalski, A., and Sherlock, A. R. 1992. "Automated Coastal Engineering System, User Guide and Technical Reference, Version 1.07," U.S. Army Engineer Waterways Experiment Station, Vicksburg, MS.

Longuet-Higgins 1953
Longuet-Higgins, M. S. 1953. "Mass Transport in Water Waves," *Philosophical Transactions of the Royal Society of London,* Ser A, Vol 245, pp 535-581.

Longuet-Higgins and Stewart 1964
Longuet-Higgins, M. S., and Stewart, R. W. 1964. "Radiation Stresses in Water Waves: A Physical Discussion with Applications," *Deep Sea Research,* Vol 2, pp 529-562.

Moore 1982
Moore, B. D. 1982. "Beach Profile Evolution in Response to Changes to Water Level and Wave Height," M.S. thesis, Department of Civil Engineering, University of Delaware, Newark.

Nairn and Southgate 1993
Nairn, R. B., and Southgate, H. N. 1993. "Deterministic Profile Modelling of Nearshore Processes; Part 2, Sediment Transport and Beach Profile Development," *Coastal Engineering,* Vol 19, pp 57-96.

National Research Council 1984
National Research Council. 1984. "Responding to Changes in Sea Level: Engineering Implications," National Research Council Committee on Engineering Implication of Change in Relative Sea Level.

Noda 1972
Noda, E. K. 1972. "Equilibrium Beach Profile Scale Model Relationships," *Journal of Waterway, Harbors and Coastal Engineering,* American Society of Civil Engineers, Vol 98, No. 4, pp 511-528.

Penland, Suter, and McBride 1987
Penland, S., Suter, J. R., and McBride, R. A. 1987. "Delta Plain Development and Sea Level History in Terrebonne Coastal Region, Louisiana," *Proceedings of Conference on Coastal Sediments '87,* American Society of Civil Engineers, pp 1689-1705.

Roelvink and Broker 1993
Roelvink, J. A., and Broker, I. 1993. "Cross-Shore Profile Models," *Coastal Engineering,* Vol 21, pp 163-191.

Roelvink and Stive 1989
Roelvink, J. A., and Stive, M. J. F. 1989. "Bar Generating Cross-Shore Flow Mechanisms on a Beach," *Journal of Geophysical Research,* Vol 94, No. C4, pp 4785-4800.

Saville 1957
Saville, T. 1957. "Scale Effects in Two-Dimensional Beach Studies," *Transactions of the Seventh General Meeting of the International Association of Hydraulic Research,* Vol 1, pp A3.1 - A3.10.

Seymour and Boothman 1984
Seymour, R. J., and Boothman, D. P. 1984. "A Hydrostatic Profiler for Nearshore Surveying," *Journal of Coastal Engineering,* Vol 8, pp 1-14.

Stive and Battjes 1984
Stive, M. J. F., and Battjes, J. A. 1984. "A Model for Offshore Sediment Transport," *Proceedings 19th Intl. Conf. on Coastal Engineering,* pp 1420-1436.

Swart 1974
Swart, D. H. 1974. "Offshore Sediment Transport and Equilibrium Beach Profiles," Publ. No. 131, Delft Hydraulics Lab, Delft, The Netherlands.

Vellinga 1983
Vellinga, P. 1983. "Predictive Computational Model for Beach and Dune Erosion During Storm Surges," *Proceedings, American Society of Civil Engineers Specialty Conference on Coastal Structures' 83,* pp 806-819.

Watanabe, Riho, and Horikawa 1980
Watanabe, A., Riho, Y., and Horikawa, K. 1980. "Beach Profile and On-Offshore Sediment Transport," *Proceedings 17th International Conference on Coastal Engineering,* pp 1106-1121.

III-3-5. Definition of Symbols

β	Ratio of the erosion time scale to the storm duration [dimensionless]
Δy	Equilibrium dry beach width [length]
$\Delta y'$	Non-dimensional equilibrium dry beach width
ε	Parameter suggested by Larson & Kraus to calibrate a sediment transport model [dimensionless]
ε	Eddy viscosity [length^2/time]
κ	Ratio of wave height to local depth within the surf zone
v_b	Instantaneous wave-induced water particle velocity at the bottom [length/time]
v_s	Steady wave-induced water particle velocity [length/time]
ρ	Mass density of water (salt water = 1,025 kg/m^3 or 2.0 slugs/ft^3; fresh water = 1,000kg/m^3 or 1.94 slugs/ft^3) [force-time^2/length^4]
σ	Angular frequency (= $2\pi/T$) [time^{-1}]
σ_H	Standard deviation of significant wave height [length]
$\overline{\tau}_b$	Average bottom shear stress [force/length^2]
$\overline{\tau}_{bs}$	Bottom shear stress [force/length^2]
τ_η	Surface wind stress [force/length^2]
τ_b	Seaward-directed shear stress at the bottom [force/length^2]
A	Sediment scale or equilibrium profile parameter (Table III-3-3) [length$^{1/3}$]
A_F	Nourishment material scale parameter [length$^{1/3}$]
A_N	Native sediment scale parameter [length$^{1/3}$]
A'	Parameter for nourished beach calculations ($= A_F / A_N$) [dimensionless]
B	Berm height [length]
$B(t)$	Instantaneous total height of the active profile above the current water level [length]
B_0	Original berm height [length]
B'	Parameter for nourished beach calculations [dimensionless]
C	Wave speed [length/time]
D	Sediment grain diameter [length - generally millimeters]
D_*	Excess energy dissipation per unit volume in the surf zone
D_F	Sediment grain diameter of beach fill material [length - generally millimeters]
D_N	Sediment grain diameter of native beach [length] - generally millimeters
D_r	Model to prototype sediment grain size scale ratio

E	Total wave energy in one wavelength per unit crest width [length-force/length]
f	Darcy-Weisbach friction coefficient [dimensionless]
g	Gravitational acceleration (32.17 ft/sec^2, 9.807m/sec^2) [length/time^2]
h	Equilibrium beach profile depth (Equation III-3-13) [length]
h	Water depth [length]
H	Wave height [length]
h_*	Depth to which nourished profile will equilibrate or closure depth [length]
$\overline{H}$	Annual mean significant wave height [length]
h_0	Asymptotic beach profile depth [length]
H_0	Deepwater wave height [length]
h_b	Breaking depth [length]
h_c	Closure depth [length]
h_{CR}	Depth over bar crest [length]
h_D	Depth to the bar base [length]
H_e	Effective significant wave height [length]
h_r	Model to prototype vertical length scale ratio
H_s	Deepwater significant wave height [length]
h_T	Depth over bar trough [length]
k	Decay constant [dimensionless]
k	Wave number (= $2\pi/L = 2\pi/CT$) [length^{-1}]
K'	Parameter used to calibrate the Kriebel & Dean simple transport relationship [dimensionless]
K''	Parameter used to calibrate a sediment transport model [dimensionless]
L_*	Width of active profile [length]
L_0	Deepwater wave length [length]
l_r	Model to prototype horizontal length scale ratio
L_r	Model to prototype length scale ratio
m_0	Beach slope [length-rise/length-run]
n	Empirical exponent used in the equilibrium beach profile equation [dimensionless]
P	Single profile parameter (Equation III-3-32) [dimensionless]
Q	Time-averaged seaward discharge due to the return flow of shoreward mass transport

q_y	Kriebel & Dean simple transport relationship (Equation III-3-45)
R	Shoreline recession [length]
R_{max}	Maximum shoreline recession [length]
R_∞	Equilibrium berm recession due to a storm surge (Equation III-3-38) [length]
S	Storm surge level [length]
S	Sea level rise [length]
s_r	Model to prototype submerged specific weight scale ratio
T	Wave period [time]
T_D	Total storm surge duration [time]
T_e	Effective wave period [time]
T_r	Model to prototype wave period scale ratio
T_S	Time-scale of the equilibrium beach response system (Equation III-3-52) [time]
V	Added volume per unit beach length [length^3/length]
V_∞	Volume of sand eroded from the berm above the initial still-water level due to a storm surge level S [length^3/length]
V'	Parameter for nourished beach calculations ($= V/(B\,W_*)$) [dimensionless]
V'_{c2}	Critical volume of sand that will just yield a finite shoreline displacement for the case of sand that is finer than the native (Equation III-3-23) [dimensionless]
V'_{c1}	Critical volume of sand delineating intersecting and nonintersecting profiles (Equation III-3-22) [dimensionless]
w	Sediment fall velocity [length/time]
W_*	Width of active nearshore zone (Equation III-3-19) [length]
W_b	Width of the surf zone [length]
w_f	Sediment fall velocity [length/time]
y	Equilibrium beach profile distance offshore (Equation III-3-13) [length]
y	Distance seaward from mean low water [length]
y_0	Offset of the shoreline between the sloping beach face and the imaginary or virtual origin of the equilibrium profile [length]
z	Elevation [length]

III-3-6. Acknowledgments

Authors of Chapter III-3, "Cross-Shore Sediment Transport Processes:"

Robert G. Dean, Ph.D., Coastal & Oceanographic Engineering, University of Florida, Gainesville, Florida.

David L. Kriebel, Ph.D., Department of Ocean Engineering, United States Naval Academy, Annapolis, Maryland.

Todd L. Walton, Ph.D., Coastal and Hydraulics Laboratory, Engineer Research and Development Center, Vicksburg, Mississippi.

Reviewers:

Kevin Bodge, Ph.D., Olsen Associates, Jacksonville, Florida.

James R. Houston, Ph.D., Engineer Research and Development Center, Vicksburg, Mississippi.

Paul A. Work, Ph.D., School of Civil and Environmental Engineering, Georgia Institute of Technology, Atlanta, Georgia.

Table of Contents

List of Tables

List of Figures

Chapter III-4
Wind-Blown Sediment Transport

III-4-1. Introduction

From an engineering perspective, the transport of sand by wind is often an important component in the coastal sediment budget. Wind transport can lead to the removal of sand or its redistribution within the littoral zone. Onshore winds carry sand from the beach and deposit it in backshore marshes, in developed backshore areas, or in natural or man-made dunes. Offshore winds carry sand from the beach into the sea or lake. In some areas, wind-blown sand is a nuisance and must be controlled. In other areas, the natural growth of protective dunes is limited by the amount of sand transported to them by wind. Wind transport of sand is a continual, natural process that is often significant in bringing about beach changes. It is important to be able to quantitatively predict how much sand will be transported by wind at a given coastal site, the direction in which that sand will be transported, and where it will be deposited. This chapter discusses wind transport prediction in coastal areas by describing the driving force (the wind) and the transport mechanisms. The overall wind regime of U.S. coastal areas, modifications to the wind field brought about by its proximity to the ground (the atmospheric boundary layer), initiation of sediment motion under wind, processes involved in sand transport by wind, and the quantitative prediction of transport rates are described.

III-4-2. Wind Regime of U.S. Coastal Areas

a. Introduction. In the Northern Hemisphere, between the equator and latitude 30° N, winds generally blow from the east (the trade winds), while between 30° N and 60° N, winds blow generally from the west (the prevailing westerlies). From autumn to spring, however, cold fronts frequently pass over the United States. During these events the normal wind regime is different. Ahead of these fronts, winds are generally southerly, while behind the fronts, winds are northerly. Southerly winds ahead of the front transport warmer tropical air northward while behind the front, northerly winds transport cold polar air southward (see Figure III-4-1).

b. Example wind environments. Four coastal areas are considered herein as example wind environments (see Figure III-4-2). The sites are: Newport, OR; Aransas Pass, TX; Westhampton Beach, NY; and New Buffalo, MI on Lake Michigan. The four sites represent a broad range of geographical parameters. Wind roses for each area are given in Figure III-4-3. Since sand transport generally occurs for wind speeds greater than about 5-m/s (approximately 10 knots), the 5 m/s wind speed is used to separate the winds into two regimes.

c. Wind direction. Winds at Newport, OR, have a major north-south component because the Cascade Range immediately behind the shoreline has a north-south orientation and because of the north-south orientation of the shoreline. Occasionally, winds may blow from land toward the sea. In Port Aransas, TX, winds come mostly from the southeast because of the influence of Bermuda high pressure systems. At Westhampton Beach, NY, winds blow mainly from the southwest around through the northwest. At New Buffalo, MI, in the Lake Michigan region, winds blow generally in the north-south direction because of the west-to-east passage of fronts. These wind conditions are in general agreement with the air mass and cold front influences shown in Figure III-4-1.

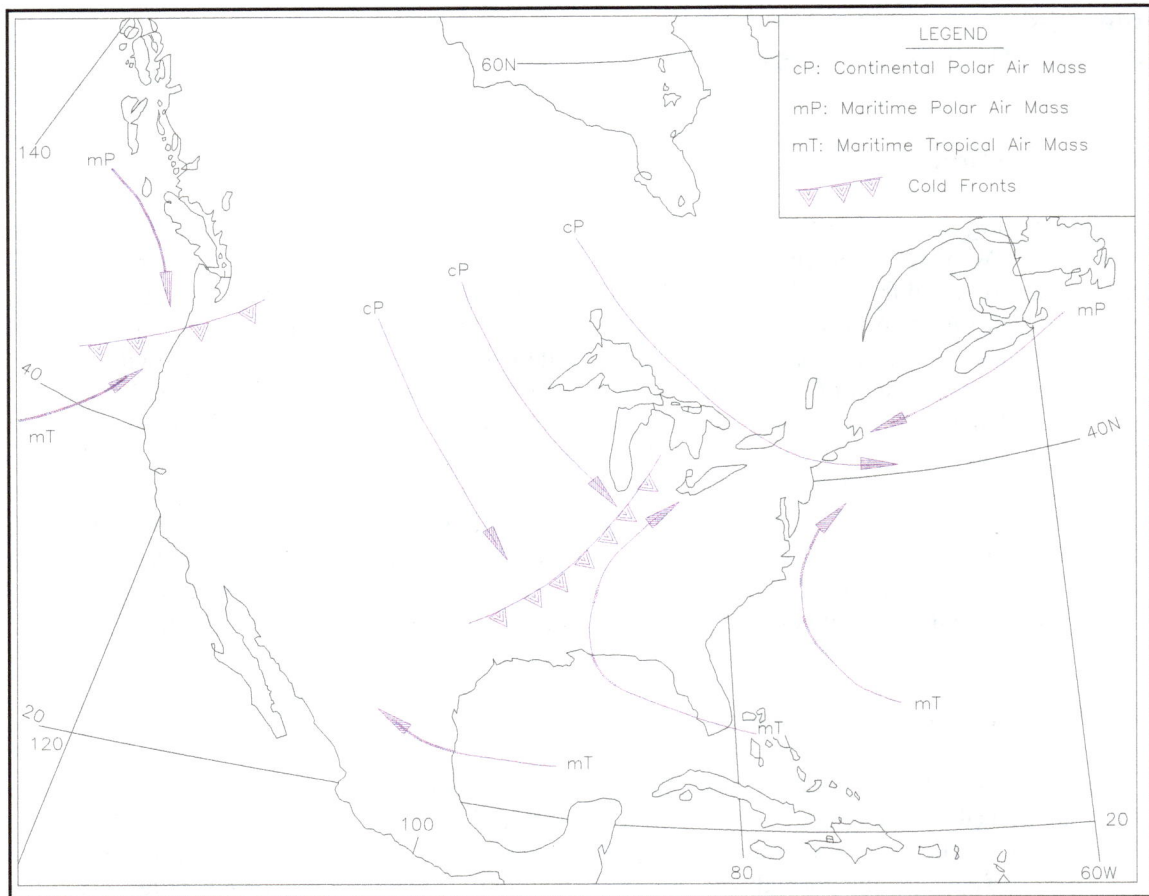

Figure III-4-1.   Air masses and fronts over the coastal areas

(1) Basic concepts.

(a) Wind velocity and the vertical gradient of wind speed near the ground are important factors in determining how much sediment will be transported by wind. In addition, sediment characteristics such as size, size distribution, packing, and moisture content play important roles. The vertical gradient of wind speed results in shearing forces within the air and eventually on the ground surface. When vegetation is present, sheer can be transferred into the uppermost soil layers. The steeper the gradient, the greater the shear stress. The velocity gradient is significantly influenced by local topography, vegetation, and land use. These factors contribute to the "roughness" of the ground surface. In the case of wind transport in coastal areas, local perturbations in the wind field may also be important in determining the eventual erosion and deposition patterns of wind-blown sand.

(b) The vertical distribution of wind speed - the velocity profile - generally follows a logarithmic distribution. An important parameter in this representation is the shear velocity defined by

$$u_* = \sqrt{\frac{\tau}{\rho_a}}$$

(III-4-1)

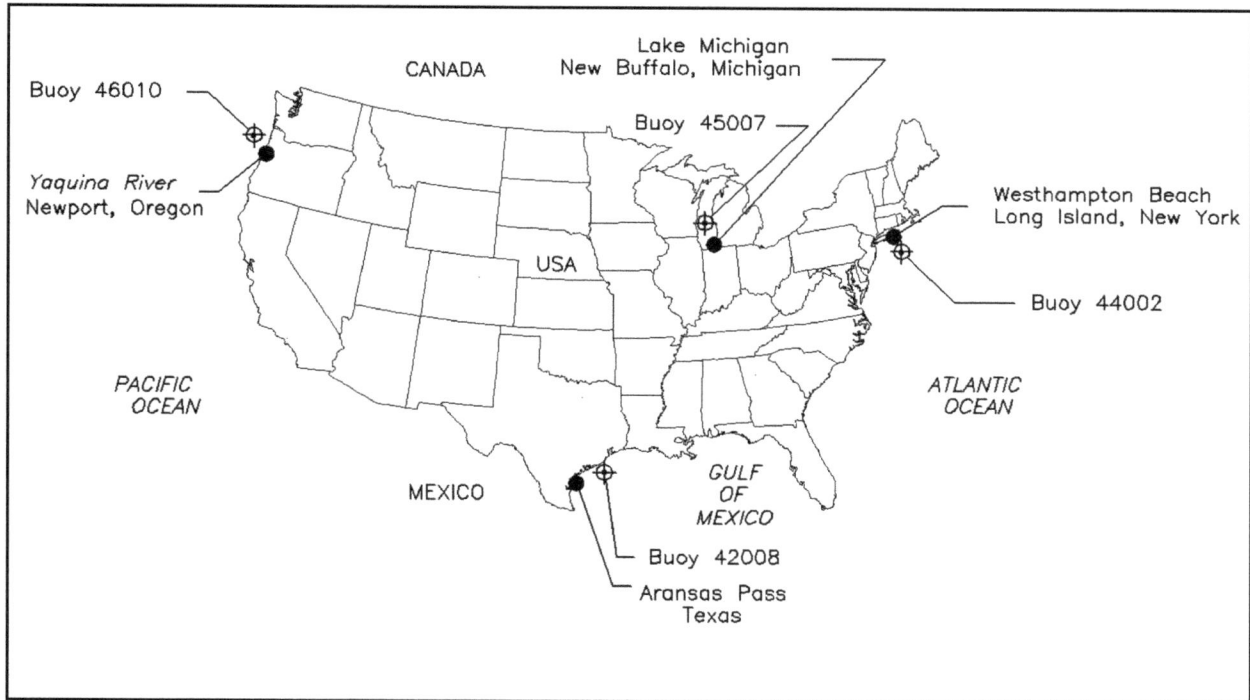

Figure III-4-2. Example wind regime areas (Solid circles denote example wind regime areas. Open circles are nearby buoys used to construct wind roses)

in which u_* = the shear or friction velocity, τ = the boundary shear stress (force per unit surface area), and ρ_a = the density of the air. The logarithmic distribution of wind velocity is given by

$$U_z = \frac{u_*}{\kappa} \ln\left(\frac{Z_o + Z}{Z_o}\right)$$

(III-4-2)

in which U_z = the average wind speed as a function of height above ground level, u_* = the shear velocity, Z = the height above ground level, Z_o = the height of a roughness element characterizing the surface over which the wind is blowing, and κ = von Karman's constant (κ = 0.4). Wind speed measurements are usually averaged over a period of several minutes to smooth out fluctuations due to gusts. Averaging time can vary depending on how the measurements are made. If readings are taken from an anemometer dial rather than from a recording, the observer may visually average the speed by observing the needle over a period of several minutes. Wind speeds are occasionally reported as the fastest mile of wind speed where the averaging time is the time it would take the wind to travel the distance of 1 mile; thus, the averaging time for a 60-mph wind would be 1 min; the averaging time for a 30-mph wind would be 2 min, etc.

(c) Since the height of a roughness element Z_o is usually small compared with Z, Equation 4-2 can be approximated by

$$U_z \approx \frac{u_*}{\kappa} \ln\left(\frac{Z}{Z_o}\right)$$

(III-4-3)

or

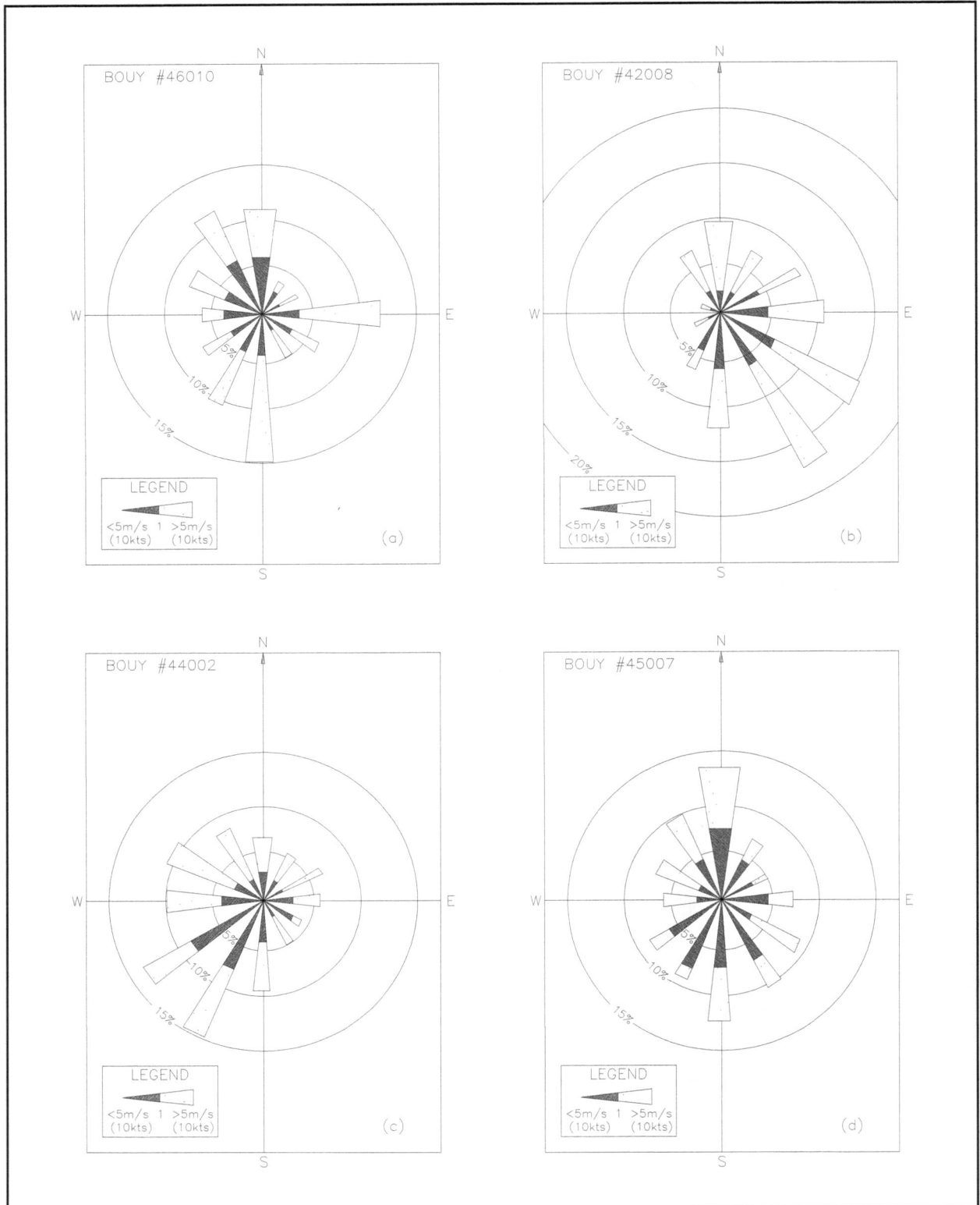

Figure III-4-3.   Wind roses based on hourly data nearby: (a) Newport, OR, (b) Aransas Pass, TX, (c) Westhampton Beach, Long Island, NY, and (d) New Buffalo, MI

$$\ln Z = \ln Z_o + \left(\frac{\kappa}{u_*} \right) U_z$$

which is of the form

$$Y = a_o + a_1 X$$

where $Y = \ln Z$ and

$$a_o = \ln Z_o \quad \text{or,} \quad Z_o = e^{a_o}$$

and

$$a_1 = \frac{\kappa}{u_*} \quad \text{or,} \quad u_* = \frac{\kappa}{a_1}$$

(d) A logarithmic velocity distribution as fit by linear regression techniques is given in Figure III-4-4 for Isle Dernieres, LA.

(e) The concept of a drag coefficient is sometimes used in wind transport calculations, where the drag coefficient is defined as

$$C_z = \left(\frac{u_*}{U_z} \right)^2 \tag{III-4-4}$$

in which C_z = the drag coefficient at height Z. Often C_z is considered a constant for a given Z; however, computations by Hsu and Blanchard (1991) for C_{10} showed a variation of from 0.5×10^{-3} to 5.0×10^{-3}, an order of magnitude. In fact, C_z varies with season and wind direction and cannot be considered constant. It is evident that the roughness characteristics of the fetch over which the wind blows depend on wind direction, particularly in coastal regions where onshore and offshore winds experience significantly different fetches. Onshore winds usually experience less friction than offshore winds because of the relative smoothness of the water surface when compared with the rougher land surface. On the other hand, shear stresses on the beach due to onshore winds will be greater than shear stresses due to offshore winds of the same speed.

(f) Because the appropriate height of a roughness element is not known a priori, it must be found by the simultaneous solution of Equation 4-2 applied at two heights above the ground. However, if Z_o is small, and the wind speed is not measured very close to the ground (well outside of the roughness elements), Equation 4-3 can be used to estimate the wind speed distribution without first finding Z_o. Equation 4-3 can be applied at two heights, Z_1 and Z_2 so that

$$\ln Z_1 = \ln Z_o + \left(\frac{\kappa}{u_*} \right) U_{z1} \tag{III-4-5}$$

and

$$\ln Z_2 = \ln Z_o + \left(\frac{\kappa}{u_*} \right) U_{z2} \tag{III-4-6}$$

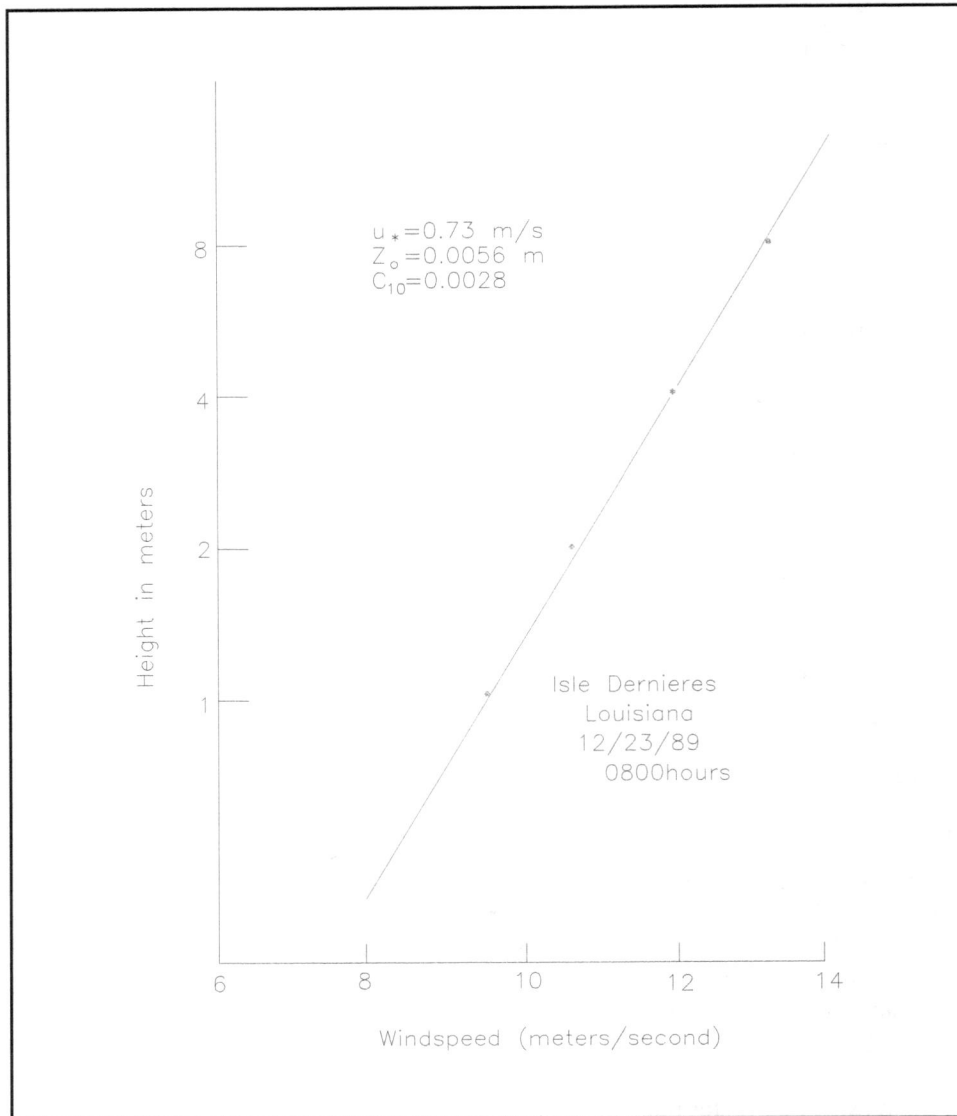

Figure III-4-4. Example logarithmic wind profile at Isles Dernieres, LA (adopted from Hsu and Blanchard (1991))

where Z_1 is below Z_2 so that U_{z1} is less than U_{z2}.

(g) Eliminating Z_o from Equations 4-5 and 4-6 gives

$$u_* = \frac{\kappa(U_{z2} - U_{z1})}{\ln\left(\dfrac{Z_2}{Z_1}\right)}$$

(III-4-7)

(h) Consequently, the shear velocity u_* can be determined from Equation 4-7 without first finding Z_o by measuring the wind velocity at two elevations. For wind transport studies, wind velocity sensors used to determine u_* should be spaced vertically as far apart as practical, say a minimum of 3 to 4 m, but preferably more.

Wind-Blown Sediment Transport

EXAMPLE PROBLEM III-4-1

Find:

The shear velocity u_* and the height of a characteristic roughness element Z_o. What is the magnitude of the shear stress acting on the ground surface?

Given:

Wind speed is measured at two elevations above the ground. At 10 m height, the average wind speed is 11.3 m/s; at 4 m, it is 9.36 m/s.

Solution:

The velocity distribution is logarithmic as given by Equation 4-2. Since there are two unknowns, Z_o and u_*, the equation must be solved at the two elevations. Substituting into the equation with $\kappa = 0.4$

$$11.3 = \frac{u_*}{0.4} \ln\left(\frac{10 + Z_o}{Z_o}\right)$$

and

$$9.36 = \frac{u_*}{0.4} \ln\left(\frac{4 + Z_o}{Z_o}\right)$$

Solving for each equation for u_* and equating

$$\frac{9.36\,(0.4)}{\ln\left(\dfrac{4 + Z_o}{Z_o}\right)} = \frac{11.3\,(0.4)}{\ln\left(\dfrac{10 + Z_o}{Z_o}\right)}$$

or

$$\ln\left(\frac{10 + Z_o}{Z_o}\right) = 1.207 \ln\left(\frac{4 + Z_o}{Z_o}\right)$$

$$\ln(Z_o + 10) - \ln Z_o = 1.207\left[\ln(Z_o + 4) - \ln Z_o\right]$$

$$\ln(10 + Z_o) = 1.207 \ln(4 + Z_o) - 0.207 \ln Z_o$$

Assuming that Z_o is small with respect to the height at which the wind speeds are measured, a first approximation to Z_o can be obtained,

(Sheet 1 of 3)

Example Problem III-4-1 (Continued)

$$\ln(10) = 1.207 \, \ln(4) - 0.207 \, \ln Z_o$$

$$2.303 = 1.673 - 0.207 \, \ln Z_o$$

$$\ln Z_o = -3.042$$

$$Z_o = 0.0477 \quad m$$

A second improved approximation can be found by substituting $Z_o = 0.0477$ into,

$$\ln(10 + 0.0477) = 1.207 \ln(4 + 0.0477) - 0.207 \ln Z_o$$

Solving for Z_o

$$Z_o = 0.05 \quad m$$

Then, substituting back into one of the wind speed equations

$$u_* = \frac{U_z \kappa}{\ln\left(\dfrac{Z + Z_o}{Z_o}\right)}$$

for the wind speed observation at 10 m

$$u_* = \frac{(11.3)(0.4)}{\ln\left(\dfrac{10 + 0.05}{0.05}\right)} = 0.852 \quad m/s$$

The assumption that Z_o is small with respect to the height at which wind speeds are measured could have been made at the outset by using Equation 4-3. Then

$$U_z = \frac{u_*}{0.4} \, \ln\left(\frac{Z}{Z_o}\right)$$

and

$$11.3 = \frac{u_*}{0.4} \, \ln\left(\frac{10}{Z_o}\right)$$

(Sheet 2 of 3)

Example Problem III-4-1 (Concluded)

with

$$9.36 = \frac{u_*}{0.4} \ln\left(\frac{4}{Z_o}\right)$$

or

$$(11.3)(0.4) = u_*\left[\ln(10) - \ln Z_o\right]$$

and

$$(9.36)(0.4) = u_*\left[\ln(4) - \ln Z_o\right]$$

Subtracting

$$0.776 = u_* \ln\left(\frac{10}{4}\right)$$

$$u_* = 0.847 \text{ m/s}$$

On substituting this value for u_* back into one of the above equations and solving for Z_0 gives

$$Z_0 = 0.048 \text{ m}$$

These are the same as the first approximations found above.

The shear stress can be found from Equation 4-1, the definition of u_*

$$u_* = \sqrt{\frac{\tau}{\rho_a}}$$

or

$$\tau = \rho_a u_*^2$$

where $\rho_a = 0.00122$ gm/cm^3 and $u_* = 0.852$ m/sec. Substituting

$$\tau = 0.00122(85.2)^2 = 8.86 \text{ dynes/cm}^2$$

The reader may wish to verify the shear velocity and characteristic roughness element height for the wind speed observation at Isle Dernieres given in Figure III-4-4.

(Sheet 3 of 3)

(i) In most sand transport studies, two-level or multilevel wind velocity measurements are rarely available and the wind speed climatology must be measured at two or more levels to establish u_*. Based on measurements at six beaches, Hsu (1977) established a relationship for u_* in terms of U_{2m}, the wind speed at the 2-m height (see Figure III-4-5). His relationship for the dry beach area is given by

$$u_* = 0.044 \, U_{2m} \qquad\qquad\qquad\text{(III-4-8)}$$

in which U_{2m} = the average wind speed at the 2-m height. Equations for u_* on tidal flats, dunes, etc. are also given by Hsu (1977) (see Figure III-4-5). Note that the coefficient in this equation depends critically on the height of a roughness element so that Equation 4-8 is valid only for beaches with roughness conditions similar to those for which the equation was developed.

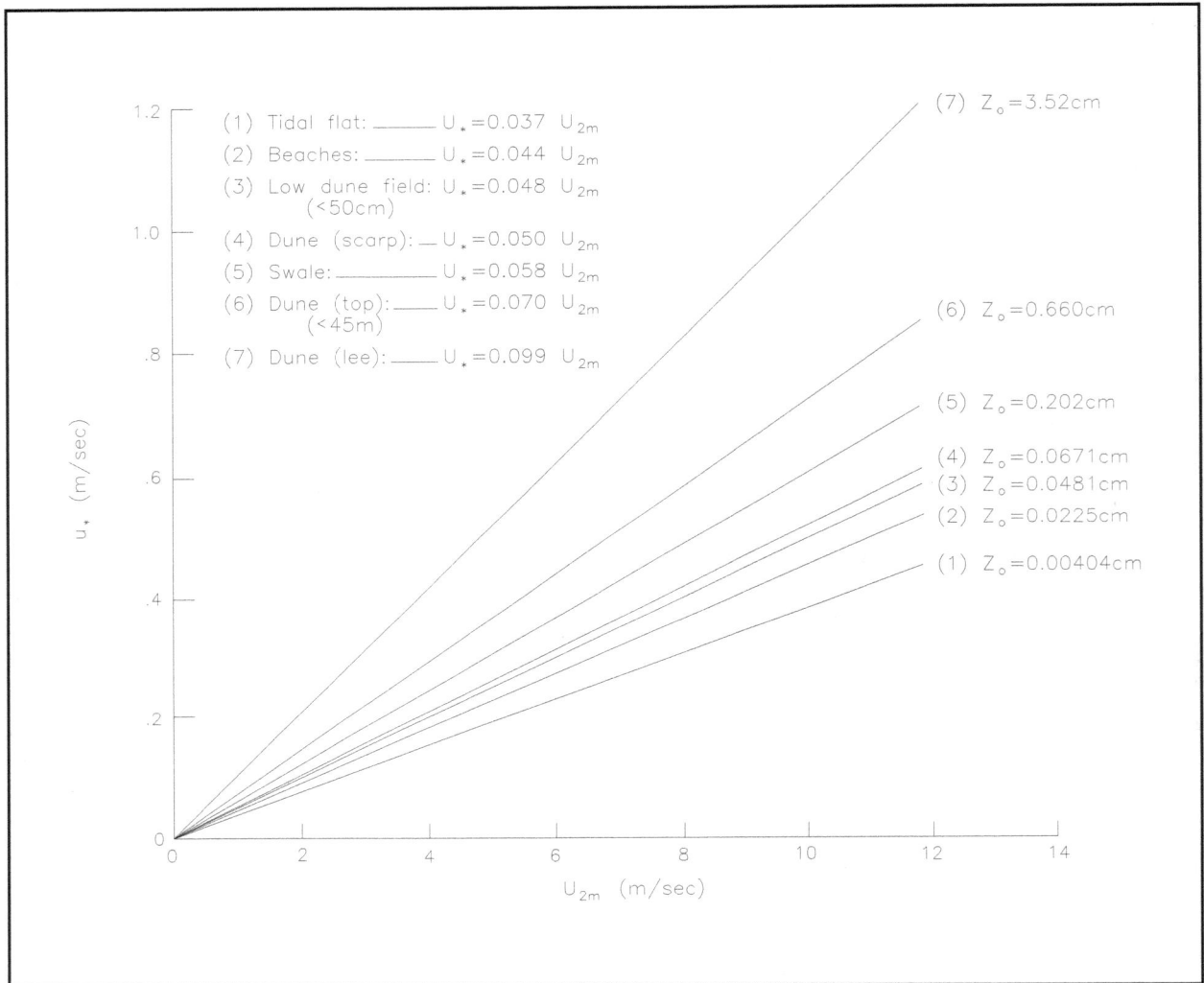

Figure III-4-5. Aerodynamic roughness element lengths Z_0 and relationship between shear velocity u_* and wind speed at the 2-m height U_{2m} in coastal environments (adopted from Hsu (1977))

EXAMPLE PROBLEM III-4-2

Find:

The drag coefficient at the 10-m, 4-m, and 2-m heights using the shear velocity calculated in the preceding problem, then estimate the shear velocity using the wind speed determined for the 2-m height. Compare the shear velocity with the value determined in the preceding example problem.

Given:

The wind speed measurements from the previous example problem.

Solution:

The definition of the drag coefficient is given by Equation 4-4. Thus, for the 10-m height,

$$C_{10m} = \left[\frac{u_*}{U_{10m}} \right]^2$$

$$C_{10m} = \left[\frac{0.852}{11.3} \right]^2 = 0.00568$$

For the 4-m height $C_{4m} = (0.852/9.36)^2 = 0.00829$. The wind speed at the 2-m height must be found from Equation 4-2 (or Equation 4-3 if the roughness element height can be assumed small). Thus,

$$U_{2m} = \frac{(0.852)}{(0.4)} \ln\left(\frac{2 + 0.05}{0.05} \right)$$

$$U_{2m} = 7.91 \quad m/s$$

and

$$C_{2m} = \left[\frac{0.852}{7.91} \right]^2 = 0.0116$$

Using this wind speed at the 2-m height to estimate the shear velocity u_* in Equation 4-8 gives

$$u_* = 0.044\, U_{2m}$$

$$u_* = 0.044(7.91) = 0.348 \quad m/s$$

This result is significantly lower than the $u_* = 0.852$ value found from the wind speed distribution because the example wind speed distribution is somewhat unusual for a beach. The characteristic roughness length for the example is larger than what is typically found on beaches. A value of Z_o typical for beaches is 0.000225 m (0.0225 cm); however, based on measurements by Hsu (1977), Z_o can be much larger in the lee of dune areas. See Figure III-4-5 for some typical values of Z_o in various coastal environments.

(2) Measurement of wind speed and direction.

(a) Wind speed is measured with an anemometer. Two types of anemometers are typically used, cup-type or propeller-type anemometers. Wind direction is measured using a vane that indicates the direction *from which* the wind is blowing. When wind speed measurements are made with a cup-type anemometer, a separate vane is needed. Propeller-type anemometers have a built-in vane that keeps the propeller directed into the wind. Wind speed measurements are usually obtained at or corrected to a standard anemometer height of 10 m above ground level, although anemometers may be mounted at other heights. The wind record from an anemometer depends significantly on the physical conditions of the surrounding terrain. The surrounding terrain determines the boundary roughness and thus the rate at which wind speed increases with height above the ground. Consequently, physical conditions characterizing the surrounding fetches determine the wind environment at a site. At coastal sites, the fetch seaward of the beach is over water and is generally characterized by relatively low roughness. The fetch landward of the beach can vary significantly from site to site since it depends on the type of vegetation, terrain, etc.

(b) Differences exist between onshore and offshore wind speeds due to differences in friction conditions over the fetch. The differences are most marked in the observed wind speeds (see Hsu (1988)). Hsu (1986) derived an equation to estimate the difference in overwater and overland wind speeds. The ratio of wind speed overwater to the wind speed overland is given by

$$\frac{U_{sea}}{U_{land}} = \left[\frac{H_{sea}}{H_{land}}\frac{C_{D,land}}{C_{D,sea}}\right]^{\frac{1}{2}}$$

(III-4-9)

where

U_{sea} and U_{land} = the wind speed over sea and land, respectively

H_{sea} and H_{land} = the height of the planetary boundary layer over the sea and land

$C_{D,sea}$ and $C_{D,land}$ = the drag coefficients over sea and land

(c) Equation 4-9 should apply for both onshore and offshore winds assuming that the geostrophic wind above the boundary layer does not change appreciably. For a given climatological area, the parameters on the right side of Equation 4-9 are generally known. Holzworth (1972) and SethuRaman and Raynor (1980) found H_{sea} = 620 m, H_{land} = 1014 m, $C_{D,sea}$ 0.0017, and $C_{D,land}$ = 0.0083, leading to U_{sea}/U_{land} = 1.7 for this area. $C_{D,sea}$ and $C_{D,land}$ were measured at the 8-m level.

(d) If the direction between either onshore near-surface winds or offshore near-surface winds is within 45 deg of shore-normal, Equation 4-9 will be valid regardless of whether the wind blows from land to sea or vice versa (Hsu (1988), pp 184-186). Equation 4-9 may, therefore, be applicable to weather systems such as land breezes, sea breezes, hurricanes near landfall, and other synoptic-scale phenomena such as cyclones (low pressure cells), anti-cyclones (high pressure cells), and monsoons. This is not the case for the passage of frontal systems and squall lines, however, since their winds do not usually blow within 45 deg of shore-normal. Typically, winds associated with the passage of fronts and squall lines do not persist for long periods of time.

(e) An empirical relationship between overland wind speed and overwater wind speed can be estimated from measured wind speeds at land and water sites by

$$U_{sea} = A + B U_{land}$$

(III-4-10)

where A and B are empirical constants.

(f) If the air-water temperature difference is small, say less than 5 °C, and if the aggregate wind estimation error at airports is less than 10 percent (Wieringa 1980), then an estimate is given by Liu, Schwab, and Bennett (1984) as

$$U_{sea} = 1.85 \ (m/s) \ + \ 1.2 \ U_{land} \tag{III-4-11}$$

where U_{sea} and U_{land} are measured in m/s.

(g) Based on many pairs of measurements of overland and overwater wind speeds, the U_{sea} was correlated with U_{land} for synoptic-scale weather systems (Hsu 1981; Powell 1982). Wind speeds used by Hsu (1981) were less than 18 m/s while Powell's (1982) data included hurricane-force wind speeds obtained during Hurricane Frederic in Alabama in 1979. The correlation between U_{sea} and U_{land} is shown in Figure III-4-6. The best-fit equation by linear regression methods is given by

$$U_{sea} = 1.62 \ (m/s) \ + \ 1.17 \ U_{land} \tag{III-4-12}$$

(h) Equation 4-11 is also plotted in Figure III-4-6. The difference between Equations 4-11 and 4-12 is not significant even though the correlation on which Equation 4-12 is based did not consider air-sea temperature differences. Equation 4-12 is recommended for use in coastal wind-blown sand transport studies where the air water temperature difference is small ($|Ta - Ts| < 5$ °C).

(i) For air-sea temperature differences exceeding 5 °C, a correction can be applied. A temperature correction applied to wind speed as given by Resio and Vincent (1977) is

$$U'_{10m} = R_T \ U_{10m} \tag{III-4-13}$$

where U'_{10m} is the wind speed corrected for the air-sea temperature difference and R_T is the correction factor applied to the overwater wind speed at the 10-m height. The correction factor R_T is shown in Figure III-4-7 and given by the equation

$$R_T = 1 \ - \ 0.06878 \ |T_a \ - \ T_s|^{0.3811} \ \ sign(T_a \ - \ T_s) \tag{III-4-14}$$

where T_a = the air temperature in degrees Celsius and T_s = the sea temperature in degrees Celsius.

(j) For wind-blown sand transport studies, wind measurement stations should be close to the study site and preferably on the beach. Land-based wind measurements at airports are generally not useful indicators of nearshore wind conditions and should not be used for coastal sand transport studies. Unfortunately, wind measurements are often not available near coastal study sites and data from nearby Coastal Marine Automated Network (C-MAN) and offshore meteorological buoys must be used. Table III-4-1 compares wind measurements at a C-MAN station (ALSN6), at an offshore buoy 44025, and at Islip Airport on central Long Island. (Wind speeds have been corrected to the standard 10-m elevation using a power-law relation. This correction is discussed below.) Wind speed measurements at Islip Airport are consistently lower than measurements at the coast. On average, $U_{44025}/U_{Islip} = 1.6$, which is consistent with the U_{sea}/U_{land} found earlier for the New York-Massachusetts area (Holzworth 1972, SethuRaman and Raynor 1980). Also, measurements at the offshore buoy are more variable than those at the C-MAN station, making the C-MAN station data preferable in this case.

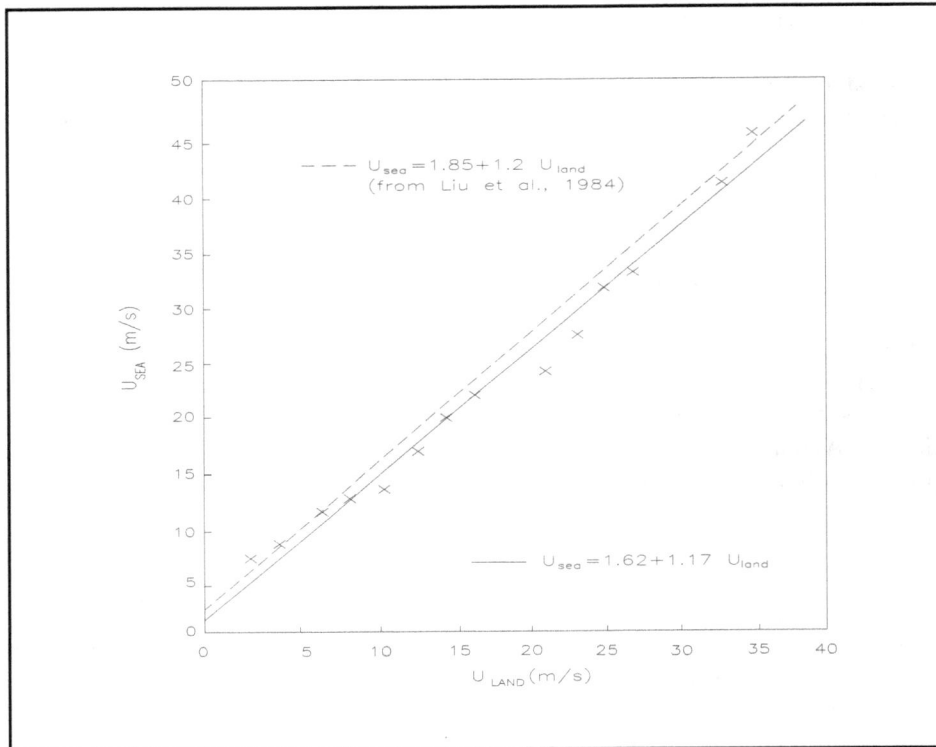

Figure III-4-6. Variation of U_{sea} with U_{land}. (Hsu 1986)

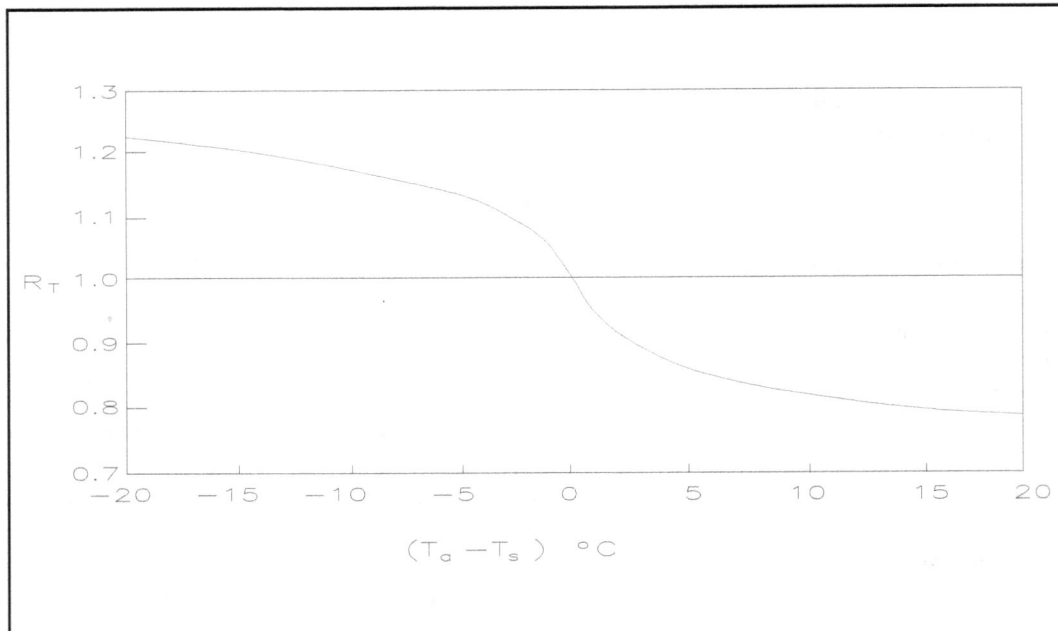

Figure III-4-7. Temperature correction factor R_T as a function of air-sea temperature difference (Resio and Vincent 1977)

Table III-4-1
Monthly Mean Wind Speeds Near Westhampton Beach, Long Island NY - Based on Simultaneous Measurements at C-MAN Station ALSN6 and Offshore Buoy 44025

		Wind Speeds in meters/sec				
		C-MAN ALSN6 @4.9 m	BUOY 44025 @ 4.9 m	C-MAN ALSN6 @ 10 m	BUOY 44025 @ 10 m	ISLIP AIRPORT @ 10 m
YEAR	MONTH					
1991	May	7.30	4.70	6.20	5.00	2.60
	Jun	6.70	5.40	5.70	5.80	2.90
	Jul	6.00	4.70	5.10	5.00	2.40
	Aug	6.80	5.00	5.80	5.40	2.80
	Sep	6.60	5.80	5.60	6.20	3.80
	Oct	7.60	6.30	6.50	6.70	4.50
1992	Nov	7.70	7.10	6.50	7.60	4.60
	Dec	8.80	7.90	7.50	8.50	4.70
	Jan	8.10	N/A	6.90	N/A	4.70
	Feb	8.70	7.30	7.40	7.80	5.10
	Mar	9.00	7.30	7.70	7.80	5.30
	Apr	7.00	5.40	6.00	5.80	4.50
	May	6.80	5.10	5.80	5.50	4.60
	Jun	7.10	4.80	6.00	5.10	4.70
AVG =		7.44	5.91	6.34	6.32	4.09
STD DEV =		0.91	1.14	0.79	1.22	0.99

(k) Wind speeds are often adjusted to a standard anemometer height by using a power-law relation. The standard anemometer height is usually 10 m above the ground. The power-law adjustment equation is given by

$$U_{ZR} = U_{ZM} \left(\frac{Z_R}{Z_M} \right)^n \tag{III-4-15}$$

in which Z_M = the anemometer height, Z_R = the reference height (usually 10 m), U_{ZM} = the wind speed at the anemometer height, U_{ZR} = the wind speed at the reference height, and n = an empirically determined exponent, ranging from 1/11 to 1/7. The following example details the calculation of this exponent as well as wind speed correction factors for air sea temperature differences.

(3) Effects of vegetation, dunes, and buildings.

(a) Air flow over a flat beach differs from air flow over dunes. There are at least three distinct wind transport zones resulting from the presence of dunes on the backbeach (Hsu 1988). Dunes produce both a stagnation zone on their windward side and a wake zone on their leeward side within which wind patterns are disturbed and no longer exhibit the typical logarithmic distribution with height above the ground. Wind blowing offshore results in a stagnation zone landward of the dune that extends landward approximately two to four times the height of the dune. In this zone, wind speeds are lower than the free stream wind speed (region of underspeed); in fact, near the ground they may even blow in the direction opposite to that of the free stream. A wake region (cavity) may extend 10 dune heights seaward of the dune. Within the wake cavity, wind speeds are also reduced and may reverse direction near ground level. Over the dune crest, wind speeds are higher than the free stream speed because the streamlines are compressed and the flow is accelerated (region of overspeed) (see Figure III-4-8).

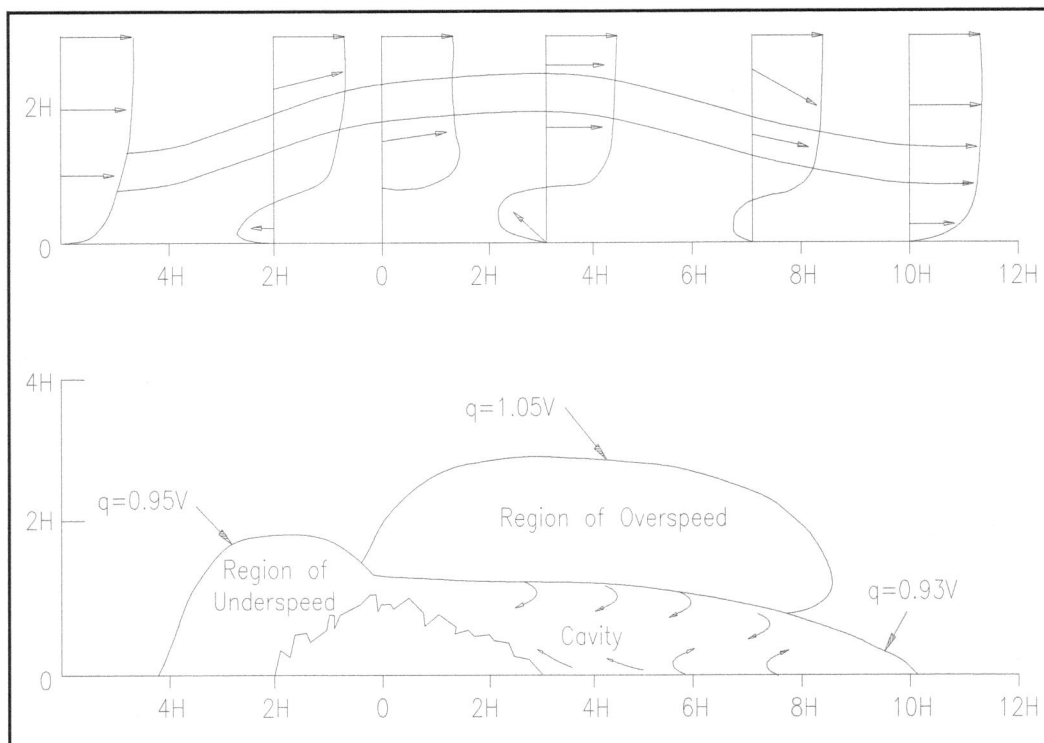

Figure III-4-8. Wind field in the vicinity of a coastal sand dune (q is the local resultant mean velocity and v is the reference velocity in the uniform stream above the dune) (Hsu 1988)

(b) Buildings and vegetation also locally modify the air flow. Generally, buildings will produce a wake in their lee. Air flow patterns near buildings will be complex, will depend on the specific building geometry, and on the presence of any surrounding buildings. Generalized air flow patterns cannot be established for the numerous building/construction configurations encountered in practice. The effect of vegetation depends on the type of vegetation, its location relative to any dunes, and how far it projects into the near-ground boundary layer. Trees usually have more effect on wind patterns than do shrubs or grasses but, depending on their location, may be less significant than shorter vegetation in affecting sand transport rates and sediment deposition patterns. Most sand transport takes place within a few centimeters of the ground and anything that modifies the air flow in this region will have some effect.

III-4-3. Transport Rates

a. Processes of sand transport by wind.

(1) The processes involved in transporting sand by wind are summarized by Raudkivi (1976). Sand grains move by bouncing along the surface, a process termed saltation, and by surface creep. Sand grains,

EXAMPLE PROBLEM III-4-3

Find:

Determine the overwater wind speed and correct the wind speed due to the air-sea temperature difference. Estimate the exponent in Equation 4-15 using the overwater wind speeds.

Given:

The winds given in Example Problem III-4-1 measured over land on a day when the air temperature is 15 °C and the water temperature is 5 °C.

Solution:

The air-sea temperature difference is $(15°-5°) = 10$ °C and the correction R_T is given by Equation 4-14

$$R_T = 1 - 0.06878 |15 - 5|^{0.3811} (+) = 0.8346$$

This correction is applied to the wind speed at the 10-m height. Therefore, from Example Problem III-4-1, $U_{10m} = 11.3$ m/s and the corrected 10-m wind speed is

$$U_{10m} = 0.8346(11.3) = 9.43 \text{ m/s}$$

Assuming that the height of the roughness elements are $Z_o = 0.05$ as found in Example Problem III-4-1, the shear velocity can be found from Equation 4-2

$$U_z = \frac{u_*}{\kappa} \ln\left(\frac{Z + Z_o}{Z_o}\right) \qquad 9.43 = \frac{u_*}{0.4} \ln\left(\frac{10 + 0.05}{0.05}\right)$$

$$u_* = 0.7113 \quad \text{m/s}$$

The wind speed at the 8-m height is then calculated as

$$U_{8m} = \frac{0.7113}{0.4} \ln\left(\frac{8 + 0.05}{0.05}\right) = 9.036 \text{ m/s}$$

Using these two wind speeds to estimate the exponent in Equation 4-15

$$\frac{U_{10m}}{U_{8m}} = \left(\frac{10}{8}\right)^n \qquad \frac{9.43}{9.036} = (1.25)^n$$

$$n = \frac{\ln(1.0436)}{\ln(1.25)} = 0.1913 = \frac{1}{5.23}$$

once dislodged, are carried up into the moving air by turbulence; they acquire energy from the moving air, and they settle through the air column due to their weight and impact on the ground. The saltating grains impact on the ground surface at a flat angle and transmit a portion of their energy to the grains on the ground. Some of the grains on the surface are dislodged and are carried upward into the flow where they continue the process of saltation; others are moved forward on the surface by the horizontal momentum of the impacting, saltating grains. This latter forward movement of sand on the ground surface is termed surface creep. Saltating grains are capable of moving much larger surface grains by surface creep due to their impact. On beaches, saltation is the more important of the two processes.

(2) The presence of sand in the air column near ground level also leads to a reduction in the wind speed near the ground and a modification of the wind speed distribution.

(3) Because saltating sand grains impact the ground surface at a relatively flat angle which ranges from about 10 to 16 deg with the horizontal, there is a tendency for small depressions in the ground surface to lead to the formation of sand ripples. Saltating sand grains have a greater tendency to impact the back face of a small depression because of the flat angle of incidence. This preferential movement on the back face of the depression moves sand up the back face to form a mound or ripple. Once small ripples are established they may grow until they reach a limiting height which is determined by the prevailing wind speed.

(4) On beaches, the most important sand features frequently present are sand dunes. Dunes are much larger than the sand features described above. They are large enough to significantly alter wind patterns and to shelter the area on their leeward side. Once established, dunes cause wind pattern changes which lead to dune growth.

    b. Sand transport rate prediction formulas.

(1) Many equations have been proposed to predict sand transport by wind; e.g., see Horikawa (1988) and Sarre (1988). Several example sand transport equations follow:

$$G = 0.036 \ U_{5ft}^{3} \qquad \text{(O'Brien and Rindlaub 1936)}$$

in which G = the dry weight transport rate in pounds per foot per day and U_{5ft} = the wind speed at the 5-ft elevation in feet per second ($U_{5ft} > 13.4$ feet per second);

$$q = B_{Bagnold} \ \frac{\rho_a}{g} \sqrt{\frac{D}{d}} \ u_*^{\,3} \qquad \text{(Bagnold 1941)}$$

in which q = the mass transport rate in gm/cm-s, $B_{Bagnold}$ = a coefficient, ρ_a = the mass density of the air = 0.001226 gm/cm^3, d = a standard grain size = 0.25 mm, D = grain size in mm, and u_* = the shear velocity in cm/sec;

$$q = Z_{Zingg} \ \frac{\rho_a}{g} \left[\frac{D}{d} \right]^{\frac{3}{4}} \ u_*^{\,3} \qquad \text{(Zingg 1953)}$$

in which Z_{Zingg} = a coefficient, and q, D, d, ρ_a, and u_* are as in the previous expression.

(2) Chapman (1990) provides an evaluation of these and several other equations. The predictive capability of the seven equations investigated by Chapman (1990), as gauged by the coefficient of determination, ranged from $r^2 = 0.63$ to $r^2 = 0.87$ (see Table III-4-2).

Table III-4-2
Predictive Capability of Sand Transport Equations (Chapman 1990)

Equation	Value of r^2
Bagnold (1941)	0.63
Horikawa & Shen (1960)	0.84
Hsu (1973)	0.87
Kad b (1964)	0.65
O'Brien & Rindlaub (1936)	0.80
Williams (1964)	0.80
Zingg (1953)	0.78

(3) In Chapman's evaluation, Hsu's (1986) equation, which like Bagnold's (1941) is based on considerations of the turbulent kinetic energy relationship, had an $r^2 = 0.87$. Hsu's equation is given by

$$q = K \left[\frac{u_*}{\sqrt{gD}} \right]^3 \qquad \text{(III-4-16)}$$

where

q = sand transport rate in gm/cm-s

u_* = shear velocity

g = acceleration of gravity

D = mean sand grain diameter

K = dimensional eolian sand transport coefficient

(4) The term in square brackets in Equation 4-16 is a dimensionless Froude number, which can be computed using any consistent units. The dimensions of K are the same as those of q; in this case, grams per centimeter per second. Values of K as a function of sand grain diameter can be obtained from Figure III-4-9. A regression equation ($r^2 = 0.87$) fitted to both laboratory and field data is given by

$$\ln K = -9.63 + 4.91 D \qquad \text{(III-4-17a)}$$

or

$$K = e^{-9 63 + 4 91 D} \qquad \text{(III-4-17b)}$$

where D is in millimeters and K in grams per centimeter per second, which plots as a straight line on semi-logarithmic graph paper (see Figure III-4-9).

(5) Equations 4-16 and 4-17a can be used to estimate sand transport rates for given wind speeds and mean sand-grain diameters within the range of empirical observations from which the equations were developed. Equation 4-17a is based on the data in Figure III-4-9, which include transport data for mean sand grain diameters up to 1.0 mm; consequently, the equations should not be used to estimate transport on beaches with mean grain diameters greater than about 1.0 mm.

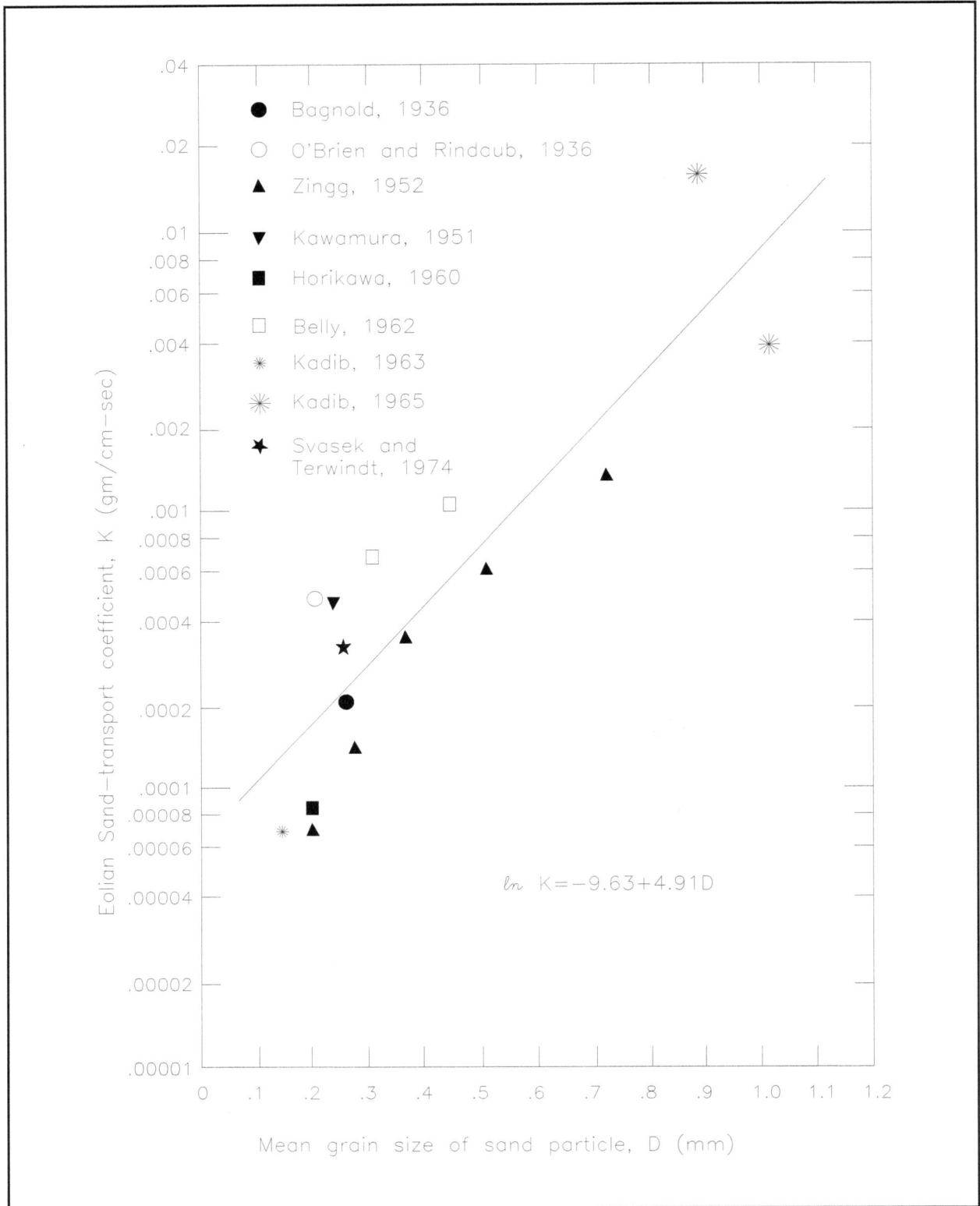

Figure III-4-9. Wind-blown sand transport coefficient as a function of mean sand-grain diameter (adopted from Hsu (1977)

(6) Equation 4-16 can be recast into a dimensionless form, which allows it to be used with any consistent set of units. The revised equation is given by

$$\frac{q}{v_a \rho_a} = K' \left[\frac{u_*}{\sqrt{gD}} \right]^3 \qquad \text{(III-4-18)}$$

in which v_a = the kinematic viscosity of the air and ρ_a = the mass density of the air. The dimensionless coefficient K' is given by

$$K' = e^{-1.00 + 4.91D} \qquad \text{(III-4-19)}$$

in which D = *the median grain diameter in millimeters*. Equation 4-18 reduces to Equation 4-16 when v_a = 0.147 cm²/sec and ρ_a = 0.00122 gm/cm³ are substituted into it.

c. *Initiation of sand transport.* Before sand will be transported by wind, the boundary shear stress must be increased above some critical or threshold value. The critical shear stress, in terms of the shear velocity, is given by Bagnold (1941)

$$u_{*t} = A_t \sqrt{\frac{(\rho_s - \rho_a)gD}{\rho_a}} \qquad \text{(III-4-20)}$$

in which ρ_s = the mass density of the sediment, ρ_a = the mass density of the air, and A_t = a dimensionless constant (A_t = 0.118). The effect of soil moisture that increases the critical shear stress is not considered in Equation 4-20.

III-4-4. Procedures for Calculating Wind-Blown Sand Transport

The steps for calculating wind-blown sand transport on beaches follow.

a. Obtain hourly average wind speed and direction data. (Wind data tabulated at intervals less frequent than 1 hr may be used in lieu of hourly data; however, hourly data are preferable.)

b. Obtain daily precipitation data and monthly evaporation records from a nearby National Weather Service (NWS) station. (These data are available in "Climatological Data" summaries published monthly for each state by the National Climatic Data Center (NCDC), Asheville, NC).

c. Obtain the density and median grain size of the beach sand at the study site.

d. Compute the critical shear velocity u_{*t} for the mean grain diameter using Equation 4-20.

e. Compute the critical wind speed at the 2-m height U_{2mt} using Equation 4-8 with the value of u_{*t} computed under Step 4 above. (This is the wind speed measured at the 2-m height that can initiate sand transport.)

f. Shear velocity u_* is relatively independent of height up to a height of about 50 m above ground level; therefore, Equation 4-7 can be used to compute the critical wind speed at any height above the ground using the U_{2mt} and u_{*t}. (For example, let Z_1 = 2 m, U_{z1} = U_{2mt}, Z_2 = the height at which the available wind

EXAMPLE PROBLEM III-4-4

Find:
Determine the volume of sand being transported per meter of beach width by the wind if the sand is dry.

Given:
The median sand-grain diameter at Westhampton Beach, Long Island, NY, is 0.26 mm. The wind velocity is measured by an anemometer located 8 m above the beach and found to be 21 m/s (see Figure III-4-10 for site location).

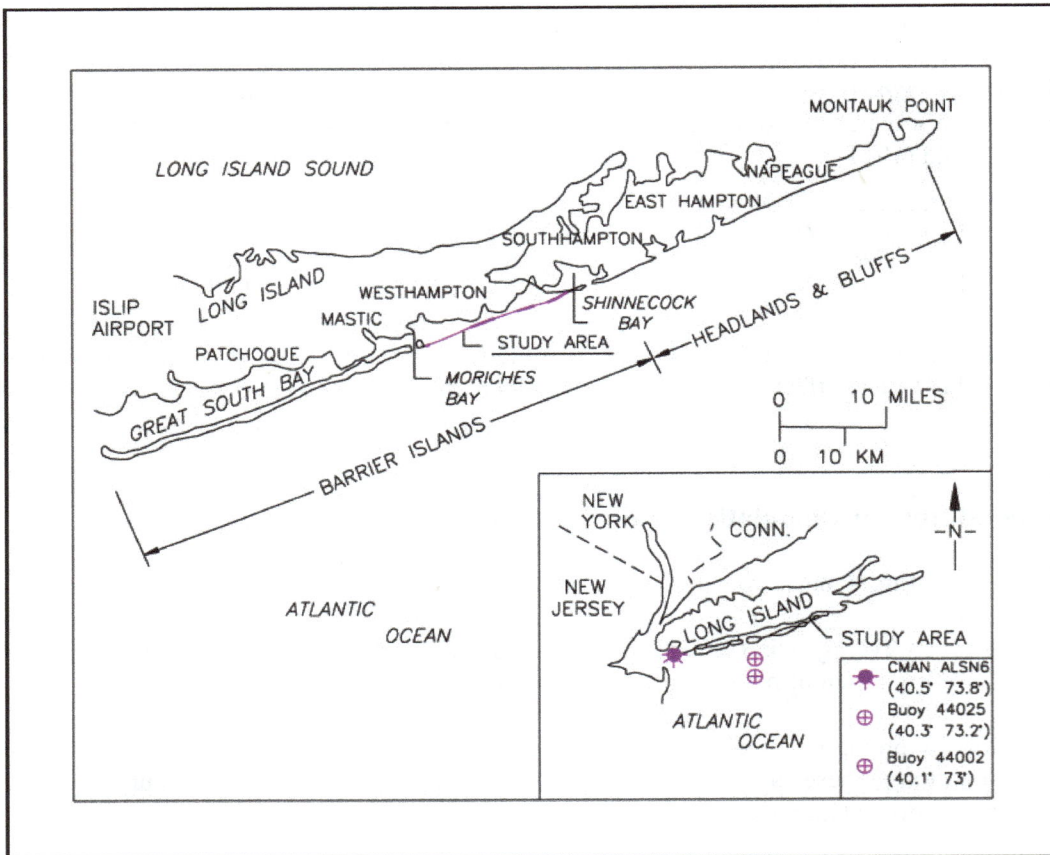

Figure III-4-10. Westhampton Beach, Long Island, NY (Hsu 1994)

Solution:
The first step is to compute the threshold wind speed that will just initiate sand transport. This is obtained from Equation 4-20

$$u_{*t} = A_t \sqrt{\frac{(\rho_s - \rho_a)gD}{\rho_a}}$$

(Sheet 1 of 5)

Wind-Blown Sediment Transport

Example Problem III-4-4 (Continued)

where $A_t = 0.118$, g = acceleration of gravity, D = sand-grain diameter, ρ_a = density of air, and ρ_s = density of the sediment. At 18 °C, $\rho_a = 1.22 \times 10^{-3}$ gm/cm³; $\rho_s = 2.65$ gm/cm³, $g = 980$ cm/sec², and $D = 0.026$ cm. Note that Equation 4-20 is valid for any consistent set of units. Substituting into the equation

$$u_{*t} = 0.118 \sqrt{\frac{(2.65 - 0.00122)(980)(0.026)}{(0.00122)}} = 27.75 \text{ cm/s} \quad (0.2775 \text{ m/s})$$

Next compute the threshold wind speed at the 2-m height. Using Equation 4-8 and the threshold shear velocity determined above

$$u_* = 0.044\, U_{2m}$$

where u_* = the shear velocity and U_{2m} = the wind velocity at the 2-m height. For threshold conditions

$$u_{*t} = 0.044\, U_{2mt}$$

where U_{2mt} is the threshold wind speed at a height of 2 m. Thus

$$U_{2mt} = 630.8 \text{ cm/sec} \quad (6.308 \text{ m/s})$$

This is the wind speed measured at the 2-m height that will initiate sand transport. Only average wind speeds that exceed 630.8 cm/s at the 2-m elevation will transport sand.

Because the wind speed is measured at the 8-m height, the threshold wind speed at that height must be determined. Equation 4-7 is given by

$$u_* = \frac{\kappa\,(U_{Z2} - U_{Z1})}{\ln\,(Z_2/Z_1)}$$

in which u_* is taken equal to u_{*t} since the shear velocity is independent of height above the ground.

Then, letting $U_{Z2} = U_{8mt}$, $U_{Z1} = U_{2mt}$, $Z_2 = 8$ m, $Z_1 = 2$ m, and noting that $u_{2t} = 630.8$ cm/s,

$$27.75 = \frac{0.4\,(U_{8mt} - 630.8)}{\ln(8/2)}$$

Solving for U_{8mt} gives $U_{8mt} = 727.0$ cm/s (7.27 m/s). Consequently, winds speeds that exceed 727.0 cm/s at the 8-m height are capable of transporting sand, while wind speeds less than 727.0 cm/s can be ignored in any wind-blown sediment transport study at this site.

(Sheet 2 of 5)

Example Problem III-4-4 (Continued)

Alternatively, Figure III-4-11a or III-4-11b can be used to find the threshold velocity for wind speed measurements at the 8-m height. The curves on Figures III-4-11a and III-4-11b were calculated using the preceding procedure. Enter the figure with the median sand-grain diameter of 0.26 mm and read from the curve for an anemometer height of 8 m. This gives $U_{8mt} = 730$ cm/s. (Figure III-4-11b is an enlargement of Figure III-4-11a for median sand sizes up to 0.25 mm.)

The amount of sediment being transported by the 21-m/s wind speed can now be calculated from Equation 4-16 or 4-18.

$$q = K \left[\frac{u_*}{\sqrt{gD}} \right]^3$$

where K = an empirical coefficient given by Equation 4-17a or 4-17b. The term in the square brackets is a dimensionless number having the form of a Froude number. Since this term is dimensionless, variables in any consistent set of units can be used to evaluate it. Equation 4-17b is

$$K = e^{-9.63 + 4.91D} \quad (\text{gm/cm-s})$$

in which D must be in millimeters. Equation 4-17a yields K in units of gm/cm-s; consequently, when used in Equation 4-16, the resulting mass transport q also will be in units of gm/cm-s. For the example,

$$K = e^{-9.63 + 4.91(0.26)} = e^{-8.354}$$

hence

$$K = e^{-8.354} \quad \text{or} \quad K = 2.355 \times 10^{-4} \text{ gm/cm-s}$$

Then

$$q \ (\text{gm/cm-s}) = (2.355 \times 10^{-4}) \left[\frac{u_* \ (\text{cm/s})}{\sqrt{(980)(0.026)}} \right]^3$$

in which u_* must be specified in cm/s since g is in centimeters per second squared and D is provided in centimeter units for dimensional consistency. Then

$$q = (1.831 \times 10^{-6}) u_*^3$$

The shear velocity u_* must now be expressed in terms of the wind speed at the 8-m height. Equation 4-7 is used again, noting that $u_* = 0.044 \ U_{2m}$, or $U_{2m} = 22.73 \ u_*$

$$u_* = \frac{0.4(U_{8m} - 22.73u_*)}{\ln(8/2)}$$

(Sheet 3 of 5)

Example Problem III-4-4 (Continued)

solving for u_*

$$u_* = 0.0382 \, U_{8m}$$

and

$$q = (1.019 \times 10^{-10}) \, U_{8m}^3$$

in which q is in grams per centimeter-second and U_{8m} is in centimeters per second.

For the given wind speed at the 8-m height, $U_{8m} = 2,100$ cm/s and

$$q = 0.9433 \, \frac{gm}{cm-s}$$

Equation 4-16 can be rewritten simply as

$$q = K^*(D, Z_M) \, U_{zM}^3$$

where $K^*(D, Z_M)$ = a coefficient in units of grams-seconds squared per centimeter to the fourth power which depends on the mean grain diameter D and the height at which the wind speed is measured Z_M. Values of K^* can be obtained from Figure III-4-12a or III-4-12b. Entering the figure with a sand-grain diameter of 0.26 mm and reading from the curve for an anemometer height of 8 m, $K^* = 1.01 \times 10^{-10}$ gm-s^2/cm^4, so that

$$q = 1.01 \times 10^{-10} \, U_{8m}^3 \quad \frac{gm}{cm-s}$$

as above (where U_Z must be specified in centimeters per second). Figure III-4-12b is an enlargement of Figure III-4-12a for sand-grain diameters up to 0.25 mm.

Figure III-4-12a indicates that transport rates decrease as the mean grain diameter increases for grain diameters up to about 0.3 mm. For larger grain diameters, transport rates increase with increasing grain diameter. Equation 4-17a, on which Figure III-4-12a is based, was derived empirically for mean grain diameters up to 1.0 mm; consequently, Figure III-4-12a is believed to be valid for grain diameters up to 1.0 mm; however, the relationships should not be used for grain diameters greater than 1.0 mm (i.e., outside the range of data utilized to develop the empirical relationship).

The volumetric sand transport rate can be obtained from the mass transport rate from

$$q_v = \frac{q}{\rho_s(1 - p)}$$

(Sheet 4 of 5)

Example Problem III-4-4 (Concluded)

in which ρ_s = the mass density of the sand and p = the porosity (the fraction of the in situ sand deposit which is pore space). Assuming a typical porosity of 0.4 for the in situ sand and using ρ_s = 2.65 gm/cm³, the volume rate of dry sand transport per unit width is

$$q_v = \frac{(0.9433)}{(2.65)(1 - 0.4)} \quad \text{or} \quad q_v = 0.5933 \text{ cm}^3/\text{cm-s}$$

Therefore

$$q_v = 5.933 \times 10^{-5} \quad \text{m}^3/\text{m-s}$$

An equation for the volume sand transport rate can be written

$$q_v = K_v^*(D.Z_M) \, U_{zM}^3$$

in which K_v^* = a coefficient in units of m² - sec²/cm³ which depends on the mean sand-grain diameter D and the height at which the wind speed is measured Z_M. The coefficient K_v^* can be obtained from Figure III-4-13 which is based on the assumption that the sediment has a mass density of ρ_s = 2.65 gm/cm² and an in situ porosity of 0.4. Entering Figure III-4-13 with the grain diameter of 0.26 mm and reading from the curve for an anemometer height of 8 m gives

$$q_v = 6.3 \times 10^{-15} U_{8m}^3$$

in which q_v = the volume transport rate in cubic meters per meter-second and U_{8m} is in centimeters per second. Therefore

$$q_v = 6.3 \times 10^{-15}(2,100)^3 = 5.83 \times 10^{-5} \quad \text{m}^3/\text{m-s}$$

which is approximately the same result obtained earlier. Note that the wind speed is specified in centimeters per second and the resulting volume transport is in cubic meters per meter-second.

While the foregoing involves a lot of computation, repeated application of the equations to wind speed and direction data is much simplified, since the equations for the given site reduce to

$$U_{8mt} = 727.0 \text{ cm/s}$$

the threshold wind velocity at the 8-m height, and

$$q = (1.019 \times 10^{-10}) \, U_{8m}^3 \quad \text{gm/cm-s}$$

the mass transport rate per unit width where U_{8m} is in centimeters per second. These equations will be used in a subsequent example.

(Sheet 5 of 5)

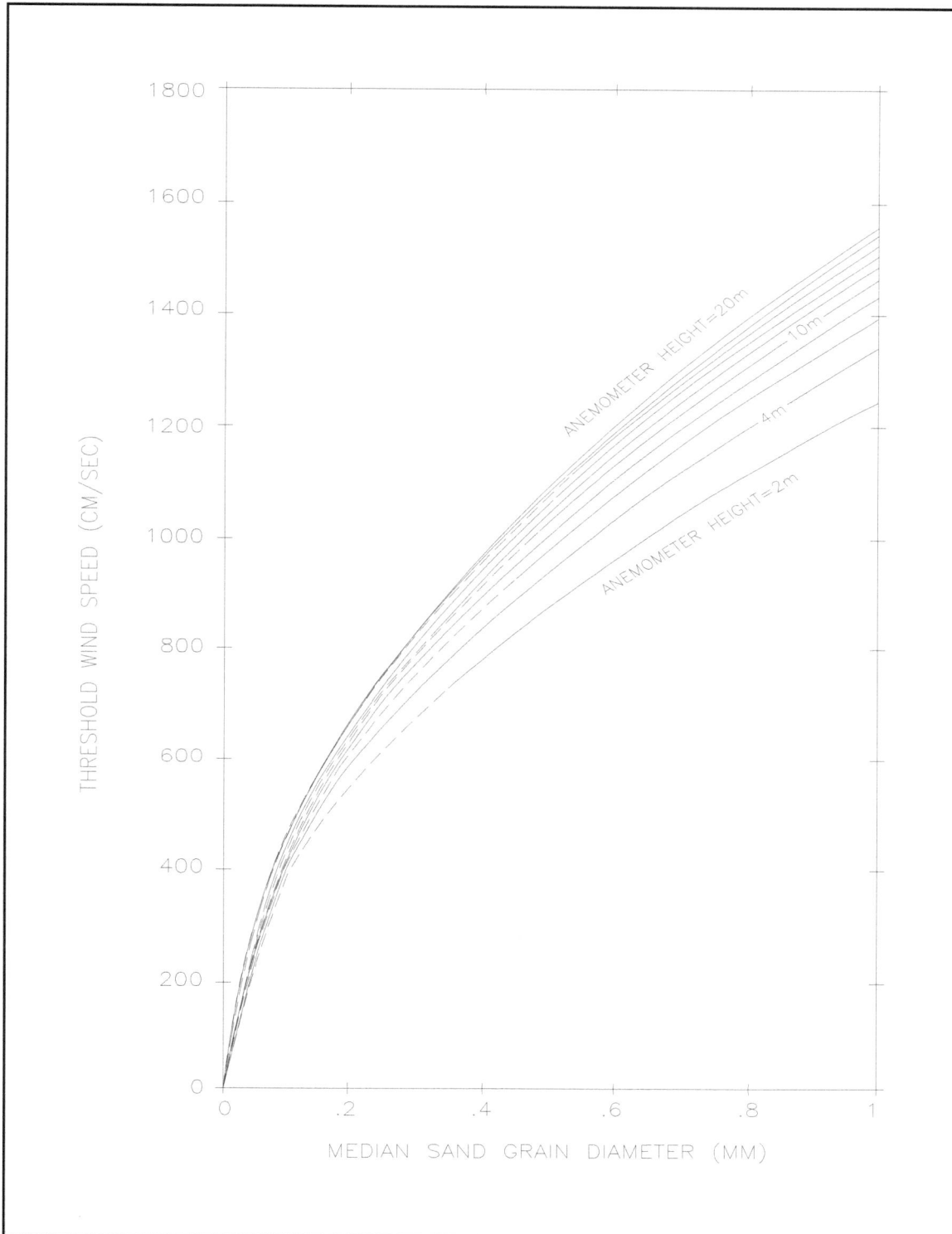

Figure III-4-11a. Threshold (critical) wind speed as a function of median sand-grain diameter and the anemometer height at which wind speed is measured (Continued)

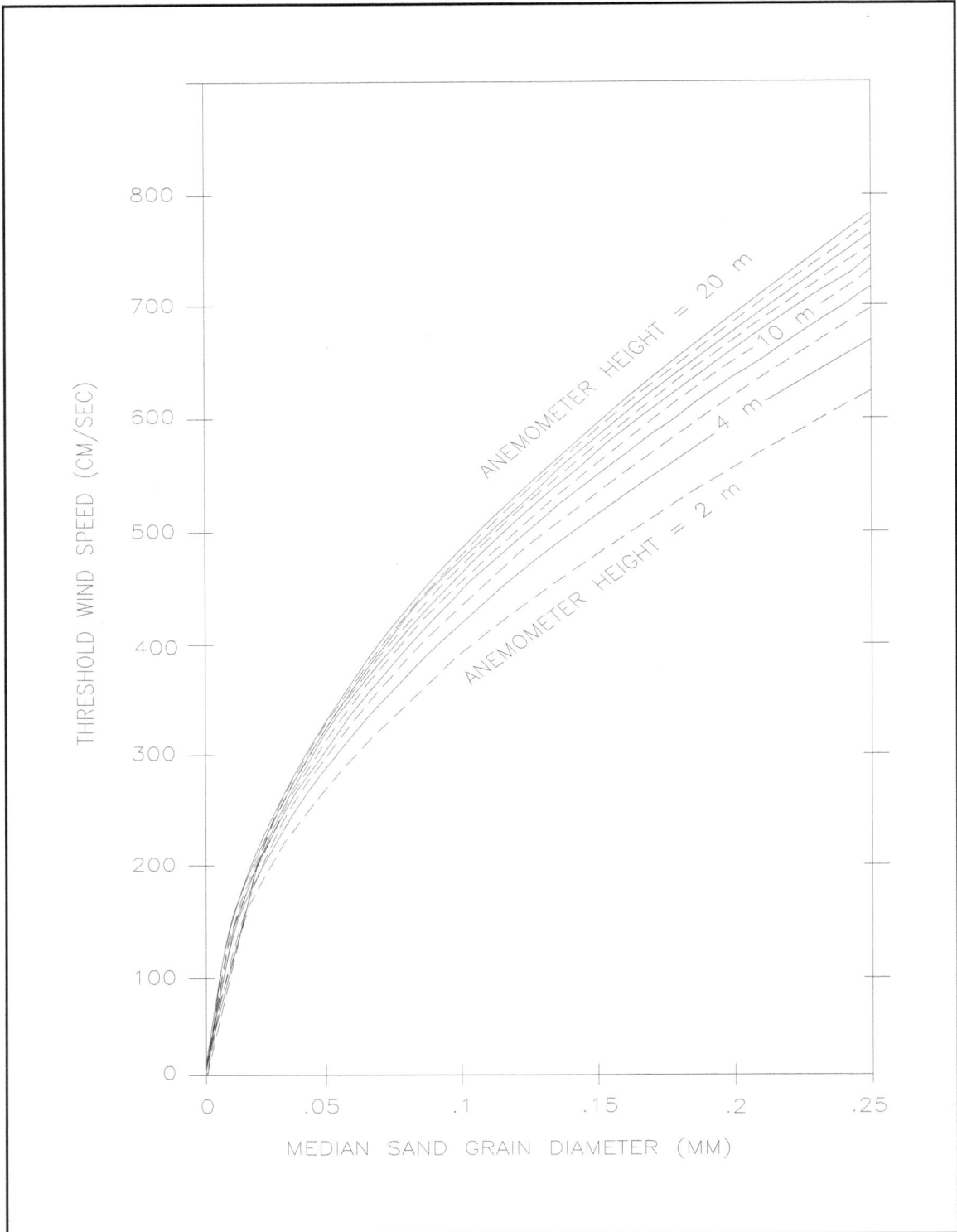

Figure III-4-11b.   (Concluded)

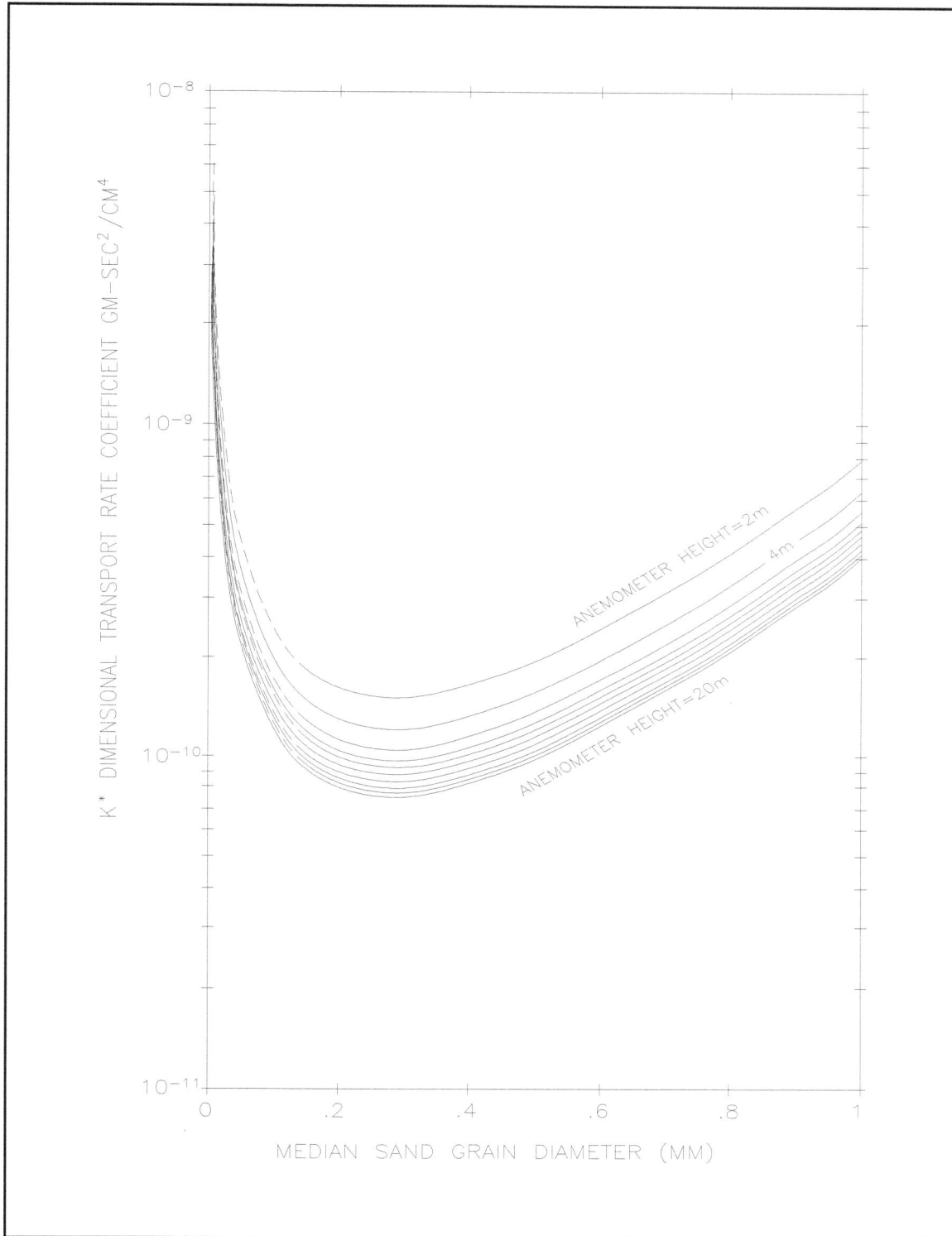

Figure III-4-12a.   Mass transport rate coefficient K^{*} as a function of median sand-grain diameter and the anemometer height at which wind speed is measured (Continued)

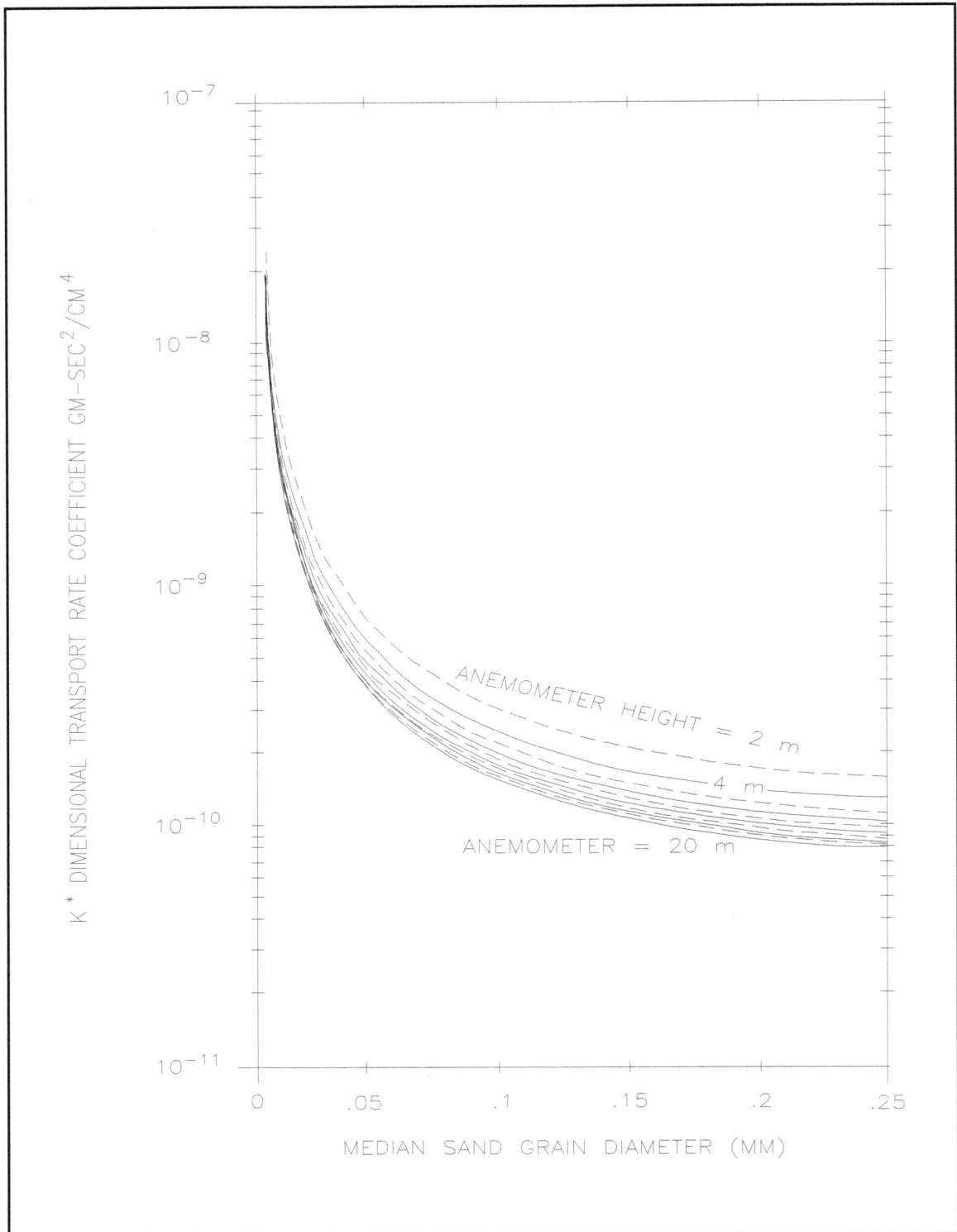

Figure III-4-12b. (Concluded)

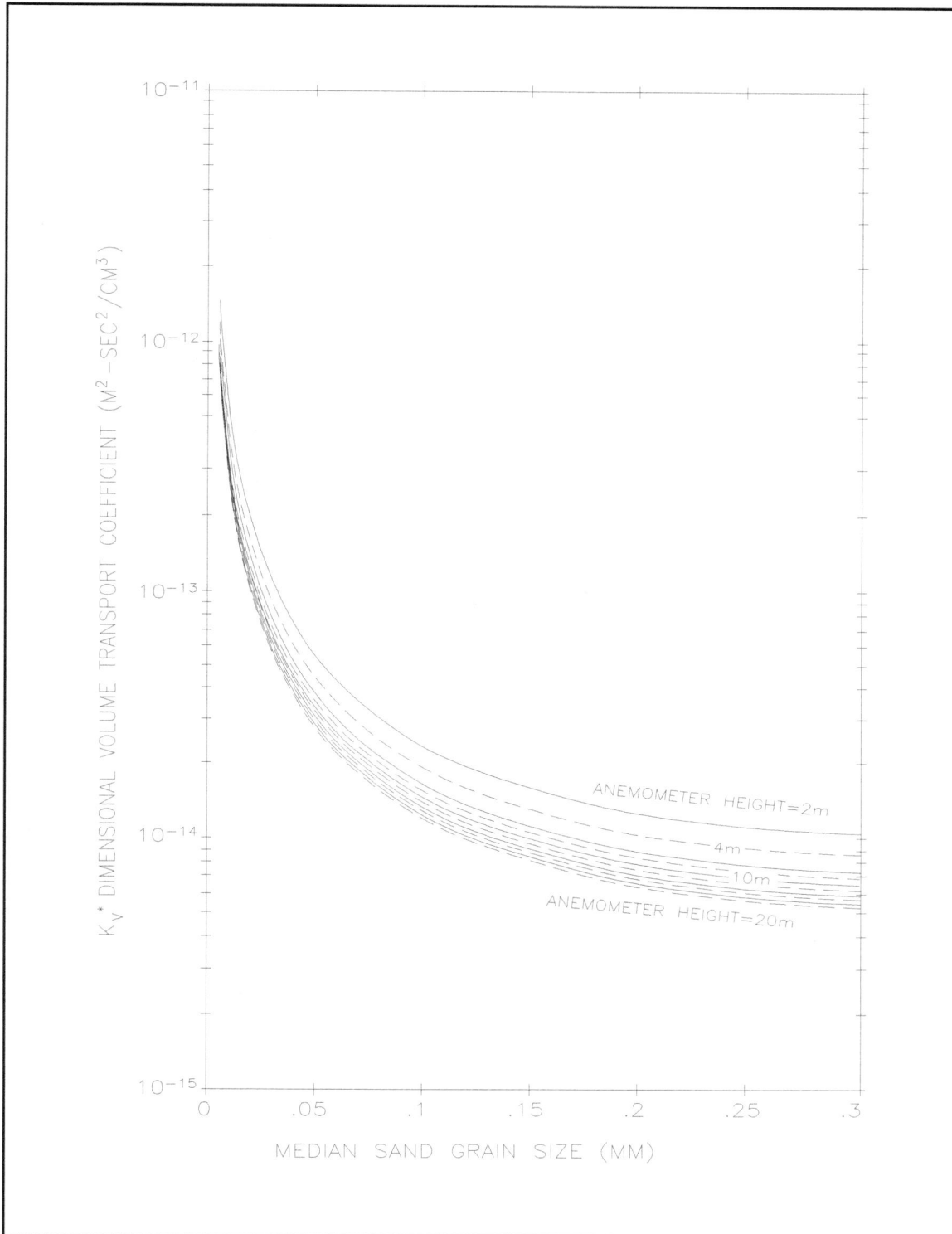

Figure III-4-13. Volume transport rate coefficient K_V^* as a function of median sand-grain diameter and the anemometer height at which wind speed is measured (assumes sediment density = 2.65 gm/cm^3 and porosity = 0.4)

EXAMPLE PROBLEM III-4-5

Find:

The total amount of sand transported by the wind during the given day and the total transported in each of the 16 point compass directions.

Given:

Hourly wind speed and direction data given in Table III-4-3 for Westhampton Beach, Long Island, NY, were measured by an anemometer located 8 m above the ground. The median sand-grain diameter at Westhampton Beach is 0.26 mm.

Solution:

The solution is given in Table III-4-4. The preceding example problem shows that for a sand-grain diameter of 0.26 mm, the threshold wind speed at the 8-m height is 7.27 m/s. Winds less than 7.27 m/s will not transport sand. Columns 1, 2, and 3 in Table III-4-4 are the given data. (Note that wind directions are the direction from which the wind is blowing.) Column 4 is constructed from column 3 and indicates the direction in which the sand will move rather than the direction from which the wind is blowing. The given winds are essentially from the south and will transport sand northward. Column 5 gives the effective winds. Winds less than the threshold value of 7.27 m/sec have been set to zero. Column 6 is computed from column 5 using the equation

$$q = (1.019 \times 10^{-10}) \ U_{8m}^{3}$$

in which U_{8m} is the wind speed at the 8-m height in centimeters per second; this equation gives q in grams per centimeter-seconds. Column 7 is obtained from column 6 by dividing by the density of the sand, 2.65 gm/cm^3, and the packing factor, $(1 - 0.4) = 0.6$. The total volume transported during each hour between observations (column 8) is 3,600 s/hr times the average transport rate during each hour. The total sand transport during the day is 0.5994 m^3/m. Transport in each compass direction is given at the bottom of the table.

Table III-4-3
Hourly Wind Data, Westhampton Beach, Long Island, NY

Time (hr)	Wind Speed (m/sec)	Wind Direction (Azimuth)	Transport Direction (Compass)
(1)	(2)	(3)	(4)*
Midnight	12.0	175	N
1	8.0	170	N
2	9.0	175	N
3	11.0	150	NNW
4	9.0	175	N
5	9.0	190	N
6	10.0	170	N
7	9.0	185	N
8	6.0	165	NNW
9	5.0	155	NNW
1	4.0	170	N
1	5.0	195	NNE
Noon	9.0	215	NE
1	14.0	200	NNE
2	12.0	215	NE
3	13.0	190	N
4	11.0	180	N
5	10.0	160	NNW
6	7.0	180	N
7	11.0	170	N
8	16.0	175	N
9	14.0	175	N
10	9.0	175	N
11	9.0	150	NNW
Midnight	7.0	145	NW

* Direction of transport (opposite of wind direction).

Table III-4-4
Hourly Transport Analysis Under Dry Conditions, Westhampton Beach, Long Island, NY

Time (hr) (1)	Speed (cm/sec) (2)	Direction Azimuth (3)	Direction Compass (4)	Effective Winds (cm/sec) (5)	Rate (g/cm-sec) (6)	Rate (m³/m-sec) (7)	Volume (m³/m) (8)
1 am	800	170	N	800	0.0522	3.281e-06	0.0258
2	900	175	N	900	0.0743	4.672e-06	0.0143
3	1100	150	NNW	1100	0.1356	8.530e-06	0.0238
4	900	175	N	900	0.0743	4.672e-06	0.0238
5	900	190	N	900	0.0743	4.672e-06	0.0168
6	1000	170	N	1000	0.1019	6.409e-06	0.0199
7	900	185	N	900	0.0743	4.672e-06	0.0199
8	600	165	NNW	0	0	0.000e+00	0
9	500	155	NNW	0	0	0.000e+00	0
10	400	170	N	0	0	0.000e+00	0
11	500	195	NNE	0	0	0.000e+00	0
Noon	900	215	NE	900	0.0743	4.672e-06	0.0084
1 pm	1400	200	NNE	1400	0.2796	1.759e-05	0.0401
2	1200	215	NE	1200	0.1761	1.107e-05	0.0516
3	1300	190	N	1300	0.2239	1.408e-05	0.0453
4	1100	180	N	1100	0.1356	8.530e-06	0.0407
5	1000	160	NNW	1000	0.1019	6.409e-06	0.0269
6	700	180	N	0	0	0.000e+00	0
7	1100	170	N	1100	0.1356	8.530e-06	0.0154
8	1600	175	N	1600	0.4174	2.625e-05	0.0626
9	1400	175	N	1400	0.2796	1.759e-05	0.0789
10	900	175	N	900	0.0743	4.672e-06	0.0401
11	900	150	NNW	900	0.0743	4.672e-06	0.0168
Midnight	700	145	NW	0	0	0.000e+00	0
TOTAL							0.5711
TOTAL NW							0
TOTAL NNW							0.0675
TOTAL N							0.4035
TOTAL NNE							0.0401
TOTAL NE							0.0600

EXAMPLE PROBLEM III-4-6

Find:

If the amount of precipitation on the given day exceeds the amount of evaporation, determine the total amount of sand transported by wind and the amount of transport in each of the 16 compass directions.

Given:

The hourly wind speeds and directions in the preceding example problem (Table III-4-3) for Westhampton Beach, Long Island.

Solution:

Because the amount of precipitation exceeds the amount of evaporation, the sand can be assumed to be wet and the threshold velocity will be greater than the threshold velocity for dry sand. The threshold velocity for wet sand is given by Equation 4-22b

$$u_{*tw} = u_{*t} + 18.75 \quad \text{cm/s}$$

in which u_{*tw} and u_{*t} are in centimeters per second. From the preceding example, $u_{*t} = 27.75$ cm/s. Thus

$$u_{*tw} = 27.75 + 18.75 = 46.5 \quad \text{cm/s}$$

The corresponding wind speed at the 2-m height is

$$u_{*tw} = 0.044 \, U_{2mtw}$$

or

$$U_{2mtw} = \frac{46.5}{0.044} = 1057 \quad \text{cm/s}$$

For winds measured at the 8-m height, as before

$$u_* = \frac{\kappa(U_{Z2} - U_{Z1})}{\ln(Z_2/Z_1)}$$

and

$$46.5 = \frac{0.4(U_{8mtw} - 1057)}{\ln(8/2)} \qquad \text{Therefore} \qquad U_{8mtw} = 1,218 \text{ cm/s}$$

The wind speed measured at the 8-m height must exceed 1,218 cm/s in order to transport sand. Winds less than 1,218 cm/s in the record are not considered in the computations. The wet sand threshold velocity can also be found from Figures III-4-14a or III-4-14b. Entering with a grain diameter of 0.26 mm, read the curve for and anemometer height of 8 m to find the threshold velocity of $U_{8mtw} = 1,220$ cm/s. Figure III-4-14b is an enlargement of Figure III-4-14a. Table III-4-5 gives the results of the transport analysis. Wet conditions reduce the total transport from 0.5994 m³/m to 0.2718 m³/m.

(Sheet 1 of 4)

Example Problem III-4-6 (Continued)

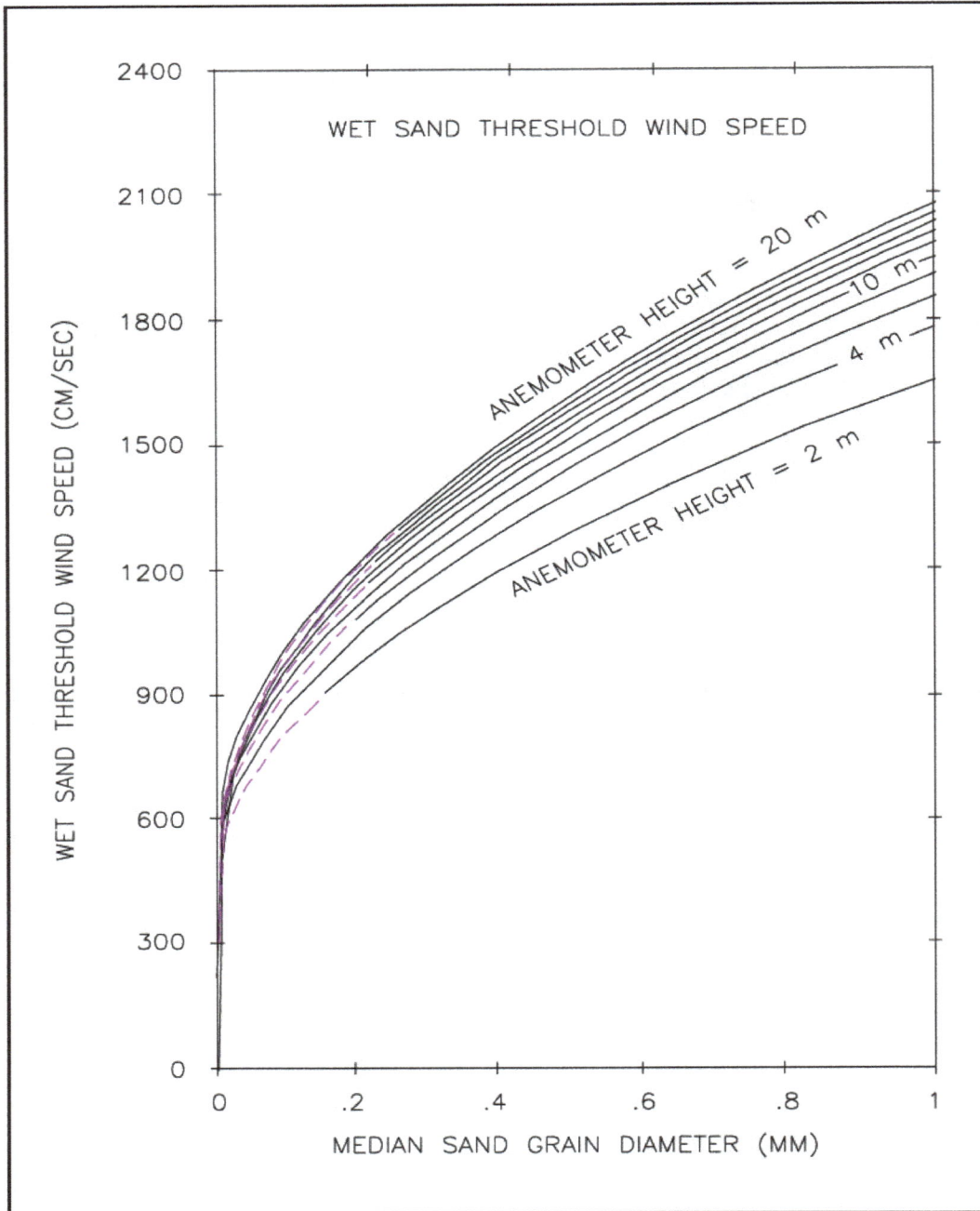

Figure III-4-14. Wet sand transport rate coefficient as a function of median sand grain diameter and the anemometer height at which wind speed is measured (Continued)

(Sheet 2 of 4)

Example Problem III-4-6 (Continued)

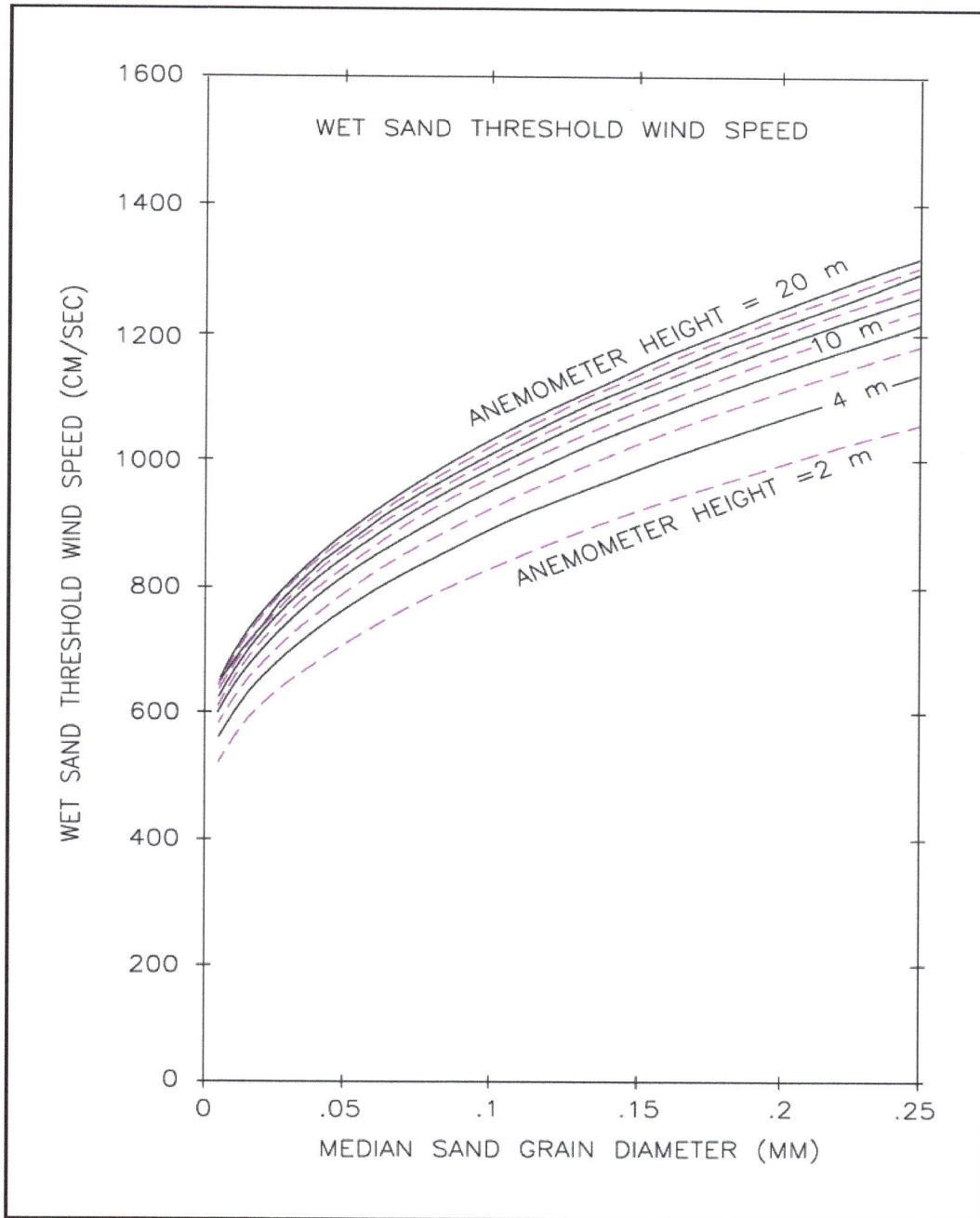

WET SAND THRESHOLD WIND SPEED

(y-axis) WET SAND THRESHOLD WIND SPEED (CM/SEC)
(x-axis) MEDIAN SAND GRAIN DIAMETER (MM)

ANEMOMETER HEIGHT = 20 m
10 m
4 m
ANEMOMETER HEIGHT = 2 m

Figure III-4-14.   (Concluded)

(Sheet 3 of 4)

Example Problem III-4-6 (Concluded)

Table III-4-5
Hourly Transport Analysis Under Wet Conditions, Westhampton Beach, Long Island, NY.

Time (hr) (1)	Speed (cm/sec) (2)	Direction Azimuth (3)	Direction Compass (4)	Effective Winds (cm/sec) (5)	Rate (g/cm-sec) (6)	Rate (m³/m-sec) (7)	Volume (m³/m) (8)
1 am	800	170	N	0	0	0.000e+00	0
2	900	175	N	0	0	0.000e+00	0
3	1100	150	NNW	0	0	0.000e+00	0
4	900	175	N	0	0	0.000e+00	0
5	900	190	N	0	0	0.000e+00	0
6	1000	170	N	0	0	0.000e+00	0
7	900	185	N	0	0	0.000e+00	0
8	600	165	NNW	0	0	0.000e+00	0
9	500	155	NNW	0	0	0.000e+00	0
10	400	170	N	0	0	0.000e+00	0
11	500	195	NNE	0	0	0.000e+00	0
Noon	900	215	NE	0	0	0.000e+00	0
1 pm	1400	200	NNE	1400	0.2796	1.759e-05	0.0317
2	1200	215	NE	0	0	0.000e+00	0
3	1300	190	N	1300	0.2239	1.408e-05	0.0253
4	1100	180	N	0	0	0.000e+00	0
5	1000	160	NNW	0	0	0.000e+00	0
6	700	180	N	0	0	0.000e+00	0
7	1100	170	N	0	0	0.000e+00	0
8	1600	175	N	1600	0.4174	2.625e-05	0.0473
9	1400	175	N	1400	0.2796	1.759e-05	0.0789
10	900	175	N	0	0	0.000e+00	0
11	900	150	NNW	0	0	0.000e+00	0
Midnight	700	145	NW	0	0	0.000e+00	0
TOTAL							0.1832
TOTAL NW							0
TOTAL NNW							0
TOTAL N							0.1515
TOTAL NNE							0.0317
TOTAL NE							0

(Sheet 4 of 4)

measurements were taken, and solve for U_{z2t} = the critical wind velocity at the Z_2 height.) Only wind speeds in excess of the computed U_{z2t} will result in sand transport.

g. If wind speeds exceed the critical value and there was no precipitation on a given day, compute the potential sand transport rate using Equation 4-16 or Equation 4-18.

h. If there was precipitation on a given day, the amount of precipitation should be compared with the amount of evaporation. If evaporation exceeds precipitation, compute the potential sand transport rate using Equation 4-16 or 4-18. (If daily evaporation data are not available, daily evaporation can be estimated by dividing monthly evaporation by the number of days in the month.)

i. If precipitation exceeds evaporation, the critical shear velocity can be computed by

$$u_{*tw} = u_{*t} + 7.5\,W \tag{III-4-21}$$

in which W = the fraction water content in the upper 5 mm of the sand (Hotta et al. 1985). Based on measurements in Holland made with a radioactive moisture probe, Svasek and Terwindt (1974) found that $W < 0.025$ (2.5 percent). Consequently

$$u_{*tw} = u_{*t} + 0.1875 \quad \text{(m/s)} \tag{III-4-22a}$$

or

$$u_{*tw} = u_{*t} + 18.75 \quad \text{(cm/s)} \tag{III-4-22b}$$

may be used in the absence of detailed soil moisture data. For days when precipitation exceeds evaporation, u_{*tw} is used as the critical shear velocity rather than u_{*t}. The computations then proceed as above.

j. If precipitation is in the form of snow, there will be no sand transport until the snow cover has melted. In the computations, days with snow and days known to have snow cover are excluded from the record. Data on snow cover are included in "Local Climatological Data" for all major National Weather Service (NWS) stations.

k. The results of a comprehensive study of potential sand transport by wind at Westhampton Beach by Hsu (1994) are shown in the form of a transport rose in Figure III-4-15. The rose gives the direction from which the sand is transported with the most being transported northward or northwestward. Wind data were obtained from C_MAN Station ALSN6 for 1989. The results of the analysis are tabulated in Tables III-4-6 through III-4-8. Table III-4-6 presents the results obtained by not considering snow cover or moisture conditions. The total potential transport for the year is 71.2 m³/m-yr. The table also shows that most transport at Westhampton Beach occurs during the winter months, with the most transport during November and December. Table III-4-7 corrects for days of snow cover. The total transport is reduced to 64.6 m³/m-yr. When the data are also corrected for moisture conditions (Table III-4-8), the total transport is further reduced to 53.2 m³/m-yr, about 75 percent of the uncorrected value of 71.2 m³/m-yr.

III-4-5. Wind-Blown Sand Transport and Coastal Dunes

a. Dunes and dune processes.

(1) Sand dunes are important coastal features that are created, enlarged, and altered by wind-blown sand. In the absence of dunes, the beach landward of the active shoreface and berm is often characterized by a

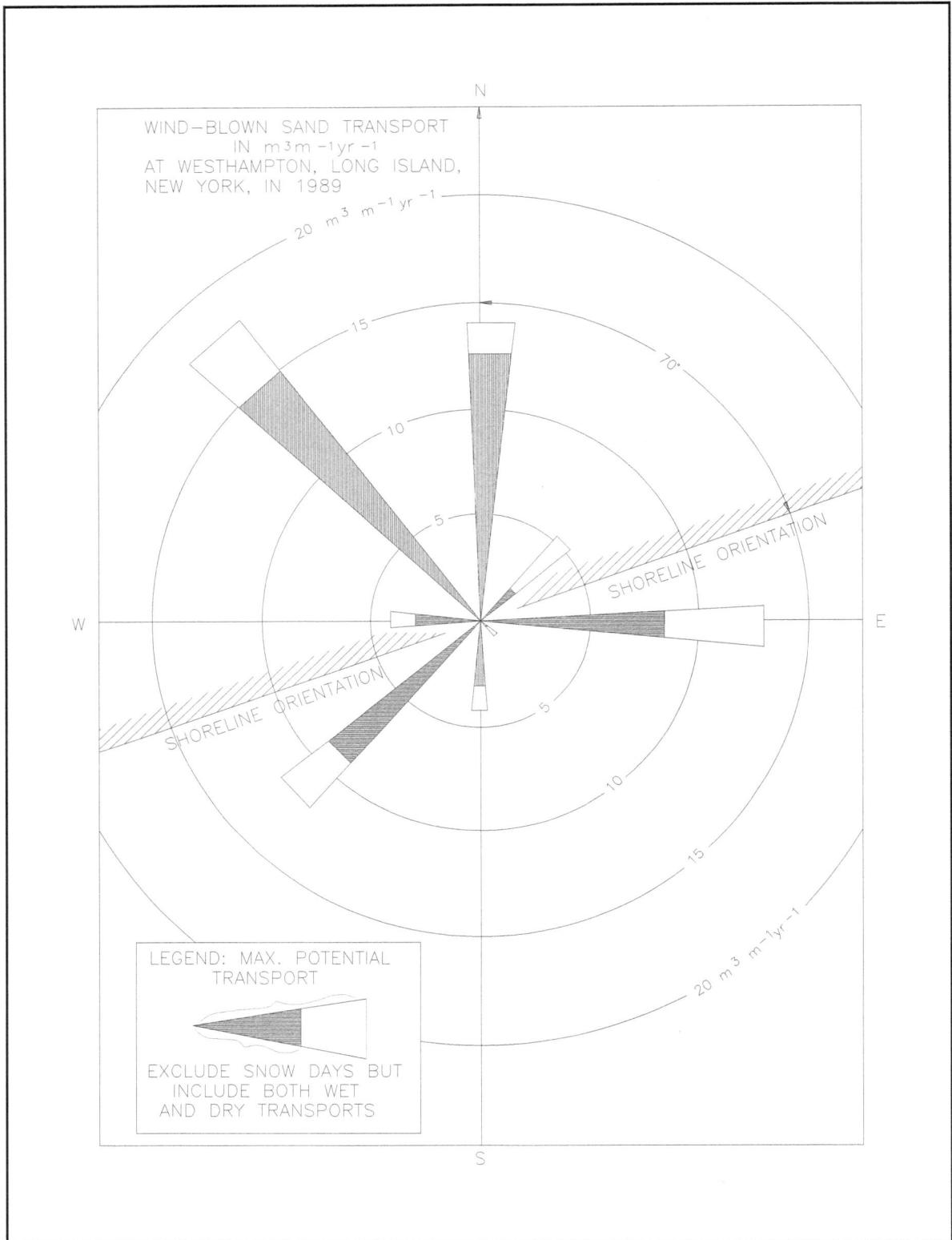

Figure III-4-15. Wind-blown sand transport rose at Westhampton Beach, Long Island, NY, for 1989 (Hsu 1994)

Table III-4-6
Monthly Sand Transport at Westhampton Beach, Long Island, NY (All Data Considered) (Hsu 1994)

						Volume, Cubic Meters Per Meter							
Direction	Jan	Feb	Mar	Apr	May	Jun	Jul	Aug	Sep	Oct	Nov	Dec	Total
N	2.66	1.06	0.80	0.91	0.33	0.16	0.07	0.09	0.85	0.69	2.88	3.37	13.88
NE	0.14	1.45	1.12	0.38	0.12	0.03	0.15	0.14	0.10	0.24	0.34	0.86	5.07
E	0.92	0.63	3.94	0.13	1.83	0.33	0.23	0.60	1.63	2.41	0.04	0.33	13.01
SE	0.10	0.00	0.16	0.09	0.05	0.01	0.18	0.06	0.14	0.20	0.03	0.00	1.03
S	0.33	0.47	0.08	0.19	0.56	0.09	0.03	0.01	0.52	0.53	1.20	0.30	4.32
SW	0.47	0.95	1.68	2.09	1.54	0.58	0.28	0.12	1.53	0.58	1.67	0.33	11.82
W	0.33	0.30	0.20	0.10	0.24	0.02	0.16	0.08	0.18	0.42	1.85	0.34	4.23
NW	1.65	1.43	0.54	1.04	0.44	0.34	0.01	0.11	0.08	1.17	4.55	6.46	17.82
Total	6.62	6.29	8.53	4.94	5.11	1.55	1.10	1.22	5.03	6.24	12.56	11.99	71.18

Table III-4-7
Monthly Sand Transport at Westhampton Beach, Long Island, NY (Snow Days Excluded) (Hsu 1994)

						Volume, Cubic Meters Per Meter							
Direction	Jan	Feb	Mar	Apr	May	Jun	Jul	Aug	Sep	Oct	Nov	Dec	Total
N	2.66	0.76	0.78	0.91	0.33	0.16	0.07	0.09	0.85	0.69	2.80	3.12	13.24
NE	0.11	0.28	0.26	0.38	0.12	0.03	0.15	0.14	0.10	0.24	0.00	0.85	2.66
E	0.22	0.60	3.35	0.13	1.83	0.33	0.23	0.60	1.63	2.41	0.04	0.07	11.45
SE	0.10	0.00	0.16	0.09	0.05	0.01	0.18	0.06	0.14	0.20	0.03	0.00	1.03
S	0.33	0.47	0.08	0.19	0.56	0.09	0.03	0.01	0.52	0.53	1.20	0.30	4.32
SW	0.47	0.95	1.68	2.09	1.54	0.58	0.28	0.12	1.53	0.58	1.62	0.33	11.77
W	0.33	0.30	0.20	0.10	0.24	0.02	0.16	0.08	0.18	0.42	1.68	0.16	3.88
NW	1.65	1.18	0.54	1.04	0.44	0.34	0.01	0.11	0.08	1.17	4.27	5.40	16.23
Total	5.90	4.54	7.06	4.94	5.11	1.55	1.10	1.22	5.03	6.24	11.64	10.23	64.58

Table III-4-8
Monthly Sand Transport at Westhampton Beach, Long Island, NY (Snow Days and Wet Days Excluded) (Hsu 1994)

						Volume, Cubic Meters Per Meter							
Direction	Jan	Feb	Mar	Apr	May	Jun	Jul	Aug	Sep	Oct	Nov	Dec	Total
N	2.54	0.64	0.59	0.73	0.25	0.16	0.07	0.08	0.77	0.69	2.72	3.12	12.36
NE	0.11	0.27	0.19	0.19	0.00	0.03	0.05	0.08	0.08	0.00	0.00	0.85	1.85
E	0.22	0.49	2.89	0.00	1.00	0.02	0.04	0.49	1.00	2.14	0.00	0.07	8.37
SE	0.02	0.00	0.11	0.01	0.01	0.01	0.15	0.06	0.14	0.04	0.00	0.00	0.55
S	0.08	0.31	0.08	0.09	0.33	0.00	0.02	0.01	0.46	0.25	1.15	0.27	3.06
SW	0.44	0.70	1.30	1.86	1.44	0.23	0.28	0.10	1.19	0.32	0.83	0.24	8.93
W	0.23	0.28	0.20	0.09	0.22	0.01	0.06	0.08	0.18	0.30	1.24	0.13	3.03
NW	1.53	1.18	0.52	0.91	0.32	0.30	0.01	0.09	0.08	1.06	3.74	5.26	15.00
Total	5.17	3.87	5.89	3.88	3.56	0.76	0.67	1.00	3.90	4.81	9.69	9.95	53.15

EXAMPLE PROBLEM III-4-7

Find:

Estimate the sand transport in each of the eight compass directions. (Note that the results of this analysis will only be approximate since no information on moisture conditions and/or snow cover is available. Also, the data are daily average wind speeds rather than hourly values.) Ideally, this type of analysis would be performed with several years of hourly wind data; however, a rough estimate can be obtained from wind rose data.

Given:

The wind rose for Atlantic City, NJ, in Figure III-4-16 and the median sand grain diameter of 0.25 mm. The winds have been corrected to offshore conditions and the standard anemometer height of 10 m.

Solution:

Data from the wind rose (Figure III-4-16) are tabulated in Table III-4-9. The transport rates at the wind speed interval boundaries are computed in Table III-4-10. An upper bound of 35 mph was selected to determine the average transport rate over the interval. This upper bound is somewhat arbitrary and was selected only 6 mph above the 29 mph lower bound since it is likely that on those days with average winds exceeding 29 mph, the speed will not exceed 29 mph by much. Column 3 of Table III-4-10 gives the threshold wind speed at the 10-m height for a median sand-grain diameter of 0.25 mm. The transport coefficient in column 4 is from Figure III-4-12b. Column 5 is the transport rate and column 6 is the average transport rate over the interval. (Note that it is important to use average transport rates rather than average wind speeds since transport rate varies with the cube of the wind speed.) For example, the average transport rate over the interval between wind speeds of 29 mph and 35 mph is 0.2944 gm/cm-s. The amounts of sand transported are computed in Table III-4-11.

The threshold velocity of 7.25 m/s is greater than the 6.26 m/s corresponding to the 6.3 m/s (14 mph) interval boundary; therefore, the number of days that the wind speed was between 7.25 m/s and 12.96 m/s (29 mph) was calculated assuming that wind speeds are uniformly distributed over the interval. Adjusted days/year were computed as $(12.96 - 7.25)/(12.96 - 6.26) = 0.852$ times the total number of days in the 6.3 m/s (14 mph) $< U <$ 13.0 m/s (29 mph) interval. Column 5 is from Table III-4-10. The mass transport (column 6) is column 5 x number of days x number of seconds/day. Column 7 is the volume transport equal to column 6 divided by (0.6) x 2.65 gm/cm^3 and converted to cubic meters per meter. The total amount of sand transported in all directions is 146.1 m^3/m.

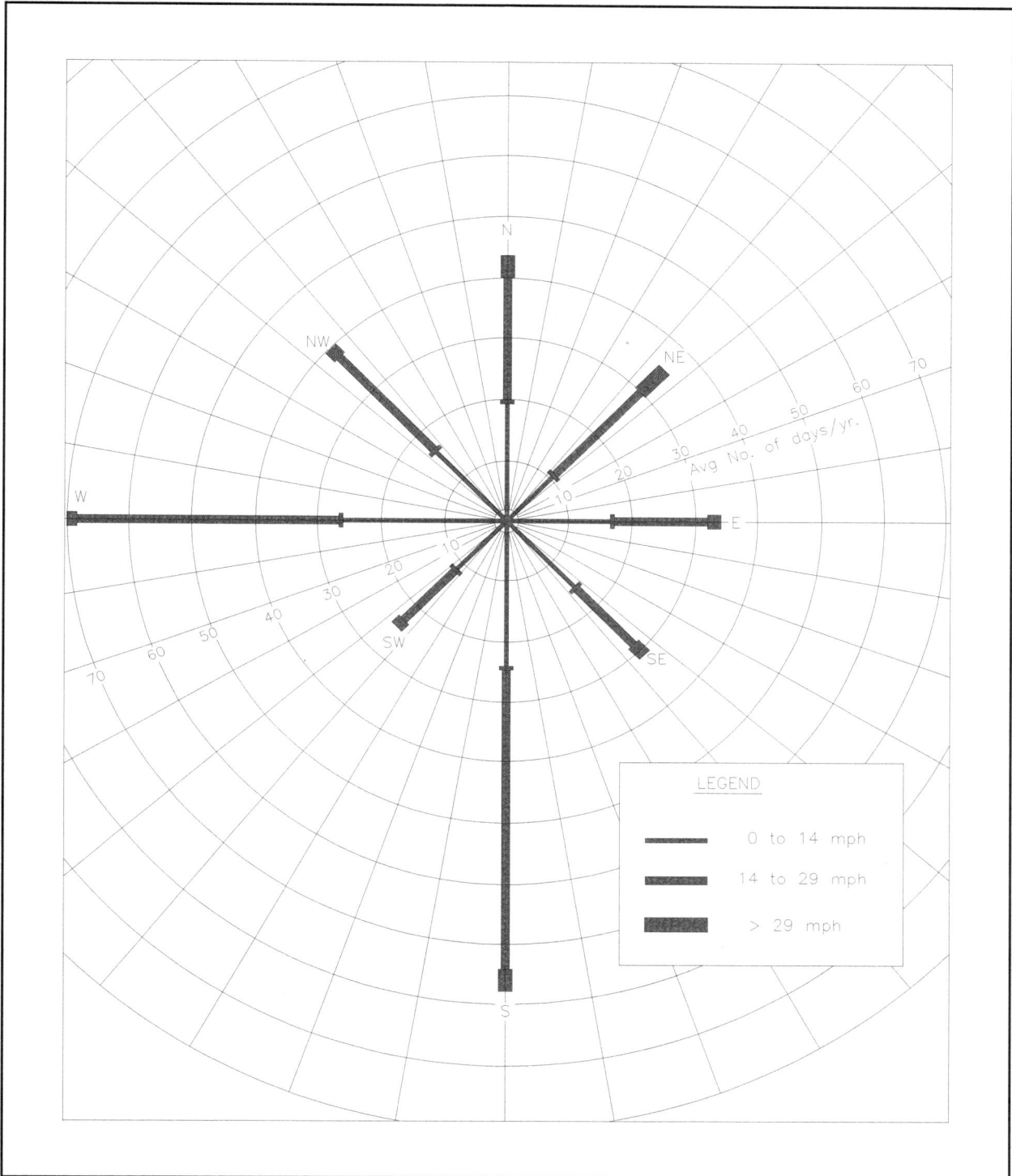

Figure III-4-16.   Wind Rose for Atlantic City, NJ (data from 1936 to 1952)

Table III-4-9
Wind Rose Data, Atlantic City, NJ (Wind Data, 1936 to 1952)

Direction	Wind Speed Interval	Days/Year
N	29 mph < U	3
N	14 mph < U < 29 mph	20
N	0 mph < U < 14 mph	20
NE	29 mph < U	5
NE	14 mph < U < 29 mph	19
NE	0 mph < U < 14 mph	11
E	29 mph < U	2
E	14 mph < U < 29 mph	15
E	0 mph < U < 14 mph	17
SE	29 mph < U	2
SE	14 mph < U < 29 mph	13
SE	0 mph < U < 14 mph	16
S	29 mph < U	3
S	14 mph < U < 29 mph	49
S	0 mph < U < 14 mph	25
SW	29 mph < U	1
SW	14 mph < U < 29 mph	23
SW	0 mph < U < 14 mph	12
W	29 mph < U	1
W	14 mph < U < 29 mph	42
W	0 mph < U < 14 mph	27
NW	29 mph < U	1
NW	14 mph < U < 29 mph	21
NW	0 mph < U < 14 mph	17
TOTAL		365

Table III-4-10
Wind Rose Data Analysis, Atlantic City, NJ (Anemometer at Standard 10-m Height)

Wind Speed (mph)	Wind Speed (m/sec)	Threshold Wind Speed * (m/sec)	Transport Coef. * $(gm\text{-}s^2/cm^4)$	Transport Rate (gm/cm-sec)	Average Rate Over Interval (gm/cm-sec)
(1)	(2)	(3)	(4)	(5)	(6)
35	15.64	7.25	9.8000e-11	0.3753	
					0.2944
29	12.96	7.25	9.8000e-11	0.2135	
					0.1254
14	6.26	7.25	9.8000e-11	0.0373 **	
					0 ***
0	0	7.25	9.8000e-11	0	

* Based on median grain size of 0.25 mm
** Calculated for threshold wind speed of 7.25 m/s.
*** Wind speed in interval does not exceed threshold wind speed.

Table III-4-11
Wind Blown Sand Transport Analysis, Wind Rose Data, Atlantic City, NJ (Wind Data, 1936 to 1952)

Wind Direction	Wind Speed Interval	Days/Year	Adjustment Days/Year	Transport Rate (gm/cm-sec)	Mass Transport (gm/cm-yr)	Volume Transport (m³/m-yr)	Direction in which sand is transported (m³/m)
(1)	(2)	(3)	(4)	(5)	(6)	(7)	(8)
N	13 m/s < U	3	3	0.2944	76,308	4.80	
N	6.3 m/s < U < 13 m/s	20	17*	0.1254	184,621	11.61	Total S = 16.41
N	0 m/s < U < 6.3 m/s	20	0	0	0	0	
NE	13 m/s < U	5	5	0.2944	127,181	8.00	
NE	6.3 m/s < U < 13 m/s	19	16*	0.1254	175,390	11.03	Total SW = 19.03
NE	0 m/s < U < 6.3 m/s	11	0	0	0	0	
E	13 m/s < U	2	2	0.2944	50,872	3.20	
E	6.3 m/s < U < 13 m/s	15	13*	0.1254	138,466	8.71	Total W = 11.91
E	0 m/s < U < 6.3 m/s	17	0	0	0	0	
SE	13 m/s < U	2	2	0.2944	50,872	3.20	
SE	6.3 m/s < U < 13 m/s	13	11*	0.1254	120,004	7.55	Total NW = 10.75
SE	0 m/s < U < 6.3 m/s	16	0	0	0	0	
S	13 m/s < U	3	3	0.2944	76,308	4.80	
S	6.3 m/s < U < 13 m/s	49	42*	0.1254	452,321	28.45	Total N = 33.25
S	0 m/s < U < 6.3 m/s	25	0	0	0	0	
SW	13 m/s < U	1	1	0.2944	25,436	1.60	
SW	6.3 m/s < U < 13 m/s	23	20*	0.1254	212,314	13.35	Total NE = 14.95
SW	0 m/s < U < 6.3 m/s	12	0	0	0	0	
W	13 m/s < U	1	1	0.2944	25,436	1.60	
W	6.3 m/s < U < 13 m/s	42	36*	0.1254	387,704	24.38	Total E = 25.98
W	0 m/s < U < 6.3 m/s	27	0	0	0	0	
NW	13 m/s < U	1	1	0.2944	25,436	1.60	
NW	6.3 m/s < U < 13 m/s	21	18*	0.1254	193,852	12.19	Total SE = 13.79
NW	0 m/s < U < 6.3 m/s	17	0	0	0	0	
Total		365	190			146.07	

* Number of days when wind speed exceeds the threshold wind speed and some transport occurs. Assumes uniform distribution of wind speeds over the 6.3 m/s < U < 13 m/s interval.

Wind-Blown Sediment Transport

"deflation plain" from which sand is removed and the profile lowered by wind. The presence of vegetation or other obstructions to the wind results in sand being trapped and leads to the creation of sand dunes. Foredunes are the first line of dunes landward of the shoreline. These dunes often result from the natural accumulation of wind-blown sand originating on the beach face; however, they may also be man-made. Whether natural or man-made, they are subject to continued growth, alteration, and movement due to natural wind transport processes. When present and if sufficiently large, foredunes can provide protection from coastal flooding, shoreline erosion, and wave damage. Dunes can also shelter the area landward of them from wind-blown sand by providing an area within which wind-blown sand accumulates. This can be an important function of stabilized dunes, since wind-blown sand is often a nuisance that must be controlled.

(2) A line of dunes can provide protection from flooding due to high-water levels and wave overtopping. Sand stored in dunes is also available as a sacrificial contribution to the increased longshore and offshore sand transport during storms. Dunes, by contributing their volume to the sand in transport, reduce the landward movement of the shoreline during storms. Dunes can also reduce wave damage in developed landward areas by limiting wave heights by causing waves to break as they propagate over the dunes. On undeveloped barrier islands, dunes provide a source of sand for overwash processes. During storms with elevated water levels, waves overtopping the dunes carry sand into the bay or lagoon behind a barrier island. Over a period of time, this natural process results in landward migration of the island.

(3) Dunes can be constructed artificially by: (a) beach nourishment, (b) grading existing sand available on the dry beach, or (c) "beach scraping" - removing sand from below the high-water line during low tide and using it to construct a protective dune. In areas where there is sufficient sand and the beach is wide, dunes can be encouraged to develop by natural accumulation by planting dune grass or other vegetation, or by installing sand fences to trap the landward-blowing sand. The rate at which dunes grow depends on the rate at which sand is transported to the dune from the fronting beach and on the effectiveness of the vegetation, fencing, and/or the dune itself in retaining sand. Because of the effect of the dune on the wind field (see Figure III-4-8) and the absence of a mobile sand supply on the landward side of coastal dunes, there is little transport from the dune to the beach, even during periods when winds blow seaward. Coastal dunes generally grow by trapping and accumulating sand on their seaward side; thus, they grow toward the source of sand - the fronting beach. If more than one line of foredunes is to be developed by natural accumulation, the landward line of dunes must be constructed before the line closest to the shoreline is constructed, otherwise the line closest to the shoreline will trap sand and keep it from reaching the more landward line.

(4) When designing a dune system intended to accumulate sand by natural wind transport processes, the dunes should be set back from the shoreline so that there is sufficient dry beach to provide a source of sand. A distance of 60 m between the toe of the dune and the high-water line has been recommended (*Shore Protection Manual* 1984). This expanse of sand is often not available on many beaches so that lesser distances must be accepted. At the very least, the toe of the foredune should be landward of the normal seasonal fluctuation of the shoreline. Narrow beaches are less effective sand sources and dune growth will be slower. Typical rates of sand accumulation in dunes are about 2.5 to 7.5 m^3/m-yr; however, rates as high as 25 m^3/m-yr have been measured (Woodhouse 1978). Savage (1963) measured accumulations of 4.5 to 6 m^3/m over a period of 7 to 8 months using a single row of sand fencing. Savage and Woodhouse (1968) measured accumulations of 39 m^3/m over a 3-year period in a multiple-fence experiment. This is an average of 13 m^3/m-yr; however, accumulation rates must have been higher than this average rate during portions of the 3-year period.

(5) Dunes that are intended to provide protection against flooding and erosion must be stabilized to prevent their deflation and/or landward migration. Stabilization is also necessary to control the landward movement of wind-blown sand into developed areas. Stabilization can be achieved using vegetation or sand fencing. Guidelines for dune creation and stabilization using vegetation are given by U.S. Army Corps of

EXAMPLE PROBLEM III-4-8

Find:

If the shoreline azimuth at Atlantic City is $\theta = 50$ deg (angle between north and the shoreline), estimate the rate at which sand might accumulate in a dune field behind the beach. (As in the preceding example, this type of analysis ideally should be done with hourly wind data.)

Given:

The transport computations from the preceding Example Problem III-4-7 based on the wind rose at Atlantic City, NJ.

Solution:

The transport rate in a given direction is the transport rate times the cosine of the angle between the wind direction and the given direction. See Figure III-4-17. For Atlantic City, shoreline alignment and compass directions are shown in Figure III-4-18. For example, an easterly wind will transport sand perpendicular to the shoreline $q_{\perp E} = q_E \cos 50°$ where $q_{\perp E}$ is the transport perpendicular to the shoreline into the dunes. For a southeasterly wind, the transport perpendicular to the shoreline is $q_{\perp SE} = q_{SE} \cos 5$ deg. In general,

$$q_{\perp} = q \cos \alpha$$

where α = angle between wind direction and offshore, directed normal to the shoreline.
For the example

$$q_{\perp E} = q_E \cos(50°)$$
$$q_{\perp SE} = q_{SE} \cos(5°)$$
$$q_{\perp S} = q_S \cos(40°)$$
$$q_{\perp SW} = q_{SW} \cos(85°)$$

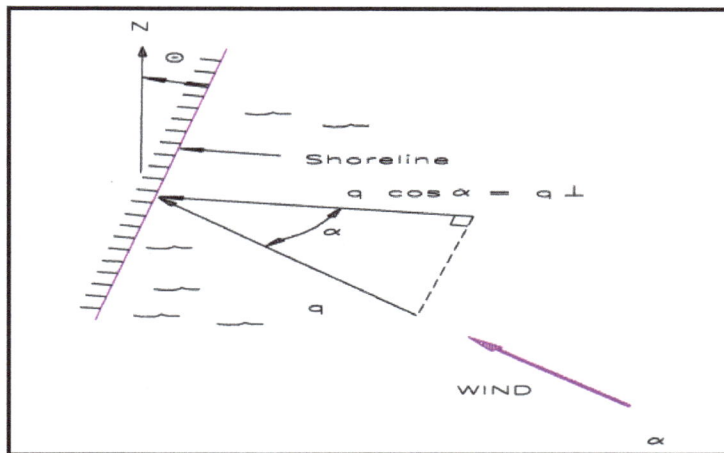

Figure III-4-17. Net sand transport at an angle α with the wind direction

(Sheet 1 of 4)

Example Problem III-4-8 (Continued)

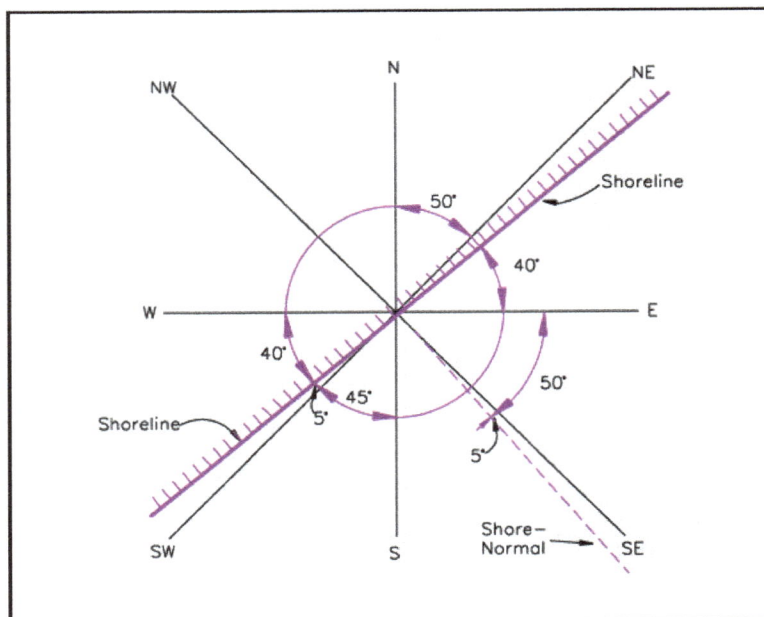

Figure III-4-18. Shoreline orientation with respect to compass directions at Atlantic City, NJ

Winds blowing generally from the dunes or backbeach towards the shoreline are not as efficient in transporting sand as are winds blowing generally onshore. This is because of sheltering provided by the dune itself and stabilization by vegetation on the dune, and because there is often only a limited supply of sand landward of the dunes. This latter condition prevails at Atlantic City, where much of the area landward of the beach is paved or developed. Figure III-4-8 also shows that in the lee of a dune, winds near ground level may be in the direction opposite of the wind direction at a higher elevation. Winds blowing seaward from the dunes are not very efficient in transporting sand from the dunes back onto the beach. Sand transport efficiency, proportional to the square of the cosine of the angle between the wind direction and a vector perpendicular to the shoreline, is assumed. See Figure III-4-19 for the efficiency factor as a function of β, the angle between the shoreline and wind direction. Figure III-4-20 gives the definition of β.

Sand transport rates in the offshore direction, perpendicular to the general orientation of the dune toe, are given by

$$q_\perp = q \cos\alpha \cos^2\beta, \qquad 180° < \beta < 360°$$

where β is the angle the wind makes with the shoreline and α is the angle the wind makes with a vector perpendicular to the shoreline. Therefore, $\alpha = \beta - 90°$ and

$$\cos(\alpha) = \cos(\beta - 90°) = -\sin(\beta)$$

(Sheet 2 of 4)

Example Problem III-4-8 (Continued)

Therefore,

$$q_\perp = -q \sin\beta \cos^2\beta, \qquad 180° < \beta < 360°$$

The $\sin \beta$ term in the equation corrects the transport from the wind direction to a direction perpendicular to the shore, while the $\cos^2 \beta$ term is the efficiency term introduced to consider the sheltering effects of the dune.

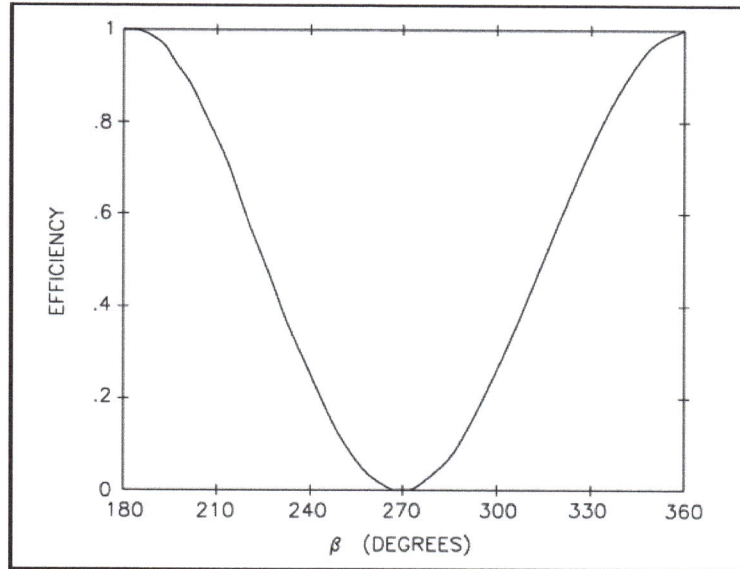

Figure III-4-19. Sand transport efficiency for winds blowing from dunes toward beach as a function of the angle between wind direction and shoreline

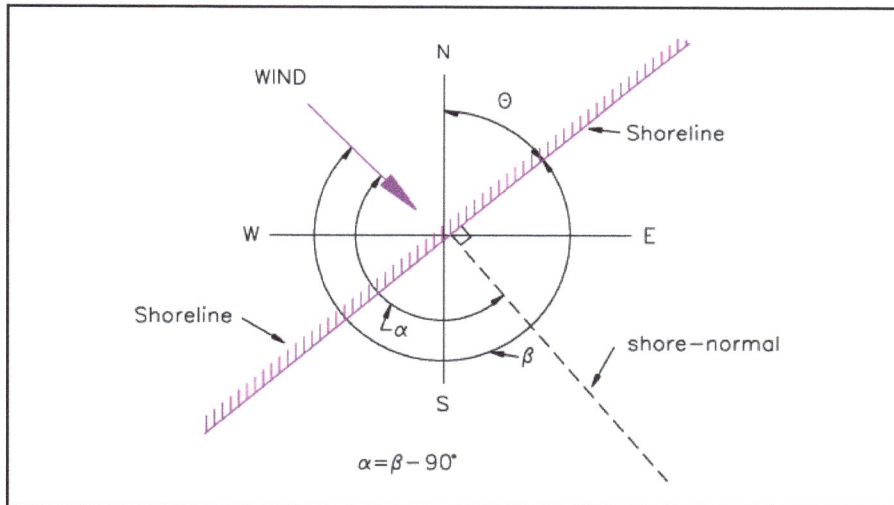

Figure III-4-20. Definition of angle ß, between wind and shoreline

(Sheet 3 of 4)

Example Problem III-4-8 (Concluded)

Results of the analysis are given in Table III-4-12. Columns 1 and 2 are obtained from Table III-4-11. Column 3 is the angle β between the given wind directions and the shoreline azimuth of $\theta = 50°$ at Atlantic City. Column 4 is $\sin(\beta)$ and column 5 is the offshore wind transport efficiency, $\cos^2(\beta)$. The efficiency is assumed equal to 1.0 for onshore winds. Column 6, the rate of dune growth due to winds from the given direction, is the product of columns 2, 4, and 5. The estimated total potential dune growth rate is 27.2 m³/m-yr. This estimate represents a potential growth rate which depends on the sand supply. The analysis does not consider the effects of snow cover and moisture to reduce transport. Woodhouse (1978) reports on a 30-year average dune accumulation rate of 13.7 m³/m-yr on the Clatsop Plains, Oregon, and suggests that higher accumulation rates probably prevailed in the early years.

Table III-4-12
Estimated Annual Dune Growth at Atlantic City, NJ

Wind Direction	Annual Transport (m³/m)	Wind Angle (β) (degrees)	COS(α) or SIN(β)	Efficiency COS²(β)	Dune Growth (m³/m/yr)
(1)	(2)	(3)	(4)	(5)	(6)
N	15.79	310	-0.7660	0.4132	-5.00
NE	18.44	355	-0.0872	0.9924	-1.60
E	11.44	40	0.6428	1.0000[1]	7.35
SE	10.34	85	0.9962	1.0000[1]	10.30
S	31.73	130	0.7660	1.0000[1]	24.31
SW	14.24	175	0.0872	1.0000[1]	1.24
W	24.68	220	-0.6428	0.5868	-9.31
NW	13.14	265	-0.9962	0.0076	-0.10
TOTAL					27.20 m³/m/yr

[1] Transport efficiency = 1.0 for angles between 0 and 180 deg.

(Sheet 4 of 4)

Engineers (1978) and Woodhouse (1978). The best type of stabilizing vegetation varies with geographical area, location on the dune, exposure of the site, whether the water body is salt or fresh water, etc. Typical plants used for dune stabilization include European beach grass (*Ammophila arenaria*) and, less frequently, American dunegrass (*Elymus mollis*) along the Pacific Northwest and California coasts; American beach grass (*Ammophila breviligulita*) along the mid- and upper-Atlantic and Great Lakes coasts; sea oats (*Uniola paniculata*) along the South Atlantic and gulf coasts, and panic grasses (*Panicum amarum* or *P. amarulum*) along the Atlantic and gulf coasts.

(6) Guidelines for dune creation and stabilization with fencing are given in the *Shore Protection Manual* (1984) and Woodhouse (1978). Typical sand fencing is either wooden-slat fencing of the type used to control snow along highways or geosynthetic fabric fencing. A wooden-slat fence is typically 1.2 m high with 3.8-cm-wide slats wired together to provide about 50 percent porosity. Stabilization by fencing requires periodic inspection of the installation, replacement of damaged or vandalized fencing, and installation of new fencing to continue the stabilization process as the dune grows. The size and shape of a dune can be controlled by stategically installing subsequent lines of fencing to encourage the dune to grow higher or wider. Most fence installations fill within 1 year of construction. Dune growth is most often limited by the available sand supply rather than by the capacity of the fence installation. Dunes created by fencing must also subsequently be planted to stabilize them for when the fencing eventually deteriorates.

(7) Stabilization by vegetation has the advantage that it is capable of growing up through the accumulating dune and of repairing itself if damaged, provided the damage is not too extensive. After planting and for a year or two thereafter, vegetation requires fertilization to establish healthy plants. Traffic through vegetated dunes must also be controlled to prevent damage to plants by excluding any pedestrians and vehicles or by providing dune walk-over structures. Dune areas with barren areas in an otherwise healthy vegetated dune, will be subject to "blowouts," local areas of deflation. Blowouts reduce a dune's capability to protect an area from flooding.

(8) An example of estimating the rate of dune growth follows. Example III-4-8 assumes that all of the sand transported to the dune is trapped by vegetation and/or sand fencing and that the fronting beach is sufficiently wide to supply sand at the potential transport rate for a given wind speed, i.e., dune growth is not limited by sand supply. Subsequent examples consider the effects of armoring and a limited sand supply.

b. Factors affecting dune growth rates.

(1) The dune growth rates presented in the preceding section and Example III-4-8 are "potential" rates, since they consider only the accumulation of sand in the dunes and not any of the processes that remove sand from the dunes. Furthermore, the trapping process is assumed to be nearly perfect, i.e., little sand is removed from the dunes when wind direction reverses and blows offshore and no sand is blown landward of the dunes. In some developed coastal communities, sand is often removed from the area landward of the dunes; hence, the rate of dune growth often underestimates wind-blown sand transport rates or, vice versa, calculated transport rates overestimate the rate of dune growth. Some typical measured dune growth rates are given in Table III-4-13.

(2) On relatively narrow beaches backed by dunes, the dunes are occasionally eroded by waves during storms having elevated water levels. The storms often leave a nearly vertical scarp with the height of the scarp and the amount of erosion depending on the severity of the storm; i.e., the height of the waves and storm surge. Less frequently, during unusual storms, the dune may be completely destroyed. Dunes erode during storms only to be rebuilt during the period between storms with sand transported by winds having an onshore component. The net rate of dune growth, therefore, is only a fraction of what would be predicted if only rates of accumulation are considered and sand losses are ignored. Dune growth rates can be estimated

Table III-4-13
Measured Dune Growth Rates (Field experiments conducted by Coastal Engineering Research Center in 1960s and 1970s)

Site	Padre Island, TX			Padre Island, TX			Ocracoke Island, NC		
Type	Wood fencing (3 fences, 3 lifts)			Wood fencing (4 fences, 3 lifts)			American Beachgrass		
Beach	60 - 90 m			60 - 90 m			180 - 210 m		
Growth Rates	Δt (mos)	ΔV (m³/m)	Rate (m³/m-yr)	Δt (mos)	ΔV (m³/m)	Rate (m³/m-yr)	Δt (mos)	ΔV (m³/m)	Rate (m³/m-yr)
	12	8.5	8.53	12	6.0	6.02	24	12.8	6.40
	12	6.8	6.77	12	10.3	10.28	27	9.8	4.35
	12	10.8	10.79	12	6.5	6.52	29	16.6	6.85
Avg Rate			8.70			7.61			5.86
Site	Core Banks, NC			Padre Island, TX			Core Banks, NC		
Type	American Beachgrass and fence			Sea oats			Sea oats		
Beach	120 - 150 m			60 - 90 m			90 m		
Growth Rates	Δt (mos)	ΔV (m³/m)	Rate (m³/m-yr)	Δt (mos)	ΔV (m³/m)	Rate (m³/m-yr)	Δt (mos)	ΔV (m³/m)	Rate (m³/m-yr)
	32	11.7	4.42	36	29.6	9.87	22	5.0	2.74
	22	11.0	4.98	60	42.6	8.53	14	2.8	2.37
	26	10.8	4.98				19	6.3	3.96
Avg Rate			5.14			9.20			3.02
Site	Clatsop Spit, OR								
Type	European beachgrass								
Beach	180 m								
Growth Rates	Δt (mos)	ΔV (m³/m)	Rate (m³/m-yr)						
	360	384	12.79						
Avg Rate			12.79						

by constructing a sediment budget for the dune that balances rates of accumulation against rates of erosion. However, estimating the occurrence and volume of dune erosion is difficult. Estimating how much sand is actually trapped by the dune is also difficult.

(3) The rate at which sand is supplied to a dune depends also on the availability of a sand source upwind of the dunes. A wide beach usually provides an adequate source. However, if the beach is narrow, there may not be a sufficient upwind supply in the sense that the over-sand fetch is insufficient for the transport rate to reach an equilibrium (Nordstrom and Jackson 1992; Gillette et al. 1996). Consequently, calculated transport rates overestimate actual transport rates.

(4) The size gradation of the beach sand also has an influence on transport rates. Sand transport analyses based on a single sand-grain diameter fail to recognize that beach sands are comprised of a gradation of sizes each of which has a different threshold wind speed that initiates motion. Grains finer than the median size are more easily transported and may be moved by lower wind speeds while coarser grains are left behind. Basing transport rates on only the median grain size fails to recognize that finer grains on the surface can be transported even when wind speeds are relatively low. Ideally, each grain size should be considered separately. In addition, as fines are removed from the beach sand surface, the remaining larger size fraction armors the surface, making it difficult to transport additional sand. Figure III-4-21(a) shows a relatively uniform distribution of sand grains where almost all grains have about the same threshold speed. Figure III-4-21(b) shows a well-graded sand prior to wind erosion, while Figure III-4-21(c) shows the same well-graded sand after the fines have been removed leaving behind the coarse fraction on the surface. Higher wind speeds are required to continue the transport process after the fine surficial materials are removed. Alternatively, processes that break up the beach surface to expose additional fine sediments to wind will reestablish the transport of fines. Such processes include pedestrian and vehicular traffic and beach raking.

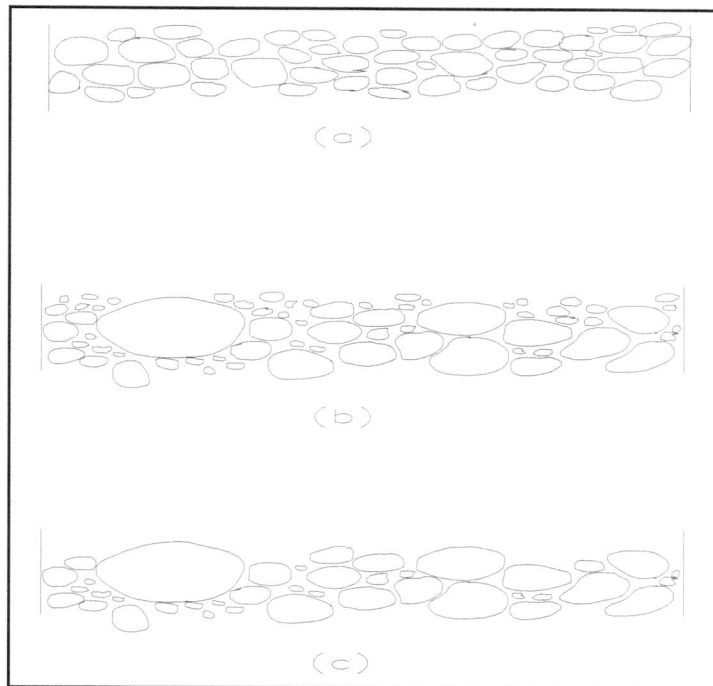

Figure III-4-21. Schematic views of sorted sand deposits: (a) well-sorted (poorly graded); (b) poorly sorted (well-graded); (c) poorly sorted after fines have been removed by wind erosion

(5) The formation of a salt crust on the beach surface by the evaporation of seawater also increases threshold wind speeds and reduces transport (Nickling and Egglestone 1981). Sand below the high tide line is most affected by this process; however, occasional saturation of the beach above the usual high tide level by storm tides will expand the affected beach area inland. Additionally, salt spray carried inland by wind can further expand the affected area. Processes that break up the crust to restore usual threshold wind speeds include pedestrian and vehicular traffic.

c. *Dune sediment budget.*

(1) Sand gains and losses by a typical dune are illustrated in Figure III-4-22. Sand is transported by onshore winds to the dune where vegetation, sand fencing, or the dune itself alter the wind field and allow the sand to accumulate. If the trapping is incomplete, the same onshore winds may transport sand inland from the dune.

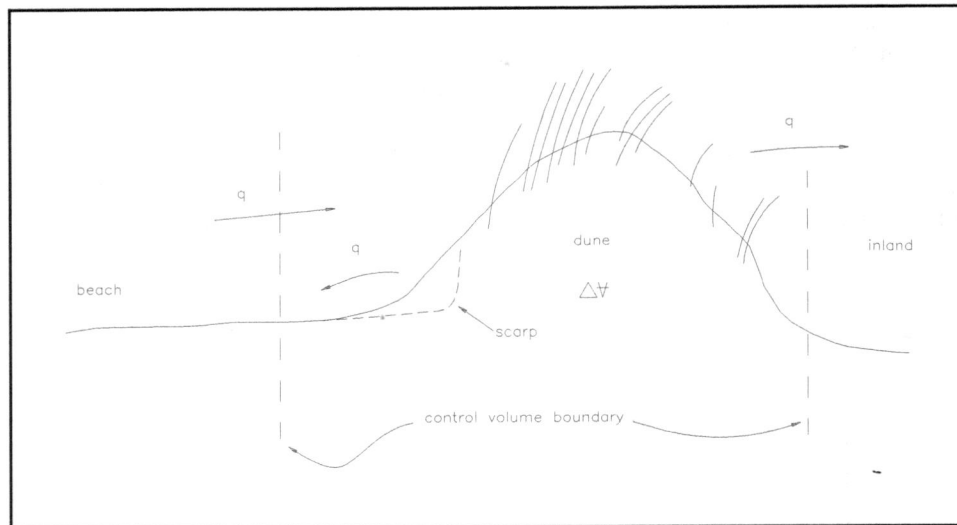

Figure III-4-22.   Definition of terms, dune sediment budget

(2) If sand accumulates at the dune's seaward side and erodes from its landward side, the dune will migrate seaward. Occasional high water levels during storms may allow waves to erode the base of the dune to create a scarp. Sand eroded from the dune by waves becomes a part of the littoral transport. Following the storm, onshore winds again rebuild the dune's seaward side. During periods of offshore winds, some sand from the dune will be blown back onto the beach. Similarly, if there is an unstabilized source of sand landward of the dune, it will contribute to building the dune. The sand budget equation for the dune is given by

$$q - q_s - q_i = \rho (1-p) \frac{\Delta \forall}{\Delta t} \qquad \text{(III-4-23)}$$

in which q = wind-borne sand transport rate (mass per unit time per unit length of beach), q_s = dune erosion by waves (mass per unit time per unit length of beach), q_i = the sand blown inland from the dune, $\Delta \forall$ = the volume change per unit length of beach, ρ = mass density of the sand, p = the porosity of the *in situ* dune sand, and t = time. Note that q and q_i include both sand gains and losses depending on their sign. In fact, q and q_i should each be multiplied by transport efficiency terms, which depend on wind direction and dune conditions such as the presence of vegetation and sand fencing. For example, onshore winds can efficiently transport sand, while offshore winds, which would carry sand out of the dunes, are less efficient if vegetation and fencing are present. The calculations in Example Problem III-4-8 consider only q with a transport efficiency between 0 and 1 depending on the wind direction relative to the axis of the dune. Net dune losses embodied in q_s and q_i are not included. Equation 4-23 might be rewritten

$$\eta \, q - q_s - \eta_i \, q_i = \rho (1 - p) \frac{\Delta \forall}{\Delta t} \qquad \text{(III-4-24)}$$

in which η = the transport efficiency for winds at the seaward side of the dune and η_i = the transport efficiency for winds at the landward side of the dune. For the calculations in Example Problem III-4-8, η was assumed to be a function of wind direction ($\eta = 1$ for winds with an onshore component and $0 < \eta < 1$ for winds with an offshore component), and η_i was assumed zero (no inland sand losses and no inland sand source). Note that η was assumed to vary as $\cos^2\beta$ with 180 deg $< \beta <$ 360 deg, where β = the angle between the wind and the dune axis (the dunes are usually parallel with the shoreline) so that $\beta = 270$ deg represents wind blowing from the back of the dune.

d. Procedure to estimate dune "trapping factor."

(1) Unfortunately, the preceding discussion assumes knowledge of quantities that are most typically unknown and a simple approach must be taken. A "dune trapping factor" can be computed to correct predicted "potential" dune growth rates in order to obtain better estimates of actual rates. The trapping factor T_f is based on the difference between the sand grain size distribution on the dune and the size distribution on the beach in front of the dune. If R_t = a theoretical growth rate based on calculations of sand transport and R_a = the actual growth rate, the relationship is given by

$$R_a = T_f R_t \tag{III-4-25}$$

(2) The value of T_f can be found using the procedure suggested by Krumbein and James (1965) to compare the characteristics of potential beach nourishment sand with native beach sand characteristics. While the Krumbein and James (1965) procedure has been superseded by better methods to estimate beach nourishment overfill ratios, in principle it presents a way to estimate what volume of sand with a given gradation is needed to produce a unit volume of sand having a different gradation. The procedure is based on the assumption that the gradations of the sand on the beach and the sand in the dunes are both log-normally distributed and can be described by their phi-mean μ_φ and phi-standard deviation σ_φ. These parameters can be estimated using the relationships

$$\mu_\varphi = \frac{\varphi_{16} + \varphi_{84}}{2} \tag{III-4-26}$$

and

$$\sigma_\varphi = \frac{\varphi_{16} - \varphi_{84}}{2} \tag{III-4-27}$$

in which φ_{16} = the phi diameter of the 16th percentile (16 percent of the grains are finer) and φ_{84} = the phi diameter of the 84th percentile. The phi diameter is given by

$$\varphi = -\log_2 d_{mm}$$

or equivalently by

$$\varphi = -3.3219 \log_{10} d_{mm} \tag{III-4-28}$$

in which φ = the diameter in phi units and d_{mm} = the grain diameter in millimeters. The phi-mean and phi-standard deviation are calculated for sand taken from the dune, $\mu_{\varphi D}$ and $\sigma_{\varphi D}$, and for sand from the beach in front of the dune, $\mu_{\varphi B}$ and $\sigma_{\varphi B}$, and the following parameters are calculated

$$\frac{\sigma_{\varphi D}}{\sigma_{\varphi B}}$$

(III-4-29)

and

$$\frac{\mu_{\varphi D} - \mu_{\varphi B}}{\sigma_{\varphi B}}$$

(III-4-30)

and used with Figure III-4-23 to determine how much beach sand is needed to produce a unit volume of dune sand. Figure III-4-23 is derived from the equation

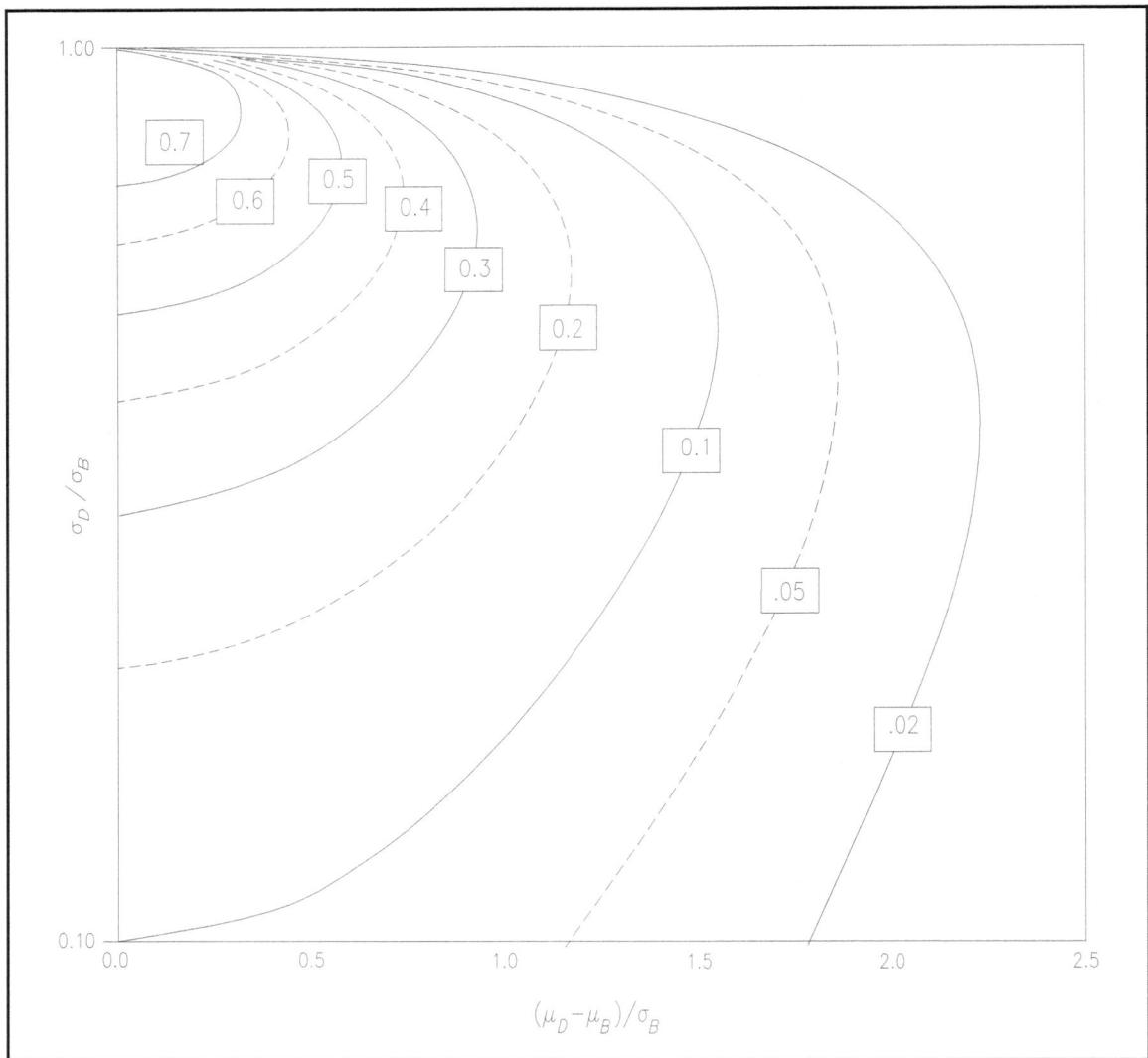

Figure III-4-23.   Dune sand trapping factor T_f

EXAMPLE PROBLEM III-4-9

Find:

Determine the dune trapping factor and estimate the rate of dune growth for the sands from the beach and dune in Atlantic City with the size characteristics given below.

Given:

The dune growth rate from Example Problem III-4-8 based on the transport rates of Example Problem III-4-7 and the following data from a sand size analysis of the beach and dune sands at Atlantic City.

Beach sand: $d_{16} = 0.159$ mm $d_{84} = 0.664$ mm

Dune sand: $d_{16} = 0.173$ mm $d_{84} = 0.403$ mm

Solution:

To use the relationships for μ_φ and σ_φ, the phi diameter must be calculated for each of the above percentiles using Equation 4-28.

For the beach sand

$$\varphi = -\log_2 d_{mm}$$

or equivalently

$$\varphi = -3.3219 \log_{10} d_{mm}$$

and, therefore

$$\varphi_{16} = -3.3219 \log_{10} (0.159) = 2.65$$

Similarly

$$\varphi_{84} = 0.59$$

For the dune sand

$$\varphi_{16} = 2.53 \quad \text{and} \quad \varphi_{84} = 1.31$$

From Equation III-4-26 and III-4-27

$$\mu_{\varphi B} = \frac{2.65 + 0.59}{2} = 1.62$$

(Continued)

Example Problem III-4-9 (Concluded)

and

$$\sigma_{\varphi B} = \frac{2.65 - 0.59}{2} = 1.03$$

Similarly, for the dune sand

$$\mu_{\varphi D} = 1.92 \quad \text{and} \quad \sigma_{\varphi D} = 0.61$$

Substituting these values into Equation 4-31 gives

$$T_f = \left(\frac{0.61}{1.03} \right) \exp \left[\frac{(1.92 - 1.62)^2}{2(0.61^2 - 1.03^2)} \right] \cong 0.55$$

The trapping factor T_f can also be found from Figure III-4-23.

Consequently, the transport rate found in example problem III-4-8 should be multiplied by 0.55. The transport rate is therefore

$$R_a = 0.55 \; R_t$$

or

$$R_a = 0.55 \; (27.2) \cong 15.0 \; \frac{m^3}{m-yr}$$

$$T_f = \left(\frac{\sigma_{\varphi D}}{\sigma_{\varphi B}} \right) \exp \left[\frac{(\mu_{\varphi D} - \mu_{\varphi B})^2}{2 \ (\sigma_{\varphi D}^2 - \sigma_{\varphi B}^2)} \right] \qquad \text{(III-4-31)}$$

which is based on a comparison of the log-normal distributions of dune and beach sands. The procedure is valid only if the dune sand is finer than the beach sand and more narrowly graded (better sorted). This is generally the case since the dune sand is almost always derived from the sand on the beach in front of the dune.

(3) Some example values of the various parameters for typical beach and dune sands along with calculated dune trapping factors are given in Table III-4-14.

Table III-4-14
Typical Beach and Dune Sand Phi Diameters and Dune Trapping Factors Calculated from Beach and Dune Sand Samples at Various U. S. Beaches

Location	φ_{16} (dune)	φ_{84} (dune)	φ_{16} (beach)	φ_{84} (beach)	$\mu_{\varphi B}$	$\mu_{\varphi D}$	$\sigma_{\varphi B}$	$\sigma_{\varphi D}$	T_f
North Street Ocean City, NJ	2.48	1.88	2.61	1.72	2.17	2.18	0.45	0.30	0.67
9th Street Ocean City, NJ	2.59	0.79	2.49	0.19	1.34	1.69	1.15	0.90	0.69
29th Street Ocean City, NJ	2.64	1.98	2.69	1.41	2.05	2.31	0.64	0.33	0.46
36th Street Ocean City, NJ	2.53	1.31	2.65	0.59	1.62	1.92	1.03	0.61	0.55
Cape May, NJ Location 1	1.86	0.60	2.06	0.20	1.13	1.23	0.93	0.63	0.67
Cape May, NJ Location 2	1.62	0.58	1.71	0.39	1.05	1.10	0.66	0.52	0.78
CERC FRF Duck, NC					0.76	1.08	1.00	0.77	0.68
Clearwater Beach, FL	2.87	2.09	2.66	1.41	2.04	2.48	0.63	0.39	0.41

e. Continuity equation for wind transport of beach sand.

(1) The continuity equation for the beach sand transported by an onshore wind is developed in Figure III-4-24.

(2) At the upwind end of the fetch, the ocean and/or saturated sand precludes any wind transport. Further shoreward, the sand is dry and transportable by the prevailing wind; however, the amount in transport has not reached its potential rate predicted by the equations. In this region the wind is deflating the beach. The amount of sand leaving the Δx-long control volume exceeds the amount entering and the difference is comprised of sand eroded from the beach. Downwind along the fetch, the sand transport rate has reached equilibrium and the amount of sand entering the control volume equals the amount leaving. Here the beach is not deflating; rather the amount of sand eroded is balanced by the amount deposited. Further along the fetch, near the base of the dune, the amount of sand entering the control volume is less than the amount removed by the wind, and sand accumulates to build the dune. The change in transport rate is often due to a decrease in local wind speed induced by vegetation, sand fencing, or by the dune itself. Writing a sediment balance for the Δx control volume yields the equation

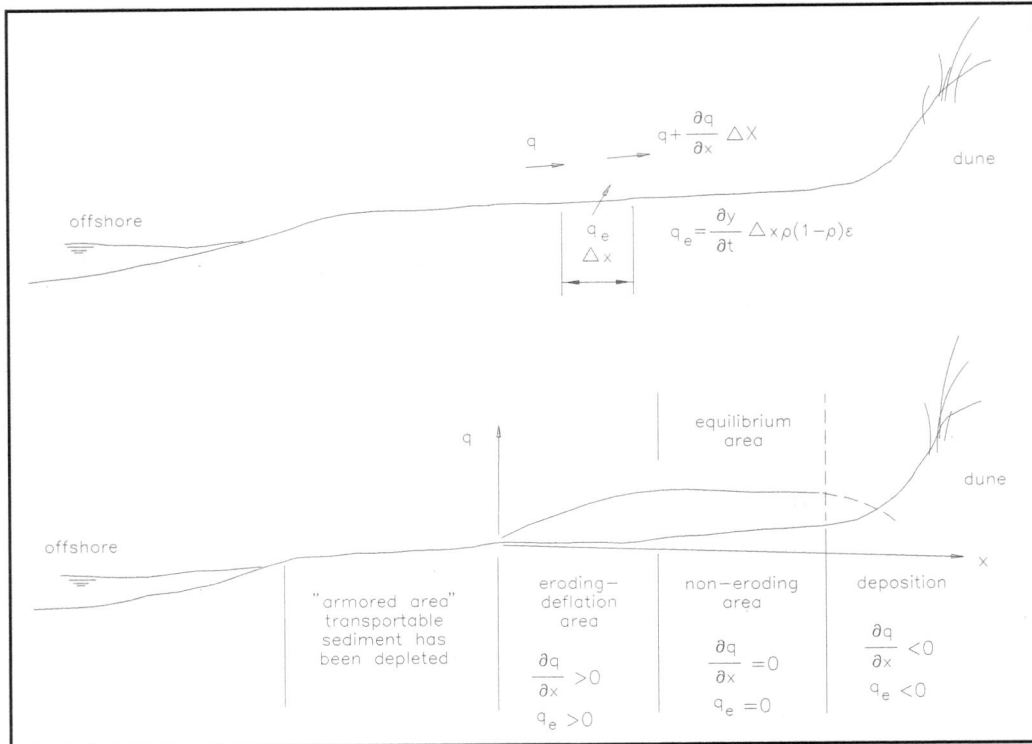

Figure III-4-24. Sand conservation equations for a beach showing deflation area, equilibrium area, and deposition area

$$q + q_e - \left(q + \frac{\partial q}{\partial x} \Delta x \right) = 0 \tag{III-4-32}$$

in which q equals the wind-blown transport into the control volume (mass per unit time per unit length perpendicular to the direction of transport), q_e equals the amount contributed to the control volume by erosion from the beach (mass per unit time per unit length), and Δx equals the width of the control volume in the direction of transport. The first two terms are the sand influx to the control volume while the term in brackets is the efflux. Simplifying Equation 4-32 gives

$$\frac{\partial q}{\partial x} = \frac{q_e}{\Delta x} \tag{III-4-33}$$

(3) The contribution from the beach q_e can be expressed as a deflation rate times the beach surface area

$$q_e = \frac{\partial y}{\partial t} \Delta x \, \rho \, (1 - p) \, \varepsilon \tag{III-4-34}$$

in which y equals the height of the beach surface above some datum, ρ equals the mass density of the sand, p equals the porosity of the *in-situ* beach sand, and ε equals a dimensionless "erodibility factor" related to the beach sand that is potentially transportable. ε is related to the sand size gradation as well as to any conditions that increase the threshold wind speed such as soil moisture content, salt crust formation, etc. Equation III-4-33 becomes

$$\frac{\partial q}{\partial x} = \frac{\partial y}{\partial t} \rho \left(1 - p\right) \varepsilon \qquad\qquad (\text{III-4-35})$$

(4) The upwind boundary condition, $q = 0$ at $x = 0$, is applied where the sand is first transportable, i.e., at the shoreline or the point where the sand is no longer wet. Since different size fractions are transported at different rates by a given wind speed, Equation 4-35 must be applied to each range of sizes present on the beach. Unfortunately, the relationship between the size gradation parameters and the amount of surficial, transportable sand is not known. Therefore, it is not possible to establish limit depths of erosion for a given wind speed blowing over sand with a given size distribution.

(5) The amount of sand available for transport at a given wind speed depends on the gradation as well as the mean or median grain size. For a fine, poorly graded sand, once transport is initiated, there is a nearly continuous supply of sizes that can be transported. For a well-graded sand subjected to winds capable of transporting the finer fractions, fine sediments are eventually removed from the surface leaving behind the coarse fraction. See Figure III-4-21. In order to continue the transport process, higher wind speeds must occur or fine sediments must be exposed at the surface again.

  f. Limited source due to gradation armoring. Beach sands generally are not uniform in size but have a gradation that is often log-normally distributed. The finer fraction at the beach surface is more easily transported than the coarser fraction. Finer sand in the gradation also has the potential to be transported at a greater rate by lower wind speeds while larger sand is transported at a slower rate, if at all. However, as finer sands are depleted from the surface layer of the beach, the remaining larger particles shelter the underlying fines and armor the beach. Then, at the upwind end of the fetch, transport of fine sand ceases and, unless the wind speed increases to remove and transport the coarser fraction, transport will cease. Further downwind, the fines leaving the control volume balance the fines entering so that transport continues. Consequently, there is a moving, upwind boundary between that portion of beach where the fines have been removed and transport has ceased, and where transport of fines continues. Each size fraction has its own boundary. As fines are removed and the coarser fraction remains behind to armor the surface, the no-transport boundary moves downwind. Note that for sand with a uniform size gradation, this armoring process does not occur while for a well-graded sand it does. Other armoring processes may occur for uniformly graded sands. Figure III-4-24 shows the location of the upwind boundary separating the armored area from the area where transport processes continue.

EXAMPLE PROBLEM III-4-10

Find:

The wind transport rate for each fraction of the sand at Atlantic City having the distribution given in Figure III-4-25. Determine the total transport by summing the transport of all fractions using the beach sand size distribution as a weight function.

Given:

The wind rose data from Atlantic City in Table III-4-9 and the beach and dune sand size distributions in Figure III-4-25.

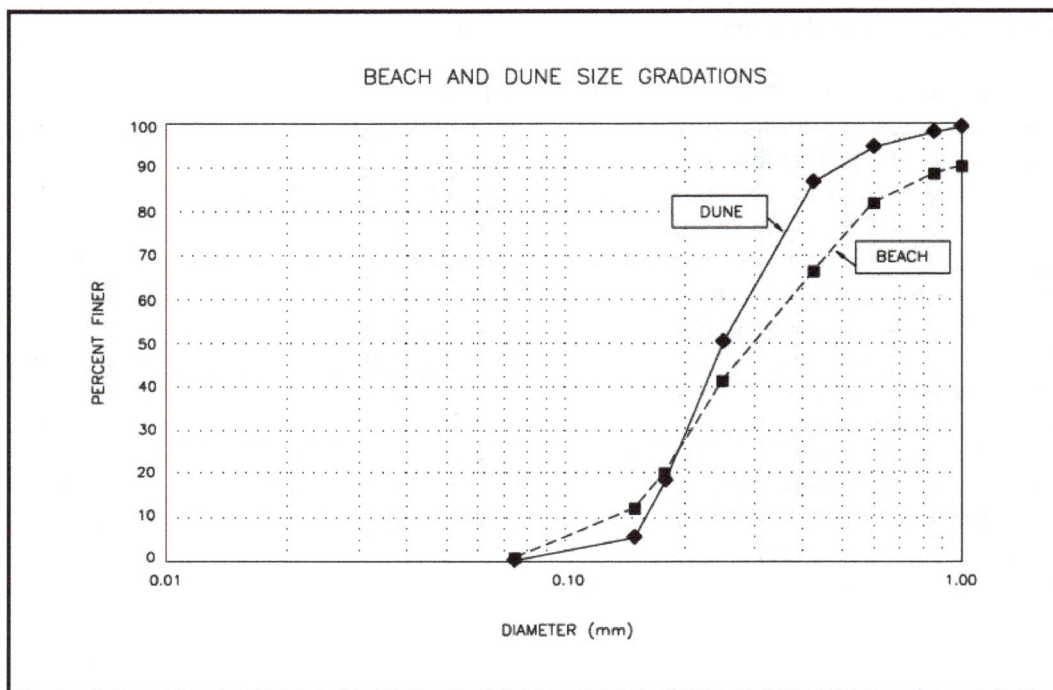

BEACH AND DUNE SIZE GRADATIONS

Figure III-4-25. Beach and dune and size gradations for Atlantic City, NJ

Solution:

The first step is to divide the size distribution for the beach sand into discrete increments and to determine the threshold wind speed for the upper limit of each size interval. See Table III-4-15. The volumetric transport coefficient is obtained from Figure III-4-13 entering with the appropriate sand grain diameter and elevation at which the wind speeds are measured (in this example assumed to be 10 m). The wind speeds in both miles per hour (mph) and in centimeters per second at the wind rose interval boundaries are given in Table III-4-16.

(Sheet 1 of 6)

Wind-Blown Sediment Transport

Example Problem III-4-10 (Continued)

Table III-4-15
Size Distribution, Threshold Wind Speeds and Volumetric Transport Coefficients for Beach Sand Size Intervals, Atlantic City, NJ

d (mm)	Fraction Finer	Fraction Within Interval	u_{thresh} (cm/sec)	Volumetric Transport Coefficient K'_v (m^2-sec^2/cm^3)
0.08	0.02	0.020	420	1.80e-14
0.10	0.05	0.030	450	1.50e-14
0.20	0.14	0.090	555	1.00e-14
0.25	0.28	0.140	650	8.00e-15
0.30	1.42	0.140	725	5.90e-15
0.35	0.51	0.090	800	5.35e-15
0.40	0.65	0.140	910	3.36e-15
0.50	0.75	0.100	1020	9.43e-15
0.60	0.83	0.080	1120	1.23e-14
0.70	0.87	0.040	1210	1.57e-14
0.80	0.89	0.015	1310	2.20e-14
0.90	0.90	0.015	1395	3.02e-14

Table III-4-17 gives the potential sand transport at the wind rose interval boundaries for each of the grain size intervals. Column 4 of the table gives the potential transport for each of the size intervals in cubic meters per meter-second. For those cases where the threshold wind speed exceeds the actual wind speed (the coarser fraction), the transport is zero. Column 5 gives the potential transport values in cubic meters per meter-year while column 6 gives the potential transport weighted by the fraction of sand within the given size interval. For example, for a wind speed of 14 mph (626 cm/sec) 77.4 m^3/m-yr of sand with a diameter of 0.2 mm could be transported. However, only 9 percent of the sand is within this size interval so that the weighted transport rate is 0.09(77.4) = 6.96 m^3/m-yr. Note that for a wind speed of 14 mph (626 cm/sec), only sand with a diameter less than about 0.2 mm will be transported. For a wind speed of 29 mph, sand with a diameter less than 0.70 mm is transported.

Table III-4-16
Wind Speeds for Atlantic City Wind Rose Intervals

Wind Speed	
(mph)	(cm/sec)
0	0
14	626
29	1,296
35	1,565

(Sheet 2 of 6)

Example Problem III-4-10 (Continued)

Table III-4-17
Transport Rates for Size Intervals at Atlantic City Wind Rose Speed Intervals

d (mm)	Fraction Within Interval	Wind Speed (cm/sec)	Potential Transport (m³/m-sec)	Potential Transport (m³/m-yr)	Transport Weighted by Sand Fraction Within Interval (m³/m-yr)
0.08	0.020	626	4.413e-6	139.3	2.79
0.10	0.030	626	3.677e-6	116.0	3.48
0.20	0.090	626	2.451e-6	77.4	6.96
0.25	0.140	626	0[1]	0[1]	0[1]
0.30	0.140	626	0	0	0
0.35	0.090	626	0	0	0
0.40	0.140	626	0	0	0
0.50	0.100	626	0	0	0
0.60	0.080	626	0	0	0
0.70	0.040	626	0	0	0
0.80	0.015	626	0	0	0
0.90	0.015	626	0	0	0
0.08	0.020	1296	3.922e-5	1237.7	24.75
0.10	0.030	1296	3.268e-5	1031.4	30.94
0.20	0.090	1296	2.179e-5	687.6	61.88
0.25	0.140	1296	1.743e-5	550.1	77.01
0.30	0.140	1296	1.286e-5	405.7	56.80
0.35	0.090	1296	1.166e-5	367.9	33.11
0.40	0.140	1296	7.321e-6	231.0	32.34
0.50	0.100	1296	2.055e-5	648.4	64.84
0.60	0.080	1296	2.680e-5	845.7	67.66
0.70	0.040	1296	3.421e-5	1079.5	43.18
0.80	0.015	1296	0[1]	0[1]	0[1]
0.90	0.015	1296	0	0	0
0.08	0.020	1565	6.895e-5	2175.8	43.52
0.10	0.030	1565	5.746e-5	1813.2	54.40
0.20	0.090	1565	3.830e-5	1208.8	108.79
0.25	0.140	1565	3.064e-5	967.0	135.38
0.30	0.140	1565	2.260e-5	713.2	99.85
0.35	0.090	1565	2.049e-5	646.7	58.20
0.40	0.140	1565	1.287e-5	406.2	56.86
0.50	0.100	1565	3.612e-5	1139.9	113.99
0.60	0.080	1565	4.711e-5	1486.8	118.94
0.70	0.040	1565	6.014e-5	1897.8	75.91
0.80	0.015	1565	8.427e-5	2659.3	39.89
0.90	0.015	1565	1.157e-4	3650.5	54.76

[1] Threshold wind speed exceeds actual wind speed and there is no transport of coarser sands.

(Sheet 3 of 6)

Example Problem III-4-10 (Continued)

Table III-4-18 is constructed from Table III-4-17 and the wind rose data. Table III-4-18 is only for winds from the north. The values in columns 2, 3, and 4 are the average transport values over the wind speed interval; hence, for example, the first value in column 3 is the average transport over the wind speed interval 14 mph < U < 29 mph, or $(2.79 + 24.75)/2 = 13.77$ m^3/m-yr. Similarly, the first value in column 4 is the average over the interval 29 mph < U < 35 mph, or $(24.75 + 43.52)/2 = 34.14$ m^3/m-sec. The values in column 5 are the weighted sums of the values across each row with the number of days that the wind was within the given speed interval used as the weight factor. Hence, the value in the first row is given by $(1.395)(20/365) + (13.77)(20/365) + (34.14)(0/365) = 0.831$ m^3/m-yr. Consequently, the total amount of sand transported northward is 14.218 m^3/m-yr. Table III-4-18 is for winds from the north only. Similar analyses must be done for the remaining seven compass directions. The results of the analyses for all eight compass directions are summarized in Table III-4-19 where, for example, column 2 is derived from column 5 of Table III-4-18. (Calculations for the remaining compass directions are omitted for brevity.)

Table III-4-18
Wind Transport of Size Intervals for Winds from North Weighted by Fraction of Year Wind was within Given Speed and Direction Interval

(Similar tables must be constructed for the other seven compass directions)

	Wind From North			
d (mm)	Avg. Transport Over Interval 0<U<14 No. Days = 20	Avg. Transport Over Interval 14<U<29 No. Days = 20	Avg. Transport Over Interval 29<U<35 No. Days = 0	Sum of Transports Weighted by No. of Days (m^3/m-yr)
0.08	1.395	13.77	34.14	0.831
0.10	1.740	17.21	42.67	1.038
0.20	3.480	34.42	85.33	2.077
0.25	0	38.5	106.20	2.110
0.30	0	28.4	78.33	1.556
0.35	0	16.55	45.66	0.907
0.40	0	16.17	44.60	0.886
0.50	0	32.42	89.42	1.776
0.60	0	33.83	93.30	1.854
0.70	0	21.59	59.55	1.183
0.80	0	0	19.95	0
0.90	0	0	27.38	0
				TOTAL N = 14.218

(Sheet 4 of 6)

Example Problem III-4-10 (Continued)

Table III-4-19
Summary of Transports for Eight Compass Directions

d (mm)	Transport (m²/yr)							
	N	NE	E	SE	S	SW	W	NW
0.08	0.831	1.226	0.818	0.739	2.225	1.007	1.781	0.951
0.10	1.038	1.533	1.022	0.923	2.780	1.259	2.226	1.188
0.20	2.077	3.066	2.044	1.846	5.560	2.517	4.452	2.376
0.25	2.110	3.456	2.164	1.953	6.041	2.717	4.721	2.506
0.30	1.556	2.551	1.596	1.441	4.456	2.004	3.483	1.849
0.35	0.907	1.487	0.930	0.840	2.597	1.168	2.029	1.077
0.40	0.886	1.453	0.909	0.820	2.537	1.141	1.983	1.053
0.50	1.776	2.912	1.822	1.645	5.087	2.288	3.975	2.110
0.60	1.854	3.039	1.902	1.716	5.308	2.387	4.148	2.202
0.70	1.183	1.940	1.214	1.095	3.388	1.524	2.647	1.405
0.80	0	0.273	0.109	0.109	0.164	0.055	0.055	0.055
0.90	0	0.375	0.150	0.150	0.225	0.075	0.075	0.075
TOTAL	14.218	23.314	14.680	13.277	40.307	18.142	31.576	16.847

Dune growth depends on the wind direction relative to the dune axis and the assumptions made regarding the ability of the dune to trap and hold sand. The procedures presented in Example Problem III-4-8 are applied to each individual grain size interval and the values summed to estimate the amount of sand trapped by the dune. Transport in each direction is multiplied by $\sin\beta$, the angle between the dune axis and the direction from which the wind is blowing, and $\cos^2\beta$, the dune trapping efficiency. These values are given at the bottom of Table III-4-20. The values in Table III-4-20 are the net contribution to dune growth by winds from the indicated directions. For example, winds from the north remove sand from the dunes since values in column 2 are negative (the wind is blowing from the dune toward the beach). For this site, most dune growth is due to southerly winds and, to a lesser extent, to southeasterly winds.

(Sheet 5 of 6)

Example Problem III-4-10 (Concluded)

For the example, the total dune growth rate is 38.63 m³/m-yr. This represents a high rate of growth and most certainly overestimates dune growth at Atlantic City since the example assumes an unlimited supply of sand in all size intervals available for transport. As in Example Problem III-4-9, the calculated dune growth rate can be adjusted by using the trapping efficiency based on a comparison of the beach sand with the dune sand. From the preceding example, $T_f =$ 0.55; hence, the estimated dune growth rate is $R_a = 0.55(38.63) = 21.4$ m³/m-yr. An alternative procedure to adjust dune growth rates based on comparing the dune sand distribution with the beach sand distribution is given in Example Problem III-4-11, which follows.

Table III-4-20
Computation of Dune Growth Rate

d (mm)	Dune Growth (m²/yr)								
	N	NE	E	SE	S	SW	W	NW	Total
0.08	-0.263	-0.106	0.526	0.736	1.704	0.088	-0.672	-0.007	2.110
0.10	-0.329	-0.133	0.657	0.919	2.130	0.110	-0.840	-0.009	2.639
0.20	-0.657	-0.265	1.314	1.839	4.260	0.219	-1.679	-0.018	5.277
0.25	-0.668	-0.299	1.391	1.946	4.628	0.237	-1.781	-0.019	5.734
0.30	-0.493	-0.221	1.026	1.435	3.144	0.175	-1.314	-0.014	4.230
0.35	-0.287	-0.129	0.598	0.836	1.989	0.102	-0.766	-0.008	2.465
0.40	-0.280	-0.126	0.584	0.817	1.944	0.100	-0.748	-0.008	2.408
0.50	-0.562	-0.252	1.171	1.638	3.897	0.199	-1.500	-0.016	2.828
0.60	-0.587	-0.263	1.222	1.710	4.066	0.208	-1.565	-0.011	5.038
0.70	-0.374	-0.168	0.780	1.091	2.595	0.133	-0.999	-0.010	3.215
0.80	0	-0.024	0.070	0.109	0.126	0.005	-0.021	-0.000	0.288
0.90	0	-0.032	0.096	0.150	0.172	0.007	-0.028	-0.001	0.396
TOTAL	-4.500	-2.017	9.436	13.226	30.925	1.581	-11.911	-0.127	
$\sin\beta$	-0.7660	-0.0872	0.6428	0.9962	0.7660	0.0872	-0.6428	-0.9962	
$\cos^2\beta$	0.4132	0.9924	1.0000	1.0000	1.0000	1.0000	0.5868	0.0076	

TOTAL DUNE GROWTH RATE = 38.631 m³/m-yr

(Sheet 6 of 6)

EXAMPLE PROBLEM III-4-11

Find:

The trapping efficiency for natural dune building in Atlantic City and the dune grain size distribution calculated from the results of Example Problem III-4-10 given the measured and calculated grain-size distributions for the dune sand at Atlantic City.

Given:

The dune growth rates based on the individual grain size intervals as calculated in Example Problem III-4-10 and the actual dune sand size distributions given in Figure III-4-25.

Solution:

The calculated amounts of sand trapped in the dune from each size interval are given in Table III-4-21. The values in column 2 of Table III-4-21 are found by summing across the rows in Table III-4-20. Thus, 2.111 m^3/m-yr of sand with a diameter less than 0.08 mm is trapped by the dunes. Similarly, 2.639 m^3/m-yr of sand with a diameter between 0.08 mm and 0.10 mm is trapped. The calculated size distribution of the dune sand is determined by normalizing the values in column 2 by dividing each value in column 2 by the sum of column 2. The resulting values are given in column 3, which represents the fraction of the sand in the dune within the given size interval. For example, 0.0547 or 5.47 percent of the dune sand is finer than $d = 0.08$ mm and 6.83 percent is between $d = 0.10$ mm and $d = 0.08$ mm, etc. Column 4 of Table III-4-21 gives the cumulative distribution.

Table III-4-21
Calculated Dune Sand Size Distribution

d (mm)	Amount Trapped in Dune (m^3/m-yr)	Normalized Amount Trapped in Dune	Cumulative Size Distribution in Dune
0.08	2.111	0.0547	0.0547
0.10	2.639	0.0683	0.1230
0.20	5.277	0.1366	0.2596
0.25	5.732	0.1484	0.4080
0.30	4.230	0.1095	0.5175
0.35	2.465	0.0638	0.5813
0.40	2.408	0.0623	0.6436
0.50	4.828	0.1250	0.7686
0.60	5.038	0.1304	0.8990
0.70	3.215	0.0832	0.9823
0.80	0.288	0.0075	0.9898
0.90	0.396	0.0102	1.0000
TOTAL	38.631	1.0000	

(Sheet 1 of 4)

Example Problem III-4-11 (Continued)

Figure III-4-26 shows the calculated sand size distribution in the dune along with the actual dune size distribution. The actual distribution differs from the calculated distribution because not all of the sand present on the beach is available for transport due to processes like gradation armoring, crust formation, etc. Some fines might be confined between larger grains and once the surficial fines are removed, the remaining fines are not available for transport. Also, the coarser fraction can only be transported by the less frequent high wind speeds.

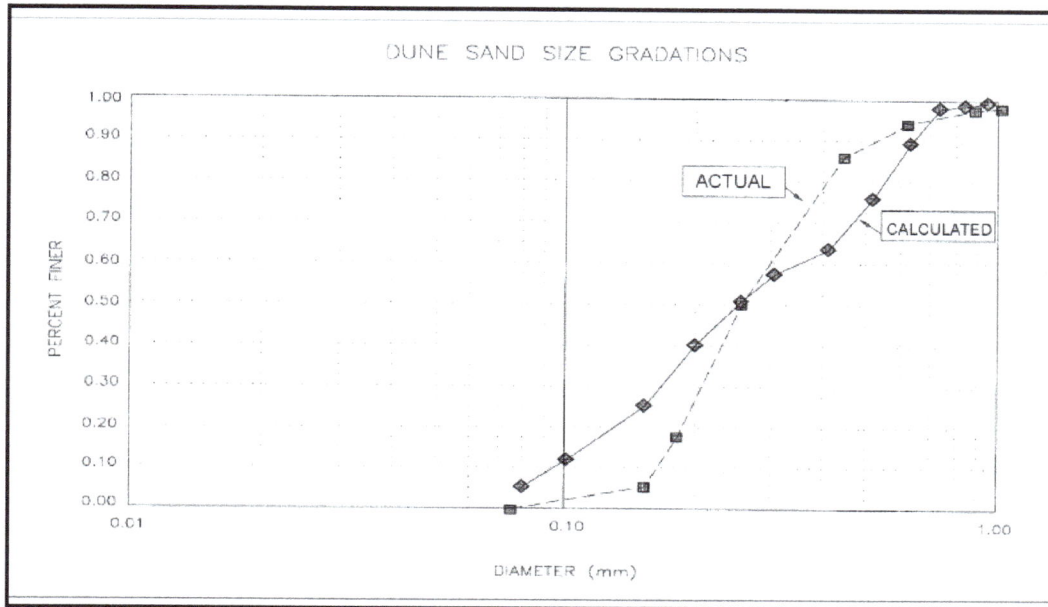

Figure III-4-26. Actual and calculated dune sand size gradation at Atlantic City, NJ

The transport rates calculated for each size interval are potential transport rates. Comparison of the calculated size distribution with the actual distribution allows a correction factor to be calculated for each size fraction. Columns 2 and 3 of Table III-4-22 present the calculated and actual size distributions, respectively. Column 4 is a calculated correction factor equal to the actual interval fraction divided by the calculated interval fraction. For example, the correction factor 0.1818 in the first row equals 0.0100/0.0547, the ratio of column 3 over column 2. Thus column 4 gives the numbers by which calculated values must be multiplied in order to obtain actual values. The amount of sand available for transport in each size interval is limited, so, in order to calculate the amount actually transported in each size interval, the correction factor values in column 4 must be normalized by dividing through by the largest value. (Column 4 would suggest that 300 percent of the amount of sand between 0.35 mm and 0.40 mm is needed to produce the observed distribution. Clearly, no more than 100 percent can be available.) The values in column 5 result. These values can be considered to be transport correction factors for each size interval. They are the values by which the transport rate in each size interval must be multiplied in order to produce the measured dune size distribution.

(Sheet 2 of 4)

Example Problem III-4-11 (Continued)

Table III-4-22
Comparison of Calculated Dune Sand Size Distribution and Actual Dune Sand Size Distribution

d (mm)	Calculated Dune Sand Size Distribution	Actual Dune Sand Size Distribution	Correction = Actual/Calculated	Normalized Correction
0.08	0.0547	0.0100	0.1818	0.0604
0.10	0.0683	0.0470	0.6912	0.2297
0.20	0.1366	0.1355	0.9891	0.3286
0.25	0.1484	0.0908	0.6135	0.2039
0.30	0.1095	0.2250	2.0455	0.6797
0.35	0.0638	0.1334	2.1175	0.7036
0.40	0.0623	0.1896	3.0095	1.0000
0.50	0.1250	0.0832	0.6656	0.2212
0.60	0.1304	0.0397	0.3054	0.1015
0.70	0.0832	0.0125	0.1506	0.0500
0.80	0.0075	0.0125	1.6026	0.5325
0.90	0.0102	0.0100	0.9804	0.3258
TOTAL	1.0000	1.0000		

Table III-4-23 gives the corrected transport rates for each size interval. Column 2 is the calculated amount in each size interval trapped in the dune. (See column 2, Table III-4-21.) Column 3 is the normalized correction factor (see column 5, Table III-4-22) and column 4 is the corrected dune growth rate for sand within the given size fraction. The sum of the values in column 4 gives the corrected dune growth rate. For the given example, the corrected dune growth rate is 12.7 m³/m-yr. This procedure is similar to the procedure used to derive the trapping factor T_f but by matching the calculated and actual dune size distributions and without assuming that the distributions are log-normal. The value of T_f can be calculated for this example by taking the ratio of the value of the total from column 4 in Table III-4-23 to the total in column 2. This ratio is $T_f = 12.68/38.63 = 0.33$, which can be computed with the value of $T_f = 0.57$ calculated in Example Problem III-4-9.

Figure III-4-27 compares the actual dune size distribution with the calculated distribution after the correction has been applied.

(Sheet 3 of 4)

Example Problem III-4-11 (Concluded)

Table III-4-23
Corrected Transport Rates for Size Intervals and Corrected Dune Growth Rate

d (mm)	Calculated Transport Rates (m³/m-yr)	Normalized Correction Factor	Corrected Transport Rates (m³/m-yr)
0.08	2.111	0.0604	0.1275
0.10	2.639	0.2297	0.6062
0.20	5.277	0.3286	1.7340
0.25	5.732	0.2039	1.1688
0.30	4.230	0.6797	2.8751
0.35	2.465	0.7036	1.7344
0.40	2.408	1.0000	2.4080
0.50	4.828	0.2212	1.0680
0.60	5.038	0.1015	0.5114
0.70	3.215	0.0500	0.1608
0.80	0.288	0.5325	0.1534
0.90	0.396	0.3258	0.1290
TOTAL =	38.627		12.6766

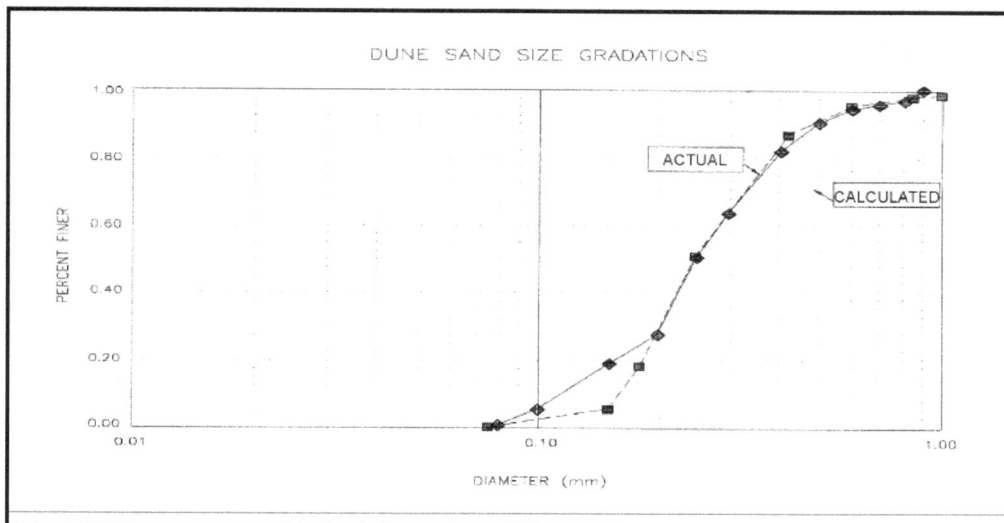

Figure III-4-27. Comparison of actual dune sand size gradation with corrected, calculated gradation

(Sheet 4 of 4)

III-4-6. References

EM 1110-2-5003
Planting Guidelines for Dune Creation and Stabilization

Bagnold 1936
Bagnold, R. A. 1936. "The Movement of Desert Sand," *Proceedings Royal Society of London*, Series A, Vol 157, pp 94-620.

Bagnold 1941
Bagnold, R. A. 1941. *The Physics of Blown Sand and Desert Dunes*, Morrow, New York, (republished in 1954 by Methuen, London).

Belly 1962
Belly, P. Y. 1962. "Sand Movement by Wind," Technical Memorandum, TM-1, Coastal and Hydraulics Laboratory, U.S. Army Engineer Waterways Experiment Station, Vicksburg, MS.

Chapman 1990
Chapman, D. M. 1990. "Aeolian Sand Transport - An Optimized Model," *Earth Surface Processes and Landforms*, Vol 15, pp 751-760.

Gillette, Herbert, Stockton, and Owen 1996
Gillette, D. A., Herbert, G., Stockton, P. H., and Owen, P. R. 1996. " Causes of the Fetch Effect in Wind Erosion," *Earth Surface Processes and Landforms*, Vol 21, pp 641-659.

Holzworth 1972
Holzworth 1972. "Mixing Heights, Wind Speeds, and Potential for Urban Air Pollution Throughout the Contiguous United States," Office of Air Programs, Publication No. AP-101, Environmental Protection Agency, Research Triangle Park, North Carolina.

Horikawa 1988
Horikawa, K. (Editor). 1988. *Nearshore Dynamics and Coastal Processes*, University of Tokyo Press, Tokyo, Japan.

Horikawa and Shen 1960
Horikawa, K., and Shen, H. W. 1960. "Sand Movement by Wind Action - on the Characteristics of Sand Traps," Technical Memorandum 119, Coastal and Hydraulics Laboratory, U.S. Army Engineer Waterways Experiment Station, Vicksburg, MS.

Hotta, Kubota, Katori, and Horikawa 1985
Hotta, S., Kubota, S., Katori, S., and Horikawa, K. 1985. "Sand Transport by Wind on a Wet Sand Surface," *Proceedings, 19th International Conference on Coastal Engineering,* American Society of Civil Engineers, pp 1265-1281.

Hsu 1973
Hsu, S. A. 1973. "Computing Eolian Sand Transport from Shear Velocity Measurements," *Journal of Geology*, Vol 81, pp 739-743.

Hsu 1977
Hsu, S. A. 1977. "Boundary Layer Meteorological Research in the Coastal Zone," *Geoscience and Man*, H. J. Walker, ed., School of Geoscience, Louisiana State University, Baton Rouge, LA, Vol 18, pp 99-111.

Hsu 1981
Hsu, S. A. 1981. "Models for Estimating Offshore Winds from Onshore Meteorological Measurements," *Boundary Layer Meteorology*, Vol 20, pp 341-351.

Hsu 1986
Hsu, S. A. 1986. "Correction of Land-Based Wind Data for Offshore Applications: A Further Evaluation," *Journal of Physical Oceanography*, Vol 16, pp 390-394.

Hsu 1988
Hsu, S. A. 1988. *Coastal Meteorology*, Academic Press, San Diego, CA.

Hsu 1994
Hsu, S. A. 1994. Personal communication, Coastal Studies Institute, Louisiana State University, Baton Rouge, LA.

Hsu and Blanchard 1991
Hsu, S. A., and Blanchard, B. W. 1991. "Shear Velocity and Eolian Sand Transport on a Barrier Island," *Proceedings, Coastal Sediments '91,* American Society of Civil Engineers, New York, Vol 1, pp 220-234.

Kadib 1964
Kadib, A. 1964. "Calculation Procedure for Sand Transport on Natural Beaches," Miscellaneous Paper 2-64, Coastal and Hydraulics Laboratory, U.S. Army Engineer Waterways Experiment Station, Vicksburg, MS.

Kadib 1966
Kadib, A. 1966. "Mechanics of Sand Movement on Coastal Dunes," *Journal of the Waterways and Harbors Division*, American Society of Civil Engineers, WW2, pp 27-44.

Kawamura 1951
Kawamura, R. 1951. "Study on Sand Movement by Wind," Report, Institute of Science and Technology, University of Tokyo. (translation, University of California, Berkeley, Hydraulic Engineering Laboratory Research Report, HEL-2-9, Berkeley, CA, 1964, pp 1-64).

Krumbein and James 1965
Krumbein, W. C., and James, W. R. 1965. "A Lognormal Size Distribution Model for Estimating Stability of Beach Fill Material," TM-16, Coastal and Hydraulics Laboratory, U.S. Army Engineer Waterways Experiment Station, Vicksburg, MS.

Liu, Schwab, and Bennett 1984
Liu, P. C., Schwab, D. J., and Bennett, J. R. 1984. "Comparison of a Two Dimensional Wave Prediction Model with Synoptic Measurements in Lake Michigan," *Journal of Physical Oceanography*, Vol 14, pp 1514-1518.

Nickling and Egglestone 1981
Nickling, W. G., and Egglestone, M. 1981. "The Effects of Soluble Salts on the Threshold Shear Velocity of Fine Sand," *Sedimentology*, Vol 28, pp 505-510.

Nordstrom and Jackson 1992
Nordstrom, K. F., and Jackson, N. L. 1992. "Effect of Source Width and Tidal Elevation Changes on Aeolian Transport on an Estuarine Beach," *Sedimentology*, Vol 39, pp 769-778.

O'Brien and Rindlaub 1936
O'Brien, M. P., and Rindlaub, B. D. 1936. "The Transport of Sand by Wind," *Civil Engineering*, Vol 6, pp 325-327.

Powell 1982
Powell, M. D. 1982. "The Transition of the Hurricane Frederic Boundary-Layer Wind Field from the Open Gulf of Mexico to Landfall," *Monthly Weather Review*, Vol 110, pp 1912-1932.

Raudkivi 1976
Raudkivi, A. J. 1976. *Loose Boundary Hydraulics*, 2nd ed., Pergamon Press, New York.

Resio and Vincent 1977
Resio, D.T., and Vincent, C. L. 1977. "Estimation of Winds Over the Great Lakes," *Journal of the Waterway, Port, Coastal and Ocean Engineering Division*, American Society of Civil Engineers, Vol 103, No. WW2.

Sarre 1988
Sarre, R. D. 1988. "Evaluation of Eolian Transport Equations Using Intertidal Zone Measurements: Saunton Sands, England," *Sedimentology*, Vol 35, pp 671-679.

Savage 1963
Savage, R. P. 1963. "Experimental Study of Dune Building with Sand Fences," *Proceedings, International Conference on Coastal Engineering*, 1963, pp 380-396.

Savage and Woodhouse 1968
Savage, R. P., and Woodhouse, W. W., Jr. 1968. "Creation and Stabilization of Coastal Barrier Dunes," *Proceedings, 11th International Conference on Coastal Engineering*, Vol 1, 1968, pp 671-700.

SethuRaman and Raynor 1980
SethuRaman, S., and Raynor, G. S. 1980. "Comparison of Mean Wind Speeds and Turbulence at a Coastal Site and Offshore Location," *Journal of Applied Meteorology*, Vol 19, pp 15-21.

Shore Protection Manual 1984
Shore Protection Manual. 1984. 4th ed., 2 Vol, U.S. Army Engineer Waterways Experiment Station, U.S. Government Printing Office, Washington, DC.

Svasek and Terwindt 1974
Svasek, J.N., and Terwindt, J. H. J. 1974. "Measurements of Sand Transport by Wind on a Natural Beach," *Sedimentology*, Vol 21, pp 311-322.

Wieringa 1980
Wieringa, J. 1980. "Representativeness of Wind Observation at Airports," Bulletin, American Meteorological Society, Vol 61, pp 962-971

Williams 1964
Williams, G. 1964. "Some Aspects of the Aeolian Saltation Load," *Sedimentology*, Vol 3, pp 257-287.

Woodhouse 1978
Woodhouse, Jr. 1978. "Dune Building and Stabilization with Vegetation," Special Report No. 3, Coastal and Hydraulics Laboratory, U.S. Army Engineer Waterways Experiment Station, Vicksburg, MS.

Zingg 1953
Zingg, A. W. 1953. "Wind Tunnel Studies of the Movement of Sedimentary Material," *Proceedings, 5th Hydraulic Conference*, State University of Iowa, Studies in Engineering, Bulletin, Vol 34, pp 111-135.

III-4-7. Definition of Symbols

$\Delta \forall$	Volume change per unit length of beach [length^3/length]
ε	Erodibility factor related to beach sand that is potentially transportable [dimensionless]
η	Transport efficiency for winds at the seaward side of a dune
η_i	Transport efficiency for winds at the landward side of a dune
κ	von Karman's constant (= 0.4)
μ_φ	Mean of a sediment sample [phi units]
$\mu_{\varphi D\,or\,B}$	Mean of a sediment sample taken from the dune or beach [phi units]
v_a	Kinematic viscosity of air [length^2/time]
ρ	Mass density of sediment grains [force-time^2/length^4]
ρ_a	Mass density of air [force-time^2/length^4]
σ_φ	Standard deviation of a sediment sample [phi units]
$\sigma_{\varphi D\,or\,B}$	Standard deviation of a sediment sample taken from the dune or beach [phi units]
τ	Boundary shear stress [force/length^2]
φ	Sediment grain diameter in phi units ($\varphi = -\log_2 d_{mm}$, where d_{mm} is the grain diameter in millimeters)
φ_x	Sediment grain diameter of the x-percentile of a sample (x-percent of the grains are finer) [phi units]
A	Empirical constant (Equation III-4-10) [dimensionless]
A_t	Dimensionless constant (= 0.118) used in calculating the critical shear stress, u_{*_t}, for sand transport by wind (Equation III-4-20)
B	Empirical constant (Equation III-4-10) [dimensionless]
$B_{Bagnold}$	Bagnold coefficient [dimensionless]
$C_{D,land}$	Drag coefficients over land [dimensionless]
$C_{D,sea}$	Drag coefficient over sea [dimensionless]
C_z	Wind drag coefficient at height Z (Equation III-4-4) [dimensionless]
d	Standard grain size (= 0.25mm)
D	Median sediment grain diameter [length - generally millimeters]
D	Mean sediment grain diameter [length - generally millimeters]
d_{mm}	Sediment grain diameter in millimeters

d_x	Sediment grain diameter of the x-percentile of a sample (x-percent of the grains are finer) [length]
e	Base of natural logarithms
g	Gravitational acceleration (32.17 ft/sec^2, 9.807m/sec^2) [length/time^2]
G	Dry weight sand transport rate [force/length/time]
H_{land}	Height of the planetary boundary layer over the land [length]
H_{sea}	Height of the planetary boundary layer over the sea [length]
K	Empirical dimensional eolian sand transport coefficient (Equation III-4-17) [force/length/time]
K'	Eolian sand transport coefficient (Equation III-4-19) [dimensionless]
n	Empirically determined exponent (ranging from 1/11 to 1/17) (Equation III-4-15) [dimensionless]
p	Porosity of the *in situ* dune sand [percent]
q	Wind-borne sand transport rate [mass/time/length]
q_i	Sand blown inland from the dune [mass/time/length]
q_s	Dune erosion by waves [mass/time/length]
q_v	Volumetric sand transport rate [length^3/length/time]
R_a	Actual dune growth rate [length^3/length-time]
R_t	Theoretical dune growth rate [length^3/length-time]
R_T	Temperature correction factor applied to wind speed at the 10-m height (Figure III-4-7) [dimensionless]
t	Time
T_a	Air temperature [degrees Celsius]
T_f	Dune trapping factor [dimensionless]
T_s	Sea temperature [degrees Celsius]
u_*	Wind shear or friction velocity [length/time]
u_{*t}	Critical or threshold velocity for sand transport by wind (Equation III-4-20) [length/time]
u_{*tw}	Critical or threshold velocity for wet sand transport by wind (Equation III-4-20) [length/time]
U_{land}	Wind speed over land [length/time]
U_{sea}	Wind speed over sea [length/time]

U_x	Average wind speed at x-elevation [length/time]
U_z	Average wind velocity as a function of height above ground level [length]
U_{ZM}	Wind speed at the anemometer height [length/time]
U'_{10m}	Wind speed at the 10-meter height corrected for the air-sea temperature difference [length/time]
W	Fraction of water content in the upper 5mm of the sand [percent]
Z	Height at which wind speed is measured [length]
Z_0	Height of a roughness element characterizing the surface over which the wind is blowing [length]
Z_M	Anemometer height [length]
Z_R	Reference height [length]
Z_{Zingg}	Zingg Coefficient [dimensionless]

III-4-8. Acknowledgments

Authors of Chapter III-4, "Wind-Blown Sediment Transport:"

S. A. Hsu, Ph.D., Coastal Studies Institute, School of Geosciences, Louisiana State University, Baton Rouge, Louisiana.

J. Richard Weggel, Ph.D., Dept. of Civil and Architectural Engineering, Drexel University, Philadelphia, Pennsylvania.

Reviewers:

Robert G. Dean, Ph.D., Coastal & Oceanographic Engineering, University of Florida, Gainesville, Florida.

Scott Douglass, Ph.D., Dept. of Civil Engineering, University of South Alabama, Mobile, Alabama.

James R. Houston, Ph.D., Engineer Research and Development Center, Vicksburg, Mississippi.

David B. King, Ph.D., Coastal and Hydraulics Laboratory (CHL), Engineer Research and Development Center, Vicksburg, Mississippi.

Joon Rhee, Ph.D., CHL

Todd L. Walton, Ph.D., CHL

Chapter 5
EROSION, TRANSPORT, AND DEPOSITION
OF COHESIVE SEDIMENTS

EM 1110-2-1100
(Part III)
30 April 2002

Table of Contents

Page

Erosion, Transport, and Deposition of Cohesive Sediments

List of Tables

List of Figures

Chapter III-5
Erosion, Transport, and Deposition of Cohesive Sediments

III-5-1. Introduction

a. Cohesive sediments are those in which the attractive forces, predominantly electrochemical, between sediment grains are stronger than the force of gravity drawing each to the bed. Individual grains are small to minimize mass and gravitational attraction, and are generally taken to be in the silt ($<70 \mu$) to clay ($<4 \mu$) range. The strength of the cohesive bond is a function of the grain mineralogy and water chemistry, particularly salinity. Thus, a coarse silt behaves like noncohesive fine sand in fresh water, but is cohesive in an ocean environment. Similarly, a fine sand exhibits cohesion in salt water. In other words, it is easier to define cohesive sediment by behavior than by size.

b. Grain size and shape nevertheless play a significant role in the lack of permeability of cohesive sediment. As grain size decreases, so does the size of the interstitial pore spaces while drainage path length increases. The small pores result in greater resistance to flow, exacerbating the effects of the long drainage path. Clay minerals tend to form flake-shaped platelets, rather than spherical particles. These platelets deposit with the smallest dimension vertical, further reducing pores and increasing vertical drainage paths. For this reason, clay is often used as an impermeable layer in hydraulic earthworks such as dikes and channels.

c. In coastal engineering terms, the principal indicator of cohesive sediment behavior is a critical shear for erosion of bed sediment τ_c, which is significantly greater than the critical shear for deposition τ_s. In other words, once the sediment has been deposited on the bed, the cohesive bond with other bed particles makes it more difficult to remove than particle mass alone would suggest.

d. The processes and states of coastal cohesive sediment listed below are shown schematically in Figure III-5-1 and Table III-5-1.

(1) Consolidated. Stiff or hard cohesive sediment that has had centuries to drain, probably compressed beneath glaciers or other overburden.

(2) Suspension. Individual grains or flocs dispersed in the water column and transported with the water.

(3) Fluid Mud. A static or moving intermediate state between suspension and deposition, analogous to bed-load transport of noncohesive sand, that can move in the direction of flow supported by the bed. Fluid mud is the result of excess pore pressure, built up by hindered settling or wave action. Water cannot escape from the sediment deposit, and builds up the excess interstitial pore pressure necessary to support the weight of sediment above it. The whole mass of sediment and trapped water behaves like a uniform dense viscous fluid, flowing downhill or in the direction of the water flow. Fluid mud layers can often be seen on echo soundings as a false bottom in depressions in the seabed.

(4) Mud. Unconsolidated cohesive sediment that has been recently deposited. 'Recently' may be a matter of a few hours to several years.

e. Processes and states in Figure III-5-1 may be skipped. For example: most coastal mud, even fluid mud, is eroded before it has undergone sufficient consolidation to be defined as 'consolidated'; many cohesive sediments do not form fluid mud, but deposit directly as stationary mud. Differences between mud and consolidated sediments occur during erosion. Transport, deposition, and consolidation are the same for both mud and consolidated cohesive sediments.

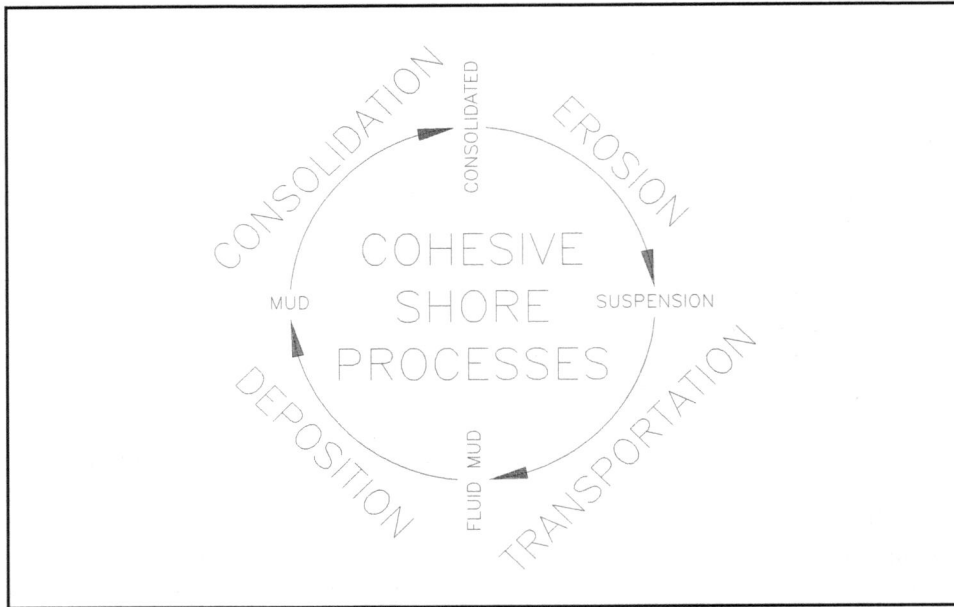

Figure III-5-1. Outline of cohesive shore processes — Any process or state may be bypassed, for example, fluid mud may be eroded without passing through (further) deposition and consolidation

Table III-5-1
Cohesive Sediment Density

Soil Description	Typical Saturated Bulk Density	
	kg/m³	lb/ft³
Suspension	1,020	64
Fluid Mud	1,100	70
Freshly Deposited Mud	1,300	80
Very Soft Consolidated	1,500	90
Soft Consolidated	1,600	100
Medium Consolidated	1,800	110
Stiff Consolidated	1,900	120
Very Stiff Consolidated	2,100	130
Hard Consolidated	2,200	140

III-5-2. Consolidated and Unconsolidated Shores

a. Consolidated shore.

(1) A shore is defined as consolidated cohesive when the erosion process is directly related to the irreversible removal of a cohesive sediment substratum (such as glacial deposits, ancient lagoon peats, tidal flat muds, valley and bay fill muds, lacustrine clays, flood deltas consisting of fine sediments, soft rock or other consolidated or over-consolidated deposits). Even when sand beaches are present, under the sand beach there lies an erodible surface that plays the most important role in determining how these shores erode, and ultimately, how they evolve in the long term. This differs fundamentally from sandy shores where erosion (or deposition) is directly related to the net loss (or gain) of noncohesive sediment from a given surface area.

Erosion on a sandy shore is a potentially reversible process (i.e., due to natural processes), while erosion on a consolidated cohesive shore is irreversible.

(2) Consolidated shores consist of consolidated or partially consolidated cohesive sediments which are usually covered by a thin veneer of sand and gravel, sometimes forming a beach at the shore (Part III-5-3 describes the techniques available for determining the properties of cohesive sediment). In essence, these shores are defined by an insufficient supply of littoral sand and gravel (i.e., noncohesive sediments). Consolidated shorelines may be associated with an eroding bluff or cliff face, or they may consist of a transgressive barrier beach perched over older sediments. The sand veneer often disguises the underlying cohesive substratum, and therefore, at many locations consolidated shores are incorrectly assumed to behave as sandy shores. The veneer thickness is usually in the range of a few centimeters to 2 or 3 m.

(3) Consolidated cohesive shores compose a large part of the Great Lakes, Arctic, Atlantic, Pacific, and U.S. Gulf coasts, a large part of the North Sea coast of England, and sections of the Baltic and Black Seas. Examples along the U.S. Atlantic coast include many of the barrier islands that are perched over older consolidated sediment; Riggs, Cleary, and Snyder (1995) estimate that 50 percent of the North Carolina coast is underlain by older estuarine peats and clays. Other examples along the U.S. east coast include the shores of Chesapeake and Delaware Bays. In many instances, the erosion of the shores associated with the Mississippi Delta and the transgressive barrier islands along the Texas coast is the result of cohesive processes. Cliff erosion along the South California coast and along large parts of the Beaufort Sea coast of Alaska are related to an insufficient supply of littoral sand, the hallmark of consolidated cohesive shores. Many other examples throughout the world, including erodible rocky coasts, are cited by Sunamura (1992). As awareness of the importance of the distinction of this shore type grows, and as sub-bottom investigations become more prevalent, more examples are identified. As Riggs, Cleary, and Snyder (1995) note, in many cases the shore is not just a 'thick pile of sand.'

(4) Consolidated cohesive shores are often difficult to identify owing to the presence of a sand beach at the shore. The existence of an eroding bluff or cliff at the shore, featuring consolidated or cohesive sediment of some form, is a sure sign of a consolidated shore. However, in many cases, cohesive shores do not feature eroding bluffs. Examples include many of the barriers along the Atlantic and U.S. Gulf coasts.

(5) There are at least six ways of visually identifying the presence of underlying consolidated cohesive sediment, which would distinguish a consolidated cohesive shore from a sandy shore. A series of photos of a transgressive shoreline along the east coast of Ghana in West Africa provide examples of the different types of evidence which may indicate the presence of cohesive sediment under a sand beach. Long-term recession rates along this 7-km section of the Ghanaian coast are in the range of 2 to 8 m/year. The six distinguishing features are:

(a) The most straightforward evidence is the presence of exposed cohesive sediment on the beach. Figure III-5-2 shows a large expanse of peat exposed on the beach in Ghana. Such exposures may be infrequent and result from severe erosion events (where the overlying sand is stripped and carried offshore) or may appear between large alongshore sand waves.

(b) Pieces of clay or peat on the beach. Figure III-5-3 shows some angular clay blocks that have been removed from the seabed and transported towards the shore, along with some pieces of rubble from old buildings that have been destroyed by erosion. In many locations, clay balls can be found along the shoreline. The more rounded clay pieces probably result from transport over a greater distance, i.e., the exposed cohesive sediment may not be located in the immediate vicinity of where the clay balls are found.

Figure III-5-2. Peat exposed on a beach along the Keta shoreline in Ghana, West Africa, May 1996

Figure III-5-3. Pieces of eroded clay (and some rubble) scattered on a beach along the Keta shoreline in Ghana, West Africa, September 1996

(c) Springs or surface runoff across a beach. Figure III-5-4 shows springs near the waterline along the beach in Ghana which result from groundwater flow over the impervious underlying clay and peat sediments. In this case, the groundwater flow gradient is driven by the presence of an enclosed lagoon with water levels higher than mean sea levels on the other side of the washover terrace.

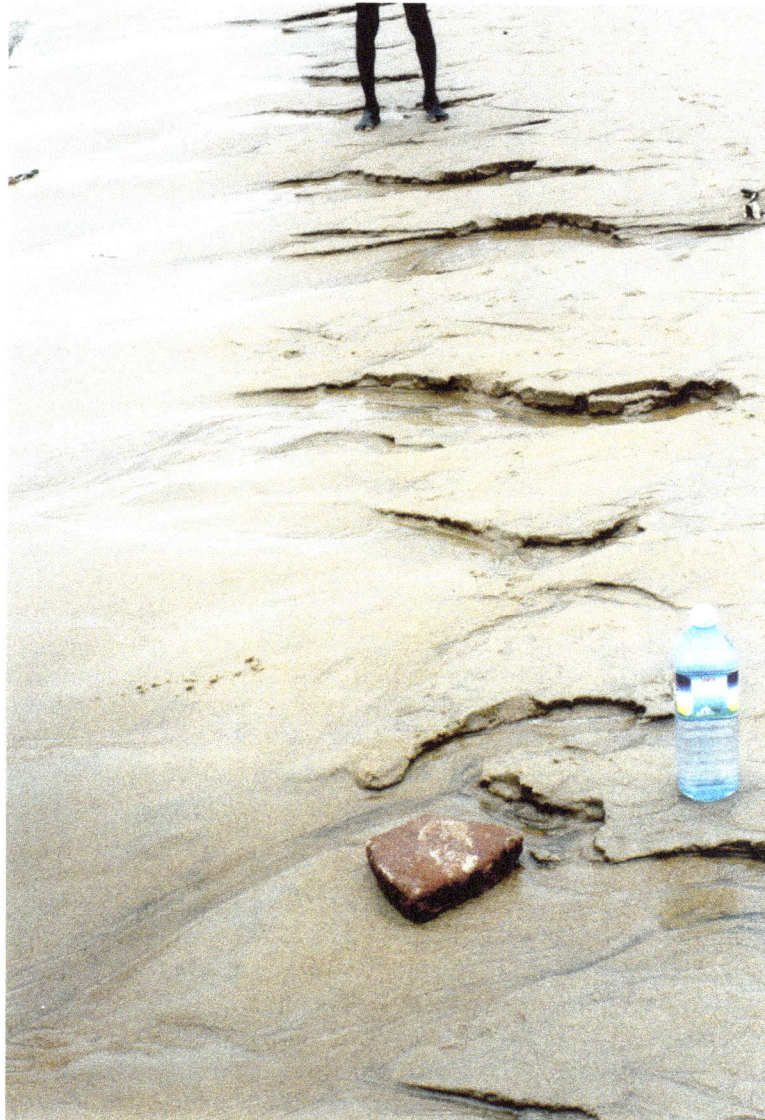

Figure III-5-4.   Springs flowing over the beach surface along the
Keta shoreline in Ghana, West Africa, September 1996

(d) Discoloration of water in the nearshore zone. Along eroding bluff shorelines, the water takes on the color of the bluff sediment in response to wave attack. Figure III-5-5 shows discoloration of the seawater near the shoreline caused by erosion of underlying lagoon sediments along the Ghana shoreline. At this location, this was the first evidence that easily erodible cohesive sediments existed in the nearshore zone.

(e) Permanent undulations in the shoreline planform may also signify the presence of cohesive or consolidated sediments in the nearshore zone with alongshore variability in erosion resistance. These features are best identified through oblique or overhead aerial photographs. An example along the Ghana

Figure III-5-5. Discolored water from erosion of exposed cohesive sediment along the Keta shoreline in Ghana, West Africa, July 1996

coast is shown in Figure III-5-6. These same undulations could be seen at the same location in several aerial photos taken over a 35-year period. Therefore, these features are distinguished from migrating alongshore sand waves which also occur here. Subsurface investigations at this location have subsequently revealed considerable alongshore variability in the stratigraphy of the underlying sediments, possibly relating to the presence of old channels into the lagoon inshore of the present beach.

(f) Finally, exposed cohesive sediments can often be identified in the troughs between offshore bars through a swimming or diving survey.

(6) At the site in Ghana, subsequent to the initial visual observations of evidence of the presence of underlying cohesive sediments, a series of subsurface investigations was completed to define these conditions. These included augers, boreholes, vibracores and sub-bottom profiling. Use of more detailed subsurface investigations to confirm visual observations and provide more detailed information is discussed in Part III-5-6.

b. Mud shore. Unconsolidated mud coasts are generally the result of more recent deposition of cohesive sediment. Deposition requires quiet or calm hydrodynamic conditions, where large waves are rare. Muddy shore naturally occurs as mud flats and coastal wetlands in protected waters, such as estuaries and other embayments with short fetches.

(1) Mud flat. Generally without vegetation, lying below high water in tidal areas (Figure III-5-7).

(2) Coastal Wetland. Salt marsh in saltwater, stabilized by vegetation, generally above low water in tidal areas, often above mud flats (Figure III-5-8) and mangrove. Marsh vegetation stabilizes the mud bed and provides shelter in which more sediment can deposit.

Erosion, Transport, and Deposition of Cohesive Sediments

Figure III-5-6.   Permanent undulations in the Keta shoreline in Ghana, West Africa, July 1996

Figure III-5-7.   Mud flat in Cumberland Basin, Bay of Fundy, Canada, showing drainage gullies

Figure III-5-8. A mud 'beach' backed with eroding salt marsh at Annapolis Royal, in the Annapolis Bay of the Bay of Fundy. Basalt revetment at top of salt marsh is an attempt to halt the erosion

(3) Mangrove. Tropical coastal wetland where the vegetation, in the form of trees, can withstand relatively severe wave attack (Figure III-5-9). Depth-limited waves will penetrate a mangrove, eroding mud from around the root system in extreme storms, and endangering individual trees. Mangrove can be found between mud flats and salt marsh, and can transmit waves, which erode the adjacent salt marsh behind it.

(4) Mud 'beach.' A sloping mud shore, exposed to wave attack often following the destruction of mangrove or other wetland vegetation (Figure III-5-8). The usual cause of wetland or mangrove destruction is property developers, municipal agencies, and individuals wishing to 'reclaim' the shore. Natural destruction occurs from the seaward edge, through undermining of the root system by waves and currents.

III-5-3. Erosion Processes on Consolidated Shores

a. Erosion processes on cohesive shores are distinctly different from those on sandy shores. There are also differences between consolidated and mud cohesive shores. On consolidated shores, the erosion process is irreversible because, once eroded, the cohesive sediment (e.g., glacial till, glacio-lacustrine deposits, ice bonded sediments, soft rock or other consolidated deposits) cannot be reconstituted in their consolidated form in the energetic coastal environment. Furthermore, since the sand and gravel content is low in these deposits (often less than 20 percent), erosion is not balanced by an equal volume of deposition within the littoral zone. The eroded fine sediments (silt and clay) are winnowed, carried offshore, and deposited in deep water in contrast to the sand fraction, which usually remains in the littoral zone.

Figure III-5-9.   Sand beach disappearing into mangrove on the island of Borneo. Sediment within the mangrove is cohesive mud

b.   Consolidated cohesive sediment is eroded by at least four mechanisms:

(1) Through abrasion by sand particles moved by waves and low currents.

(2) Through pressure fluctuations associated with turbulence generated at various scales such as wave-breaking-induced turbulence that reaches the lake or seabed and large-scale eddies that may develop in the surf zone.

(3) Through chemical and biological influences.

(4) Through wet/dry and freeze/thaw cycles where exposed to the atmosphere.

c.   Sand can also provide a protective cover to the underlying cohesive substratum. However, only when the sand cover is sufficient to protect the cohesive substratum at all times will the shore revert to a sandy classification (i.e., truly a 'thick pile of sand').

d.   On consolidated cohesive shores, the rate of lake or seabed downcutting determines the long-term rate at which the bluff or cliff retreats at the shoreline. In other words, while subaerial geotechnical processes may dictate when and where a slope failure will occur, the frequency of failures over the long term is determined by the rate at which the nearshore profile is eroded (i.e. the downcutting rate). Subaqueous and subaerial erosion processes on cohesive shores are discussed in detail in Part III-5-7. In addition, the geomorphology of cohesive shores and the relationship to erosion processes is the topic of Part III-5-5.

III-5-4. Physical and Numerical Modeling

a. Laboratory or physical model experiments have been used in two ways: (1) to improve the understanding of the fundamental principles of cohesive shore erosion processes, and (2) to develop a measure of the erodibility of specific samples of cohesive sediment.

b. In the former category, scale model tests have been performed in wave flumes as described by Sunamura (1975, 1976 and 1992) for erodible rocky coasts similar to consolidated cohesive behavior, by Nairn (1986) for a section of the Lake Erie shoreline using an artificial clay, and by Skafel and Bishop (1994) and Skafel (1995) using intact samples of till removed from the Lake Erie shoreline. The tests described by Skafel and Bishop (1994) included an assessment of the relationship between wave properties (e.g., wave height, orbital velocity, type and fraction of broken waves) and local erosion rate and the relationship between sand cover and erosion rates. Laboratory experiments on erosion and deposition of mud are described in Part III-5-6c(9). The annular flume, used in both the laboratory and the field (e.g., Amos et al. 1992, Krishnappan 1993) may be regarded as a full-scale model of the response of a mud bed to shear.

c. The primary difficulty in the use of physical model experiments of cohesive shores is the scaling of the cohesive material. At present, it is not possible to accurately scale cohesive sediment with respect to its erosion resistance properties. Therefore, model tests must be interpreted qualitatively, or full-scale tests must be conducted using low wave energy conditions. Nevertheless, the noted tests have been extremely valuable in advancing the understanding of cohesive shore erosion processes both inside and outside the surf zone.

d. A technique of assessing the erodibility of intact samples of consolidated cohesive sediment in unidirectional flow conditions has been developed and applied by Kamphuis (1990), and, more recently, for the assessment of cohesive sediment samples removed from the southeast shoreline of Lake Michigan (Parson, Morang, and Nairn 1996). These tests are typically performed for both clear water and sand in flow conditions to elucidate the importance of sand as an abrasive agent. This approach for defining the erodibility of cohesive sediment samples is discussed in more detail in Part III-5-7b.

e. The development and application of numerical models for describing erosion processes on cohesive shores is not far advanced owing to the complexity of the processes involved. Most numerical models may be described as little more than numerical frameworks for interpolating or extrapolating observed behavior of cohesive sediments in water. Essentially, a numerical model of cohesive shore erosion must define the near-bed flow conditions within the surf zone, the movement of any overlying noncohesive sediment cover, as well as the erosion resistance properties of the cohesive sediment (which change with time due to exposure of sediment layers and subaerial drying). Numerical modeling of cohesive shores is summarized in Part III-5-12.

f. The best 3-D numerical mud models (e.g. Le Hir 1994) treat the water column as a continuum: from stationary consolidating bed, through fluid mud, to sediment maintained in suspension by turbulence. There are nevertheless bed sediment 'modules,' which use the equations presented in Parts III-5-7, 9, and 10 to calculate erosion and deposition, supplying sediment and new bathymetry to numerical hydrodynamic models that transport the sediment by advection and dispersion (Part III-5-8a).

g. All mud models need to track sediment layering - the composition and state of each sediment layer at each grid point in the model — more a bookkeeping function than numerical modeling.

h. Part III-5-13 provides some detailed guidance on the specific coastal engineering and management issues associated with cohesive shores. The success of engineering and management techniques on cohesive shores is dependent on a recognition of the fundamental differences between sandy and cohesive shore processes. Engineering techniques that may have been successful on the more familiar sandy shores, may be unsuitable or inappropriate along cohesive shores.

III-5-5. Geomorphology of Consolidated Shores

This section provides a review of the geomorphology of cohesive shores and the relationship between the land forms and the erosion processes discussed in the previous sections. The discussion is subdivided under two headings: "Controlling Factors" and "Profile Types." This section focuses on the geomorphology of consolidated cohesive sediment shores. For more detail on erosion processes along rocky coasts, refer to Sunamura (1992).

a. Controlling factors. The primary controlling factors discussed in the following paragraphs influence the geomorphology of cohesive shores:

(1) Lag deposits.

(a) Some consolidated cohesive sediment units have cobbles or boulders within their composition. For example, along the Great Lakes, glacial till may be either fine-grained or stony (i.e., containing gravel and cobbles). During the evolution of stony till shores, the cobbles and boulders that are left behind after the removal of the finer clay, silt, and sand build up to form a protective armor for the underlying till. In these cases, an erosion-resistant nearshore shelf will usually have formed. The depth of the shelf at any location is such that the lag deposit remains immobile, and therefore the depth is determined by the local wave climate and the grain size of the lag deposit. Generally, for the Great Lakes, the depth of this shelf is approximately 2 m below low water datum. Along sea and ocean coasts with large tidal ranges and longer waves, lag deposits may occur at much greater depths (e.g., lag deposits over clay have been found in water depths of 10 m below datum along the North Sea coast of England). The shelf creates what has been referred to as a convex profile, in contrast to the concave profile associated with situations where lags are not present.

(b) The armored shelf acts to dissipate wave energy, and therefore reduce or even prevent bluff erosion. Boyd (1992) gives several examples where a bluff is protected from erosion by a nearshore shelf and notes that the reduced wave energy may also allow a stable beach to exist at the shore, providing additional protection to the bluff toe. Natural headlands along an eroding cohesive shoreline often owe their existence to the presence of a lag-protected nearshore shelf.

(c) An example of a site where a lag deposit has resulted in an erosion-resistant foreshore and a stable shoreline is located near the town of Goderich on the Canadian shore of Lake Huron. A nearshore profile for this site is shown in Figure III-5-10. The cobble-protected shelf has a depth of 1.75 m and is about 200 m wide. The stratigraphy at the site consists of a stony till unit below the average lake level and a fine-grained till unit above the average lake level. At nearby sites where the fine-grained till unit dips below the average lake level to depths of greater than 2 m, and the nearshore shelf is no longer present, the bluff recession rates range from 0.3 m/year to over 1 m/year.

(2) Different stratigraphic units.

(a) Along most cohesive shores, 3-D variations in contact surfaces between stratigraphic units are common. This results from the complex geomorphologic conditions that formed the underlying geology, which, depending on the location, may include some combination of: glacial, lacustrine, estuarine, or fluvial

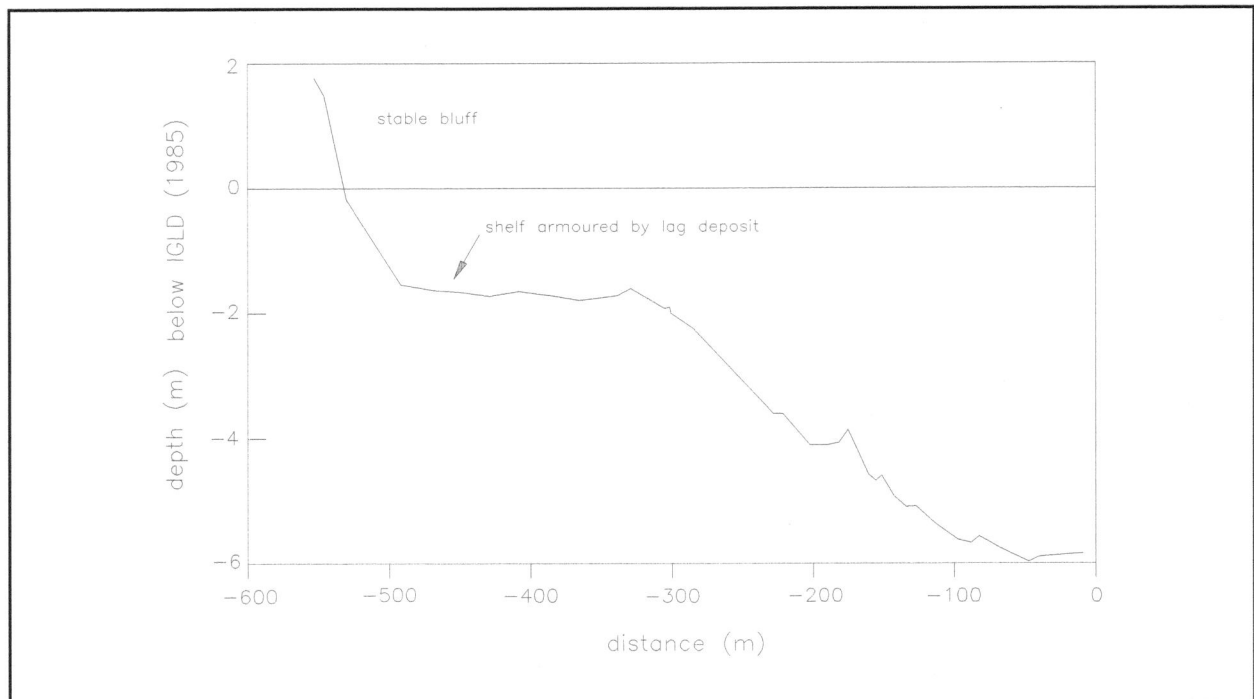

Figure III-5-10. A convex consolidated cohesive profile with a shelf protected by lag deposits located near Goderich, Ontario, on Lake Huron

processes. Therefore, as the shore erodes in time, the recession rates can change, as either more or less erosion resistant, units are encountered further inland. One example of the influence of erosion resistance is East Point located along the Scarborough Bluffs east of Toronto on Lake Ontario. The plan and stratigraphy cross section of this feature are shown in Figure III-5-11. Clearly, East Point is an expression of the more erosion-resistant Leaside (or northern) till unit, which dips below the average lake level at this location. Leaside till is relatively hard and includes boulder pavements (Boyce, Eyles, and Pugin 1995). Along the neighboring shore, the less erosion-resistant Scarborough clay unit exists below the lake level.

(b) Recognition of the variation in erosion resistance of different till units resulted in an unlikely finding during the investigation of the influence of a harbor structure at Port Burwell on downdrift erosion along the north central shore of Lake Erie (Figure III-5-12). A long harbor jetty here intercepts almost all of the sand moving along the shore from west to east, therefore depriving the downdrift shore of sediment. On a sandy shore, this would be a clear case of downdrift erosion due to sediment supply starvation. However, the investigation described by Philpott (1984) revealed that cohesive shores along the north central shore of Lake Erie seldom have (and probably never had) enough sand to halt the downcutting of the underlying till. Updrift of the harbor, the trapping of large quantities of sand eventually halted the nearshore profile downcutting, and the bluff position was stabilized. However, this still did not fully explain why recession rates updrift of the harbor fillet beach were generally lower (about 1 m/year) than those on the downdrift shoreline to the east (in the range of 2 to 4 m/year).

(c) Figure III-5-12 also describes the bluff face stratigraphy at Port Burwell. There is a change in the subaqueous stratigraphic unit at the harbor mouth, where Port Stanley till exists below the lake level on the updrift side and Waterlain till forms the nearshore profile on the downdrift side. Based on a comprehensive investigation, including laboratory testing of the erodibility of the different till units, it was concluded that Waterlain till was less erosion-resistant than the Port Stanley till. It is not coincidental that the change in till

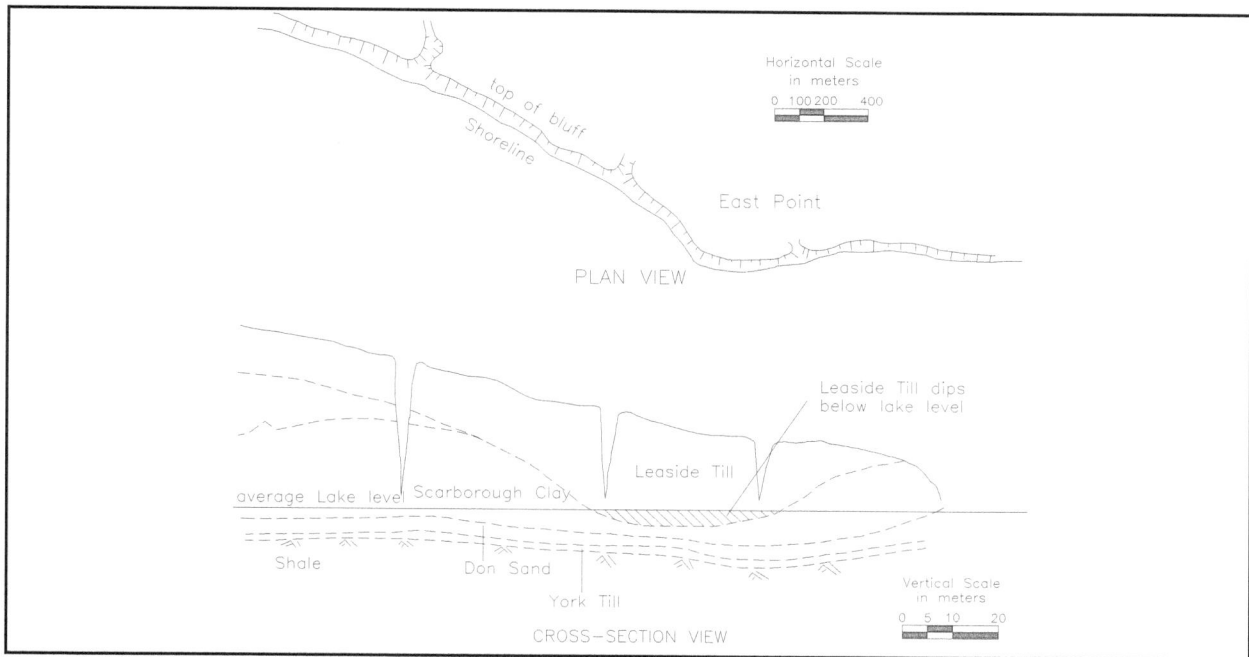

Figure III-5-11. Plan and cross section of East Point along the Scarborough Bluffs (located east of Toronto on Lake Ontario) showing the influence of the erosion-resistant leaside (or northern) till on the local geomorphology

occurred at the harbor mouth, as often creeks or rivers follow the interface between different stratigraphic units.

(d) Riggs, Cleary, and Snyder (1995) provide several examples of both headlands (local areas of slowly retreating or stable shoreline) and local sections of rapidly retreating shoreline, that are a direct result of variability in the erosion resistance of different stratigraphic units that make up the shoreface. They found that headlands result from the presence in the nearshore of more erosion-resistant Pleistocene or older sediments. In contrast, rapidly retreating sections of shore consisted of sand-poor, valley-fill sediment or compact peats and clays deposited in modern estuarine environments.

(3) Quantity and mobility of sand cover.

(a) In natural situations, a protective sand cover can build up over an erodible substrate and protect it from further erosion. Investigations of Great Lakes sites have shown that approximately 200 m³/m of sand cover (measured from the top of the beach out to the 4-m contour) is required to halt the downcutting process (Nairn 1992). However, even half as much as this quantity can afford at least some protection to the underlying cohesive substratum.

(b) Another important factor is the volatility or mobility of sand cover. If the overlying noncohesive bed forms are rapidly and frequently changing position, the underlying cohesive substratum will be exposed to erosive situations more frequently. Nairn and Parson (1995) have indicated that on the Great Lakes, the shift of bar position in response to changing lake levels has an important influence on the exposure of the underlying till (in the troughs between the bars) to erosion. On an eroding cohesive bluff shore along the North Sea coast of England, Pringle (1985) identified alongshore migrating areas of reduced beach cover and exposed glacial till called 'Ords.' These Ords have an important role in exposing the underlying glacial till to erosion. Figure III-5-13 shows an Ord on the Holderness coast at low tide with glacial till exposed (partly

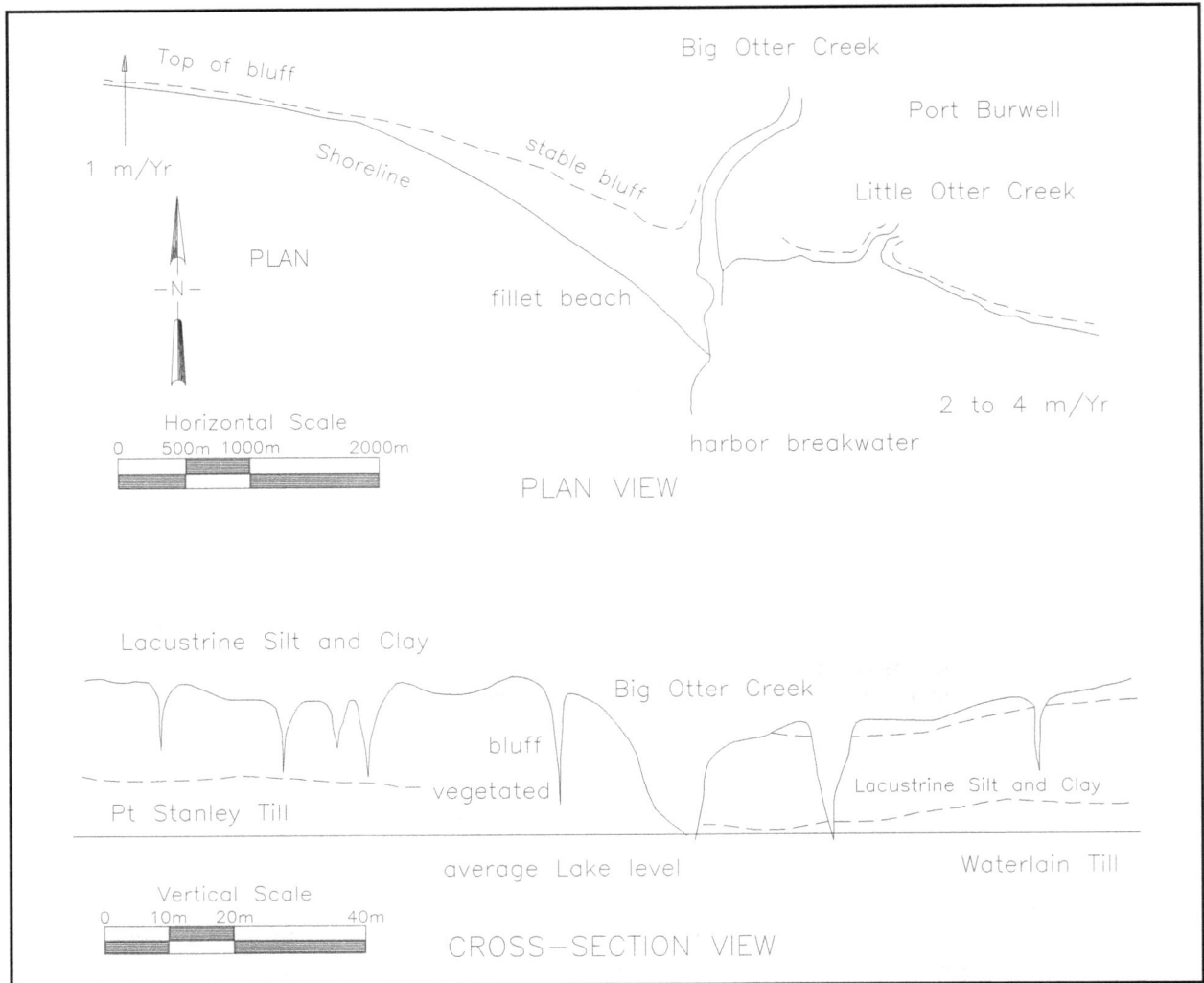

Figure III-5-12. Plan and cross section of the Port Burwell area on the north central shore of Lake Erie showing the influence of a fillet beach and stratigraphy changes on the geomorphology of a cohesive shore

covered with cobbles) between the upper coarse sand beach and a bar consisting of finer sand. The closeup photo of the Ord (Figure III-5-14) shows erosion of the exposed till around a 10-cm-high pedestal protected by a rock cap. Pringle (1985) found that the migration rate for these features was approximately 500 m/year, which is similar to the rate of migration for large sand waves or rhythmic features along Long Island, NY (Thevenot and Kraus 1995). An Ord could be defined as the area between two migrating sand waves. In summary, both cross-shore and alongshore variations in sand cover thickness resulting from migration of bars and sand waves or Ords, respectively, have an important influence on the rate of cohesive sediment erosion in the nearshore zone.

(c) Deposition of large quantities of sand over a cohesive substrate can occur at a change in shoreline orientation where the potential alongshore sediment transport rate rapidly decreases, or at a natural obstruction to alongshore transport, such as at a rock headland. Other instances where sand may eventually build up to protect a profile include sites where the alongshore transport of sediment is intercepted at a harbor jetty. The Port Burwell fillet beach shown in Figure III-5-6, as discussed above, protects the nearshore cohesive substratum and has stopped the bluff recession behind the fillet beach.

Figure III-5-13. Bluff erosion along the Holderness coast of the North Sea. The underlying cohesive profile is exposed at low tide in a trough (referred to as an "Ord") between the upper beach and first bar

Figure III-5-14. Close-up of the exposed cohesive profile on the Holderness coast (Figure III-5-13). A rock-capped pedestal of cohesive sediment, about 10 cm in height, has developed through erosion of the adjacent seabed

(d) Many areas along the Great Lakes shores that once had sufficient sand cover to protect an underlying cohesive substratum from downcutting are now coming under attack as the sediment supply has been reduced through human influences. Reductions to the sediment supply occur through entrapment of sediment at structures which protrude into the lake (including harbor jetties and land reclamation projects that have been created for many purposes, such as power plants, marinas, and docking facilities) as well as through protection of previously eroding sections of shoreline. Shabica and Pranschke (1994) describe one such area north of Chicago on Lake Michigan where the sand cover has decreased from 560 m³/m in 1975 to 190 m³/m in 1989. If the depletion of sediment cover continues at this site, the previously very low rates of shoreline recession (less than 0.2 m/year, or 8 in./year) may accelerate.

(4) Local wave and water level conditions. The characteristics of the local wave and water level conditions represent the fourth controlling factor on the geomorphology of consolidated cohesive coasts. Both the intensity and the directionality of the waves can influence the rate of erosion at a particular shore site. Other factors being equal, greater wave energy translates to higher downcutting rates and more rapid shoreline erosion. Directionality of the waves can have a secondary influence on downcutting rates by affecting the mobility of the sand cover over the underlying till. Large swings in wave direction can result in a more dynamic system with respect to the sediment cover. Fluctuations in water level also have an important role in cohesive shore erosion processes as explained by Stewart and Pope (1993) and Fuller (1995). While direct erosion at the bluff toe may be accelerated during high-water conditions, low water leads to acceleration of the nearshore downcutting process (which in turn allows more waves to reach the bluff toe).

b. Profile types. Boyd (1981, 1992) completed an extensive review of nearshore profile shapes for consolidated cohesive shores on the Great Lakes. These essentially fall into two categories: concave profiles and convex profiles. Figure III-5-15 provides a schematic description of these two profile types.

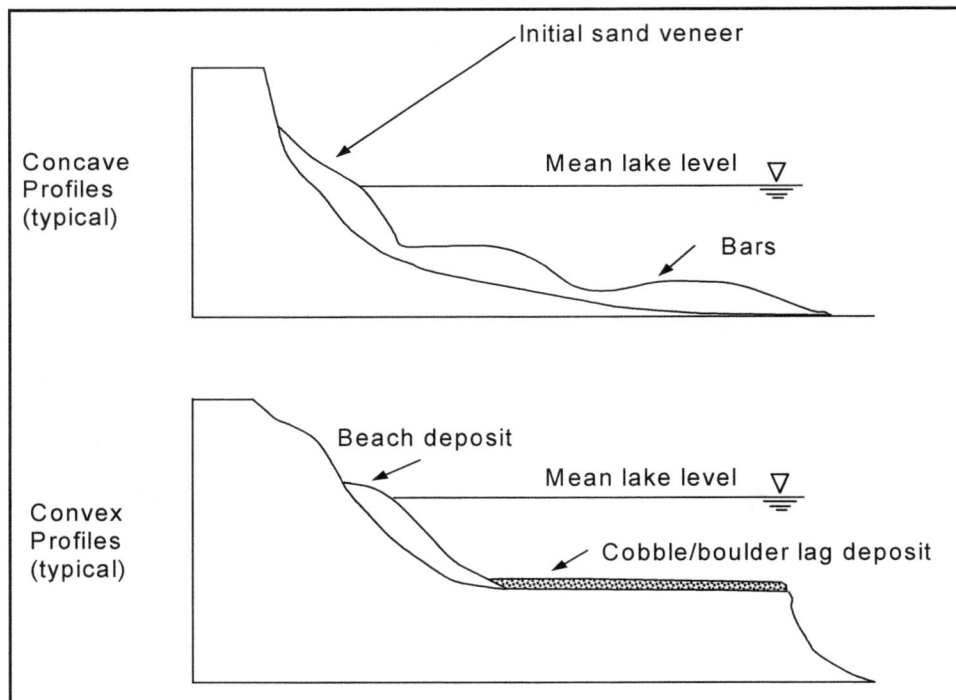

Figure III-5-15. Distinctions between concave and convex consolidated cohesive profiles

(1) Concave profiles. Concave profiles develop in fine-grained sediment with a relatively uniform erosion resistance from the closure point up to the top of the bluff or cliff. These profiles have an exponential form similar to sandy shore profiles described by Dean (1977). Sand cover over these cohesive profiles can range from perhaps as little as 25 to over 200 m^3/m measured between the bluff toe and the 4-m depth contour. The sand cover can be in the form of bars, and, in areas with a sand cover in the high end of the range, a substantial beach at the shore. Stewart and Pope (1993) found that a reduction in the range of water level fluctuations would not reduce the long-term erosion rates for cohesive shores with concave profiles. As explained above, lower water levels result in accelerated lowering of the nearshore profile, which essentially has the same effect as high water levels — allowing waves to reach the bluff toe.

(2) Convex profiles. As noted above, convex profiles develop at locations where potential lag deposits exist within the eroding material. These profiles are characterized by a nearshore shelf, which on the Great Lakes has a depth approximately 2 m below low water datum. At other locations, this depth will be determined by the median grain size of the lag deposit, the wave climate, and the range of water level fluctuations. Long-term erosion rates along these shores are less than rates for concave cohesive shores (having limited sand cover) with the same wave exposure. With the exception of high-water periods, the erosion-resistant nearshore shelf acts to dissipate wave energy before it reaches the shoreline. However, during high water periods, these shorelines are more vulnerable to erosion when waves are able to attack the bluff toe. Therefore, in contrast to cohesive shores with concave profiles, shores with convex profiles would benefit from a reduction in the range of water level fluctuations (Stewart and Pope 1993). Finally, the fish habitat function of cohesive shores with a convex profile shape is much more important owing to the surficial substrate (cobbles and boulders with limited sand cover) and the proximity to a deepwater drop-off at the edge of the shelf.

III-5-6. Sediment Properties and Measurement Techniques

a. Introduction.

(1) In the case of noncohesive sand and gravel, sediment mobility can be estimated just by knowing the grain size and shape, specific gravities of the sediment and water, and the viscosity or temperature of the water (i.e., physical properties). The mobility of cohesive sediment is a more complex phenomenon. Cohesion (particle attraction) is governed by the electrochemistry of the sediment mineral and water; its state of consolidation; and in many cases, by the presence of organisms like diatoms, which can bind the sediment particles together with mucus.

(2) The extent of data requirements will vary depending on the nature of the coastal engineering problem and the nature of the shore. This section presents an overview of the range of possible field and laboratory investigations that can be used to characterize the conditions associated with erosion, transport, deposition, and consolidation on a cohesive shore. This discussion focuses on a characterization of the specific geologic conditions related to the cohesive shore. For the measurement of environmental conditions (e.g., waves and water levels), refer to Part II-3.

b. Consolidated shore erosion. Developing an understanding of consolidated shore erosion requires information on profile shape (beach and nearshore profile techniques are summarized in Part III-3-2), presence or absence of lag deposits, bluff and nearshore stratigraphy, erodibility of the one or more cohesive units in the active nearshore erosion zone (i.e., between high water and the depth of closure), and the sand cover thickness and stability. Available techniques to assess the characteristics listed above are as follows:

(1) Field sampling and geotechnical analyses.

(a) Testing may be performed in situ or in a laboratory on samples extracted from the field. Extraction techniques for seabed or lake bed cohesive sediments include: dredging; coring; box coring; and cutting samples (the latter using a chainsaw with a trenching chain). As noted in Part III-5-6, it is important to retrieve intact samples that, to the extent possible, preserve the natural structure of the cohesive sediment.

(b) None of the available standard geotechnical test procedures provide a direct measure of the erosion resistance of a cohesive sediment in the coastal environment. Nevertheless, the more important characteristics that provide an indirect assessment of erodibility include: grain size analysis (including clay content); liquid and plastic limits; water content; undrained shear strength; bulk density; and consolidation pressure. Techniques for establishing these parameters are presented in the USACE Engineering Manual "Laboratory Soils Testing" (EM 1110-2-1906). Undrained shear strength can be determined in the field using a cone penetrometer or a vane shear apparatus.

(c) Borehole information can be valuable for assessing variations in stratigraphy both above and below the water level.

(2) Laboratory erodibility experiments.

(a) There are no standard and accepted approaches for establishing the erodibility of cohesive sediment in the coastal environment based on geotechnical properties. Therefore, to quantify the relationship between erodibility and shear stress applied under a given flow condition, it is usually necessary to perform laboratory experiments. Experiments may not be required where direct techniques have been applied to determine the erodibility of similar sediment. It is advisable that these experiments be performed with intact, and to the extent possible, undisturbed samples of cohesive sediment in order to preserve the natural structure of the soil.

(b) Four laboratory techniques for assessing erodibility are briefly reviewed in this section. These provide an example of the range of techniques that are available.

(c) Arulanandan, Loganatham, and Krone (1975), and more recently Zeman (1986), describe the use of a rotating cylinder apparatus to assess the erodibility of intact and undisturbed samples of cohesive sediment. This technique is also mentioned for testing the erodibility of mud. In this approach, a long cylindrical sample is mounted inside a larger transparent cell. The cell is then filled with water and rotated. During rotation, the torque transmitted to the inside stationary cylinder is measured to quantify the shear stress applied to the sample. At the end of the test, erosion rates are determined by the loss in mass of the sample. A disadvantage of this approach is the inability to introduce sand to the flow to assess the important influence of sand abrasion.

(d) Another small-scale laboratory technique for testing intact and undisturbed samples is described by Rohan et al. (1986). This procedure is based on an adaptation of the standard pinhole test. Water is circulated through a hole drilled through the axis of a cylindrical sample. The head loss caused by friction in the sample is measured using differential manometers in order to assess the shear stress applied to the soil by the flow. Depending on the size of the hole bored in the sample, it is possible that this technique could be adapted to assess the influence on erosion of sand in the flow.

(e) At a larger scale, intact and undisturbed cohesive sediment samples can be placed in a drop section in the floor of a unidirectional flow flume or tunnel. The sample is then exposed to different flow conditions with and without the presence of sand, and the erosion of the sample surface is surveyed intermittently to determine erosion rates for the different conditions. This technique is also frequently used for mud. Kamphuis (1990) describes the use of a tilting tunnel in which Pitot tubes were used to determine a velocity profile upstream and downstream of the sample in order to determine the shear stress applied to the sample by the flow. Cornett, Sigouin, and Davies (1994) describe a similar approach using a tilting flume for the analysis of samples extracted from the bed of Lake Michigan near St. Joseph Harbor (Parson, Morang, and Nairn 1996). In this case, a laser doppler velocimeter measured the velocity profile near the bed in order to establish the shear stress applied to the sample by the flow (Figure III-5-16). This figure shows a sand veneer migrating over the till sample in the unidirectional flow flume. In both the Kamphuis (1990) and the Cornett, Sigouin, and Davies (1994) tests, the maximum flows generated were in the range of 3 to 3.5 m/sec (10 to 12 ft/sec). Results from experiments using this technique to estimate erodibility are presented in Part III-5-7b.

Figure III-5-16. Laser doppler velocimeter (LDV) used to determine shear stress exerted on the till bed in a unidirectional flow flume test. This test features sand in the flow acting as an abrasive

(f) The most realistic approach that can be taken to assess erodibility in a laboratory setting is to create a nearshore profile with intact and undisturbed cohesive sediment samples in a wave flume or basin. This approach was used by Skafel and Bishop (1994) to complete important research into the erosion processes on cohesive shores. Intact samples, measuring 1 m by 0.35 m by 0.45 m, encased in an open-ended steel box were extracted from the top of a bluff on Lake Erie and placed directly in a wave flume. The open-ended steel box was pushed slowly into the till by a 20-ton hydraulic ram and the till at the inner end of the box was cut away using a chainsaw with a trenching chain. The box was then removed with a crane. The till boxes were installed in the flume to create the desired profile shape. In these tests, the effects of sand cover in the form of migrating bars or a patchy veneer were tested. Also, the influence of breaking waves on the erosion of the cohesive sediment was assessed.

(3) Field techniques for assessing surface and subsurface conditions.

(a) One of the most important pieces of information in characterizing a cohesive shore profile is the sand cover thickness across the underlying cohesive profile (i.e., measured from the bluff toe out to a depth of at least 4 m). In addition, where the cohesive profile is exposed, it is also important to determine whether or not a protective lag deposit exists. As with any coastal engineering site investigation, beach and nearshore profiles are essential information. In this section, a variety of techniques for characterizing the surface and subsurface conditions, with particular focus on the sand cover thickness, are presented, ranging from the simplest to the most sophisticated.

(b) The simplest technique of estimating the thickness of the sand cover across the profile involves the following tasks:

• Complete a beach and nearshore profile from the toe of the bluff out to the depth of closure (between the 5- and 10-m water depth).

• Through the use of a steel probe or test pits, attempt to determine the thickness of sand cover near the waterline.

• Estimate the shape of the underlying cohesive profile (as a smooth exponential form) joining points between the toe of the bluff, the position of the till at the waterline (if determinable), and the troughs between the bars on the profile. Typically, the till will be exposed or only thinly covered in the troughs. If repeated profiles are available at a site, these may provide additional information on the position of the underlying till if the position of the troughs between bars shifts between surveys.

(c) In order to complement the simple technique described above, a diving inspection could be completed across the profile. The diver could use an underwater video to document conditions, and a steel probe to estimate sand cover thickness at different locations. Depending on the extent of sand cover, the till may be exposed in some areas. Alternatively, a frame-mounted video camera lowered from a boat or a remotely operated vehicle with video could be used. Video is also valuable in assessing whether or not a lag deposit exists where the cohesive layer is exposed.

(d) In place of a simple steel probe, a jet probe could be used to survey the thickness of the sand cover on the land and underwater. A jet of either water or air can be used to penetrate the sand cover (the latter is only applicable underwater). Shabica and Pranschke (1994) describe the use of a hydraulic probe consisting of an extendible 20 mm diameter pipe through which water is pumped at 2.8 kg/cm^2 (40 psi).

(e) A technique based on electrical resistivity has been used to establish the sand thickness across the subaerial section of beach for sections of the Holderness shoreline. This method is particularly useful at locations with large tidal ranges that allow for significant sections of the profile to be surveyed at low tide.

(f) Ground-penetrating radar was used to survey the thickness of the sand cover for several profiles downdrift of St. Joseph Harbor (Parson, Morang, and Nairn 1996). The limitation of this technique is that it can only be used in a freshwater environment.

(g) Sub-bottom profiling, or high-frequency seismic imaging, is another geophysical technique that is capable of establishing the thickness of sand cover over an underlying cohesive profile. Side-scan sonar is an acoustic technique that provides an image of the seabed or lake bed surficial conditions. While this procedure would not be capable of determining the thickness of sand cover, it could provide useful surficial information such as the extent of exposed gravel and cobble lag deposits. These methods are described in Part IV-5 in greater detail.

c. Erosion, transport, and deposition of mud. In the case of mud, it is useful first to examine the following hydrodynamic sediment properties that will be required by the equations presented later:

(1) Cohesion. The cohesive bond is predominantly electrochemical, increasing with the electrical conductivity of the ambient water and proximity of the particles. Conductivity increases with salinity. The bond between particles may be enhanced, particularly at rest on the bed, by biological 'glues' such as the mucus excreted by diatoms, worm tubes, and feces (Paterson 1994).

(2) Critical shear for erosion.

(a) As water flows over the mud bed, as either steady flow or oscillatory flow under tides and waves, it exerts a shear stress τ on the bed due to viscosity and turbulence (described in greater detail in Part III-6). Not only is shear a real physical stress on the bed sediment, but it also serves as empirical shorthand for the level of turbulence in the flow. Thus, it is a useful parameter in describing suspended load sediment transport; as well as fluid mud (bed load), erosion, and deposition.

(b) At the level of a stationary particle on the bed, shear forces are balanced by the forces of gravity, interparticle friction, and cohesion. Shear is augmented by lift and drag, making the force balance

$$\text{SHEAR} + \text{LIFT} + \text{DRAG} < \text{GRAVITY} + \text{FRICTION} + \text{COHESION}$$

a vector sum, the same as that for noncohesive sand and gravel, but with the addition of cohesion. As flow increases, the left-hand side of this balance increases approximately as the square of velocity, until

$$\text{SHEAR} + \text{LIFT} + \text{DRAG} = \text{GRAVITY} + \text{FRICTION} + \text{COHESION}$$

and the formerly stationary particle leaves the bed and begins to move. The shear stress at which this occurs is known as the *critical shear for erosion* or *erosion threshold* τ_c. τ_c is still shorthand for the entire left-hand side of the balance, not shear alone.

(c) The sediment 'particle' may be an individual grain; but more likely a floc, made up of several grains held together by cohesion. Cohesion plays the major role in the right-hand side of the force balance, and failure (erosion) will occur where cohesion is weakest.

(d) Critical shear τ_c is not a particularly useful concept in fluid mud. At the water/fluid mud interface, the applied shear stress is balanced by a shear strain (flow) of the fluid mud, rather than GRAVITY + FRICTION + COHESION. Also, the fluid mud is essentially a thick, viscous, laminar, boundary layer, protecting the stationary bed from any SHEAR + LIFT + DRAG approaching τ_c. Erosion of fluid mud is better described by densimetric Froude Number entrainment between two fluids (Part III-5-7e).

(3) Erosion rate at twice critical shear. Both the Parthenaides and Krone Equations (Parts III-5-7d and III-5-9c) are 'excess shear' fits to observed erosion and deposition, respectively. For example, the Parthenaides Equation (Part III-5-7d) correlates observed erosion rates with

$$\text{Dimensionless Excess Shear} = (\tau_c - \tau) / \tau_c; \text{ negative } (\tau_c < \tau) \text{ for erosion}$$

The Parthenaides coefficient M_p (in units of $kg/m^2/sec$) (see Equation III-5-1), is the correlation coefficient between erosion rate and excess shear, when dimensionless excess shear = -1; that is, when $\tau = 2\tau_c$.

(4) Critical shear for deposition.

(a) A critical shear stress for deposition τ_s Pa (lbf/ft^2) is not obvious at first glance. In noncohesive sediment, the critical shear for deposition is only slightly less than that for erosion: a noncohesive particle will come to rest almost as soon as the shear is too small to move it. But the process of deposition of cohesive sediment flocs is quite different; τ_s is generally on the order of one fourth of τ_c

(b) High shear near the bed breaks up large flocs before they can settle. Then, the resulting smaller flocs and individual particles are resuspended. The critical shear for deposition τ_s is that through which large flocs can pass without being broken up. Note that τ_s is not shorthand for something more; τ_s really is the shear stress in the bottom boundary layer which cannot overcome cohesion in the settling flocs.

(5) Sediment, fluid mud, and water densities.

(a) Important densities and specific gravities:

ρ_w = specific gravity (mass density) of water — 1,000 kg/m^3 (62.4 lb/ft^3) in fresh water, up to 1,030 kg/m^3 (64.3 lb/ft^3) in seawater

ρ_s = specific gravity (mass density) of sediment mineral (no voids): generally 2,000 kg/m^3 (125 lb/ft^3) to 2,700 kg/m^3 (170 lb/ft^3) depending on mineral

(b) Bulk sediment density voids filled with ambient water:

- Of freshly deposited flocs (may be fluid mud if < 1,100 to 1,200 kg/m^3 (70 to 75 lb/ft^3), corresponding to a mass concentration in excess of about 20 kg/m^3 (20 ppt)).

- Of existing bed surface, and layers, e.g., 1,400 kg/m^3 (90 lb/ft^3).

- Of fully consolidated sediment, generally < 2,000 kg/m^3 (125 lb/ft^3).

(6) Grain size and settling velocity.

(a) Settling velocity is a more important hydrodynamic property of cohesive sediment than grain size. Settling velocity is a measure of the sediment's behavior in suspension; grain size only allows us to guess the settling velocity.

(b) The first thing to know about cohesive sediment grain size is that it is *not* a measurable physical constant. True, the size of individual dispersed grains may be inferred from measurements of their settling velocities in distilled (free of dissolved chemicals) water: generally on the order of 10-μ or less. The settling velocity of a 10-μ sphere, 2,500 kg/m^3 (156 lb/ft^3), in water of 20 °C (68 °F) is 0.06 mm/sec (0.002 in./sec).

(c) But cohesive sediment of this size in natural, often salt, water does not stay dispersed for long. Grains stick together when they come close enough for the cohesive forces to overcome the fluid shear and gravity keeping them apart. Aggregations of cohesive sediment grains are called 'flocs.' Flocs are larger than individual grains, of course, but because of water trapped within the floc, they are also less dense than the pure mineral. Depending on the relationships among floc size, shape, and density, the result is a *floc settling velocity* that may be more or less than that of individual grains. The settling velocity must be determined with the natural sediment in the natural water.

(d) Mud may also be biologically cohesive (Paterson 1994), for example, due to mucus excreted by diatoms. Biological cohesion is even more difficult to predict than electrochemical, providing yet another reason for using natural sediment and natural water in determining floc size and settling velocities.

(7) Degree of consolidation.

(a) The degree of consolidation u is defined as the ratio of the bulk density of the sediment to the bulk density of the 'fully consolidated' sediment, measured under Part III-5-6b(5). Consolidation of cohesive sediment is the compaction of the soil mass accompanied by drainage of the interstitial water, just as with noncohesive sediment. The principal difference is the length and cross-sectional area of the drainage path. In cohesive sediments, the path length is long and the area is small (i.e., low permeability); slowing down drainage and consolidation. Drainage is through the bed surface, into the ambient water, so that a good relative measure of the length of the drainage path P is the depth of burial below the surface. Cross-sectional area of the drainage path must be inferred from measurements of permeability.

(b) Overburden speeds up consolidation but increases the length of the drainage path, especially when the overburden is also cohesive. Nevertheless, consolidation starts at the bottom of a sediment layer and follows the draining water upward, giving even a freshly deposited layer a density gradient, denser at the bottom to less dense at the surface, until the entire layer is fully consolidated. The strength of the sediment represented by the critical shear for erosion τ_c increases with density and consolidation.

(8) Field measurement techniques. Many of the cohesive sediment field and laboratory measurement techniques are the same as those for noncohesive sediment (Part III-1). Nevertheless, some accommodation must be made for mud:

(a) Bed sampling. Much depends on knowledge of the composition and density of surficial sediments, which can be gained from laboratory analysis of surface samples obtained in the field. It is unreasonable to expect that undisturbed mud samples can be collected. In fact, it is difficult to contain most surficial mud in the commonly used Shipek or Ponar grab samplers because the samples leak out. Underwater samples may have to be obtained by divers, and all samples that include entrapped water should be transported in sealed jars and stored at 4 °C (39 °F).

(b) Boreholes and cores. The techniques give information on subsurface sediment layers. Blow counts and cone penetration tests give a relative measure of the strength and density of the layers (but not of the critical shears for erosion and deposition of mud, see below), and cores taken from the layers are as close to undisturbed samples as is possible in cohesive sediment. Boreholes should extend to bedrock or similar hard, impenetrable layers, with cores and cone penetration tests in each major layer.

(c) Suspended sediment sampling. This type of testing is needed to determine composition and quantity of sediment in suspension. Generally, the technique is to pump and filter 4 L of suspension and transport the filter and contents to the laboratory for subsequent analysis. Alternatively, 1 L of suspension may be sealed in jars and transported to the laboratory for filtering and analysis there; this liter may be obtained by pumping or from any of the proprietary suspension samplers. Filters should be no larger than 10 microns. Sampling should be carried out at a minimum of four elevations over the depth, with special attention to the near-bed or fluid mud layer.

(d) Settling tube. Field settling tubes, e.g., the 'Owen Tube' (Eisma, Dyer, and van Leussen 1994) measure the settling velocity of cohesive sediment flocs in 'live' natural water, even in a natural level of turbulence. Typically an undisturbed sample of sediment-water suspension is captured in a horizontal tube. The tube is immediately turned into the vertical position, and the settling velocity of the flocs is determined from density changes that occur in the suspension at various depths in the tube and over various times.

(e) Piezometers. These instruments measure rate of drainage of excess pore pressure from natural muds, and thus permeability and rates of consolidation. Lancelot (Christian, Heffler, and Davis 1993) is a piezometer that first creates excess pore pressure on its insertion in the mud bed, and then measures the rate of decay or drainage.

(f) Optical techniques. These techniques are primarily used as a substitute for suspended sediment sampling. Two basic techniques are used: measuring light transmitted through a known illuminated volume of suspension in a turbidity meter or transmissometer (e.g. Bartz, Zaneveld, and Pak 1978); or measuring light reflected from the suspension by an 'Optical Backscatterance (sic) Sensor' or OBS (e.g. Sternberg, Shi, and Downing 1989). Both require calibration against natural sediment in known concentrations in natural water.

(g) Acoustic techniques. Many novel applications are still under development, mostly in the high-frequency (MHz) range, where for example, suspended sediment concentration and grain size profiles can be measured (e.g. Hay and Sheng 1992). At lower frequencies, echo sounding detects the elevation of the bed and of the surface of fluid mud; and in the side-scan mode, detects bed forms such as ripples and dunes, and their orientation (Hay and Wilson 1994). At still lower (seismic) frequencies, sound penetrates the bed and detects the interfaces between sediment layers of different densities, creating sub-bottom profiles.

(h) Radioactivity techniques. In these techniques, Gamma rays or X-rays are passed through a sediment/water suspension or bed layer (e.g. Sills 1994). The energy passing can be related by calibration to the mass density of the suspension or layer. These techniques are particularly useful in characterizing fluid mud layers, as and where they occur. Radioactivity techniques are also used in laboratory consolidation columns.

(i) Direct shear techniques. There are several field devices that apply a variable shear stress to the surface of a cohesive sediment bed (Gust 1994), and measure the variable rate of erosion (increase in suspended sediment in the water column) and deposition (decrease in suspended sediment in the water column). Results can be used directly in the Parthenaides and Krone equations (Equations 5-1 and 5-4) respectively. The prototype for all such devices is the annular flume, described under Part III-5-6b(9) "Laboratory Measurement Techniques," and sketched in Figure III-5-17. The Sea Carousel described by Amos et al. (1992) is an example of field adaptation of the annular flume.

(j) Correlation with shear strength. Although the critical shear for erosion τ_c would seem to be a function of the shear strength of soft cohesive soil (measured by vane, cone, or penetrometer), the form of that function is not yet known and certainly not linear. Even measuring mechanical surface shear directly on tidal mud flats (Faas et al. 1992) produces mechanical yield stress an order of magnitude larger than the hydrodynamic τ_c.

(9) Laboratory measurement techniques.

(a) Grain size analysis. Standard ASTM D422 laboratory techniques should be applied to determine the physical size of individual grains in bed, core, and suspended sediment samples. Although no ASTM standard has been published, the pipette technique (removing a known volume of suspension from a known

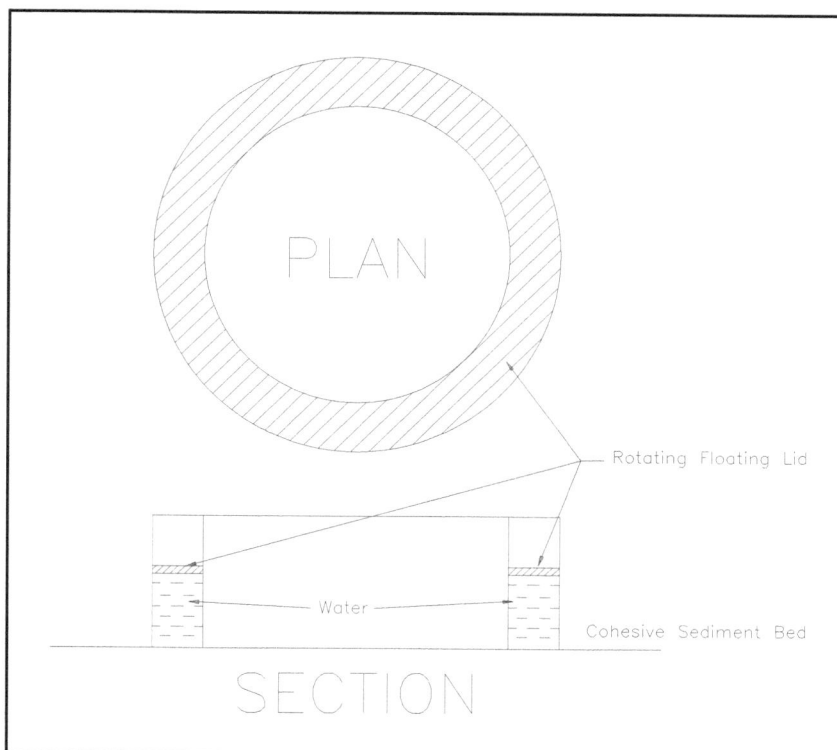

Figure III-5-17.   Prototype direct shear device, the annular flume

elevation in the settling column, and filtering or drying to determine sediment concentration), is an alternative to the hydrometer technique.   Total dry weight of suspended samples is also required to give field concentration in mg/l (ppm). Both hydrometer and pipette techniques measure settling velocity, and infer grain size from it.   The settling column also measures settling velocity in the manner of a large-scale (typically greater than 1 m (40 in.) deep, 0.3 m (12 in.) diameter) hydrometer or pipette test (Gibbs 1972). Sensitive differential pressure transducers record the variations in suspended sediment concentration with depth and time, from which settling velocity distribution in the sample can be computed.   Like the consolidation column below, settling columns need to be well-isolated from vibration and temperature changes to prevent artificial flocculation of the settling particles. For clay particles (<4 m), it will be necessary to use a nonstandard particle counter, e.g., Coulter counter.   Nonstandard (natural water) hydrometer, pipette, or settling column tests should be used to estimate settling velocity of the flocs and bulk density of deposited sediment.

(b) Consolidation column. A consolidation column is a cylinder containing 2 to 3 m of natural sediment and natural water in a vibration-free environment to ensure natural rates of consolidation (Sills 1994). Variations in pore pressure with time and depth (overburden) are measured with piezometers, and in density, with gamma ray or observed volume. Estimates of permeability (length and diameter of drainage paths, and variation with bulk density) come out of the same measurements, using Equation 5-6, for example.

(c) Direct shear techniques. There are several laboratory devices that apply a variable shear stress to the surface of a cohesive sediment bed and measure the variable rate of erosion (increase in suspended sediment in the water column) and deposition (decrease in suspended sediment in the water column). Results can be used directly in the Parthenaides and Krone Equations of Parts III-5-7d and III-5-9c, respectively.   The prototype for all such devices is the annular flume (e.g., Krishnappan 1993), sketched in Figure III-5-17. An annular flume is simply an endless channel in which the shear or velocity of rotation of the lid can be varied and the changes in the mud bed can be inferred from changes in suspended sediment concentration, measured using one of the techniques described above. The shear force driving the circulation can be measured directly as the force needed to rotate the lid. Laboratory annular flumes generally rotate the flume and the lid in opposite directions, to minimize secondary radial circulation of the water, and to obtain a more uniform distribution of shear on the mud bed. An increase in suspended sediment corresponds to a decrease in the mass of sediment on the bed m and a decrease in suspended sediment to an increase in the mass of the bed. It is then possible to extrapolate these measured shear stresses to τ_c and τ_s, at which erosion or deposition ceases; and to interpolate the Parthenaides coefficient M_p and fall velocity w (Example Problem III-5-1).

(10) Calibration techniques. Ideally, one should be able to measure all the hydrodynamic properties of a cohesive sediment, and proceed directly to a model of the shore. The model, however, will always need calibration and verification against measurements of erosion and deposition, to fine tune measurements and confirm that the model represents the shore. With some of the hydrodynamic properties unknown, the model can be used to choose between high and low values of the unknown properties, using "design of experiments" techniques (Willis and Crookshank 1994). This still requires intelligent estimates of high and low values of the unknown properties and good field measurements of erosion, deposition, or transport.

III-5-7. Erosion Processes

a. Shear stress.

(1) The formulae for predicting the movement of cohesive sediment predict rates of erosion and deposition, not transport. Try putting a typical cohesive floc size into a noncohesive sediment transport formula (Part III-6) and you will predict virtually infinite transport rates, in which the predicted density of 'sediment in suspension' may exceed its real density on the bed. Noncohesive sediment formulae are generally based on transport limitations: assuming there is a sediment supply to match the transport potential. Cohesive sediment formulae are generally based on supply limitations, and assume the flow can transport all eroded sediment. They define the sediment exchange between the bed and the water column.

(2) The cohesive sediment formulae are also less theoretically based. They form a simple numerical framework for interpolating and extrapolating observed hydrodynamic behavior of cohesive sediment. Generally erosion or deposition is correlated with the excess shear stress.

b. Erodibility of consolidated sediments.

(1) There have been many studies of the erosion resistance of cohesive soils to flowing water. Very few of these investigations have considered the much more complex flow conditions encountered in the coastal zone. Nevertheless, a basic understanding, such as it is, of the complex process of erosion of consolidated cohesive soil provides a basis for assessing the erosion resistance of cohesive soils in the coastal environment.

(2) The erodibility of cohesive soils is controlled by the bonds between cohesive particles. Many tests of remolded cohesive sediments have found that the most important parameters in describing the erodibility of cohesive sediments include consolidation and physio-chemical conditions, both of which influence the degree of bonding between particles. Aside from the difficulty of obtaining intact cohesive samples, the reason for testing remolded and reconsolidated samples is to establish the erodibility of recompacted clays, used as construction materials. Extensive investigations into the relationships between the properties of these 'homogenous' cohesive sediments (e.g., consolidation and other geotechnical parameters), and the physio-chemical properties of the fluid (including temperature, pH, and salt or cation content) have found a direct relationship between these parameters and erodibility (Croad 1981; Arulanandan, Loganatham, and Krone 1975).

(3) More recent investigations of intact consolidated sediment samples have found that the natural structure of the material, including the presence of fissures, fractures, and seams of noncohesive materials such as silt and fine sand, is the most important factor in determining erodibility (Lefebvre, Rohan, and Douville 1985; Hutchinson 1986). The natural structure is uniquely defined by the environmental conditions during the original deposition and subsequent weathering of the sediment (including overconsolidation during glacial periods for some sediments). Conventional geotechnical parameters such as clay content and shear strength do not provide a direct measure of the influence of the natural structure of consolidated sediments as it relates to erodibility. Nevertheless, Kamphuis (1987) suggested that the presence of fissures in samples may be indirectly reflected in the undrained shear strength, consolidation pressure, and clay content.

(4) A technique for assessing the hydraulic erodibility of natural and engineered earth materials including both soil and rock is described by Annandale (1996). This empirical method, which provides a relationship between threshold stream power for erosion and an erodibility index, was developed from field observations of spillway performance downstream of dams. The erodibility index is determined as a scalar product of indices representing the following material properties: (1) mass strength, (2) block/particle size, (3) discontinuity/interparticle bond shear strength, and (4) shape of the material units and their orientation relative to the flow velocity.

(5) For clay materials as for mud, the erodibility index is primarily a function of the shear strength of the soil. Stream power is calculated as the product of near-bed velocity and shear stress. This approach was applied to the scour of weak rock in the presence of waves and currents to investigate scour potential around bridge piers (Anglin et al. 1996).

(6) In summary, in the absence of a reliable and standardized technique for assessing the natural structure of consolidated cohesive sediment, as it relates to erodibility in the coastal environment, more empirical approaches must be followed, such as establishing erodibility coefficients from laboratory tests or field data.

(7) One such example of a direct empirical technique for estimating erodibility is described by Kamphuis (1990) and Parson, Morang, and Nairn (1996). In these tests, intact (undisturbed) samples of consolidated sediment were placed in a drop section of the floor of a high-velocity unidirectional flow flume or tunnel with transparent walls or windows. The shear stress over the samples was determined indirectly by measuring the vertical profile of velocity just above the bed. The average erosion rate was then determined by measuring the volumetric erosion experienced on the surface of a sample within a test period. Rates were determined for velocities in the range of 0.5 to 3 m/sec. Further details on this test procedure are presented in Part III-5-6a. All results determined by using this technique for various types of consolidated sediment (including mudstone, till, and lacustrine clay) are summarized in Figure III-5-18 (from Parson, Morang, and Nairn (1996)). It was found that shear stresses in the range of 0 to 18 Pa resulted in erosion rates in the range of 0 to 8 mm/hr.

(8) In a further extension of this type of testing, Kamphuis (1990) found that the erosion rate increased dramatically when sand was added to the flow. The results of all of the erodibility tests using this technique with sand in flow are presented in Figure III-5-19 (from Parson, Morang, and Nairn (1996)). A comparison of the clear water erosion results of Figure III-5-18 and the sand in flow results of Figure III-5-19, indicates that for the same shear stress and sediment sample, the erosion rate is increased by a factor of 3 to 8 when

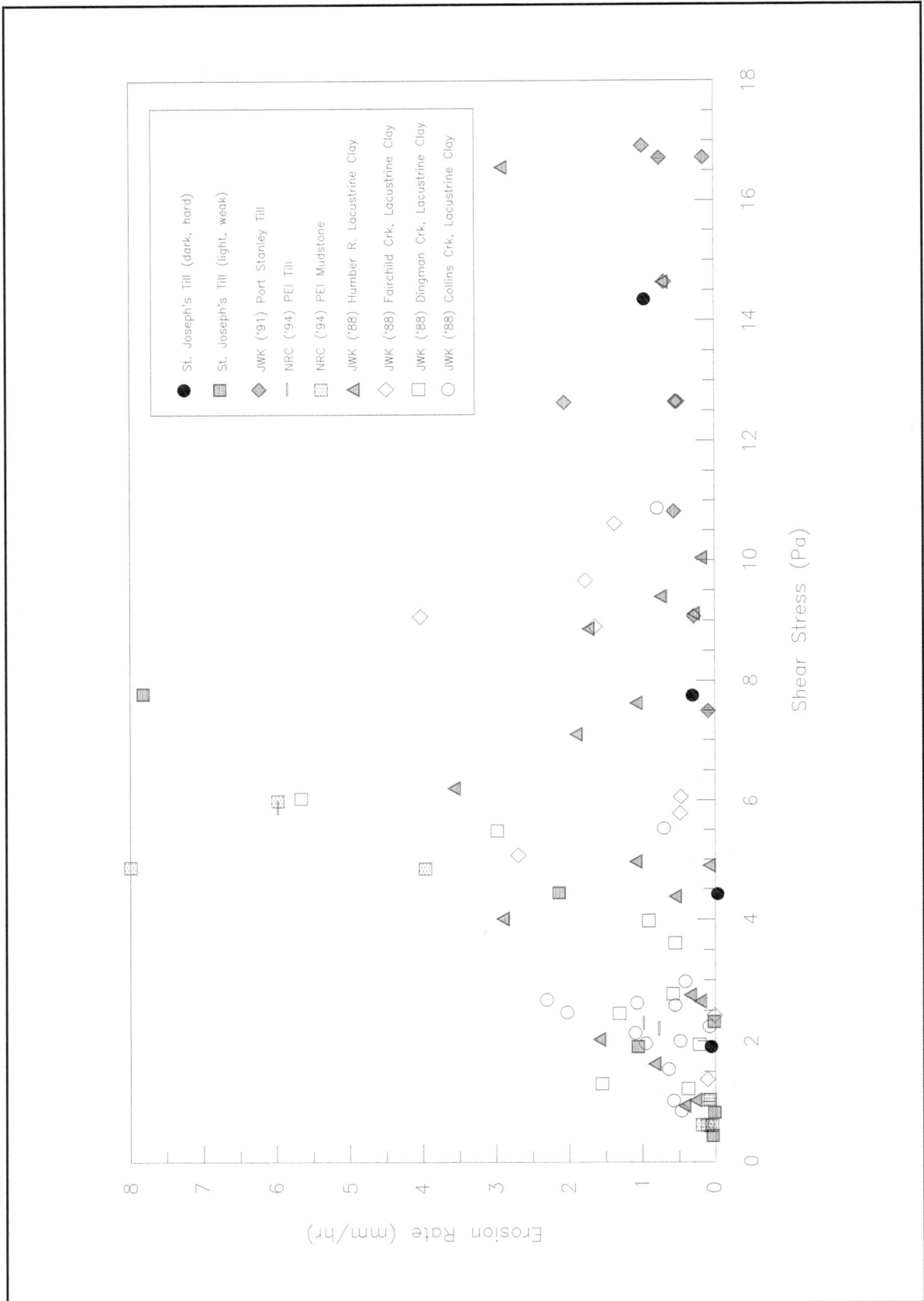

Figure III-5-18.   Clear-water erosion rates from unidirectional flow flume and tunnel tests for various materials

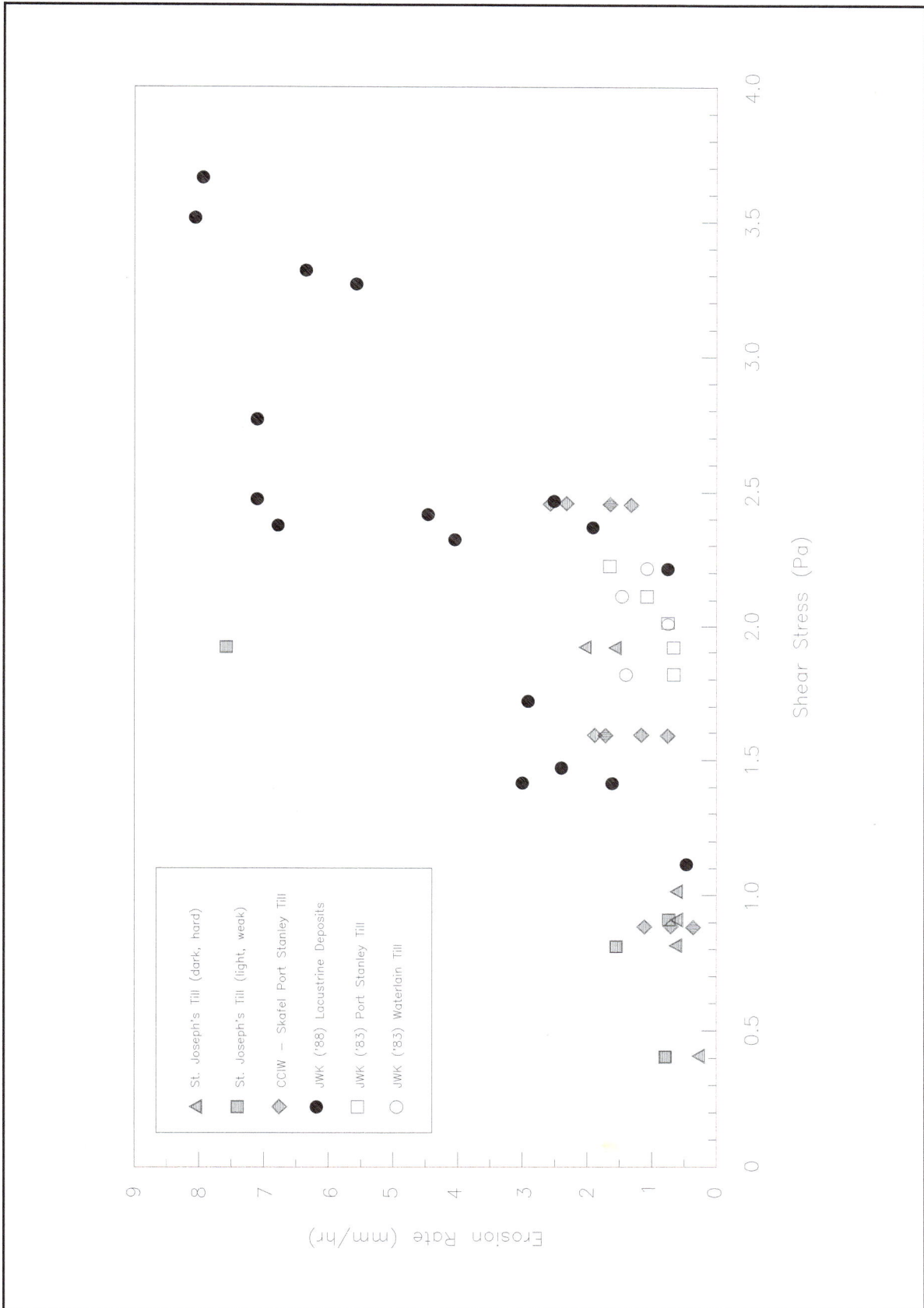

Figure III-5-19. Sand in flow erosion rates from unidirectional flow flume and tunnel tests for various materials

sand is introduced into the flow. It should be noted that the sand-grain size was selected so that sand transport occurred in a saltating (bed load) manner in the tests.

(9)   A close review of the results presented in Figure III-5-18 reveals two populations of erosion response. The less erosion-resistant group consists of soft cohesive sediments as well as sediments characterized with a high degree of fracturing and/or the presence of silt seams. In contrast, the more erosion-resistant group consisted of firm, homogenous cohesive sediments. The shear strengths for the material tested ranged from 50 to 200 kPa.

(10)   Some of the most recent research into the erodibility of cohesive sediments in the coastal environment has focused on the potential for softening of the exposed surface layer. Davidson-Arnott and Langham (1995) report that at the western end of Lake Ontario, previously overconsolidated till featured shear strengths in the range of 10 to 100 kPa. The softer sediment (10- to 20-kPa shear strength range) was associated with deeper water areas where the till was always exposed (i.e., the noncohesive sediment cover was insignificant) and where the frequency of erosion events was low. In contrast, in shallower areas (depths less than 3 m or 10 ft), the shear strengths were found to be in the range of 30 to 80 kPa. In the shallower depths, the lake bed is exposed to erosion events more frequently and sand cover thicknesses were greater. Using an adapted micro shear vane for a location in a water depth of 3.5 m, they found that the shear strength was lower in the upper 0.1 m of the lake bed and much lower in a 10-mm-thick surface layer. In fresh water, the softening process may be a result of cyclic loading by wave action, whereas in a seawater environment, factors such as salinity and biological activity (such as burrowing organisms) may also be important (Hutchinson 1986).

(11)   Finally, an indirect approach to determining the erodibility of consolidated sediment at a specific site is through the use of numerical models to 'back-calculate' the erodibility coefficient through an analysis of environmental conditions (wave action and water levels) and the observed shoreline or lake bed erosion over a given period of time. This approach is discussed in more detail in Part III-5-12.

c. Subaqueous erosion of consolidated sediments.

(1) In this section, the underwater erosion process is described. In the previous section, we explained that the erosion of hard cohesive soils consists of the destruction of bonds between clay particles and the natural structure or framework created through consolidation of the soil matrix. Erosion of consolidated sediment is irreversible. Once the sediment, which often consists of 80 to 90 percent fines, is eroded, it cannot be reconstituted in its consolidated form in the littoral zone. The eroded fines (silt and clay) are winnowed, carried offshore, and deposited in deep water in contrast to the small fraction of sand and gravel, which remains in the littoral zone. Therefore, the erosion of cohesive sediment is fundamentally different from the erosion of noncohesive sediment. In the latter case, for every volume of eroded sand, a large portion of that material will be deposited somewhere in the littoral zone (in some specific instances, onshore or offshore losses of sand from beaches can occur). Therefore, on sandy shores, the process of erosion is reversible.

(2) An extensive study of nearshore profiles on the north central shore of Lake Erie described by Philpott (1984) revealed that the profile shape remained relatively constant over an 80-year interval despite dramatic shore recession. This led Philpott (1984) to conclude that the controlling process in bluff or cliff recession on cohesive shores is not restricted to wave action at the toe (as proposed by Sunamura (1992) for eroding rocky coasts) but by the erosion of the nearshore profile by waves. Boyd (1992) cites many earlier references that also suggest that the nearshore has a controlling influence on shoreline recession. The shoreward shift of the dynamic equilibrium profile implies that erosion or downcutting is proportional to the gradient of the nearshore profile and is, thus, greatest close to shore. Davidson-Arnott (1986) describes field measurements of downcutting for a till profile (through the deployment of micro-erosion meters across a

transect) at a site near Grimsby on Lake Ontario. The results confirm the hypothesis on downcutting, i.e., the rates increase towards the shore in a manner related to the local bed slope, thus allowing for the preservation of the profile shape as it shifts shoreward with time. The downcutting hypothesis has now been confirmed by many other field investigations including 9 years of profile retreat data at Maumee Bay State Park in Ohio (Fuller 1995). Hutchinson (1986) and Sunamura (1992) also note that the rate of lowering or downwasting of the intertidal platform on erodible rocky coasts probably determines the long-term rate of cliff retreat in most instances.

(3) In general, it has also been shown that the underlying cohesive profile, for cases where the properties of the cohesive sediment are uniform along the profile, follows an equilibrium profile shape as defined by Dean (1977). Kamphuis (1990) went on to show that the specific exponential shape of the equilibrium profile at a cohesive shore site was related to the grain size of the overlying noncohesive sediment. In other words, the profile shape could be determined using the grain size of the sand veneer to define the sediment scale parameter A (see Equation 3-15). The shape of the profile can also be influenced by other factors such as variable stratigraphy and the presence of lag deposits as discussed in Part III-5-5. In these cases, a smooth equilibrium profile with an exponential shape will not exist, at least not over the full profile. Riggs, Cleary, and Snyder (1995) note that in most instances along the North Carolina coasts, the complexity of the underlying stratigraphy is such that the profile rarely resembles the equilibrium form found on truly sandy shores.

(4) The downcutting process is illustrated in Figure III-5-20 for a cohesive shore site located east of Toronto along the Scarborough Bluffs. The bluff face has retreated approximately 30 m (100 ft) in a 37-year period. The underlying cohesive profile shape in 1952 is very similar to that in 1989; it has simply shifted shoreward by 30 m. Therefore, the long-term bluff or cliff retreat rate is equivalent to the profile retreat rate. This figure also shows that there can be a significant quantity of sand covering an underlying cohesive profile. The position of the underlying cohesive profile shown in Figure III-5-20 was estimated based on observations that the cohesive sediment is usually exposed or very thinly covered in the troughs between the bars. Also, it is known that the till is exposed at the toe of the bluff (i.e., at the back of the beach).

(5) Figure III-5-20 demonstrates that there is not a cross-shore balance of erosion and deposition. All of the eroded material from the cohesive profile and the bluff is either winnowed offshore (clay and silt fractions) or transported alongshore (sand and gravel fractions).

(6) The profile retreat model for cohesive shores implies that: the amount that the driving forces for erosion exceed the resisting forces is inversely proportional to the water depth. In other words, the most active subaqueous erosion occurs at the shoreline. In general, it may be assumed that the erosion resistance of the cohesive sediment is consistent across the profile (if anything, the sediment may be less erosion-resistant in deeper water due to the increased role of softening). Therefore, the driving force for erosion must increase in the shoreward direction. These observations provide important evidence on the nature of the driving forces for cohesive profile erosion.

(7) Coakley, Rukavina, and Zeman (1986) proposed that outside the surf zone, the downcutting process is driven by shear stresses generated by the orbital motion under waves. Outside the surf zone, this driving force is inversely proportional to water depth. However, considering that wave heights and the related orbital velocity decreases in the surf zone, this mechanism cannot explain the inverse relationship between depth and driving force in this zone. Nairn, Pinchin, and Philpott (1986) proposed that the complex combination of driving forces in the surf zone may be represented by the rate of energy dissipation (described by the rate of wave height decay). Using a model of wave energy dissipation for random waves, it was shown that the rate of energy dissipation was directly proportional to the rate of downcutting in the surf zone. In the surf

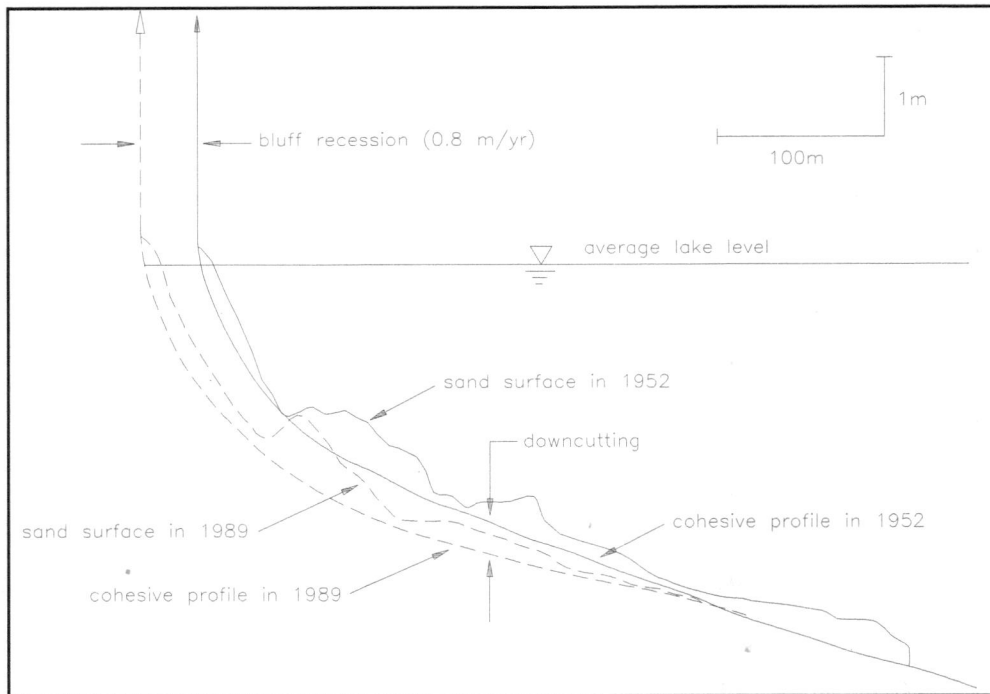

Figure III-5-20.   Bluff retreat and profile downcutting over a 37-year period at Scarborough Bluffs, located east of Toronto on Lake Ontario

zone, the rate of energy dissipation provides a good indicator of the possible driving forces such as intensity of flows (e.g., undertow and alongshore currents) as well as the intensity of near-bed turbulence (i.e., plunging breakers, which generate high near-bed turbulence, are associated with rapid wave decay). The erosive nature of plunging breakers (and the associated near-bed turbulence) on cohesive profile erosion was demonstrated through laboratory experiments performed in a wave flume with intact cohesive samples (Skafel 1995).

(8) Sand acts as an abrasive agent in the erosion of consolidated shores. Referring to Figure III-5-20, it is also likely that the presence of alongshore bars can have the opposite effect of protecting the underlying cohesive sediment from exposure and subsequent downcutting. As these bars migrate with changing wave and water level conditions, different areas of the underlying cohesive profile become exposed in the troughs between the bars. The influence of the quantity and mobility of sand cover over a cohesive profile is explored in greater detail in Part III-5-2.

(9) At the boundary between the action of subaqueous and subaerial erosion processes, the bluff or cliff toe can in some instances experience notching. This notching typically occurs in more competent materials such as rock, frozen sediments, and harder cohesive sediments, (i.e., in materials that are capable of withstanding failure when undercut).

(10)   In summary, the fundamental principles of consolidated cohesive shore erosion are:

(a) The erosion of consolidated cohesive sediment is irreversible.

(b) The long-term rate of shoreline retreat is directly related to the rate of nearshore downcutting and the associated profile retreat.

Erosion, Transport, and Deposition of Cohesive Sediments

(c) The local rate of downcutting is proportional to the gradient of the nearshore slope at any location across the profile.

(d) In addition to acting as an abrasive agent that accelerates consolidated cohesive sediment erosion, sand can also serve to protect an underlying cohesive profile from erosion.

d. Subaqueous erosion of mud. The Parthenaides Equation for mud erosion (Parthenaides 1962, Mehta et al. 1989) is an excess shear equation describing the erosion rate of a cohesive sediment:

$$\frac{dm}{dt} = M_p \frac{(\tau_c - \tau)}{\tau_c}$$

(III-5-1)

where

m = mass of sediment on the bed, kg/m^2 (lb/ft^2)

t = time, sec

τ = bed shear, Pa (lbf/ft^2)

τ_c = critical shear for erosion, Pa (psf) (lbf/ft^2)

M_p = 'Parthenaides Coefficient,' erosion rate at twice τ_c, kg/m^2/sec (lb/ft^2/sec)

e. Fluid mud.

(1) This assumes a well-defined interface between bed and water column: sediment on the bed remains at rest; while that in the water column moves with the water. Often, cohesive sediment forms an intermediate layer of fluid mud (denser than water; less dense than the bed; still capable of motion, but slower than the ambient flow). Fluid mud layers are frequently found near the shoreline, where wave activity can 'pump up' excess pore pressures within the fluid mud mass, slowing down drainage and consolidation. Hydrographers generally define fluid mud as having a density of less than 1,100 to 1,200 kg/m^3 (70 to 75 lb/ft^3).

(2) Erosion processes in fluid mud are similar to mixing processes at a salt-wedge interface. A layer of lighter fluid (water with suspended sediment) flowing above the denser fluid mud eventually induces waves in the interface when a critical Richardson Number or densimetric Froude Number is exceeded. Wind-wave activity in the upper layer increases the interfacial waves, diverting energy from the surface waves. As the difference in flow rates increases, the interfacial wave energy increases until breaking occurs, putting some fluid mud back in suspension and entraining clearer water in the fluid mud.

(3) The densimetric Froude Number is defined as:

$$F_r = \sqrt{\frac{(V_w - V_{fm})^2 \rho_{fm}}{g h_{fm}(\rho_{fm} - \rho_w)}}$$

(III-5-2)

where

V_w = mean velocity in the water layer, m/sec (fps), above

V_{fm} = mean velocity in the fluid mud layer, m/sec (fps)

ρ_w = density of the water layer, kg/m^3 (lb/ft^3)

ρ_{fm} = density of the fluid mud layer, kg/m^3 (lb/ft^3)

g = acceleration due to gravity, m/sec^2 (ft/sec^2)

h_{fm} = thickness (depth) of fluid mud layer, m (ft)

(4) This reduces to the more familiar Froude Number at the water surface, where the upper layer is air of relatively negligible density and the lower fluid mud layer is the water. It defines the ratio of the velocity differential at the interface (V_w - V_{fm}) to the celerity of a small gravity wave in the interface. When that ratio is unity, the equivalent of a hydraulic jump (breaking wave) may be expected in the interface, with resultant entrainment of fluid mud in the water (erosion of the fluid mud) balanced by entrainment of water in the fluid mud.

(5) At the interface between the fluid mud and the bed, the fluid mud protects the bed from erosion. Shear that develops between the mud and the bed is generally too low to entrain stationary particles. The processes at this interface are entirely deposition, as water drains from the fluid mud.

f. Subaerial erosion processes.

(1) Subaerial erosion processes on cohesive shores do not necessarily have anything to do with the air or wind, although strengthening of mud flats has been noted by Amos et al. (1992) due to evaporation at low water in the macrotidal Bay of Fundy. On consolidated shores, the primary subaerial erosion process is slumping of oversteep bluffs or cliffs.

(2) As stated earlier, the long-term bluff or cliff retreat rate is determined by the rate of profile downcutting. In a review of shoreline erosion data from the Lake Erie shoreline, Kamphuis (1987) points out that cliff height does not exert much influence on the process (in fact, a distinct lack of correlation was noted) because erosional debris from a shore cliff is quickly swept away, winnowed offshore, and deposited in deep water. Exceptions to this generalization include locations where the debris is not easily removed from the toe of the cliff (e.g., in the case of eroded rock cliffs or blocks of frozen sediment along Arctic shores). The primary reason for slope failures along a cohesive shore is the oversteepened nature of the slope owing to the ongoing profile and toe erosion. Nevertheless, even though subaerial processes do not determine the long-term rate of shoreline recession on cohesive shores (i.e., the frequency of slope failures), these processes are critical in determining when and where a failure will occur.

(3) Slope stability is a function of the balance between the downward force of gravity and the strength of the geologic materials in a bluff or cliff. The strength of the geologic materials depends on the cohesion of particles and the presence or absence of groundwater. The stratigraphy of a bluff or cliff can have a significant influence on slope stability. Weak clay layers can provide slip planes for slope failures or serve to confine groundwater flows, which may appear as springs at the bluff face. Where groundwater exits the bluff face, seepage erosion can occur. Also, depending on the sequence of layering, groundwater flows can act to increase pressures within the slope and contribute to instability. This will often occur when seepage pathways at the bluff face are blocked by talus from a slide further up the slope. In some instances, the

combination of surface runoff and seepage can lead to the development of large gullies. Slope failures may be classified as: falls and topples; rotational (i.e., circular) and translational slides; and, spreads and flows. The type of failure is a function of the geologic conditions at the site. There are a variety of methods for assessing slope stability.

(4) Figures III-5-21 to III-5-23 show some examples of eroding bluffs along the Great Lakes. Recent rotational failures are evident in Figure III-5-21 along the north central shore of Lake Erie. The development of a gulley is shown in Figure III-5-22 from another location on the north central shore of Lake Erie. Figure III-5-23 shows an eroding shale bluff along the western Lake Ontario shoreline.

Figure III-5-21. A rotational bluff failure along the north central shore of Lake Erie

(5) In addition to slope stability, surface erosion of the cliff or bluff face can have a secondary influence on the overall erosion of the feature. Surface erosion results from runoff, seepage, rain, and spray from wave action. This would be one of the key processes leading to the erosion of the shale bluff shown in Figure III-5-23.

(6) Edil and Bosscher (1988) present a Great Lakes perspective overview of forces and resistance influencing cohesive shore slope erosion which result in mass movement (including sliding, flow, and creep) and particle movement (including wave, wind, ice, rill and sheet erosion and sapping through seepage flow).

(7) Kuhn and Osborne (1987) investigated the recession of cohesive cliffs on the California coast. The short-term cliff recession is partly related to subaerial processes such as drainage of precipitation and the effects of urbanization at the clifftop. At these locations the cliff base is well protected by a substantial beach deposit which restricts the downcutting of the subaqueous profile. Notably, almost all of the sand at these sites is supplied by nearby rivers and not by cliff recession. This is in contrast to the Great Lakes and barrier islands on the U.S. east coast, where very little sand is supplied by rivers in most areas.

Figure III-5-22. Gully erosion of a bluff along the north central shore of Lake Erie

Figure III-5-23. An eroding shale bluff along the west Lake Ontario shoreline

Erosion, Transport, and Deposition of Cohesive Sediments

(8) A cliff stabilization project for the Scarborough Bluffs (east of Toronto on Lake Ontario) is discussed by Parker, Matich, and Denney (1986) (Figure III-5-24). It has been recognized that arresting the foreshore downcutting at the base of the bluffs is not sufficient on its own to control the erosion of an oversteep slope; proper drainage systems must be implemented to prevent gullies from developing and to address the problem of piping through sand and silt lenses. Parker, Matich, and Denney (1986) also note that the toe of the bluff must be surcharged in some locations to prevent large slip failures from occurring. One function of the land base created at the toe of the slope in Figure III-5-24 is to provide sufficient area to construct a surcharge berm. The surcharge provided by extensive shingle beaches along the south coast of England is known to contribute to the stability of cliffs that rise behind the beaches (Fleming and Summers 1986).

Figure III-5-24. Shore protection consisting of a wide berm protected by a revetment along the base of the Scarborough Bluffs located east of Toronto on Lake Ontario

Bioengineering, which consists of promoting the growth of vegetation (e.g., through the placement of mats consisting of bundled twigs to enhance rooting), may also help to stabilize an oversteep slope.

(9) Another extensive review of subaerial processes was made by Hutchinson (1986); other processes that he identified included: freeze/thaw cycles; alternate wetting and drying; and mechanical and hydrodynamic effects of micro-geological features such as erratic cobbles and boring organisms. Hutchinson (1986) provides sample test results indicating that seawater penetrated the pores of glacial tills from the Holderness coast of England (along the North Sea) to depths of at least 0.85 m. He goes on to suggest that an increase in the concentration of NaCl in the pore water from the intrusion of the seawater may increase the net attractive forces between clay particles and increase the degree of aggregation. The degree to which this effect will occur will depend on the clay content and the chemical properties of the cohesive sediment. Hutchinson (1986) concludes that the opposite effect may occur along freshwater shores, where the intrusion of fresh water may dilute the salt or cation content, thus decreasing the net attractive forces between clay particles, and increasing the susceptibility to erosion.

III-5-8. Transport Processes

a. Advection and dispersion.

(1) Cohesive sediments are not transported as bed load, except in the form of fluid mud (see Part III-5-8b below). They almost always are transported in suspension: advected (carried with the ambient water at the flow velocity) and dispersed (moved from areas of high sediment concentration to low by mixing, such as turbulence). Advection and dispersion are described in Part II-6-2.

(2) But sediment, even a floc, is denser than the ambient water and hence settles as it is being advected and dispersed. This results in a downward bias of the vertical dispersion relation

$$S - Cw = -D_z \frac{dC}{dz} \tag{III-5-3}$$

where

S = vertical (upward) dispersion of sediment, kg/m²/sec (lb/ft²/sec)

w = settling velocity of floc, m/sec (ft/sec)

D_z = vertical dispersion coefficient, m²/s (ft²/sec)

C = suspended sediment concentration, kg/m³ (lb/ft³)

z = vertical dimension, m (ft)

b. Fluid mud. Fluid mud, as its name suggests, flows and is a mechanism for transport of cohesive sediment. It flows down slopes by gravity, sometimes referred to as a 'turbidity current,' which is why fluid mud is often found in the bottoms of dredged cuts. Fluid mud is also dragged along by the shear of the water flowing above it. How it flows is determined by its rheology; which in turn, must be measured for each fluid mud combination of sediment and water. Fluid mud may have an apparent yield point, remaining stable until a critical slope or shear is exceeded. More likely, its flow velocity will vary in a nonlinear way with slope or shear, from stiffest at lowest shear to most fluid at highest shear.

III-5-9. Deposition Processes

a. Flocculation.

(1) Cohesive sediments rarely settle as individual grains in nature. Collisions between sediment grains are encouraged by differences in settling velocity, turbulence, Brownian motion, and electrochemical attraction or cohesion.

(2) When cohesive grains collide they tend to stick together, or cohere. To determine settling velocity in the laboratory, cohesive grains can be kept apart in distilled water containing a dispersing agent to neutralize the electrochemical bond.

(3) The process by which individual cohesive particles agglomerate while settling is called *flocculation*; and the resulting large particles with entrapped water, *flocs*. The settling velocity of a floc is a function of its size, shape, and relative density. A floc usually settles faster than its constituent particles; but because of

the entrapped water, its density is less than that of the sediment mineral, and the settling velocity of the floc may actually be slower than that of an individual clay particle. The size and shape of flocs, and their settling velocity, are hydrodynamic sediment properties which must be measured or determined by model calibration as described in Part III-5-3.

b. Shear stress. The principle of excess shear is also used for correlating observed rates of cohesive sediment deposition with flow. But for deposition, 'excess' is the amount by which the shear τ is less than a critical shear for deposition τ_s.

c. Krone Equation. The Krone Equation for mud deposition (Krone 1962, Mehta et al. 1989) is as follows:

$$\frac{dm}{dt} = Cw\frac{(\tau_s - \tau)}{\tau_s} \qquad \text{(III-5-4)}$$

where

m = mass of sediment on the bed, kg/m^2 (lb/ft^2)

t = time, sec

τ = bed shear, Pa (lbf/ft^2)

τ_s = critical shear for deposition, Pa (lbf/ft^2)

C = suspended sediment concentration above the bed, kg/m^3 (lb/ft^3)

w = settling velocity of sediment floc, m/sec (ft/sec)

Comparison of the Krone and Parthenaides Equations (Part III-5-4d) suggests that the Krone Equation may be less empirical, more theoretical. There is no obvious empirical coefficient to match the Parthenaides coefficient M_p, but the unknown coefficient hidden in the Krone Equation is w, the settling velocity of the flocs.

d. Fluid mud. Deposition takes place at both interfaces: that between water and fluid mud; and also that between fluid mud and the stationary bed. At the water/fluid mud interface, the process can still be predicted by the Krone Equation, except the density of the deposited sediment is less than 1,100 to 1,200 kg/m^3 (70 to 75 lb/ft^3). At the fluid mud/bed interface, the deposition process is one of consolidation: as the fluid mud drains, it consolidates to the point where it is too dense to remain fluid.

III-5-10. Consolidation

a. Strength versus consolidation.

(1) The critical shear for erosion τ_c is a function of the consolidation or density of the bed. Think of the cohesion between sediment particles varying inversely, like the force of gravity, with distance between particles. The closer the particles are to each other, the stronger the cohesive bond and the greater the shear force needed to separate them.

(2) The Migniot Equation expresses the exponential relationship between mud density and critical shear for erosion as:

$$\tau_c = N\rho_s^{\ M} \tag{III-5-5}$$

where

ρ_s = bulk density of sediment on the surface of the bed, kg/m^3 (lb/ft^3)

M, N = constants to be determined for the sediment and water.
M is dimensionless and tends to be less than 1; N equals 10^{-1} or 10^{-2} in the SI system of units.

b. Degree of consolidation. The Terzaghi consolidation relation, developed for building settlement calculations, also serves to illustrate the consolidation of coastal cohesive sediment.

$$u = 0.964 \left(\frac{tC_v}{P^2} \right)^{0.415} \tag{III-5-6}$$

where

u = degree of consolidation, dimensionless ratio

C_v = a consolidation coefficient to be determined, m^2/sec (ft^2/sec) (values on the order of 1 x 10-5 m^2/sec are not uncommon)

P = length of the drainage path, m (ft) (generally depth of burial)

III-5-11. Wave Propagation

a. Roughness and shear.

(1) The predominant nearshore wave transformation associated with muddy beds is wave energy dissipation or attenuation. Refraction, diffraction, and reflection all pretty much obey the rules set out in Part II-3-3, but wave attenuation is generally much greater over mud beaches than over sand and gravel. As more wave energy is absorbed by the mud, less reaches the breaker line.

(2) This energy dissipation can only partially be accounted for through the traditional mechanisms of bed roughness and friction. In fact, a mud bed is usually smoother (less rough) than sand.

b. Fluid mud.

(1) The predominant mechanism of wave attenuation is in the thick, viscous boundary layer of fluid mud (Lee 1995, Lee and Mehta 1994). Part of the wave energy goes to 'pumping up' excess pore pressures maintaining the mud in a fluid state. But more is converted to work done in moving the fluid mud against viscous shear.

Example Problem III-5-1

FIND:

Apply the Parthenaides and Krone Equations to the annular flume test results in the table, to determine values of:

M_p = Parthenaides coefficient for this combination of sediment and water

τ_c = Critical shear for erosion at this sediment density

w = Settling velocity of sediment flocs for this combination of sediment and water

τ_s = Critical shear for deposition for this combination of sediment and water

For the purposes of this example, ignore changes in bed elevation and water depth due to erosion and deposition.

GIVEN:

Table III-5-2
Example Problem III-5-1, "Annular Flume Test Results"

Applied Shear (Pa)	Duration (sec)	Starting Concentration (kg/m³)	End Concentration (kg/m³)
0	600	0	0
0.2	600	0	0
0.4	600	0	0
0.6	600	0	1
0.8	600	1	3.33
1.0	600	3.33	7
0.8	600	7	7
0.6	600	7	7
0.4	600	7	7
0.2	600	7	1
0	50	1	0

Bulk (dry) density of sediment on the bed = 1,500 kg/m³

Water density = 1,020 kg/m³

Area of annular flume bed = 1 m²

Water depth = 0.2 m

Total water volume, therefore = 0.2 m³

(Sheet 1 of 3)

Example Problem III-5-1 (Continued)

SOLUTION:

Since the bed area of the annular flume has been carefully chosen to be 1 m^2, the rate of change of total sediment suspension will directly give us the quantity dm_b/dt. Otherwise, we would have had to multiply both sides of both equations by the area of the bed.

Our results are not given as total sediment in suspension, but as concentration in kg/m^3. We need to multiply them by the volume of water in the annular flume, 0.2 m^3, to get total sediment numbers equivalent to dm/dt on the bed.

So, for example, when the applied shear was 0.6 Pa

$$dm/dt = (V \times \Delta C) / (A \times \Delta T) = (0.2 \text{ m}^3 \times -1 \text{ kg/m}^3) / (1 \text{ m}^2 \times 600 \text{ sec})$$
$$= -3.33 \times 10^{-4} \text{ kg/m}^2\text{-sec}$$

Similarly for 0.8 Pa, -7.77 x 10^{-4} kg/m^2/sec, and for 1.0 Pa, -1.22 x 10^{-3} kg/m^2-sec.

This concludes the erosion data for the Parthenaides Equation. By plotting erosion rate versus shear (Figure III-5-25), we can extrapolate back to the shear at which the erosion rate is zero, $\tau_c = 0.45$ Pa. From the same plot we can read M_p, the erosion rate at $2\tau_c = 0.9$ Pa, $= 10^{-3}$ kg/m^2-sec.

In the same way, the deposition results can be used as shear is reduced, with the Krone Equation for
    0.2 Pa and $C_{avg} = $ 4.0 kg/m^3, dm/dt = $ 2 \times 10^{-3}$ kg/m^2-sec
and for
    0.0 Pa and   0.5 kg/m^3,   dm/dt = 4 x 10^{-3} kg/m^2-sec

Note that we can use the mean sediment concentration C_{avg}, since deposition rate is a linear function of concentration. In high concentration, settling velocity also becomes an inverse function of concentration.

The result at 0.0 Pa gives the floc settling velocity directly

$$w = (dm/dt) / (C_{avg} \times \tau_s / \tau_s) = (4 \times 10^{-3} \text{ kg/m}^2\text{-sec}) / (0.5 \text{ kg/m}^3) = 8 \times 10^{-3} \text{ m/sec}$$

Substituting this in the Krone Equation at 0.2 Pa, yields

$$\tau_s = (\tau \times C_{avg} \times w) / (C_{avg} \times w - dm/dt)$$
$$= (0.2 \text{ Pa} \times 4.0 \text{ kg/m}^3 \times 8 \times 10^{-3} \text{ m/sec}) / (4.0 \text{ kg/m}^3 \times 8 \times 10^{-3} \text{ m/sec} - 2 \times 10^{-3} \text{ kg/m}^2\text{-sec})$$
$$= 0.21 \text{ Pa}$$

A real set of data, whether from field or laboratory annular flume, will be more difficult to work with. Complications to be expected are:

The variation of erosion rate with bed shear will not be linear, but will require some judgment and nonlinear regression to determine τ_c.

(Sheet 2 of 3)

Example Problem III-5-1 (Concluded)

You will never again have the luxury of somebody timing the settlement of the last floc, giving w directly. Instead, you will have to solve two or more equations with two or more unknowns, one of which is w.

Bed sediment density, and hence τ_c, varies with depth of erosion due to consolidation.

A natural bed will be layered, with discontinuities in consolidation and sedimentology over depth.

Water volume and depth vary with depth of erosion.

Direct measurement of bed shear is often difficult, particularly in the field laboratory where rotation of the lid of the annular flume is resisted by fluid shear, both in the flume and on the upper surface.

The area of the bed in the direct shear device (annular flume in this example) is seldom a convenient 1 m^2 or 1 ft^2.

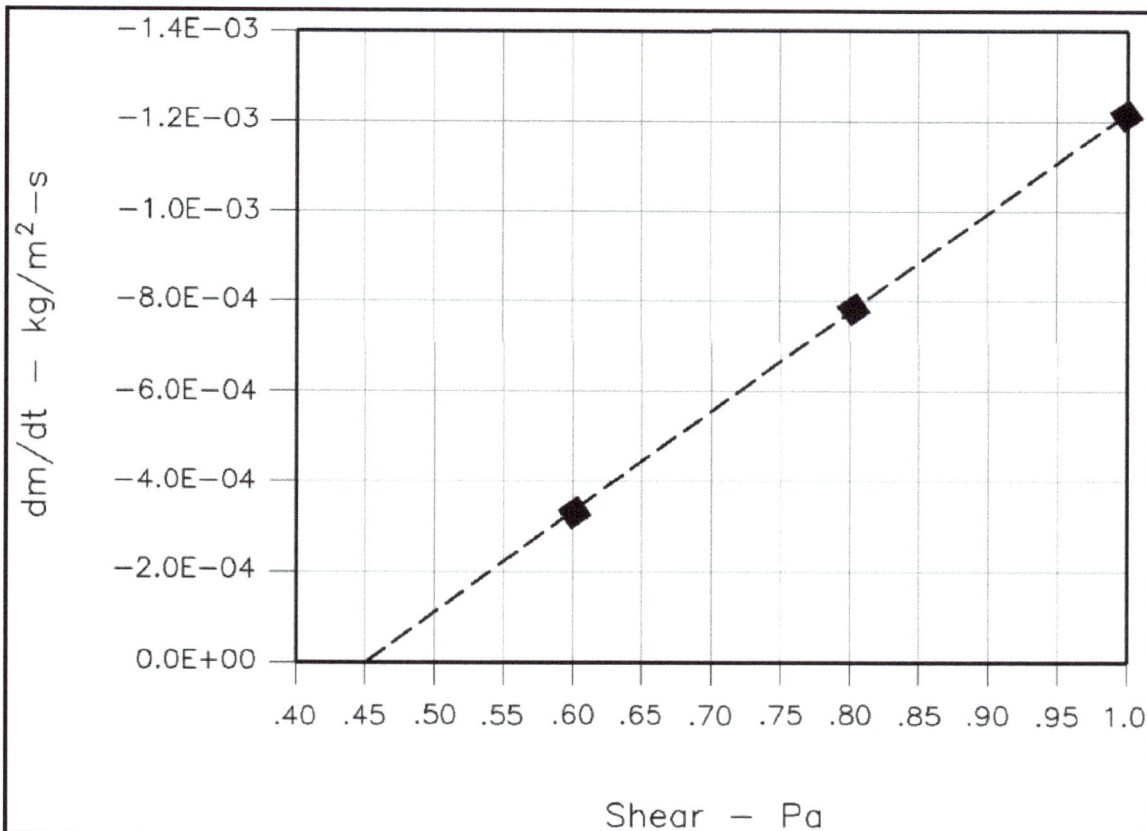

Figure III-5-25. Plot of erosion rate versus shear for example problem

(Sheet 3 of 3)

Figure III-5-26 (Lee 1995) summarizes the interactions between a mud shore and waves:

(a) At low water or under calm seas, the profile formed by the previous wave condition consolidates.

(b) Breaking waves result in mass erosion at the breaker line, with high turbidity in the surf zone; lower turbidity and surface erosion occur seaward of the breaker line, to the point where bed shear is less than the critical shear for erosion.

(c) Turbidity in the surf zone reaches that of fluid mud, flowing generally downhill offshore and collecting in troughs, under the influence of the wave orbital motion within the fluid mud.

(d) Return to (1), consolidation of the new profile under calm seas or at low water.

(3) Lee (1995) and Lee and Mehta (1994) found that wave height decays exponentially across the profile, i.e.

$$H_y = H_0 e^{-k_i y}$$
(III-5-7)

where

y = distance along wave ray, m (ft), positive towards shore

H_0 = incident wave height at $y = 0$, m (ft)

H_y = wave height at y, m (ft)

k_i = wave height attenuation coefficient, m^{-1} (ft^{-1})

(4) The wave height attenuation coefficient k_i characterizes the fluidization potential of the mud. It is obviously a function of the rheology of the bed; and also probably of the wave period T and incident wave height H_0. Lee (1995) and Lee and Mehta (1994) believe k_i to be a function of the local wave height, and talk of a spatial mean $\overline{k}_i$ across the profile. In practice, $\overline{k}_i$ is another calibration coefficient, like the Parthenaides coefficient, to be determined by fitting to observations of wave attenuation and profile development. A high $\overline{k}_i$ indicates a thick fluid mud layer, which Lee (1995) and Lee and Mehta (1994) associate with offshore flow on an eroding profile; a low $\overline{k}_i$, with an accreting profile. Lee (1995) quotes values of $\overline{k}_i$ in the range $0.0001 \leq k_i \leq 0.05$.

III-5-12. Numerical Modeling

a. *Introduction.*

(1) In this section, the role of numerical modeling is discussed both as an engineering tool and a planning tool. The development of numerical modeling of cohesive shore erosion is not far advanced. Nevertheless, approaches that have been utilized are briefly summarized. Numerical models are generally applied to simulate a future outcome given a known set of input conditions. For cohesive shore applications, there are at least four possibilities where numerical models can provide valuable information as follows:

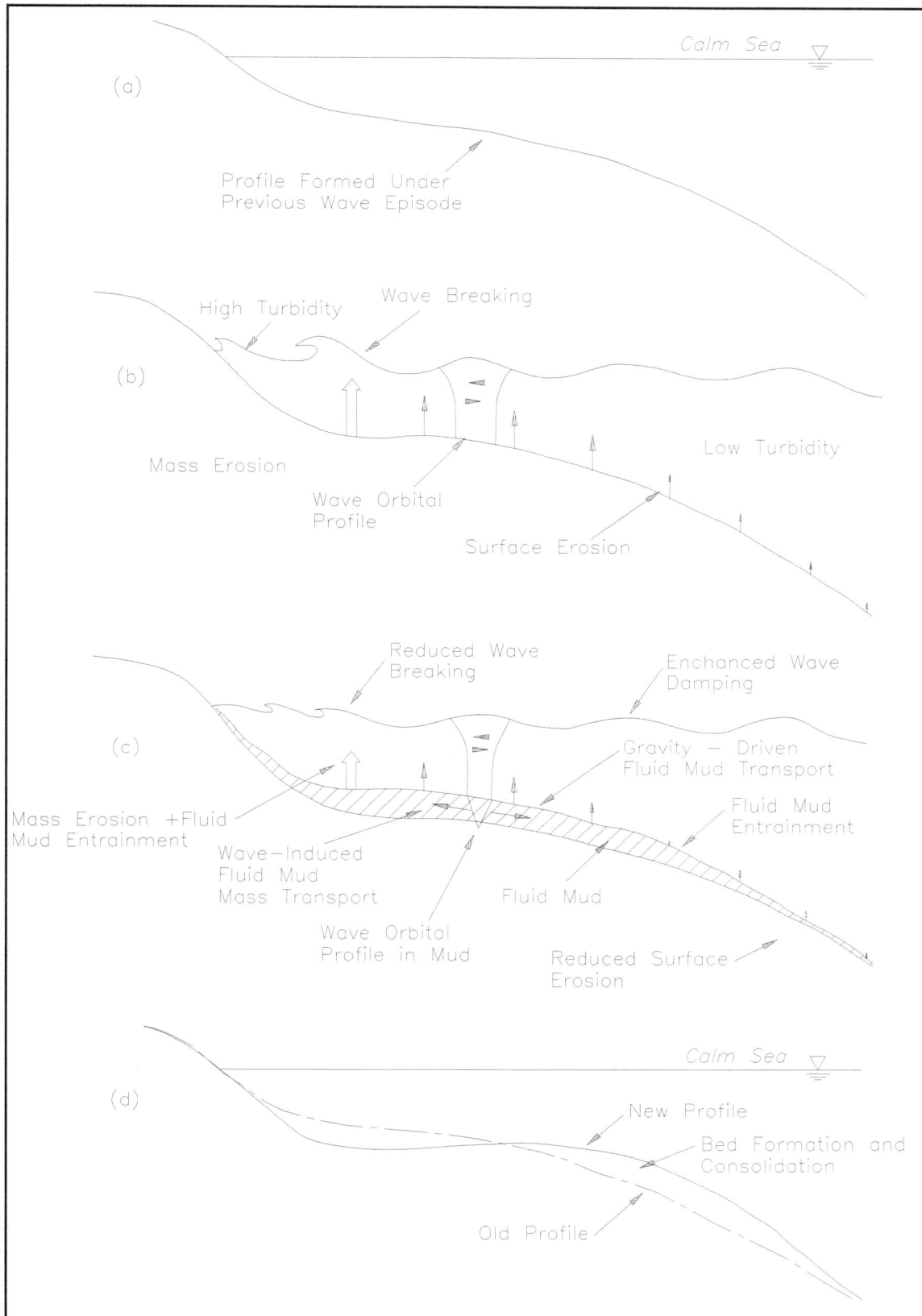

Figure III-5-26. Mud beach processes (after Lee (1995))

(a) To forecast long-term erosion rates (i.e., on the order of 50 years for developing shorelines). 'Developing shorelines' refer to areas where the evolution of a shore profile is at an early stage (e.g., newly created reservoirs). In these cases, the future rate of shoreline recession will probably differ from that experienced in the past owing to the evolving shape of the nearshore profile. Future erosion rates are required to plan setbacks and development around the new water bodies. In addition, the changing nature of the littoral zone (slopes and surficial substrate) may be important to the management of fish habitat.

(b) To forecast long-term erosion rates where either the environmental or the geologic conditions are changing due to natural or human causes. Natural influences that may change include increased storminess, sea level rise, changes to sand supply, or the changing geology encountered as the shoreline erodes. Human influences may include changes to water level fluctuations by water level regulation and reduction in the sand cover over the cohesive profile due to interruptions or reduction in sand supply to the littoral zone.

(c) To forecast the performance and potential impacts of coastal engineering works. For example, downcutting will continue offshore of the toe of a revetment constructed on a cohesive shore, and this outcome must be forecast and addressed in the design of the structure.

(d) Based on known historic recession rates and environmental conditions, numerical models can be applied to 'back-calculate' erodibility coefficients to describe the erosion resistance of a particular type of cohesive sediment. In turn, this can be used to assess future erosion rates at another site with the same sediment type but with different environmental conditions (i.e., the wave and water level climate).

(2) Therefore, there are many situations where numerical modeling would serve as an important tool for coastal engineering and coastal zone planning in cohesive shore environments.

b. Simulating erosion of consolidated sediment.

(1) Two approaches to modeling consolidated cohesive shore erosion are described in this section. Both rely on a significant degree of empiricism.

(2) The first approach is described by Penner (1993) for estimating future bank recession rates on Western Canadian lakes and reservoirs. This procedure consists of using a wave hindcast to determine the amount of wave energy that reaches a bluff face (accounting for losses over the nearshore profile). From this information and a knowledge of an erodibility coefficient, the future rate of shore recession can be determined. The erodibility coefficient relates volumetric erosion to the wave energy that reaches the bluff toe. This coefficient may be determined through back-calculation, or model calibration, for a location with known history of erosion (with the wave climate calculated by a wind-wave hindcast) or it may be estimated based on published values by others who have used this technique in the past (e.g., Penner 1993, Mollard 1986; Kachugin 1970; Newbury and McCullough 1984).

(3) Since this method essentially ignores the role of profile downcutting, it is primarily applicable to profiles that are at an early stage of development (i.e., with a very steep profile and narrow nearshore zone). In these cases, most of the wave energy is dissipated at the bluff face, in contrast to the more developed shoreline profiles on the Great Lakes and ocean coasts where the majority of wave energy is dissipated across the profile.

(4) This method has been applied to estimate the future shoreline position around new reservoirs in order to establish setback and ownership requirements. It has been refined to consider variations in erodibility of the materials encountered through time as the shoreline recedes. A table of erodibility coefficients from Penner (1993) is reproduced in Table III-5-3 below. The empirical coefficients relate wave energy that reaches the bluff toe to the weight of eroded sediment.

(5) Table III-5-3 reveals some interesting comparisons to the previous discussions of erosion processes. The high erodibility coefficient for the case of sand cover over the till confirms the importance of sand as an abrasive agent. For cases with a sand or gravel overlay, the erodibility coefficient is 2-10 times higher than the coefficients for soft and hard till. This compares well with the results of the laboratory experiments where erosion rates were increased by a factor of 3 to 8 with the introduction of sand to the flow for a range of hard and soft consolidated cohesive sediments (Part III-5-4b). Also, the category of a till bluff with a lag-protected nearshore corresponds to the slowly eroding convex profile type discussed above.

Table III-5-3
Erodibility Coefficients (from Penner (1993))

Material Type	Erodibility Coefficient (m^2/tonne)
Till with sand or gravel surficial overlay	0.0004
Soft till	0.0002
Hard till	0.00004
Till with a dense cover of surficial boulders	0.00002
Sandstone, siltstone, or shale covered with till	0.00005 to 0.0002

(6) A second approach to the numerical modeling of cohesive shore erosion is based on the conceptual model proposed by Nairn, Pinekin, and Philpott (1986). This approach relates downcutting to the shear stress generated by orbital velocities under unbroken waves and to the rate of wave energy dissipation for broken waves. The predicted downcutting determines the profile retreat rate which is assumed to determine the bluff or cliff retreat rate above water. A numerical model of coastal processes (Nairn and Southgate 1993) is used to determine the orbital velocity, the rate of wave energy dissipation, and the fraction of broken waves at locations across a nearshore profile. Erodibility coefficients relating downcutting rates to the conditions associated with broken and unbroken waves are determined in one of two ways. First, these coefficients can be back-calculated or calibrated based on a known profile retreat rate and the associated environmental conditions (i.e., the wave and water level conditions corresponding to the period of known profile retreat).

(7) Alternatively, intact cohesive sediment samples can be extracted from the study site and tested in a unidirectional flow flume as discussed in Part III-5-6a. The erodibility coefficients can then be determined based on known coefficients for sediments with a similar laboratory response. However, the latter approach is only valid for locations with similar sand cover conditions. The reason for this is that in long-term applications, the model ignores the presence of sand cover, and therefore the erodibility coefficient also accounts for the mobility and quantity of sand cover.

(8) A more detailed version of the model described in Nairn and Southgate (1993) has been developed to assess short-term erosion on cohesive profiles (i.e., over a period of weeks). This model accounts for the movement of limited quantities of sand over a hard surface. Erosion is activated in the model only where sediment transport occurs over an exposed segment of cohesive sediment. This model has been successfully tested against the laboratory wave flume data of Skafel and Bishop (1994) on cohesive profile erosion (Nairn and Southgate 1993).

(9) The numerical model described in Nairn and Southgate (1993) has been applied to assess the influence on erosion of modifications to the water level regulation on the Great Lakes. Stewart and Pope (1993) determined the potential future long-term (i.e., 50 to 100 years) profile retreat (and the associated shoreline erosion) rates for several different test sites. At each of these sites, the model was calibrated against a known profile retreat rate to determine the erodibility coefficients. These test sites featured a variety of wave exposures and profile types. For a potential lake level regulation scenario with a reduced range of water

level fluctuations (i.e., lower highs and higher lows), the long-term erosion on cohesive shores with convex profiles was found to be reduced, but the long-term erosion on concave profiles was unchanged.

   c. Simulating erosion and deposition of mud.

   (1) The numerical models described in *a.* above may also be used for erosion of mud shore, given a suitable choice of erodibility coefficient or critical shear for erosion. Nevertheless, these do not handle the other cohesive shore processes (transport, deposition, and consolidation), for which a full hydrodynamic model is required.

   (2) Willis and Crookshank (1994) describe a bed sediment 'module' designed to be called as a subroutine to a full 1-, 2-, or 3-D model of wave and current hydrodynamics on the shore. This module applies the equations presented in this chapter to compute the exchange of sediment between the bed and the water column, at each grid point in the model, at each time-step. Because of its 'add-on' nature, it cannot deal with the intermediate state of fluid mud, for which an integrated water-sediment 3-D model is required (e.g. Le Hir 1994).

III-5-13. Engineering and Management Implications

The differences between processes on sandy and cohesive shores are fundamental to successful planning and engineering on cohesive shores. This section provides a discussion of some of the important issues.

   a. Setbacks and cliff stablilization.

   (1) The two most important issues in the planning and management of cohesive shores relate to implementing setbacks for development and to managing human influences on the sediment supply. Successful coastal management can sometimes avoid the need for costly shore protection structures.

   (2) Many jurisdictions along U.S. shorelines impose a setback for new development consisting of some multiple of the average annual recession rate (e.g., 30 to 100 times the average annual recession rate). The purpose of the setback is to avoid the need for shore protection within the life of the new development, recognizing the irreversible and inevitable erosion that occurs along cohesive shores (and some sandy shores as well). However, this procedure may not be fully reliable, for several reasons. Most importantly, the environmental and geologic conditions that determine the recession rate can change through time both naturally (e.g., increased storminess, higher water levels, changes in erodibility as the shore erodes) and through human influences (e.g., by reducing the sand supply to the littoral zone).

   b. Vegetation.

   (1) Where possible, every effort should be made to replace vegetation (salt marsh, grass, mangrove) whose loss might have triggered or allowed mud shore instability (Part III-5-1b). The root system provides fiber reinforcement to the soil. Above the bed, the body of the vegetation protects the mud from shear due to currents and waves and creates a calm layer close to the bed to encourage deposition and settlement.

(2) Vegetation is required to halt mud erosion, but most vegetation cannot germinate and establish itself on an eroding or unstable shore. Some less natural form of shore protection is needed to create the conditions necessary for the vegetation to establish. This may be as simple as geotextile or willow mats imitating the root system until the natural roots can develop. But (temporary) bunds may be required to protect an area of germinating vegetation, at least until the plants can withstand currents and waves.

(3) Vegetation is also useful on consolidated cohesive shores. Brambles, for example, can help stabilize an eroding bluff or cliff against subaerial erosion while discouraging at least human traffic. However, this type of solution will only be effective in the long term if seabed or lake bed downcutting is halted.

c. Interruption to sediment supply and downdrift impacts.

(1) Another important planning requirement on cohesive shores is the management of sediment budget; specifically, human influences on sand supply. The erosion of cohesive shores results in a contribution of noncohesive sediment to the littoral sediment budget. This contribution consists of the sand and gravel-size fraction of the bluff and nearshore bottom material. There are three human influences that can reduce the sand supply:

(a) The construction of shore protection eliminates the contribution of the noncohesive fraction from the erosion of the bluff (and possibly part of the nearshore) for the protected reach of shore. On an individual-project basis, this reduction in the total supply to the littoral sediment budget may be small. However, the cumulative impact of many shore protection projects along many kilometers may be significant (as explained below).

(b) The construction of harbors, lakefills, or other projects which protrude into the lake result in large quantities of sand being trapped (in fillet beaches) and diverted offshore (once the fillet beaches have reached full capacity). Often, this sand is permanently lost from the active littoral system.

(c) Dredging or sand mining. Dredging is often required at harbor structures where sediment has been diverted offshore and deposited in navigation channels. The sand-sized portion of the dredged sediment is lost from the littoral system if it is placed outside the littoral zone (i.e., offshore, on land, or in a confined disposal facility).

It is also important to note here that sand supply can be reduced or increased through natural changes. For example, as the shore retreats, stratigraphic units containing more or less sand may be exposed to erosion.

(2) The impact of reducing the sand supply depends on the type and characteristics of the downdrift shore. Depending on the quantity of the natural sand cover along the downdrift cohesive shore, the reduction in sand supply may or may not increase the erosion rate. If the sand quantity is high (greater than approximately 150 m^3/m), the erosion rate will be increased. On the other hand, if the sand cover is already low (less than approximately 50 m^3/m), the erosion rate will probably be unaffected. Invariably, at the downdrift end of a reach of eroding consolidated shore, there is an accreting or stable sand beach deposit that receives the noncohesive fraction eroded from the bluffs. The cutoff or reduction in sand supply to this beach may trigger the onset of its erosion.

d. Remedial measures for cohesive shore erosion.

(1) In order to arrest shoreline recession on a consolidated cohesive shore, it is necessary to stop the downcutting across the profile out to a depth where the future rate of downcutting will not adversely affect the coastal protection system. Three examples of how this can be achieved are as follows:

(a) Through beach nourishment (possibly in conjunction with retaining structures such as detached breakwaters) to increase the sand cover volume to a level sufficient to protect the underlying cohesive profile under most conditions (this volume is approximately 200 m^3/m for Great Lakes shorelines).

(b) Through the construction of offshore breakwaters.

(c) Through the construction of revetment with the toe excavated a sufficient distance into the lake bed or seabed. As an alternative to excavating the toe, a toe protection berm could be placed with sufficient thickness and width to accommodate settling through erosion offshore of the toe of the berm. In most cases, this approach will lead to the eventual disappearance of beaches due to ongoing downcutting offshore of the toe of the revetment and the related steepening of the profile to an extent where sand beaches cannot be maintained naturally. Another critical issue affecting the future performance of revetments on consolidated cohesive shores is the flanking erosion of adjacent unprotected shores. The ends of the revetment may have to be periodically extended back to the shore to address flanking.

(2) One key constraint that must be considered in the design of coastal structures on consolidated cohesive shores is that the nearshore profile will continue to erode offshore of the structure. This has two potentially adverse implications for the stability and performance of the structure:

(a) The toe of the structure may be destabilized if it has not been designed to accommodate ongoing downcutting.

(b) With the deepening of the nearshore profile, the structure will be exposed to larger waves in the future. Both of these outcomes should be considered in design.

(3) Figures III-5-27 and III-5-28 demonstrate the consequences of inappropriate shore protection structures along consolidated shores. The cohesive foundations of the gravity seawall in Figure III-5-27 were undermined, causing the wall to topple. Figure III-5-28 shows that the sheet-pile wall and groin system were ineffective at preventing downcutting and arresting bluff retreat.

(4) For design purposes, the rate of foreshore lowering at any point on the profile can be determined using the profile retreat model. The required information is the approximate shape of the underlying cohesive profile and the shoreline retreat rate. Caution should be applied in using this technique since both the profile shape and the retreat rate could change in the future due to: changing stratigraphy in the foreshore, increased storminess, changes in water levels, and changes to the overlying sand cover.

e. Foundations.

(1) The foundation of structures on cohesive coasts should be assessed with respect to the stability of the underlying cohesive sediment. For locations with high bluffs or cliffs, presumably the loading associated with the structure will be much less than that associated with the bluff itself (i.e., prior to erosion).

(2) Protection of the toe of the bluff and the nearshore profile, on its own, may not be sufficient to stabilize an oversteepened bluff face. The bluff or cliff will usually be in an oversteepened condition prior to the implementation of protection (except in the case where a slope failure has recently occurred). One or more of the following three actions could be taken to address this problem:

(a) Allow the slope to naturally stabilize through surface erosion and failures, and to eventually be colonized by natural vegetation.

Figure III-5-27.   A toppled concrete seawall along the Lake Michigan coast of Berrien County. Failure probably resulted from undermining of the underlying glacial till foundation, April 1991

Figure III-5-28.   A steel sheet-pile wall and groin field has been ineffective at protecting this section of cohesive shore along the Berrien County shore of Lake Michigan, south of the town of St. Joseph, April 1994

Example Problem III-5-2

FIND:

A revetment is being designed to protect an eroding bluff. Find the total downcutting expected over a period of 25 years at the toe of a shore protection structure placed at a water depth of 2 m below low water on a cohesive profile.

Given:

The following information is available:

(1) The long-term bluff retreat rate is 1.0 m/year.

(2) The median grain size of the sand overlying the till is 0.2 mm, 200 μ.

Procedure:

The following steps are taken to develop a solution:

(1) Determine the shape of the underlying cohesive profile. In the absence of direct field information, the cohesive profile shape may be assumed to follow the equilibrium profile for the overlying sand as suggested by Kamphuis (1990) (Part III-5-5a).

(2) Determine the distance offshore to the toe of the revetment (i.e., at the 2-m depth contour) based on the equilibrium profile.

(3) Considering that the bluff retreat rate is equivalent to the profile retreat rate, find the depth of water for the distance determined in step 2 plus 25 m (i.e., 25 years at 1 m/year).

(4) Expected downcutting is the difference between the initial 2-m depth and the depth determined in step 3 above.

Solution:

(1) Using the information provided in Part III-3-3 and Figure III-3-17, the relationship between distance offshore (y) and depth of water (h) for a grain size of 0.2 mm is described by:

$$h = A\,y^{2/3}$$

(2) Therefore, $y = (h/A)^{1.5} = (2/0.1)^{1.5} = 89$ m

The toe of the revetment is located 89 m seaward of the shoreline after construction.

(3) After 25 years, the profile will have shifted inshore by a distance of 25 m. Of course, the erosion inshore of the revetment associated with this shift will be prevented by the presence of the revetment. However, offshore of the toe, the downcutting will continue at the historic rate. Therefore, the new depth at the toe will be the depth associated with a distance of 89 + 25 m from the shoreline (i.e., the distance from shore to the toe of the revetment had the revetment not been constructed). The new depth is calculated as follows:

$$h = A\,y^{0.67} = (0.1)\,114^{0.67} = 2.39 \text{ m}$$

(Continued)

Erosion, Transport, and Deposition of Cohesive Sediments

+---+

Example Problem III-5-2 (Concluded)

(4) Therefore, the expected downcutting over a 25-year period at the toe of the revetment in a water depth of 2 m is 0.39 m (i.e., 2.39 - 2.0).

In reality there are factors that may make this either a conservative (high) or nonconservative (low) estimate. If the revetment is located along a shore with steady sand transport, it is possible that the revetment toe may be protected from downcutting most of the time by the presence of a bar at the base of the revetment (this outcome depends very much on the site conditions, including: grain size, profile shape, and wave climate). If this is the case, the 0.39 m of lowering over 25 years may be an overestimate. On the other hand, if the cohesive sediment at the toe of the revetment is not protected by a sandbar, local scour could occur in addition to the erosion associated with profile lowering. Local scour could easily exceed 0.39 m. It is important to note that with the lowering of the offshore profile through the downcutting process, the local scour will increase with time as larger waves are able to reach the revetment. Also, this solution assumed that the nearshore profile was well described by an equilibrium profile shape. This may not be the case if the nearshore stratigraphy consists of geologic units with different erosion resistances.

The lowering at the toe of a shore protection structure built on a cohesive shore should also be considered in the design of the structure. Specifically, armor unit stability and overtopping potential should be estimated for the expected future depths associated with the design life of the structure.

+---+

(b) Accelerate the stabilization process by constructing a drainage system and through bioengineering to encourage and promote colonization of the slope by native vegetation.

(c) Control the local groundwater flows to minimize destabilizing factors at the bluff face.

(3) In some cases, the natural stable slope, with its toe at the back of the shore protection, may not be acceptable due to the proximity of development to the top of the bluff or cliff. In these cases, additional measures may be taken to achieve a steeper stable slope either through buttressing at the toe, through the implementation of drainage systems in the slope, or other measures to address the potential causes of slope failures at the site.

(4) What size of seawall can soft mud support? Even an earth dike a couple of meters high can result in an overnight slip circle failure through the 'beach' foundation, and longer-term subsidence will need to be compensated for in the design. Geotechnical considerations are of prime importance in the design of shore protection on soft shores; not just bearing (compressive) strength, but shear strength and consolidation rates as well (Part VI-5). Mud might not be able even to support heavy, land-based construction equipment. On many soft shores, mud has built up over centuries to thicknesses in excess of 30 m (100 ft), so that trenching to firm foundation may be uneconomical.

III-5-14. References

Amos et al. 1992
Amos, C. L., Daborn, G. R., Christian, H. A., Atkinson, A., and Robertson, A. 1992. "In-situ Erosion Measurements on Fine-Grained Sediments from the Bay of Fundy," *Marine Geology*, Vol 108, pp 175-196.

Anglin et al. 1996
Anglin, C. D., Nairn, R. B., Cornett, A., Dunaszegi, L., and Doucette, D. 1996. "Bridge Pier Scour Assessment for the Northumberland Strait Crossing, Canada," *Proceedings of the 25th International Conference on Coastal Engineering*, American Society of Civil Engineers, Orlando, FL.

Annandale 1995
Annandale, G. W. 1995. "Erodibility," *Journal of Hydraulic Research,* IAHR, Vol 33, No. 4, pp 471-494.

Arulanandan, Loganatham, and Krone 1975
Arulanandan, K., Loganathan, P., and Krone, R. B. 1975. "Pore and Eroding Fluid Influences on Surface Erosion of Soil," *Journal of Geotechnical Engineering*, American Society of Civil Engineers, Vol 101(GGT1), pp 51-66.

Bartz, Zaneveld, and Pak 1978
Bartz, R., Zaneveld, J. R. V., and Pak, H. 1978. "A Transmissometer for Profiling and Moored Observations in Water," *Photo-Optical Instrumentation Engineering*, Ocean Optics V, pp 102-108.

Boyce, Eyles, and Pugin 1995
Boyce, J. I., Eyles, N., and Pugin, A. 1995. "Seismic Reflection, Borehole and Outcrop Geometry of Late Wisconsin Tills at a Proposed Landfill Near Toronto, Ontario," *Canadian Journal of Earth Sciences,* Vol 32, pp 1331-1349.

Boyd 1981
Boyd, G. L. 1981. "Canada/Ontario Great Lakes Monitoring Programme Final Report," Department of Fisheries and Oceans Canada, Ocean and Aquatic Sciences Manuscript Report No 12.

Boyd 1992
Boyd, G. L. 1992. "A Descriptive Model of Shoreline Development Showing Nearshore Control of Coastal Landform Change: Late Wisconsinan to Present, Lake Huron, Canada," unpublished Ph.D. diss., Department of Geography, University of Waterloo, Ontario, Canada.

Christian, Heffler, and Davis 1993
Christian, H. A., Heffler, D. E., and Davis, E. E. 1993. "Lancelot — An in-situ Piezometer for Soft Marine Sediments," *Deep-Sea Research*, Vol 40, No. 7, pp 1509-1520.

Christie et al. 1995
Christie, M. C., Dyer, K. R., Fennessy, M. J., and Huntley, D. A. 1995. "Field Measurements of Erosion across a Shallow Water Mudflat," *Proceedings Coastal Dynamics '95*, American Society of Civil Engineers, pp 759-770.

Coakley, Rukavina, and Zeman 1986
Coakley, J. P., Rukavina, N. A., and Zeman, A. J. 1986. "Wave-Induced Subaqueous Erosion of Cohesive Tills: Preliminary Results," *Proceedings of the Symposium on Cohesive Shores*, National Research Council Canada, Associate Committee on Shorelines, pp 120-136.

Cornett, Sigouin, and Davies 1994
Cornett, C., Sigouin, N., and Davies, M. H. 1994. "Erosive Response of Northumberland Strait Till and Sedimentary Rock to Fluid Flow," Technical Report TR-1994-22, Institute for Marine Dynamics, National Research Council Canada.

Croad 1981
Croad, R. N. 1981. "Physics of Erosion of Cohesive Soils," Department of Civil Engineering, University of Auckland, New Zealand, Report 247.

Davidson-Arnott 1986
Davidson-Arnott, R. G. D. 1986. "Rates of Erosion of Till in the Nearshore Zone," *Earth Processes and Landforms,* Vol 11, pp 53-58.

Davidson-Arnott and Langham 1995
Davidson-Arnott, R. G. D., and Langham, D. R. J. 1995. "The Role of Softening in Erosion of a Nearshore Profile on a Cohesive Coast," *Proceedings of the Canadian Coastal Conference*, Canadian Coastal Science and Engineering Association, Vol 1, pp 215-224.

Dean 1977
Dean, R. G. 1977. "Equilibrium Profiles, U.S. Atlantic and Gulf Coasts," University of Delaware, Ocean Engineering Report No 12.

Edil and Bosscher 1988
Edil, T. B., and Bosscher, P. J. 1988. "Lake Shore Erosion Processes and Control," *Proceedings of the 19th Annual Conference of the International Erosion Control Association*, New Orleans, LA.

Eisma, Dyer, and van Leussen 1994
Eisma, D., Dyer, K. R., and van Leussen, W. 1994. "The In-situ Determination of the Settling Velocities of Suspended Fine-grained Sediment — A Review," *Proceedings of INTERCOH '94, 4th Nearshore and Estuarine Cohesive Sediment Transport Conference*, John Wiley & Sons, Paper No. 2.

Faas et al. 1992
Faas, R. W., Christian, H. A., Daborn, G. R., and Brylinsky, M. 1992. "Biological Control of Mass Properties of Surficial Sediments: An Example from Starrs Point Tidal Flat, Minas Basin, Bay of Fundy," *Nearshore and Estuarine Cohesive Sediment Transport*, A. J. Mehta, ed., Coastal and Estuarine Sciences No. 42, American Geophysical Union, pp 360-377.

Fleming and Summers 1986
Fleming, C. A., and Summers, L. 1986. "Artificial Headlands on the Clay Cliff," *Proceedings, Symposium on Cohesive Shores*, National Research Council Canada, Associate Committee on Shorelines, pp 262-276.

Fuller 1995
Fuller, J. A. 1995. "Shore and Lakebed Erosion: Response to Changing Levels of Lake Erie at Maumee Bay State Park, Ohio," Ohio Department of Natural Resources, Division of Geological Survey, Open-File Report 95-662.

Gibbs 1972
Gibbs, R. J. 1972. "The Accuracy of Particle Size Analysis Utilizing Settling Tubes," *Journal of Sedimentary Petrology*, Vol 42, No. 1, March, pp 141-145.

Gust 1994
Gust, G. 1994. "Interfacial Hydrodynamics and Entrainment Functions of Currently Used Erosion Devices," *Proceedings of INTERCOH '94, 4th Nearshore and Estuarine Cohesive Sediment Transport Conference*, John Wiley & Sons, Paper No. 15.

Hay and Sheng 1992
Hay, A. E., and Sheng, J. 1992. "Vertical Profiles of Suspended Sand Concentration and Size from Multifrequency Acoustic Backscatter," *Journal of Geophysical Research*, Vol 97, No. C10, October, pp 15,661-15,677.

Hay and Wilson 1994
Hay, A. E., and Wilson, D. J. 1994. "Rotary Sidescan Images of Nearshore Bedform Evolution During a Storm," *Marine Geology*, Vol 119, pp 57-65.

Hutchinson 1986
Hutchinson, J. 1986. "Behavior of Cohesive Shores," Keynote Paper, *Proceedings Symposium on Cohesive Shores*, National Research Council Canada, Associate Committee on Shorelines, pp 1-44.

Jiang and Mehta 1996
Jiang, F., and Mehta, A. J. 1996. "Mudbanks of the Southwest Coast of India; V: Wave Attenuation," *Journal of Coastal Research*, Vol 12, No. 4, pp 890-897.

Kachugin 1970
Kachugin, E.G. 1970. "Studying the Effect of Water Reservoirs on Slope Processes and their Shores," U.S. Army Cold Regions Research and Engineering Laboratory, Draft Translation 732, p 1980.

Kamphuis 1987
Kamphuis, J. W. 1987. "Recession Rate of Glacial Till Bluffs," *Journal of Waterway, Port, Coastal and Ocean Engineering*, American Society of Civil Engineers, Vol 13, No. 1, pp 60-73.

Kamphuis 1990
Kamphuis, J. W. 1990. "Influence of Sand or Gravel on the Erosion of Cohesive Sediment," *Journal of Hydraulic Research*, Vol 28, No. 1, pp 43-53.

Krishnappan 1993
Krishnappan, B. G. 1993. "Rotating Circular Flume," *Journal of Hydraulic Engineering*, American Society of Civil Engineers, Vol 119, No. 6, pp 758-767.

Krishnappan 1995
Krishnappan, B. G. 1995. "Cohesive Sediment Transport," Preliminary Proceedings of 'Issues and Directions in Hydraulics,' Iowa Institute of Hydraulic Research.

Krone 1962
Krone, R. B. 1962. "Flume Studies of the Transport of Sediment in Estuarial Shoaling Processes," University of California, Hydraulic Engineering Laboratory and Sanitary Engineering Research Laboratory.

Kuhn and Osborne 1987
Kuhn, G. G., and Osborne, R. H. 1987. 'Sea Cliff Erosion in San Diego County, California,' *Proceedings of Coastal Sediments '87*, American Society of Civil Engineers, pp 1839-1853.

Lee 1995
Lee, S.-C. 1995. "Response of Mud Shore Profiles to Waves," unpublished Ph.D. diss., University of Florida, Coastal and Ocean Engineering.

Lee and Mehta 1994
Lee, S.-C., and Mehta, A. J. 1994. "Equilibrium Hypsometry of Fine Grained Shore Profiles," *Proceedings of INTERCOH '94, 4th Nearshore and Estuarine Cohesive Sediment Transport Conference*, John Wiley and Sons, Paper No. 41.

Lefebvre, Rohan, and Douville 1985
Lefebvre, G., Rohan, K., and Douville, S. 1985. "Erosivity of Natural Intact Structured Clays: Evaluation," *Canadian Geotechnical Journal*, Vol 22, No. 4, pp 508-517.

Le Hir 1994
Le Hir, P. 1994. "Fluid and Sediment 'Integrated' Modelling: Application to Fluid Mud Flow in Estuaries," *Proceedings of INTERCOH '94, 4th Nearshore and Estuarine Cohesive Sediment Transport Conference*, John Wiley and Sons, Paper No. 40.

Mehta et al. 1989
Mehta, A. J., Carey, W. P., Hayter, E. J., Heltzel, S. B., Krone, R. B., McAnally, W. H., Jr., Parker, W. R., Schoellhamer, D., and Teeter, A. M. 1989. "Cohesive Sediment Transport: Part 1, Process Description; Part 2, Application," *Journal of Hydraulic Engineering*, American Society of Civil Engineers, August, Vol 115, No. 8, pp 1076-1112.

Mollard 1986
Mollard, J. D. 1986. "Shoreline Erosion and Slumping Studies on Prairie Lakes and Reservoirs," *Proceedings, Symposium on Cohesive Shores*, National Research Council Canada, Associate Committee on Shorelines, pp 277-291.

Nairn 1986
Nairn, R. 1986. "Physical Modeling of Wave Erosion on Cohesive Profiles," *Proceedings from the Symposium on Cohesive Shores,* Associate Committee for Research on Shoreline Erosion and Sedimentation, pp 210-225.

Nairn 1992
Nairn, R. B. 1992. "Designing for Cohesive Shores," *Proceedings Coastal Engineering in Canada.* J. W. Kamphuis, ed., Department of Civil Engineering, Queen's University, Kingston, Canada.

Nairn, Pinchin, and Philpott 1986
Nairn, R. B., Pinchin, B. M., and Philpott, K. L. 1986. "A Cohesive Coast Development Model," *Proceedings, Symposium on Cohesive Shores*, National Research Council Canada, Associate Committee on Shorelines, pp 246-261.

Nairn and Parson 1995
Nairn, R. B., and Parson, L. E. 1995. "Coastal Evolution Downdrift of St. Joseph Harbor on Lake Michigan," *Proceedings of Coastal Dynamics '95*, American Society of Civil Engineers, pp 903-914.

Nairn and Southgate 1993
Nairn, R. B., and Southgate, H. N. 1993. "Deterministic Profile Modelling of Nearshore Processes: Part II, Sediment Transport Processes and Beach Profile Development," *Coastal Engineering*, Vol 19, pp 57-96.

Newbury and McCullough 1984
Newbury, R. W., and McCullough, G. K. 1984. "Shoreline Erosion and Restabilization in the Southern Indian Lake Reservoir," *Canadian Journal of Fisheries and Aquatic Sciences,* Vol 41, pp 558-566.

Parker, Matich, and Denney 1986
Parker, G. F., Matich, M. A. J., and Denney, B. E. 1986. "Stabilization Studies — South Marine Drive Sector, Scarborough Bluffs," *Proceedings, Symposium on Cohesive Shores*, National Research Council Canada, Associate Committee on Shorelines, pp 356-377.

Parson, Morang, and Nairn 1996
Parson, L. E., Morang, A., and Nairn, R. B. 1996. "Geologic Effects on Behavior of Beach Fill and Shoreline Stability for Southeast Lake Michigan," Technical Report CERC-96-10, U.S. Army Engineer Waterways Experiment Station, Vicksburg, MS.

Parthenaides 1962
Parthenaides, E. 1962. "A Study of Erosion and Deposition of Cohesive Soils in Salt Water," unpublished Ph.D. diss., University of California, CITY.

Paterson 1994
Paterson, D. M. 1994. "Biological Mediation of Sediment Erodibility: Ecology and Physical Dynamics," *Proceedings of INTERCOH '94, 4th Nearshore and Estuarine Cohesive Sediment Transport Conference*, John Wiley and Sons, Paper No. 17.

Penner 1993
Penner, L. A. 1993. "Shore Erosion and Slumping on Western Canadian Lakes and Reservoirs — A Methodology for Estimating Future Bank Recession Rates," Environment Canada, Regina, Saskatchewan.

Philpott 1984
Philpott, K. L. 1984. "Comparison of Cohesive Coasts and Beach Coasts," *Proceedings, Coastal Engineering in Canada*, J. W. Kamphuis, ed., Queen's University, Kingston, Ontario.

Pringle 1985
Pringle, A. 1985. "Holderness Coast Erosion and the Significance of Ords," *Earth Surface Processes and Landforms*, Vol 10, pp 107-124.

Riggs, Cleary, and Snyder 1995
Riggs, S. R., Cleary, W. J., and Snyder, S. W. 1995. "Influence of Inherited Geologic Framework on Barrier Shoreface Morphology and Dynamics," *Marine Geology*, Vol 126, No. 1/4, pp 213-234.

Rohan et al. 1986
Rohan, K., Lefebvre, G., Douville, S., and Milette, J. P. 1986. "A New Technique to Evaluate Erosivity of Cohesive Material," *Geotechnical Testing Journal*, Vol 9, No. 2.

Schiereck and Booij 1995
Schiereck, G. J., and Booij, N. 1995. "Wave Transmission in Mangrove Forests," *Proceedings International Conference on Coastal and Port Engineering in Developing Countries (COPEDEC)*, Rio de Janeiro, September, pp 1969-1983.

Shabica and Pranschke 1994
Shabica, C., and Pranschke, F. 1994. "Survey of Littoral Drift Sand Deposits along the Illinois and Indiana Shores of Lake Michigan," *Journal of Great Lakes Research*, Vol 20, No. 1, pp 61-72.

Sills 1994
Sills, G. C. 1994. "Hindered Settling and Consolidation in Cohesive Sediments," *Proceedings of INTERCOH '94*, 4th Nearshore and Estuarine Cohesive Sediment Transport Conference, John Wiley and Sons, Paper No. 10.

Skafel 1995
Skafel, M. G. 1995. "Laboratory Measurement of Nearshore Velocities and Erosion of Cohesive Sediment (Till) Shorelines," *Coastal Engineering*, Vol 24, pp 343-349.

Skafel and Bishop 1994
Skafel, M. G., and Bishop, C. T. 1994. "Flume Experiments on the Erosion of Till Shores by Waves," *Coastal Engineering*, Vol 23, pp 329-348.

Sternberg, Shi, and Downing 1989
Sternberg, R. W., Shi, N. C., and Downing, J. P. 1989. "Suspended Sediment Measurements," *Nearshore Sediment Transport,* R. J. Seymour, ed., Plenum, pp 231-257.

Stewart and Pope 1993
Stewart, C. J., and Pope, J. (1993). "Erosion Processes Task Group Report," Working Committee 2, Land Use and Management. International Joint Commission, Great Lakes Water Level Reference Study.

Sunamura 1975
Sunamura, T. 1975. "A Laboratory Study of Wave-Cut Platform Formation," *Jour. Geology*, Vol 83, pp 389-397.

Sunamura 1976
Sunamura, T. 1976. "Feedback Relationship in Wave Erosion of Laboratory Rocky Coast," *Jour. Geology*, Vol 84, pp 427-37.

Sunamura 1992
Sunamura, T. 1992. *Geomorphology of Rocky Coasts*. John Wiley and Sons, England.

Thevenot and Kraus 1995
Thevenot, M. M., and Kraus, N. C. 1995. "Alongshore Sand Waves at Southampton Beach, New York: Observation and Numerical Simulation of Their Movement," *Marine Geology*, Vol 126, No. 1/4, pp 249-270.

Willis and Crookshank 1994
Willis, D. H., and Crookshank, N. L. 1994. "Modelling Multiphase Sediment Transport in Estuaries," *Proceedings of INTERCOH '94*, 4th Nearshore and Estuarine Cohesive Sediment Transport Conference, John Wiley and Sons, Paper No. 37.

Zeman 1986
Zeman, A. J. 1986. "Erodibility of Lake Erie Undisturbed Tills," *Proceedings, Symposium on Cohesive Shores*, National Research Council Canada, Associate Committee on Shorelines, pp 150-169.

III-5-15. Definition of Symbols

ρ_{fm}	Density of the fluid mud layer [force/length^3]
ρ_s	Mass density of sediment grains [force-time^2/length^4]
ρ_w	Mass density of water (salt water = 1,025 kg/m^3 or 2.0 slugs/ft^3; fresh water = 1,000kg/m^3 or 1.94 slugs/ft^3) [force-time^2/length^4]
τ	Bed shear [force/length^2]
τ_c	Critical shear for erosion or erosion threshold [force/length^2]
τ_s	Critical shear stress for deposition [force/length^2]
C	Suspended sediment concentration [force/length^3]
C_v	Consolidation coefficient [length^2/time]
D_z	Vertical dispersion coefficient [length^2/sec]
F_r	Densimetric Froude number (Equation III-5-2)
g	Gravitational acceleration (32.17 ft/sec^2, 9.807m/sec^2) [length/time^2]
h	Equilibrium beach profile depth (Equation III-3-13) [length]
H_0	Incident wave height [length]
h_{fm}	Thickness of fluid mud layer [length]
H_y	Wave height at location y [length]
$\overline{k_i}$	Spatial mean wave attenuation coefficient [length^{-1}]
k_i	Wave height attenuation coefficient [length^{-1}]
m	Mass of sediment on the bed [force/length^2]
M	Constant in the Mignoit equation (Equation III-5-5) which expresses the exponential relationship between mud density and critical shear for erosion [dimensionless]
M_p	Parthenaides coefficient [force/length^2/time]
N	Constant in the Mignoit equation (Equation III-5-5) which expresses the exponential relationship between mud density and critical shear for erosion [dimensionless]
P	Length of the drainage path [length]
S	Vertical (upward) dispersion of sediment (Equation III-5-3) [force/length^2/time]
t	Time
T	Wave period [time]

u	Degree of consolidation, ratio of bulk density of a sediment to bulk density of a fully consolidated sediment [dimensionless]
V_{fm}	Mean velocity in the fluid mud layer [length/time]
V_w	Mean velocity in the water layer [length/time]
w	Sediment fall velocity [length/time]
y	Distance along wave ray [length]
y	Equilibrium beach profile distance offshore (Equation III-3-13) [length]
z	Vertical dimension [length]

III-5-16. Acknowledgments

Authors of Chapter III-5, "Erosion, Transport, and Deposition of Cohesive Sediments:"

Robert B. Nairn, Ph.D., Baird and Associates, Oakville, Ontario, Canada.
David H. Willis, Ph.D., David H. Willis and Associates Ltd., Ottawa, Ontario, Canada.

Reviewers:

James R. Houston, Ph.D., Engineer Research and Development Center, Vicksburg, Mississippi.
Andrew Morang, Ph.D., Coastal and Hydraulics Laboratory (CHL), Engineer Research and Development Center, Vicksburg, Mississippi.
Todd L. Walton, Ph.D., CHL

Chapter 6
SEDIMENT TRANSPORT OUTSIDE
THE SURF ZONE

EM 1110-2-1100
(Part III)
30 April 2002

Table of Contents

List of Figures

Chapter III-6
Sediment Transport Outside the Surf Zone

III-6-1. Introduction

a. Coastal engineers often regard the seaward boundary of the surf zone as the deepwater limit of significant wave and current effect. From the outer break point shoreward to the beach, waves and their associated currents are recognized as major sources of sediment resuspension and transport. Seaward of this point there is a region from roughly 2 to 3 m depth to approximately 20 to 30 m depth within which the importance of waves and currents on sediment transport processes is not well understood. Sediment transport within this region, usually referred to as the inner shelf, is the central focus of this chapter.

b. During severe coastal storms some material removed from the beach is carried offshore and deposited on the inner shelf in depths at which, under normal wave conditions, it is not resuspended. The fate of this material has, for some time, provided a troublesome set of questions for coastal engineers. Will this sediment be returned to the surf zone and perhaps even the beach or will it be carried further offshore to the deeper waters of the continental shelf? Furthermore, what are the time scales of these sediment motions, regardless of their directionality? Similar questions are relevant for sediment deposited on the inner shelf from human activities e.g., "what is the fate and transport of dredged material from inlets or navigation channels as well as contaminated sediment resulting from discharge of sewage?"). In order to answer these and similar questions related to sedimentary processes in inner shelf waters, it is necessary to establish a quantitative framework for the analysis and prediction of sediment transport processes outside the surf zone.

c. To fully appreciate the complexity of this problem it is necessary to recognize that sediment transport has been extensively studied for decades and yet it is still not possible to predict transport rates with any degree of certainty. A majority of these studies have been carried out in laboratory flumes under highly idealized conditions of steady two-dimensional flow over uniform noncohesive sand. Few studies have addressed the added complications of bed form drag, sediment concentration influence on flow, and cohesive sediment movement. On the inner shelf, as on most of the active seabed, sediment transport is a nonlinear, turbulent, two-phase flow problem complicated by bed forms, bottom material characteristics, current variability, and by the superposition of waves. In addition, transport can be comprised of bed load as well as suspended load, the quantitative separation of which is of considerable complexity.

d. The fundamental approach to predicting sediment transport is to relate the frictional force exerted by the fluid and consequent bed shear stress τ_b to the sediment transport rate q. There are two ways to address prediction of sediment transport. Empirical approaches use measurements of fluid velocity and depth as well as bottom roughness and grain size to determine proportionality relationships over a wide variety of conditions. Theoretical approaches attempt to use turbulent flow dynamics to determine the proportionality values directly. A predictive theory of turbulent flow does not exist and stochastic analysis, though useful, cannot provide an understanding of sediment transport mechanics in turbulent flows. Validating experiments must therefore be used to determine coefficients arising from assumptions incorporated into any theoretical prediction formulations. A major difficulty encountered by either the empirical or theoretical approach is the inability to adequately measure sediment transport rates. Consequently, existing formulations for predicting sediment transport rates show large discrepancies, even when they are applied to the same data (White, Milli, and Crabbe 1975; Heathershaw 1981; Dyer and Soulsby 1988).

e. The purpose of this chapter is to present an approach to the quantitative analysis and prediction of sediment transport outside the surf zone. This approach will incorporate hydrodynamics of wave-current boundary layer flow, fluid-sediment interaction near the bottom, and sediment transport mechanics.

III-6-2. Combined Wave and Current Bottom Boundary Layer Flow

a. Introduction.

(1) In inner shelf waters and over the entire continental shelf during extreme storm events, near-bottom flow will be determined by the nonlinear interaction of waves and slowly varying currents. This superposition of flows of different time scales and hence different boundary layer scales results in the wave bottom boundary layer being nested within the current boundary layer. The high turbulence intensities within the thin wave bottom boundary layer cause the current to experience a higher bottom resistance in the presence of waves than it would if waves were absent. Conversely, the wave bottom boundary layer flow will also be affected by the presence of currents, although far less so than the current is affected by the presence of waves.

(2) A 10-s wave 1 m in height produces a near-bottom orbital velocity exceeding 0.15 m/s in depths less than 30 m. Due to the oscillatory nature of the wave orbital velocity, the bottom boundary layer has only a limited time, approximately half a wave period, to grow. This results in the development of a thin layer, a few centimeters in thickness, immediately above the bottom, called the wave bottom boundary layer, within which the fluid velocity changes from its free stream value to zero at the bottom boundary. The high-velocity shear within the wave bottom boundary layer produces high levels of turbulence intensities and large bottom shear stresses.

(3) In contrast to the wave motion, a current, wind-driven or tidal, will vary over a much longer time scale, on the order of several hours. Hence, even if the current is slowly varying, the current bottom boundary layer will have a far greater vertical scale, on the order of several meters, than the wave bottom boundary layer. Consequently the velocity shear, turbulence intensities, and bottom shear stress will be much lower for a current than for wave motion of comparable velocity.

(4) The simple eddy viscosity model proposed by Grant and Madsen (1986) is adopted throughout most of this chapter in order to obtain simple closed-form, analytical expressions for combined wave-current bottom boundary layer flows and associated sediment transport. A review of alternative models, employing more sophisticated turbulence closure schemes is given in Madsen and Wikramanayake (1991). Alternative wave-current interaction theories applicable to sediment transport are: prescribed mixing length distribution (Bijker 1967), momentum deficit integral (Fredsoe 1984), and turbulent kinetic energy closure (Davies et al. 1988). All four theories are compared in Dyer and Soulsby (1988). A detailed discussion of the eddy viscosity model approach is given in Madsen (1993).

(5) Although limitations of analyses will be pointed out in each section, some major limitations applicable throughout this chapter are stated here. (a) The hydrodynamic environment is limited to *nonbreaking* wave conditions described by *linear wave theory* and near-bottom *unidirectional* steady currents. The former limitation places the applicability of results derived in this chapter well outside the surf zone, whereas the latter precludes the use of formulas presented here for flows exhibiting appreciable turning of the current velocity vector with height above the bottom. (b) Only sediment that can be characterized as *cohesionless* is considered.

(6) Most flows that transport sediment are turbulent boundary layer shear flows and the forces exerted on the sediment bed are governed by the turbulence characteristics. Over a horizontal bottom, these flows are characterized by their large scale of variation in the horizontal plane relative to the scale of variation in the vertical plane. This disparity in length scales makes it possible to neglect vertical acceleration within the boundary layer (Schlichting 1960). Figure III-6-1 shows the turbulent boundary layer structure and mean velocity profile for a two-dimensional horizontal flow in the xz-plane, x being horizontal and z vertical. The turbulent boundary layer is made up of three sublayers; the viscous sublayer, the turbulence generation layer,

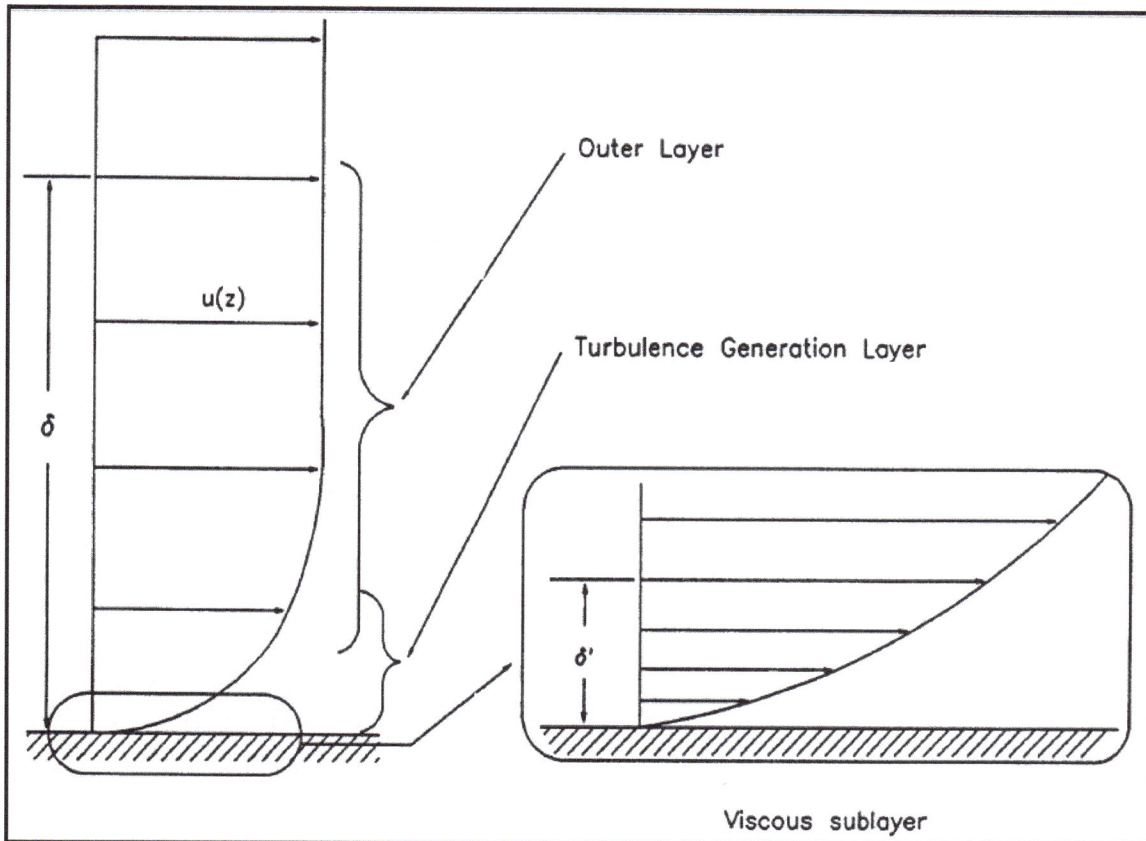

Figure III-6-1. Turbulent boundary layer structure and mean velocity profile

and the outer layer. In the viscous sublayer, turbulent fluctuations in velocity are present, but velocity fluctuations normal to the boundary must tend towards zero as the bottom is approached. Consequently, molecular transport of fluid momentum dominates turbulent transport very near the bottom and the shear stress τ can be modeled approximately by the laminar boundary layer relationship

$$\tau = \rho \, v \, \frac{\partial u}{\partial z} \tag{III-6-1}$$

where ρ is fluid density, ν the kinematic viscosity of the fluid, and u is horizontal velocity.

(7) The turbulence generation layer is characterized by very energetic small-scale turbulence and high fluid shear. Turbulent eddies produced in this region are carried outward and inward toward the viscous sublayer. If the bottom roughness elements (sediment grains and/or bed forms) have a height greater than the viscous sublayer, the turbulence generation layer extends all the way to the bottom.

(8) The outer layer makes up most of the turbulent boundary layer and is characterized by much larger eddies, which are more efficient at transporting momentum. This high efficiency of momentum transport produces a mean velocity profile that is much gentler than in the turbulence generation layer, Figure III-6-1.

(9) Turbulent boundary layer shear flow models can be simply developed by artificially choosing a constant v_T (called the eddy viscosity), which is much larger than its laminar (molecular) value and that reflects the size of the eddy structure associated with the turbulent flow. This assumption results in a simple conceptual eddy viscosity model for turbulent shear stress that may be expressed as

$$\tau = \rho \, v_T \, \frac{\partial u}{\partial z} \qquad\qquad\qquad\text{(III-6-2a)}$$

with the turbulent eddy viscosity

$$v_T = \kappa \, u_* \, z \qquad\qquad\qquad\text{(III-6-2b)}$$

where κ is known as von Karman's constant and $u_* = (\tau_b/\rho)^{1/2}$ is called the shear velocity, where τ_b is the shear stress at the bed ($z = 0$). A complete derivation of this model is given in Madsen (1993).

(10) Experimental determinations have shown relatively little variation of κ, leading to von Karman's "universal" constant assumed to have the value 0.4.

b.   *Current boundary layer.*

(1)   Currents on the inner shelf can be considered to flow at a steady velocity compared to the orbital velocity of waves. Defining the shear stress due to currents τ_c, the current velocity profile can be expressed as

$$u_c = \frac{u_{*c}}{\kappa} \, \ln\frac{z}{z_o} \qquad\qquad\qquad\text{(III-6-3)}$$

where $u_{*c} = (\tau_c/\rho)^{1/2}$ denotes the current shear velocity. Equation 6-3 is the classic <u>logarithmic velocity profile</u> expressed in terms of z_o, the value of z at which the logarithmic velocity profile predicts a velocity of zero. For a smooth bottom $z_o = z = 0$, but for a rough bottom, the actual location of the boundary is not a single value of z. Hence $z = 0$ becomes a somewhat ambiguous definition of the "theoretical" location of the bottom with some portions of the boundary actually located at $z > 0$ and others at $z < 0$. Therefore, application of Equation 6-3 in the immediate vicinity of a solid bottom is purely formal and its prediction of $u_c = 0$ at $z = z_o$ is of no physical significance.

(2)   From the extensive experiments by Nikuradse (1933), the value of z_o is obtained as

$$z_0 = \begin{cases} \dfrac{v}{9u_*} & \text{for smooth turbulent flow} \\[2ex] \dfrac{k_n}{30} & \text{for fully rough turbulent flow} \end{cases} \qquad\qquad\text{(III-6-4)}$$

in which k_n is the equivalent Nikuradse sand grain roughness, so called because Nikuradse in his experiments used smooth pipes with uniform sand grains glued to the walls, and therefore found it natural to specify the wall roughness as the sand grain diameter. Smooth or fully rough turbulent flow are delineated by Madsen (1993).

$$\frac{k_n u_*}{\nu} \geq 3.3 \quad \text{for fully rough turbulent flow}$$

$$\frac{k_n u_*}{\nu} \leq 3.3 \quad \text{for fully smooth turbulent flow} \tag{III-6-5}$$

(3)   For turbulent flows over a plane bed consisting of granular material it is natural to take $k_n = D =$ diameter of the grains composing the bed. For flow over a bottom covered by distributed roughness elements (e.g., resembling a rippled bottom), the value $k_n = 30\,z_0$ is referred to as the equivalent Nikuradse sand grain roughness, with z_0 obtained by extrapolation of the logarithmic velocity distribution above the bed, to the value $z = z_0$ where u_c vanishes.

(4)   Figure III-6-2 is a semilogarithmic plot of a current velocity profile obtained over a bottom consisting of 1.5-cm-high triangular bars at 10-cm spacing (Mathisen 1993). It is noticed that velocity measurements over crests and midway between crests (troughs) of the roughness elements deviate from the expected straight line within the lower 2.5 to 3 cm and further than ≈ 10 cm above the bottom. The near-bottom deviations reflect the proximity of the actual bottom roughness elements, which make the velocity a function of location, i.e., the flow is nonuniform immediately above the bottom roughness features. The nonlogarithmic velocity profile far from the boundary is associated with the flow being that of a developing boundary layer flow in a laboratory flume, i.e., the flow above $z \approx 10$ cm is essentially a potential flow unaffected by boundary resistance and wall turbulence.

(5)   The well-defined variation of u_c versus $\log z$ (Figure III-6-2) can be applied to the case 3 cm $<$ $z <$ 10 cm. By extrapolation to $u_c = 0$, one obtains the value $z = z_0 \approx 0.7$ cm or $k_n = 30\,z_0 \approx 21$ cm, since the flow is rough turbulent. This example clearly illustrates that k_n, the equivalent Nikuradse sand grain roughness, is a function of bottom roughness configuration and does not necessarily reflect the physical scale of the roughness protrusions. Thus, the data shown in Figure III-6-2 were obtained for a physical roughness scale of 1.5 cm, the height of the triangular bars, and result in an equivalent Nikuradse roughness of 21 cm. This large value of k_n is, of course, associated with the much larger flow resistance produced by the triangular bars.

(6)   It is often convenient to express the bottom shear stress associated with a boundary layer flow in terms of a current friction factor f_c defined by

$$\tau_c = \frac{1}{2} f_c \rho \left(u_c \left(z_r \right) \right)^2 \tag{III-6-6}$$

(7)   The somewhat cumbersome notation used in Equation 6-6 is chosen deliberately to emphasize the fact that the value of the current friction factor is a function of the reference level, $z = z_r$, at which the current velocity is specified. From Equation 6-6, the current shear velocity is obtained

$$u_{*c} = \sqrt{\frac{\tau_c}{\rho}} = \sqrt{\frac{f_c}{2}}\, u_c(z_r) \tag{III-6-7}$$

and introducing this expression in Equation 6-3, with $z = z_r$ and $\kappa = 0.4$ leads to an equation for the current friction factor

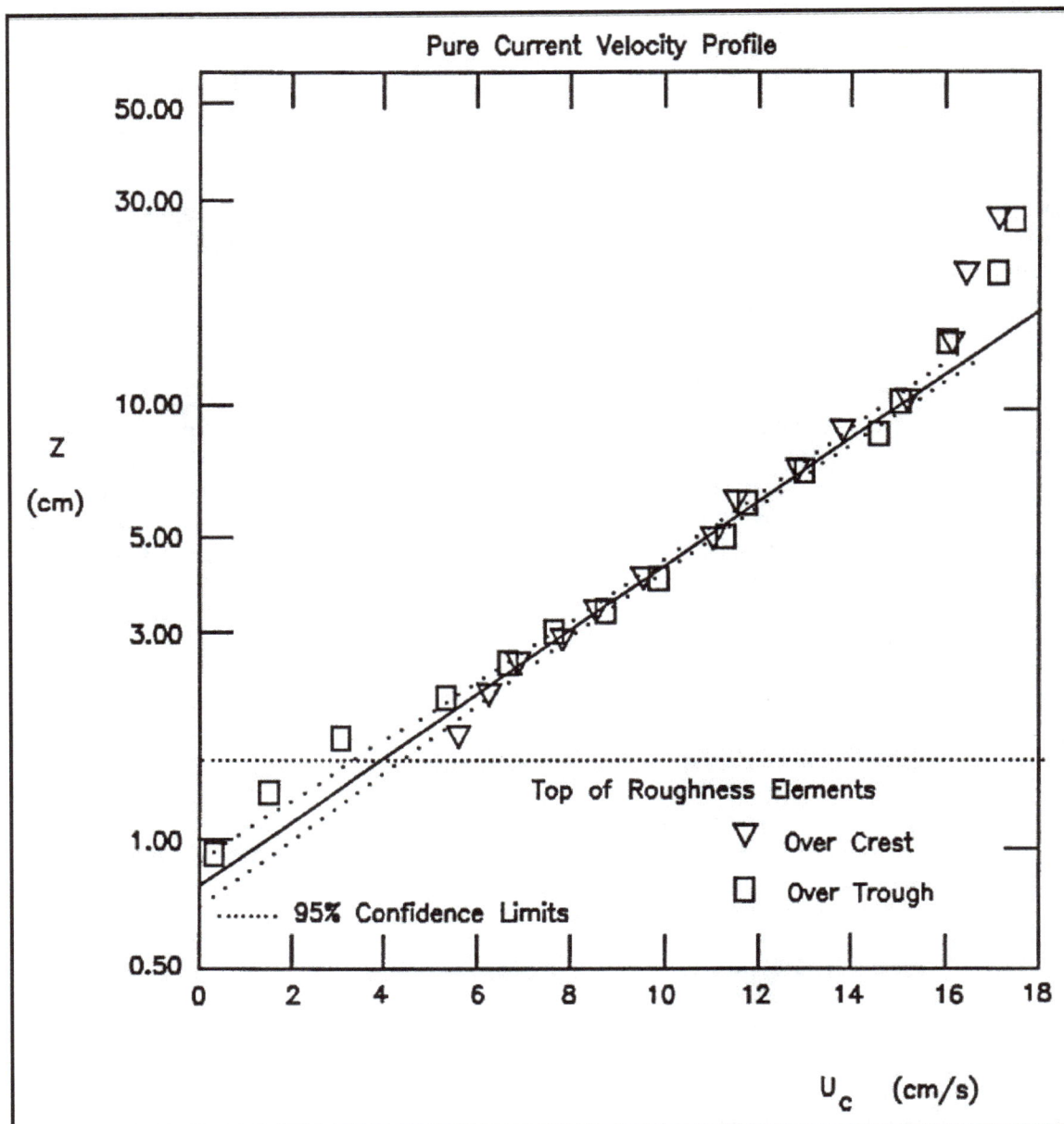

Figure III-6-2. Measured turbulent velocity profile for flow over artificial two-dimensional roughness elements (Mathisen 1993)

$$\frac{1}{4\sqrt{f_c}} \approx \log_{10} \frac{z_r}{z_0}$$

(III-6-8)

in terms of the reference elevation and the boundary roughness scale.

(8)   For rough turbulent flows $z_0 = k_n/30$ and Equation 6-8 is an explicit equation for f_c in terms of the relative roughness z_r/k_n. For smooth turbulent flows $z_0 = \nu/(9u_{*c})$ and Equation 6-8 leads to an implicit equation for f_c (Madsen 1993)

EXAMPLE PROBLEM III-6-1

FIND:

The current friction factor f_c, the shear velocity u_{*c}, and the bottom shear stress τ_c, for a current over a flat bed.

GIVEN:

The current is specified by its velocity $u_c(z_r) = 0.35$ m/s at $z_r = 1.00$ m. The bottom is flat and consists of uniform sediment of diameter $D = 0.2$ mm. The fluid is seawater ($\rho \approx 1,025$ kg/m^3, $\nu \approx 1.0 \times 10^{-6}$ m^2/s).

PROCEDURE:

1) Start with Equation 6-4 *assuming* rough turbulent flow: $z_0 = k_n/30$.

2) Solve Equation 6-8 for f_c.

3) Obtain u_{*c} from Equation 6-7.

4) Check rough turbulent flow assumption by evaluating $k_n u_{*c}/\nu$ and using Equation 6-5.

5) *If* $k_n u_{*c}/\nu \geq 3.3$, flow *is* rough turbulent. Results (f_c and u_{*c}) obtained in steps 2 and 3 constitute the solution and τ_c is obtained from Equation 6-6 or from the definition $\tau_c = \rho\, u_{*c}^2$.

6) *If* $k_n u_{*c}/\nu < 3.3$, flow *is smooth turbulent* and the solution is obtained by the following iterative procedure (steps 7 and 8).

7) Obtain $z_0 = \nu/(9u_{*c})$ using u_{*c} from step 3.

8) With z_0 from step 7 return to step 2 to obtain a new f_c, and a new u_{*c} from step 3 and reenter (step 7).

9) When consecutive values of u_{*c} agree to two significant digits, iteration is complete. The last values of f_c and u_{*c} constitute the solution and τ_c is obtained from Equation 6-6 or from the definition $\tau_c = \rho\, u_{*c}^2$

(Continued)

Example Problem III-6-1 (Concluded)

SOLUTION:

It is first *assumed* that the flow will turn out to be classified as rough turbulent, in which case Equation 6-4 gives

$$z_0 = \frac{k_n}{30} = \frac{D}{30} = 6.67 \times 10^{-6} \text{ m} = 6.67 \times 10^{-4} \text{ cm}$$

where $k_n = D$ is chosen, since the bed is flat. With this z_0 value, Equation 6-8 is solved for f_c with $z_r = 1.00$ m, giving

$$f_c = 2.33 \times 10^{-3}$$

and the corresponding shear velocity is obtained from Equation 6-7 with $u_c(z_r) = 0.35$ m/s

$$u_{*c} = 1.19 \times 10^{-2} \text{ m/s} = 1.19 \text{ cm/s}$$

But is the flow really rough turbulent as *assumed*? To check this assumption, the boundary Reynolds number is computed using the rough turbulent estimate of u_{*c}

$$\frac{k_n u_{*c}}{\nu} = \frac{D u_{*c}}{\nu} = 2.38 < 3.3$$

From Equation 6-5 it is seen that the flow should be classified as smooth turbulent. (If $k_n u_{*c}/\nu$ had been greater than 3.3, the problem would have been solved.) Therefore, an estimate

$$z_0 = \frac{\nu}{9 u_{*c}} \approx 9.3 \times 10^{-6} \text{ m}$$

is obtained from Equation 6-4 using the rough turbulent u_{*c}. With this value of z_0, Equation 6-8 gives the *current friction factor*

$$f_c = 2.47 \times 10^{-3}$$

and, from Equation 6-7, the *shear velocity* and the *bottom shear stress*

$$u_{*c} = 1.23 \times 10^{-2} \text{ m/s} = 1.23 \text{ cm/s} \quad \text{and} \quad \tau_c = \rho u_{*c}^2 = 0.155 \text{ N/m}^2 \approx 0.16 \text{ Pa}$$

Since the smooth turbulent u_{*c} obtained here is close to the rough turbulent estimate, returning with it to update $z_0 = \nu/(9 u_{*c})$ and repeating the procedure does not change f_c and u_{*c} given above, so they do indeed represent the solution. An alternative solution strategy would be to compute f_c from the implicit Equation 6-9 once it was recognized that the flow was smooth turbulent.

$$\frac{1}{4\sqrt{f_c}} + \log_{10}\frac{1}{4\sqrt{f_c}} = \log_{10}\frac{z_r u_c(z_r)}{\nu} + 0.20 \tag{III-6-9}$$

which shows f_c's dependency on a Reynolds number, $\dfrac{z_r u_c(z_r)}{\nu}$, when the flow is classified as smooth turbulent.

c. *Wave boundary layers.*

(1) Introduction. Wave boundary layers on the inner shelf are inherently unsteady due to the oscillatory nature of near-bottom wave orbital velocity. However, for typical gravity wave periods on the inner shelf, it is assumed that the boundary layer thickness, here denoted by δ_w, will be sufficiently small so that flow at the outer edge of the boundary layer may be predicted directly from linear wave theory. An extreme complication of the unsteady nature of waves arises in the application of the simple turbulent eddy viscosity model. Specifically, wave turbulent eddy viscosity becomes a time-dependent variable $u_* = (|\tau_b|/\rho)^{1/2} = u_*(t)$. However, the effect of a time-varying eddy viscosity is surprisingly small when compared with results obtained from a simple time-invariant eddy viscosity model in which $u_* = u_{*wm} = (|\tau_{wm}|/\rho)^{1/2}$ is determined from the maximum bottom wave shear stress τ_{wm}, during one wave cycle (Trowbridge and Madsen 1984). Therefore, the eddy viscosity model gives a wave shear stress

$$\tau_w = \rho\nu_t\frac{\partial u_w}{\partial z} = \rho\,\kappa\,u_{*wm}\,z\,\frac{\partial u_w}{\partial z} \tag{III-6-10}$$

where u_w is the wave orbital velocity (Grant and Madsen 1979, 1986). Using Equation 6-10, applying simple linear periodic wave theory to the boundary conditions, and simplifying by taking the limiting form of the solution for the velocity profile gives

$$u_w = \frac{2}{\pi}\sin\varphi\;u_{bm}\ln\frac{z}{z_o}\cos(\omega t + \varphi) \tag{III-6-11}$$

in which u_{bm} is the maximum near-bottom wave orbital velocity, ω is $2\pi/T$, and φ, the phase lead of near-bottom wave orbital velocity, is given by

$$\tan\varphi = \frac{\dfrac{\pi}{2}}{\ln\dfrac{\kappa u_{*wm}}{z_o\omega} - 1.15} \tag{III-6-12}$$

From the near-bottom velocity profile, the bottom shear stress may be obtained from

$$\tau_w = \tau_{\omega m}\cos(\omega t + \varphi) \tag{III-6-13}$$

Expressing the maximum bottom shear stress in terms of a *wave friction factor f_w* defined by Jonsson (1966)

$$\tau_{wm} = \frac{1}{2}f_w\rho u_{bm}^2 \tag{III-6-14}$$

or equivalently

$$u_{*wm} = \sqrt{\frac{\tau_{wm}}{\rho}} = \sqrt{\frac{f_w}{2}}\, u_{bm} \qquad\qquad\text{(III-6-15)}$$

Using Equations 6-10 to 6-15 and simplifying (Madsen 1993) results in an implicit equation for the wave friction factor

$$\kappa\sqrt{\frac{2}{f_w}} = \sqrt{\left(\ln\frac{\kappa\sqrt{\frac{f_w}{2}}\, u_{bm}}{z_0\omega} - 1.15\right)^2 + \left(\frac{\pi}{2}\right)^2} \qquad\qquad\text{(III-6-16)}$$

With $\kappa = 0.4$ and recognizing that

$$A_{bm} = \frac{u_{bm}}{\omega} \qquad\qquad\text{(III-6-17)}$$

is the bottom excursion amplitude predicted by linear wave theory, Equation 6-16 may be approximated by the following implicit wave friction factor relationship (Grant and Madsen 1986)

$$\frac{1}{4\sqrt{f_w}} + \log_{10}\frac{1}{4\sqrt{f_w}} = \log_{10}\frac{A_{bm}}{k_n} - 0.17 + 0.24\left(4\sqrt{f_w}\right) \qquad\qquad\text{(III-6-18)}$$

for rough turbulent flow when $k_n = 30z_0$.

For smooth turbulent flow, $30z_0 = 3.3\nu/u_{*wm}$ replaces k_n in Equation 6-18, which may then be written

$$\frac{1}{4\sqrt{4f_w}} + \log_{10}\frac{1}{4\sqrt{4f_w}} = \log_{10}\sqrt{\frac{RE}{50}} - 0.17 + 0.06\left(4\sqrt{4f_w}\right) \qquad\qquad\text{(III-6-19)}$$

in which

$$RE = \frac{u_{bm}\,A_{bm}}{\nu} \qquad\qquad\text{(III-6-20)}$$

is a wave Reynolds number. For completeness, the wave friction factor for a *laminar wave boundary layer* is given by (Jonsson 1966)

$$f_w = \frac{2}{\sqrt{RE}} \qquad\qquad\text{(III-6-21)}$$

(2) Evaluation of the wave friction factor.

(a) From knowledge of the equivalent roughness k_n and wave condition A_{bm} and u_{bm}, three formulas for f_w have been presented. The choice of which f_w to select is quite simply *the largest* of the three values. In this context it is mentioned that the wave friction factors predicted for smooth turbulent and laminar flow,

Equations 6-19 and 6-21, are identical for RE $\approx 3\times10^4$, which may be taken as a reasonable value for the transition from laminar (RE $< 3\times10^4$) to turbulent flow.

(b)   The evaluation of wave friction factors from Equations 6-18 and 6-19 can proceed by iteration. The iterative procedure is illustrated for Equation 6-18 by writing it in the following form

$$\frac{1}{x^{(n+1)}} = \left(\log_{10} \frac{A_{bm}}{k_n} - 0.17 \right) - \log_{10} \frac{1}{x^{(n)}} + 0.24 x^{(n)} \qquad \text{(III-6-22)}$$

with the superscript denoting the iteration step $\left(x = 4\sqrt{f_w} \right)$ and iteration started by choosing $x^{(0)} = 0.4$.

(c)   It was assumed, in conjunction with the presentation of Equation 6-11 for the orbital velocity profile within the wave boundary layer, that this expression represented an approximation valid only for small values of $z_0\omega/(\kappa u_{*wm})=(0.12/\sqrt{f_w})(k_n/A_{bm})$, i.e., for relatively large values of A_{bm}/k_n. For this reason and also because the iterative solution of the wave friction factor Equation 6-18 is rather slowly converging for $A_{bm}/k_n < 100$, Figure III-6-3 gives the exact solution for f_w and φ for $A_{bm}/k_n < 100$.

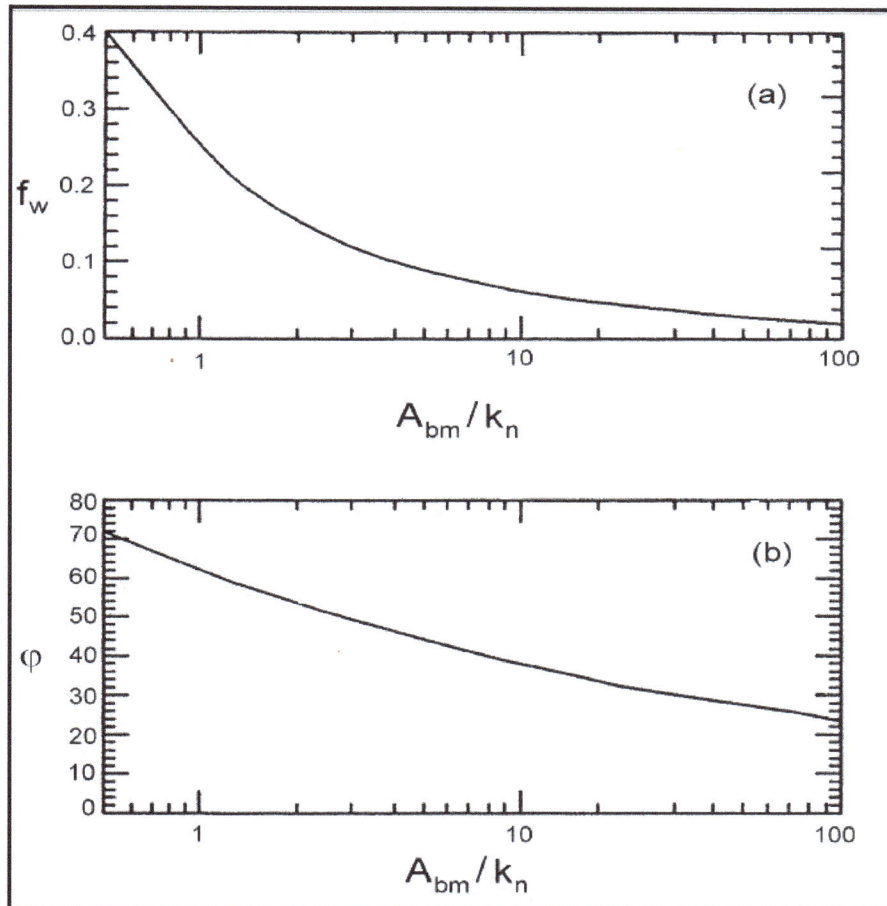

Figure III-6-3.   Wave friction factor diagram (a) and bottom friction phase angle (b)

(d)   For smooth turbulent boundary layer flows, Equation 6-19 may be arranged in a fashion similar to Equation 6-22, and the iterative solution proceeds as outlined above with $x = 4\sqrt{4f_w}$.

(3) Wave boundary layer thickness. From the logarithmic approximation given by Equation 6-11, which is valid only near the bottom and for large values of $\kappa u_{*wm}/(z_0\omega)$, the location for which $|u_w| = u_{bm}$ is obtained as $z \approx \kappa u_{*wm}/(\pi\omega)$. Although use of Equation 6-11 for the prediction of boundary layer thickness exceeds its validity, the results suggest

$$\delta_w = \frac{\kappa u_{*wm}}{\omega} = \kappa \sqrt{\frac{f_w}{2}} \, A_{bm} \qquad \text{(III-6-23)}$$

in agreement with conclusions reached based on the complete solution to the boundary layer problem (Grant and Madsen 1986).

(4) The velocity profile.

(a) With the limitation of the validity of the velocity profile given by Equation 6-11 in mind, it is seen from Equation 6-16 that the velocity profile within the wave boundary layer may be expressed in the following extremely simple form, with an upper limit of validity given by the z value for which $|u_w| = u_{bm}$, or approximately $z = \delta_w/\pi$, i.e.,

$$u_w = \frac{\sqrt{\frac{f_w}{2}} \, u_{bm}}{\kappa} \ln \frac{z}{z_0} \cos(\omega t + \varphi) = \frac{u_{*wm}}{\kappa} \ln \frac{z}{z_0} \cos(\omega t + \varphi) \qquad \text{(III-6-24)}$$

(b) Figure III-6-4 compares the experimentally obtained periodic orbital velocity amplitude within the wave boundary layer (Jonsson and Carlson 1976, Test No. 1) with the prediction afforded by the exact solution obtained from the theory presented here and the approximate solution given by Equation 6-24. Clearly, the approximate velocity profile compares favorably with both the exact and the experimental profiles near the bottom, as it should. Although some differences become apparent further from the boundary, the approximate solution provides an adequate representation of the measured profile, particularly when its intended use for the prediction of sediment transport rates is kept in mind.

(5) Extension to spectral waves.

(a) The preceding analysis of wave bottom boundary layer mechanics was based on the assumption of simple periodic waves. In reality, wind waves are more realistically represented by the superposition of several individual wave components of different frequencies and directions of propagation, i.e., by a directional wave spectrum, rather than by a single periodic component. Madsen, Poon, and Graber (1988) presented a bottom boundary layer model for waves described by their near-bottom wave orbital velocity spectrum. This reference provides the details of an analysis, the basis of which is an assumed time-invariant eddy viscosity similar to the one discussed earlier for a simple periodic wave. The analysis utilizes the near-bottom maximum orbital velocity $u_{bm,i}$ and radian frequency ω_i of the individual i^{th} wave components of the wave spectrum to calculate a single set of <u>representative periodic wave characteristics</u> (see Madsen (1993) for details).

(b) Given the widespread use of the significant wave concept in coastal engineering practice it is important to emphasize that the <u>representative periodic wave characteristics</u> defined by Madsen (1993) <u>are not those of the significant wave,</u> but rather those of a wave with the significant period and the

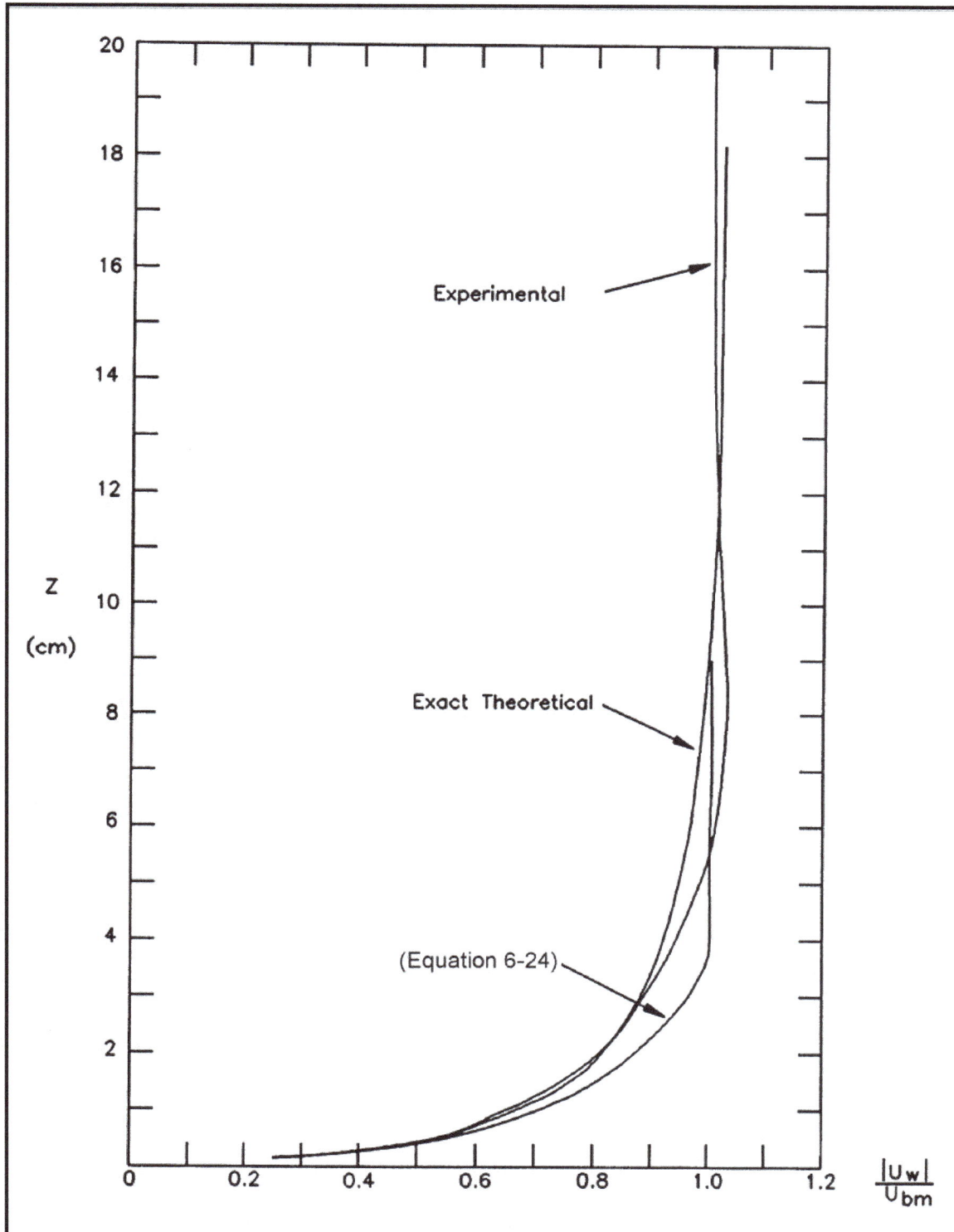

Figure III-6-4. Comparison of present model's prediction of wave orbital velocity within the wave boundary layer with measurements by Jonsson and Carlson (1976, Test No. 1)

EXAMPLE PROBLEM III-6-2

FIND:

The wave friction factor f_w, the maximum shear velocity u_{*wm}, the maximum bottom shear stress τ_{wm}, and the bottom shear stress phase angle φ, for a pure wave motion over a flat bed.

GIVEN:

The wave is specified by its near-bottom maximum orbital velocity $u_{bm} = 0.35$ m/s and period $T = 8.0$ s. The bottom is flat and consists of uniform sediment of diameter $D = 0.1$ mm. The fluid is seawater ($\rho \approx 1,025$ kg/m³, $\nu \approx 1.0 \times 10^{-6}$ m²/s).

PROCEDURE:

1) Compute bottom orbital excursion amplitude from linear wave theory: $A_{bm} = (H/2)/\sinh(2\pi h/L) = u_{bm}/\omega$, using $H = H_{rms}$ = root-mean-square wave height = $H_s/\sqrt{2}$, $\omega = 2\pi/T$ with T = significant wave period, and h = water depth. (This step is not needed here, since u_{bm} and T are specified)

2) *Assume* rough turbulent flow and solve Equation 6-18 using the iterative procedure illustrated by Equation 6-22 for f_w.

3) Obtain u_{*wm} from Equation 6-15.

4) Check rough turbulent flow assumption by evaluating $k_n u_{*wm}/\nu$ and using Equation 6-15.

5) *If* $k_n u_{*wm}/\nu \geq 3.3$, flow *is* rough turbulent. Results (f_w and u_{*cw}) obtained in steps 2 and 3 constitute the solution. τ_{wm} is obtained from Equation 6-14 or 6-15 and φ is obtained by solving Equation 6-12 with $z_0 = k_n/30$.

6) *If* $k_n u_{*wm}/\nu < 3.3$ flow *is smooth turbulent* and f_w is obtained by solving Equation 6-19 using the iterative procedure illustrated by Equation 6-22 with A_{bm}/k_n replaced by $\sqrt{RE/50}$ and $x = 4\sqrt{4f_w}$.

7) With f_w from step 6, Equation 6-15 gives u_{*wm} and τ_{wm} is obtained from Equation 6-14 or Equation 6-15.

8) With u_{*wm} from step 7, the value of $z_0 = \nu/(9u_{*wm})$ is obtained from Equation 6-15 for smooth turbulent flow.

9) The phase angle φ is calculated from Equation 6-12 using z_0 from step 8.

(Sheet 1 of 3)

Example Problem III-6-2 (Continued)

SOLUTION:

From the given wave characteristics, it follows that

$$\omega = 2\pi/T = 0.785 \text{ s}^{-1} \; ; \; A_{bm} = u_{bm}/\omega = 0.446 \text{ m} = 44.6 \text{ cm}$$

Assuming the wave boundary layer to be rough turbulent, $k_n = D = 30z_0$, and the wave friction factor is obtained by solving Equation 6-18 using the iterative procedure illustrated by Equation 6-22. For $k_n = D = 0.1$, one obtains

$$\log_{10} \frac{A_{bm}}{k_n} - 0.17 = 3.48$$

and starting with $x^{(0)} = 0.4$, Equation 6-22 gives $x^{(1)} = 0.315$ followed by $x^{(2)} = 0.327$, $x^{(3)} = 0.325$, and finally $x^{(4)} = 0.326 = 4\sqrt{f_w}$. Thus,

$$f_w = (0.326/4)^2 = 6.64 \times 10^{-3}$$

and from Equation 6-15

$$u_{*wm} = \sqrt{\frac{f_w}{2}}\, u_{bm} = 2.02 \times 10^{-2} \text{ m/s} = 2.02 \text{ cm/s}$$

This is the true solution only if the flow is rough turbulent as *assumed*, i.e., only if $k_n u_{*wm}/\nu > 3.3$. Here, $k_n = 0.1$ mm and the u_{*wm} obtained above gives

$$\frac{k_n u_{*wm}}{\nu} = 2.02 < 3.3$$

so the flow is *not rough* but smooth turbulent, i.e., f_w should be obtained from Equation 6-19 and not from Equation 6-18. Equation 6-19 may be rearranged to take the form analogous to Equation 6-22

$$\frac{1}{x^{(n+1)}} = \left(\log_{10} \sqrt{\frac{RE}{50}} - 0.17 \right) - \log_{10} \frac{1}{x^{(n)}} + 0.06 x^{(n)}$$

(Sheet 2 of 3)

Example Problem III-6-2 (Concluded)

with $x = 4\sqrt{4f_w}$ and

$$RE = \frac{u_{bm}A_{bm}}{\nu} = 1.56 \times 10^5$$

Therefore

$$\left(\log_{10}\sqrt{\frac{RE}{50}} - 0.17\right) = 1.58$$

and starting iteration with $x^{(0)} = 0.0$ gives $x^{(1)} = 08.29$, $x^{(2)} = 0.646$, $x^{(3)} = 0.7$,... and finally $x^{(6)} = 0.686 =$

$4\sqrt{4f_w} = 8\sqrt{f_w}$, or the *wave friction factor*

$$f_w = 7.35 \times 10^{-3}$$

which results in the *maximum shear velocity*

$$u_{*wm} = \sqrt{\frac{f_w}{2}}\, u_{bm} = 2.12 \times 10^{-2} \text{ m/s} = 2.12 \text{ cm/s}$$

and a *maximum bottom shear stress*, Equation 6-32,

$$\tau_{wm} = \rho\left(u_{*wm}\right)^2 = 0.461 \text{ N/m}^2 = 0.46 \text{ Pa}$$

The *phase angle* φ of the bottom shear stress is obtained from Equation 6-12 with

$$z_0 = \frac{\nu}{9u_{*wm}} = 5.24 \times 10^{-6} \text{ m}$$

obtained from Equation 6-4 for smooth turbulent flow

$$\tan \varphi = 0.242 \rightarrow \varphi = 13.6° \approx 14°$$

(Sheet 3 of 3)

root-mean-square wave height. Thus, for surface waves specified in terms of their significant height H_s and period T_s, the near-bottom representative periodic orbital velocity is obtained from a wave with root mean square height $H_{rms} = H_s/\sqrt{2}$, and wave period $T = T_s$.

(6) Dissipation. Time-averaged rate of energy dissipation in the wave bottom boundary layer is obtained from Kajiura (1968) as

$$D_p = \frac{1}{4} \rho \ (f_w \cos\varphi) \ u_{bm}^3 \tag{III-6-25}$$

for periodic waves and can be extended to spectral waves using the representative periodic wave properties as given above (Madsen, Poon, and Graber 1988).

  d. *Combined wave-current boundary layers.*

(1) Introduction.

(a) The simple conceptual model of near-bottom turbulence suggests that the eddy viscosity immediately above the bottom is a function of time whenever the flow is unsteady. In the context of a pure wave motion, it was, however, found that assuming a time-invariant eddy viscosity, based on the maximum shear velocity, resulted in a greatly simplified analysis without sacrificing accuracy appreciably (Trowbridge and Madsen 1984). In the case of combined wave-current bottom boundary layer flows, the eddy viscosity immediately above the bottom is therefore scaled by the maximum combined wave-current shear velocity. Since the vertical extent of wave-associated turbulence is limited by the wave bottom boundary layer thickness, the wave contribution to the turbulent mixing must vanish at some level above which only the current shear velocity contributes to turbulent mixing.

(b) For the general case of combined wave-current flows with the current at an angle φ_{wc} to the direction of wave propagation, the maximum combined bottom shear stress τ_m may be obtained from

$$\tau_m = |\boldsymbol{\tau}_{wm} + \boldsymbol{\tau}_c|$$

$$= \sqrt{(\tau_{wm} + \tau_c \ |\cos\varphi_{wc}|)^2 + (\tau_c \sin\varphi_{wc})^2} \tag{III-6-26}$$

$$= \tau_{wm}\sqrt{1 + 2\frac{\tau_c}{\tau_{wm}} \ |\cos\varphi_{wc}| + \left(\frac{\tau_c}{\tau_{wm}}\right)^2}$$

(c) Since the boundary layer flow for combined waves and currents at an angle is three-dimensional, one should in principle solve for the horizontal velocity vector's two components. However, when the x-direction is chosen as the wave direction, unsteady flow will exist only in this direction. In this case it can be shown that all formulas derived for the pure wave case in Part III-6(2)c are valid also for waves in the presence of a current when u_{*wm} is replaced by u_{*m} obtained from Equation 6-26 with $u_{*m} = (\tau_m/\rho)^{1/2}$. For example, the velocity profile is given by Equation 6-11 with u_{*wm} replaced by u_{*m}. As mentioned previously, $u_{*wm} \approx u_{*m}$, which shows that the wave motion is only weakly influenced by the presence of a current.

(d)   The steady flow is entirely in the direction of the current, i.e., at an angle φ_{wc} to the wave direction. Invoking the "law of the wall," the current velocity is (for $z < \delta_{cw}$ where δ_{cw} is defined in Equation 6-38)

$$u_c = \frac{u_{*c}}{\kappa} \frac{u_{*c}}{u_{*m}} \ln \frac{z}{z_0} \tag{III-6-27}$$

and z_0 is given by Equation 6-4 with u_{*m} replacing u_*.

(e)   For $z > \delta_{cw}$ the solution is

$$u_c = \frac{u_{*c}}{\kappa} \ln \frac{z}{z_{0a}} \tag{III-6-28}$$

where z_{0a} denotes an arbitrary constant of integration, referred to as the *apparent bottom roughness*. Matching the two solutions at $z = \delta_{cw}$ and introducing the expression for z_{0a} in Equation 6-28 gives an alternative form of the solution for $z > \delta_{cw}$

$$u_c = \frac{u_{*c}}{\kappa} \left(\ln \frac{z}{\delta_{cw}} + \frac{u_{*c}}{u_{*m}} \ln \frac{\delta_{cw}}{z_0} \right) \tag{III-6-29}$$

(f)   These expressions clearly reveal the pronounced effect that waves may have on current velocity profiles. First of all, the velocity gradient inside the wave-current boundary layer is reduced by a factor u_{*c}/u_{*m} relative to its value in absence of waves. This, of course, is a consequence of the increased turbulence intensity within the wave boundary layer arising from the waves. In the extreme case of $u_{*c}/u_{*m} \approx 0$, u_c remains nearly zero throughout the wave boundary layer. Thus, currents in the presence of waves experience an enhanced bottom roughness (i.e., for the same current velocity at a specified level above the bottom the 0current shear velocity and shear stress will increase with wave activity).

(g)   Figure III-6-5 compares a current velocity profile in the presence of waves predicted by the present wave-current interaction theory with the measured current profile (Bakker and van Doorn 1978). Two theoretical predictions are shown, one in which δ_{cw} is based on Equation 6-23 with $u_{*wm} = u_{*m}$ and another in which δ_{cw} was increased by a factor of 1.5. From the comparison, it is concluded that the definition of δ_{cw} given by Equation 6-23 may be adopted for the wave-current interaction theory presented here. This wave-current interaction theory may be applied to a wave described by its near-bottom orbital velocity spectrum by using the representative periodic wave with the direction of wave propagation chosen as the peak direction (Madsen 1993).

(2)   Combined wave-current velocity profile. Having chosen the x-axis as the direction of wave propagation, the wave orbital velocity profile immediately above the bottom is given by Equation 6-11 where the phase angle φ defined by Equation 6-12 is evaluated with $u_{*wm} = u_{*m}$.

The current velocity vector is given by

$$\boldsymbol{u}_c = u_c(z) \{\cos \varphi_{wc}, \sin \varphi_{wc}\} \tag{III-6-30}$$

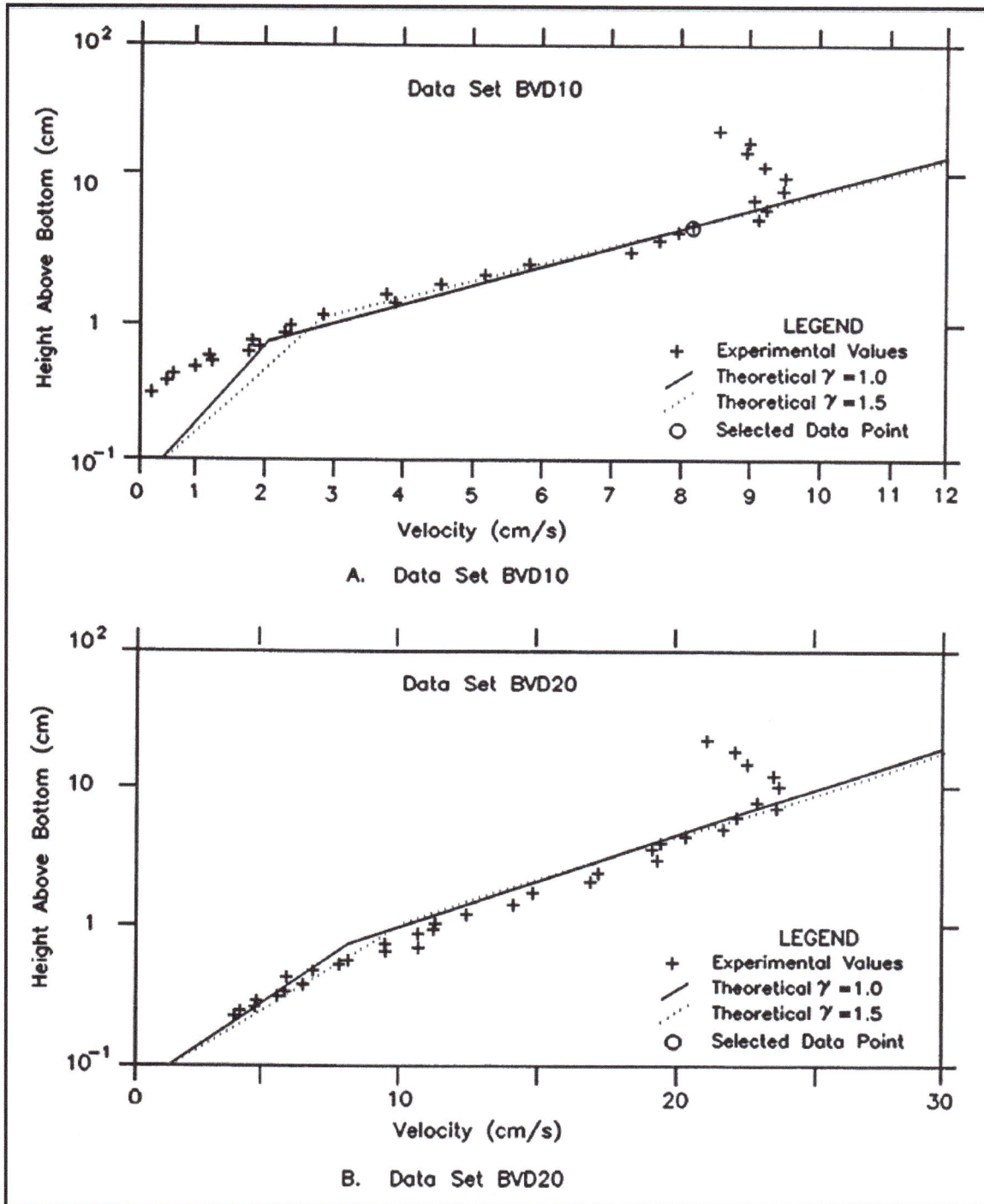

Figure III-6-5. Comparison of current profile in the presence of waves predicted by present model with measured current profile

where $u_c(z)$ is given by Equations 6-27 and 6-29.

(3) Combined wave-current bottom shear stress. Similarly, the time-varying bottom shear stress associated with combined wave-current flows may be obtained from

$$\tau_b(t) = \{\tau_{wm} \cos (\omega t + \varphi) + \tau_c \cos \varphi_{wc}, \ \tau_c \sin \varphi_{wc}\} \qquad \text{(III-6-31)}$$

(4) Methodology for the solution of a combined wave-current problem.

(a) For a wave motion specified by u_{bm} and ω ($A_{bm} = u_{bm}/\omega$) and a bottom roughness specified by its equivalent Nikuradse roughness k_n, the pertinent formulas for the solution of a combined wave-current bottom boundary layer flow problem are the relative strengths of currents and waves

$$\mu = \frac{\tau_c}{\tau_{wm}} = \frac{u_{*c}^2}{u_{*wm}^2} \qquad \text{(III-6-32)}$$

which appears in the factor relating maximum wave and maximum combined bottom shear stresses (Equation 6-26)

$$C_\mu = \sqrt{1 + 2\mu \cos \varphi_{wc} + \mu^2} \qquad \text{(III-6-33)}$$

and wave friction factor formulas, given by Equations 6-18 and 6-19 for pure waves, become for combined wave-current flows (for derivation see Grant and Madsen (1986))

$$\frac{1}{4\sqrt{\dfrac{f_{cw}}{C_\mu}}} + \log_{10} \frac{1}{4\sqrt{\dfrac{f_{cw}}{C_\mu}}} = \log_{10} \frac{C_\mu A_{bm}}{k_n} - 0.17 + 0.24 \left(4\sqrt{\frac{f_{cw}}{C_\mu}}\right) \qquad \text{(III-6-34)}$$

for rough turbulent flow, i.e., $k_n > 3.3\nu/u_{*m}$, and

$$\frac{1}{4\sqrt{\dfrac{4f_{cw}}{C_\mu}}} + \log_{10} \frac{1}{4\sqrt{\dfrac{4f_{cw}}{C_\mu}}} = \log_{10} \sqrt{\frac{C_\mu^2 \, \text{RE}}{50}} - 0.17 + 0.06 \left(4\sqrt{\frac{4f_{cw}}{C_\mu}}\right) \qquad \text{(III-6-35)}$$

for smooth turbulent flow, i.e., $k_n < 3.3\nu/u_{*m}$.

(b) The solution of Equation 6-34 or Equation 6-35 proceeds in the iterative manner illustrated by Equation 6-22 except $x = 4\sqrt{f_{cw}/C_\mu}$ replaces $x = 4\sqrt{f_w}$. Also, Figure III-6-3 may be used with $C_\mu A_{bm}/k_n$ replacing A_{bm}/k_n as entry, and the value of f_w obtained from the ordinate being interpreted as the value of f_{cw}/C_μ.

(c) The wave friction factor in the presence of currents f_{cw} is defined by

$$\frac{\tau_{wm}}{\rho} = u^2_{*wm} = \frac{1}{2} f_{cw} u^2_{bm} \tag{III-6-36}$$

and when Equation 6-33 is introduced in Equation 6-26, the maximum combined shear stress reads

$$\frac{\tau_m}{\rho} = u^2_{*m} = C_\mu u^2_{*wm} \tag{III-6-37}$$

(d)   Finally, the wave boundary layer thickness is given by the modified form of Equation 6-23, i.e.,

$$\delta_{cw} = \frac{\kappa u_{*m}}{\omega} \tag{III-6-38}$$

(e)   Current specified by τ_c, i.e., $\dfrac{\tau_c}{\rho} = u^2_{*c}$ and φ_{wc} are known.

(f)   Starting with $\mu = 0$, $C_\mu = 1$, the wave friction factor is obtained from Equation 6-34. The values of u^2_{*wm} and u^2_{*m} are obtained from Equations 6-36 and 6-37. Rough turbulent flow is checked by calculating $k_n u_{*m}/\nu$. If not greater than 3.3, flow is smooth turbulent and the wave friction factor is recalculated from Equation 6-35, followed by evaluation of u^2_{*m} and u^2_{*wm} as before. In most cases, flow is rough turbulent and the check of rough or smooth turbulent flow is only necessary during the first iteration.

(g)   With the pure wave estimate of $u^2_{*wm} = u^2_{*m}$ in hand, one obtains updated values of μ and C_μ from Equations 6-32 and 6-33. The wave friction factor is obtained from the appropriate relationship, Equation 6-34 or Equation 6-35, by solving for f_{cw}/C_μ followed by multiplication with C_μ. Values of u^2_{*wm} and u^2_{*m} are obtained from Equations 6-36 and 6-37. Upon return to Equation 6-32 with the latest value of u^2_{*wm} the procedure may be repeated until f_{cw} remains essentially unchanged from one iteration to the next. As a reasonable convergence criterion, it is recommended to calculate f_{cw} with no more than three significant figures.

(h)   The wave-current interaction problem is now solved and, following evaluation of the wave boundary layer thickness δ_{cw} from Equation 6-38, the current velocity profile may be obtained from Equation 6-27 for $z \leq \delta_{cw}$ and from Equation 6-29 for $z \geq \delta_{cw}$.

(i)   Current specified by $\boldsymbol{u}_c$ at $z = z_r$ (i.e., $u_c(z = z_r)$) and φ_{wc} are known.

(j)   Again $\mu = 0$ and $C_\mu = 1$ are used for the first iteration which proceeds as outlined under step e above. For this case, however, the first iteration is carried through determination of δ_{cw} from Equation 6-38. At this point, Equation 6-29, assuming $z_r > \delta_{cw}$, may be regarded as a quadratic equation in the unknown current shear velocity

$$\kappa u_c(z_r) = \ln \frac{z_r}{\delta_{cw}} u_{*c} + \frac{\ln \dfrac{\delta_{cw}}{z_0}}{u_{*m}} u^2_{*c} \tag{III-6-39}$$

with the solution

EXAMPLE PROBLEM III-6-3

FIND:

The current shear velocity and bottom shear stress, u_{*c} and τ_c, the maximum wave shear velocity and bottom shear stress, u_{*wm} and τ_{wm}, as well as the maximum combined shear velocity and bottom shear stress, u_{*m} and τ_m, for a combined wave-current flow over a flat bed.

GIVEN:

The wave is specified by its near-bottom maximum orbital velocity $u_{bm} = 0.35$ m/s and period $T = 8$ s. The current is specified by its magnitude $u_c(z_r) = 0.35$ m/s at $z_r = 1.00$ m and direction $\varphi_{wc} = 45°$ relative to the direction of wave propagation. The bottom is flat and consists of uniform sediment of diameter D = 0.2 mm. The fluid is seawater ($\rho = 1,025$ kg/m³, $\nu \approx 1.0 \times 10^{-6}$ m²/s).

PROCEDURE:

1) Compute wave characteristics as in step 1 of Example Problem III-6-2.

2) *Assume* rough turbulent flow and solve the wave-current interaction following the iterative procedure described in subsection III-6(2)d(4)i.

3) Check if flow is rough turbulent by evaluating $k_n u_{*m}/\nu$ with u_{*m} obtained in step 2.

4) If $k_n u_{*m}/\nu \geq 3.3$, flow *is* rough turbulent and values of u_{*c}, u_{*wm}, and u_{*m} obtained in step 2 constitute the solution. Bottom shear stresses are obtained from the definition of shear velocities: $\tau = \rho u_*^2$.

5) If $k_n u_{*m}/\nu < 3.3$ flow *is smooth turbulent* and the combined wave-current flow is solved as in step 2 except for Equation 6-35 replacing Equation 6-34 when the wave friction factor in the presence of a current is evaluated.

6) With u_{*c}, u_{*wm}, and u_{*m} from step 5, $\tau = \rho u_*^2$ is used to obtain bottom shear stresses.

SOLUTION:

For the given wave parameters, $\omega = 2\pi/T = 0.785$ s⁻¹, $A_{bm} = u_{bm}/\omega = 0.446$ m, and $k_n = D = 2\times10^{-4}$ m since the bottom is flat. Proceeding as outlined in Part III-6(2)d subsection (4)a, starting with $\mu = 0$, Equation 6-33 gives $C_\mu = 1$. *Assuming rough* turbulent flow, Equation 6-34 reduces to the exact form of the pure wave friction factor Equation 6-18 whose solution is obtained iteratively using Equation 6-22 (see Example Problem III-6-2) as

$$f_{cw} = f_w = 7.86\times10^{-3}$$

The maximum shear velocity is given by Equation 6-37, i.e.,

$$u_{*m} = \sqrt{C_\mu}\, u_{*wm} = u_{*wm} = \sqrt{\frac{f_{cw}}{2}}\, u_{bm} = 2.19\times10^{-2} \text{ m/s} = 2.19 \text{ cm/s}$$

(Continued)

Sediment Transport Outside the Surf Zone

Example Problem III-6-3 (Concluded)

since C_μ is currently assumed to be 1. From $k_n u_{*m}/\nu = 4.38$ it follows that the flow indeed is *rough turbulent as assumed*. (If $k_n u_{*m}/\nu < 3.3$, one would have to return and compute f_{cw} from Equation 6-35.) It therefore follows from Equation 6-38 that

$$\delta_{cw} = 1.12 \times 10^{-2} \text{ m} = 1.12 \text{ cm}$$

With these values and $z_0 = k_n/30 = D/30 = 6.67 \times 10^{-6}$ m, Equation 6-40 is solved to give a first estimate of the current shear velocity

$$u_{*c} = 1.47 \times 10^{-2} \text{ m/s} = 1.47 \text{ cm/s}$$

Using the first estimates of $u_{*c} = 1.47$ cm/s and $u_{*wm} = 2.19$ cm/s, Equation 6-32 gives

$$\mu = \left(\frac{u_{*c}}{u_{*wm}} \right)^2 = 0.45$$

and with $\varphi_{wc} = 45°$, Equation 6-33 gives $C_\mu = 1.36$.

To solve Equation 6-34 for the wave friction factor in the presence of a current f_{cw}, it is rearranged in a manner identical to Equation 6-22, i.e.,

$$\frac{1}{x^{(n+1)}} = \left(\log_{10} \frac{C_\mu A_{bm}}{k_n} - 0.17 \right) - \log_{10} \frac{1}{x^{(n)}} + 0.24 x^{(n)}$$

with $x = 4\sqrt{f_w/C_\mu}$. Solving iteratively starting with $x^{(0)} = 0.4$ gives $x^{(4)} = 0.342 = 4\sqrt{f_w/C_\mu}$, or

$$f_{cw} = (x/4)^2 C_\mu = 9.94 \times 10^{-3}$$

The value of the wave shear velocity is now obtained from Equation 6-36

$$u_{*wm} = \sqrt{\frac{f_{cw}}{2}} \, u_{bm} = 2.47 \times 10^{-2} \text{ m/s} = 2.47 \text{ cm/s} \quad \text{and} \quad \tau_{wm} = \rho u_{*wm}^2 = 0.624 \text{ Pa}$$

followed by

$$u_{*m} = \sqrt{C_\mu} \, u_{*wm} = 2.88 \times 10^{-2} \text{ m/s} = 2.88 \text{ cm/s} \quad \text{and} \quad \tau_m = \rho u_{*m}^2 = 0.850 \text{ Pa}$$

from Equation 6-37 and

$$\delta_{cw} = \frac{\kappa u_{*m}}{\omega} = 1.47 \times 10^{-2} \text{ m} = 1.47 \text{ cm}$$

With these values in hand, Equation 6-40 gives a second estimate of the current shear velocity

$$u_{*c} = 1.63 \times 10^{-2} \text{ m/s} = 1.63 \text{ cm/s} \quad \text{and} \quad \tau_c = \rho u_{*c}^2 = 0.272 \text{ Pa}$$

EXAMPLE PROBLEM III-6-4

FIND:

The current shear velocity u_{*c}, the maximum wave shear velocity u_{*wm}, the maximum combined shear velocity u_{*m}, and the wave boundary layer thickness δ_{cw} for a combined wave-current turbulent boundary layer flow.

GIVEN:

The wave is specified by its near-bottom maximum orbital velocity $u_{bm} = 0.35$ m/s and period $T = 8.0$ s. The current is specified by its magnitude $u_c = 0.35$ m/s at $z_r = 1.00$ m and direction $\varphi_{wc} = 45°$ relative to the direction of wave propagation. The equivalent Nikuradse roughness of the bottom is $k_n = 4.4$ cm. The fluid is seawater ($\rho = 1,025$ kg/m³, $\nu \approx 10^{-6}$ m²/s).

PROCEDURE:

Identical to Example Problem III-6-3. However, the iterative solution of the wave friction factor in Equation 6-34 is started with an initial value obtained from Figure III-6-3 to speed up convergence.

SOLUTION:

For the given wave parameters $\omega = 2\pi/T = 0.785$ s^{-1} and $A_{bm} = u_{bm}/\omega = 0.446$ m, the procedure to follow is identical to that used in Example Problem III-6-3, except that here

$$\frac{A_{bm}}{k_n} = \frac{0.446}{0.044} \approx 10$$

is much smaller. Therefore, during the first iteration (with $\mu = 0$ and $C_\mu = 1$) the wave friction factor is obtained from Figure III-6-3

$$f_w = f_{cw} \approx 0.058$$

instead of solving Equation 6-34. The pure wave approximation therefore gives

$$u_{*wm} = u_{*m} = \sqrt{\frac{f_w}{2}}\, u_{bm} = 5.96 \text{ cm/s}$$

and

$$\delta_{cw} = \frac{\kappa u_{*wm}}{\omega} = 3.04 \text{ cm}$$

With these values, the first estimate of the current shear velocity is obtained from Equation 6-40

$$u_{*c} = 2.84 \text{ cm/s}$$

(Continued)

Example Problem III-6-4 (Concluded)

An updated value of μ is then

$$\mu = \left(\frac{u_{*c}}{u_{*wm}} \right)^2 = 0.23$$

and from Equation 6-33

$$C_\mu = 1.17$$

Recognizing that Equation 6-34 is identical to the pure wave friction factor (Equation 6-18) if $C_\mu A_{bm}/k_n$ replaces A_{bm}/k_n and f_{cw}/C_μ replaces f_w, the value of f_{cw}/C_μ may be read off Figure III-6-3 when $C_\mu A_{bm}/k_n = 11.7$ is used as an entry. In this manner

$$f_{cw}/C_\mu \approx 0.053 \rightarrow f_{cw} \approx 0.062$$

is obtained. Clearly, the accuracy of the reading obtained from Figure III-6-3 is not impressive, so the value is regarded only as a rough estimate. From Equation 6-36, Equation 6-37, and Equation 6-38, it follows that

$$u_{*wm} = 6.16 \text{ cm/s}, \ u_{*m} = 6.67 \text{ cm/s}, \ \delta_{cw} = 3.40 \text{ cm}$$

and from Equation 6-40

$$u_{*c} = 2.94 \text{ cm/s}$$

With these values $\mu = (u_{*c}/u_{*wm})^2 = 0.23$, i.e., the iteration is complete.

So the solution is $u_{*c} = 2.94$ cm/s, $\tau_c = \rho u_{*c}^2 = 0.89$ Pa, $u_{*wm} = 6.16$ cm/s, $\tau_{wm} = \rho u_{*wm}^2 = 3.89$ Pa, $u_{*m} = 6.67$ cm/s, $\tau_m = \rho u_{*m}^2 = 4.56$ Pa, and $\delta_{cw} = 3.40$ cm.

EXAMPLE PROBLEM III-6-5

FIND:

The current velocity $u_c(\delta_{cw})$ at $z = \delta_{cw}$, the apparent roughness z_{0a}, and the phase angle of the wave-associated bottom shear stress for a combined wave-current bottom boundary layer flow.

GIVEN:

Wave, current, and bottom roughness are as specified in Example Problem III-6-4.

PROCEDURE:

1) Solve the wave-current interaction problem following the procedure described for Example Problem III-6-4.

2) With u_{*m}, from step 1 δ_{cw} is obtained from Equation 6-38.

3) $u_c (z = \delta_{cw})$ is obtained from Equation 6-27 with $z = \delta_{cw}$ obtained in step 2 and $z_0 = k_n/30$ since flow is rough turbulent. If flow had been smooth turbulent, i.e., $k_n u_{*m}/v < 3.3$, Equation 6-27 is used with $z_0 = v/(9u_{*m})$ as given by Equation 6-5.

4) Equation 6-28 is solved for z_{0a} using u_{*c} from step 1, $z = \delta_{cw}$ from step 2, and $u_c = u_c(z = \delta_{cw})$ from step 3.

5) The phase angle φ is obtained from Equation 6-12 with u_{*wm} replaced by u_{*m} from step 1 and $z_0 = k_n/30$.

SOLUTION:

With $u_{*c} = 2.94$ cm/s, $u_{*m} = 6.67$ cm/s, and $\delta_{cw} = 3.40$ cm from Example Problem III-6-4, Equation 6-27 gives (with $z_0 = k_n/30 = 4.4/30$ cm)

$$u_c(\delta_{cw}) = \frac{u_{*c}}{\kappa} \frac{u_{*c}}{u_{*m}} \ln \frac{\delta_{cw}}{z_0} = 10.2 \text{ cm/s}$$

By matching this with the expression for the current velocity profile outside the boundary layer (Equation 6-28) one obtains

$$u_c(\delta_{cw}) = \frac{u_{*c}}{\kappa} \ln \frac{\delta_{cw}}{z_{0a}} \quad \text{and} \quad \text{solving for } z_{0a} \text{ gives} \quad z_{0a} = 0.85 \text{ cm/s}$$

The phase of the wave-associated bottom shear stress is obtained from Equation 6-12 with u_{*wm} replaced by u_{*m}, i.e.,

$$\tan \varphi = 0.79 \rightarrow \varphi = 38°$$

$$u_{*c} = u_{*m} \frac{\ln \dfrac{z_r}{\delta_{cw}}}{\ln \dfrac{\delta_{cw}}{z_0}} \left[-\frac{1}{2} + \sqrt{\frac{1}{4} + \kappa \frac{u_c(z_r)}{u_{*m}} \frac{\ln \dfrac{\delta_{cw}}{z_0}}{\left(\ln \dfrac{z_r}{\delta_{cw}} \right)^2}} \right]$$ (III-6-40)

(k)   Having now obtained an estimate of u_{*c}, returning to Equation 6-32 with u_{*wm} known from the first iteration, the value of μ is updated. With this updated value of μ, the entire procedure is repeated until a new value of u_{*c} is obtained. For this iteration, convergence of μ to no more than two significant digits is recommended.

(l)   The theoretical predictions shown in Figure III-6-5 were obtained in this manner using the indicated data point to specify the current.

III-6-3. Fluid-Sediment Interaction

a.   Introduction.

(1)   In Part III-6-2, the purely hydrodynamic problem of turbulent bottom boundary layer flows was treated. There it was found that the fluid-bottom interaction could be represented by a bottom shear stress τ_b, which, for the general case of combined wave-current flows, is given by Equation 6-31. Since the wave component $\tau_{wm}\cos(\omega t + \varphi)$ and the current component τ_c are both associated with logarithmic velocity profiles immediately above the bottom, Equations 6-24 and 6-27, respectively, the combined velocity immediately above the bottom is logarithmic.

(2)   For a bottom consisting of a granular material, the physical flow condition in the vicinity of the individual grains is sketched in Figure III-6-6. Recalling that the logarithmic velocity profile in this region is merely an extrapolation from flow conditions obtained further from the boundary, it is evident that the actual resistance experienced by the flow is not a uniformly distributed force per unit horizontal area, but rather a sum of drag forces on individual grains (roughness elements). Thus, in reality

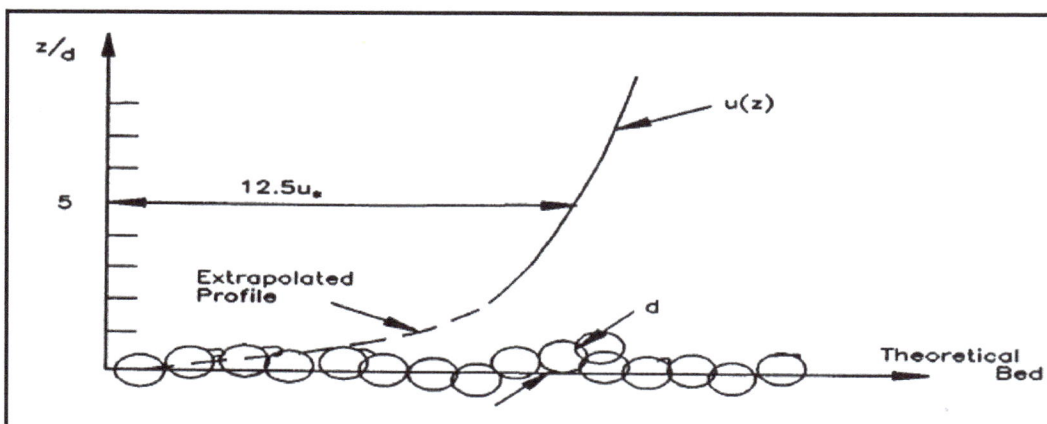

Figure III-6-6.   **Sketch of turbulent flows with logarithmic profile over granular bed**

$$\tau_b = \sum_{\substack{\text{unit} \\ \text{area}}} \bar{F}_{D,\text{grain}}$$

(3)   Not until a level somewhat above the individual roughness elements is it reasonable to assume that eddies, forming around individual grains and ejected into the flow, have merged to produce a turbulent shear flow of the type considered in the conceptual model of turbulent shear stresses presented in Part III-6-2.

(4)   In addition to the spatial variation of the actual interaction between a turbulent flow and a granular bottom, it should also be kept in mind that a turbulent flow, even for a *steady* current, is inherently *unsteady*. The logarithmic velocity profile represents the time-averaged velocity profile with high-frequency turbulent fluctuations being removed by the averaging process. The drag force on individual grains is therefore not constant but varies with time due to turbulent fluctuations. The drag forces on individual grains are therefore average values as suggested by the overbar notation.

(5)   It is evident from the description above that the complex nature of interaction between a flowing fluid and the individual grains on the fluid-sediment interface defies rigorous mathematical treatment unless a hydrodynamic model is employed which resolves the flow structure around individual grains including temporal variation associated with turbulent fluctuations. Since such a detailed hydrodynamic model is not available at present, we are forced to adopt heuristic, physically based conceptual models consistent with the level of our flow description to "derive" quantitative relationships for fluid-sediment interaction, and rely on empirical evidence to determine "constants" that are necessary to render the relationships predictive.

b.   The Shields parameter.

(1)   From the preceding discussion of the nature of near-bottom turbulent flows, it follows that the bottom shear stress may be taken as an expression for the drag force, i.e., the mobilizing force, acting on individual sediment grains on the bed surface. For a sediment of diameter D there are approximately $1/D^2$ grains per unit area, so

$$\bar{\bar{F}}_{D,\text{grain}} \propto \tau_b D^2 \qquad \qquad \text{(III-6-41)}$$

where the double overbar indicates a temporal, as well as a spatial, average.

(2)   For a *cohesionless sediment* the individual sediment grains rest and stay on the bottom due to their submerged weight and resist horizontal motion due to the presence of neighboring grains. Thus, for a cohesionless sediment, the stabilizing force is associated with the submerged weight of the individual grains

$$W_{grain} \propto (\rho_s - \rho)gD^3 \qquad \qquad \text{(III-6-42)}$$

where ρ_s denotes the density of the sediment material ($\rho_s \approx 2{,}650$ kg/m^3 for quartz).

(3)   The ratio of mobilizing (drag) and stabilizing (submerged weight) forces is of fundamental physical significance in fluid-sediment interaction for *cohesionless sediment*. This ratio is known as the *Shields parameter* (Shields 1936)

$$\psi = \frac{\tau_b}{(\rho_s - \rho)gD} = \frac{\tau_b}{(s - 1)\rho gD} = \frac{u_*^2}{(s - 1)gD} \qquad \qquad \text{(III-6-43)}$$

in which $s = \rho_s/\rho$ is the density of the sediment relative to that of the fluid.

c. Initiation of motion.

(1) Introduction.

(a) For a turbulent flow over a flat bed consisting of cohesionless sediment of diameter D the equivalent Nikuradse sand grain roughness is $k_n = D$. From the discussion of turbulent boundary layer flows it was found that the roughness scale experienced by the flow, z_0 as given by Equation 6-5, depends on whether the flow is characterized as smooth or rough turbulent. The parameter determining the characteristics of the near-bottom flow and hence the mobilizing force acting on individual sediment grains is the *boundary Reynolds number*

$$\text{Re}_* = \frac{u_* k_n}{\nu} = \frac{u_* D}{\nu} \tag{III-6-44}$$

(b) From simple physical considerations it therefore follows that the conditions of neutral stability of a sediment grain on the fluid-sediment interface may be expressed as a *critical value of the Shields parameter*

$$\psi_{cr} = \frac{\tau_{cr}}{(s-1)\rho g D} = \frac{u_{*cr}^2}{(s-1)g D} = f(\text{Re}_*) \tag{III-6-45}$$

where $f(\text{Re}_*)$, which is obtained from models or from experimental data, denotes some function of Re_*. Figure III-6-7 shows the traditional Shields diagram given by Equation 6-45 with supporting data obtained from uniform steady flow experiments (Raudkivi 1976). It shows that $\psi_{cr} \approx 0.06$ is approximately constant for $\text{Re}_* > 100$, i.e., for fully rough turbulent flow, whereas the critical Shields parameter increases steadily from a minimum of 0.035 for values of $\text{Re}_* < 10$, i.e., essentially corresponding to smooth turbulent flow conditions.

(c) In establishing the critical condition for initiation of motion, also referred to as threshold or incipient motion condition (expressed formally by Equation 6-45), it is important to keep in mind the nature of fluid-sediment interaction. For values of $\psi > \psi_{cr}$, mobilizing forces exceed stabilizing forces and sediment motion occurs. This does not mean that for $\psi < \psi_{cr}$ by a small amount, sediment does not move at all. For turbulent flow conditions in the vicinity of $\psi \approx \psi_{cr}$, the arrival of particularly strong turbulent eddies from the turbulence generation layer may momentarily increase the drag on one or a few sediment grains and, if the response time of the individual grains is shorter than the duration of this pulse, a few grains may be dislodged (i.e., a movement of the sediment may occur). The empirical curve on a ψ versus Re_* diagram that expresses the critical condition (i.e., the Shields curve or the Shields criterion) should therefore be interpreted as representing a "gray area" around the curve itself. For flow conditions within this gray area, sediment movement may take place but it becomes increasingly sporadic as ψ drops below the curve defining ψ_{cr}.

(2) Modified Shields diagram.

(a) In order to predict the flow condition that will cause sediment motion for a given sediment (i.e., s and D known) the traditional Shields curve given by Equation 6-45 or Figure III-6-7 is somewhat cumbersome to use, since the flow characteristic of interest (u_{*cr}) is involved in both parameters. This problem can be circumvented by recognizing that the Shields curve defines a unique relationship between ψ_{cr} and Re_*. Thus, from the definition of ψ_{cr} (Equation 6-45), one obtains

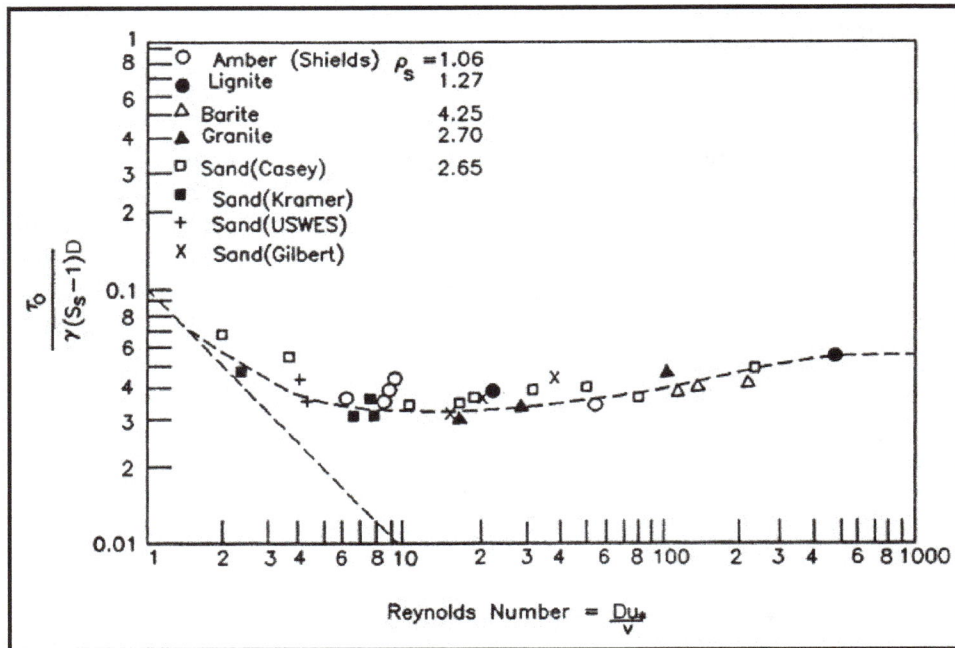

Figure III-6-7. Shields diagram for initiation of motion in steady turbulent flow (from Raudkivi (1967))

$$u_{*cr} = \sqrt{(s - 1) g D} \ \sqrt{\psi_{cr}} \tag{III-6-46}$$

which can be introduced in the definition of Re$_*$ to obtain a parameter (Madsen and Grant 1976)

$$S_* = \frac{D}{4v}\sqrt{(s - 1) g D} = \frac{\text{Re}_*}{4\sqrt{\psi_{cr}}} \tag{III-6-47}$$

(b)   The factor of 4 appears in the definition of S_* to render the numerical values of S_* comparable with the Re$_*$ values in the traditional Shields diagram. This is done merely for convenience and has no physical significance.

(c)   By taking corresponding values of Re$_*$ and ψ_{cr} from the traditional Shields diagram, one can obtain the value of the sediment-fluid parameter S_*, which in turn can be used to replace Re$_*$ in the Shields diagram. In this manner the *modified Shields diagram* (Madsen and Grant 1976) shown in Figure III-6-8 can be constructed.

(d)   The modified Shields diagram terminates at the lower end of S_* at a value of 1 which, for quartz sediments in seawater, corresponds to a sediment diameter of $D \approx 0.1$ mm, i.e., a very fine sand. Although cohesion may become important for diameters below 0.1 mm, clean silty sediments, i.e., without too much organic material and/or clay, can be considered cohesionless and therefore governed by a Shields-type initiation of motion criteria. Raudkivi (1976) presents limited experimental data on threshold conditions obtained for low values of Re$_*$ (in the range 0.03 to 1). A best-fit line to the portion of these data obtained for grain-shaped sediments gives the following criterion

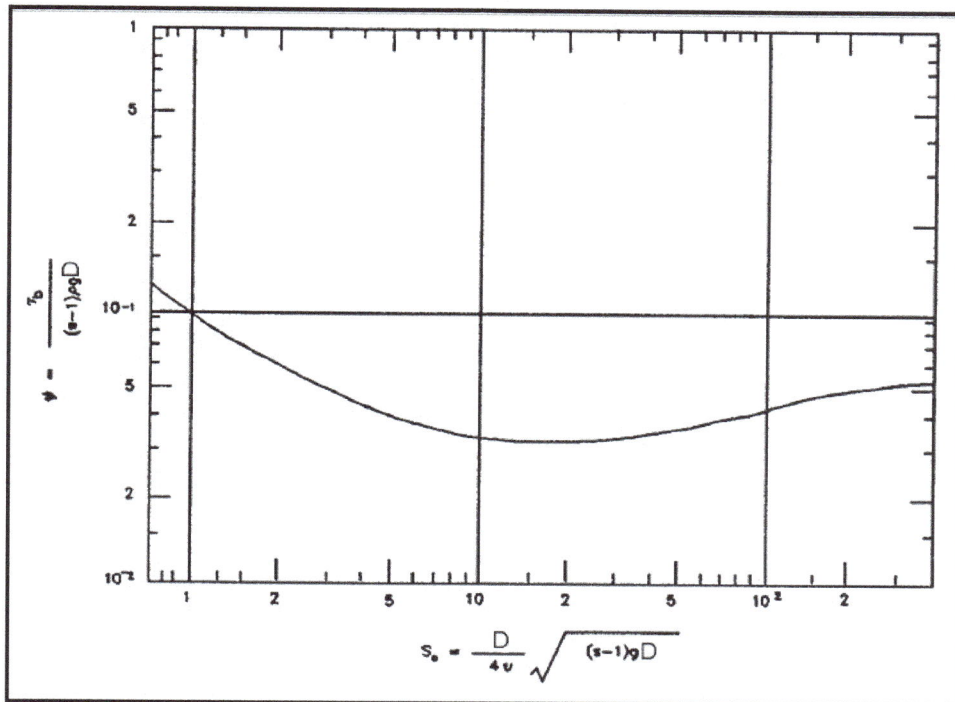

Figure III-6-8. Modified Shields diagram (Madsen and Grant 1976)

$$\psi_{cr} = 0.1\text{Re}_*^{-1/3} \quad \text{for} \quad \text{Re}_* < 1 \tag{III-6-48}$$

which upon the modification, i.e., using Equation 6-47, may be written as

$$\psi_{cr} = 0.1S_*^{-2/7} \quad \text{for} \quad S_* < 0.8 \tag{III-6-49}$$

(e) For large values of S_*, the flow corresponding to initiation of motion is fully rough turbulent, $\text{Re}_* > 100$, and the value of $\psi_{cr} \approx 0.06$ may be used for S_* values larger than the range covered in Figure III-6-9.

(3) Modified Shields criterion.

(a) The modified Shields criterion for initiation of motion of a cohesionless sediment shown in Figure III-6-8 was obtained from steady turbulent flow experiments. The fact that this criterion is applicable also for unsteady turbulent boundary layer flows, as demonstrated by the results presented in Figure III-6-9 (Madsen and Grant 1976), is not surprising when the nature of the onset of sediment movement is considered. For flow conditions in the vicinity of $\psi \approx \psi_{cr}$, the sporadic movement of a few grains is, as discussed above, associated with high-frequency turbulent fluctuations in the mobilizing force acting on a grain. Since the near-bottom mean velocity profile is logarithmic for waves as well as for currents, it is physically reasonable to expect the near-bottom turbulence to be similar, i.e., scaled by the instantaneous value of the bottom shear velocity. Provided therefore that the response time of the individual sediment grains is short relative to the time scale of turbulent fluctuations, which are expected to be the same for waves and currents if the shear velocities are the same, initiation of motion will be affected by unsteadiness only if the wave period is on the order of the time scale of the turbulent fluctuations. This not being the case, the effects of unsteadiness

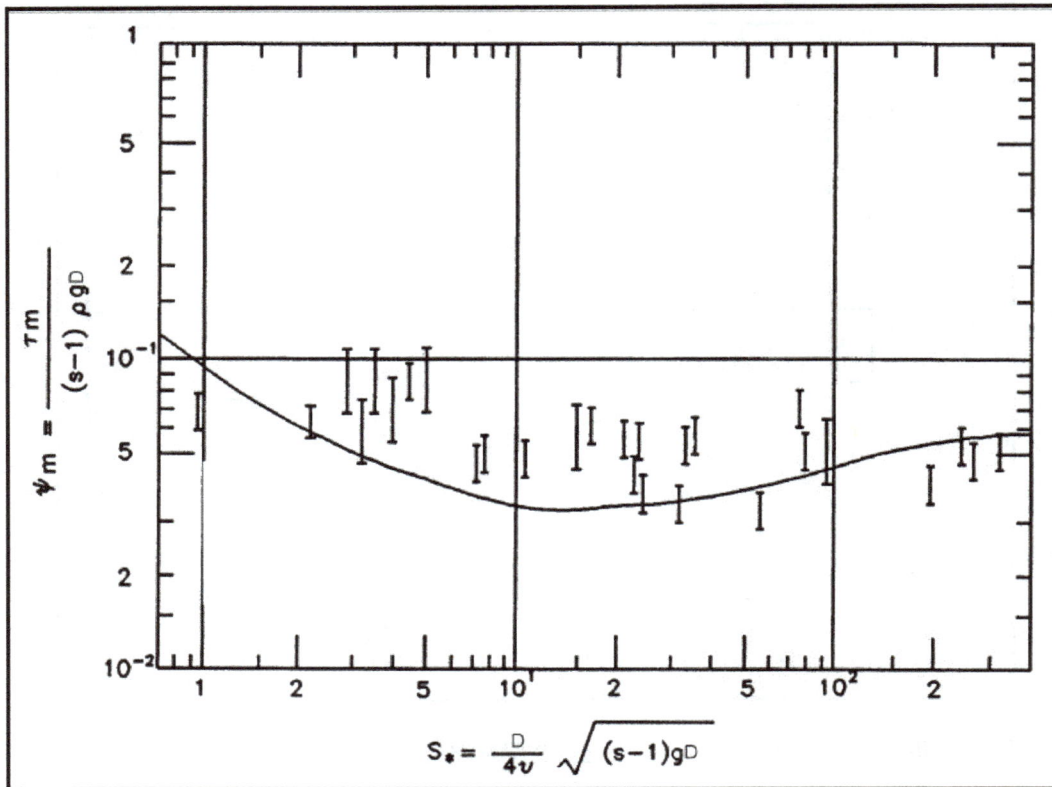

Figure III-6-9. Comparison of Shields curve with data on initiation of motion in oscillatory turbulent flows (after Madsen and Grant ((1976))

of the mean flow are negligible and sediment grains react essentially instantaneously to the applied shear stress, i.e., initiation of motion is obtained from

$$\psi_m = \frac{u_{*m}^2}{(s - 1)gD} = \psi_{cr} \qquad \qquad \text{(III-6-50)}$$

in which u_{*m} denotes the maximum shear velocity, i.e., $u_{*m} = u_{*c}$, u_{*wm}, and u_{*m} for pure current, pure wave, and combined wave-current flows, respectively.

(b)   The effect of bottom slope on the initiation of motion of sediment grains on the fluid-sediment interface may be accounted for by considering a modification of the simple force balance (Madsen 1991) between fluid drag, gravity, and frictional resistance against movement along a plane bottom inclined at an angle β to horizontal in the direction of flow, where β is taken positive if the bottom is sloping upward in the direction of flow. The resulting force balance suggests that the critical Shields parameter for flow over a sloping bottom may be expressed as

$$\psi_{cr,\beta} = \psi_{cr}\left\{\cos \beta \left(1 + \frac{\tan \beta}{\tan \varphi_s}\right)\right\} \qquad \qquad \text{(III-6-51)}$$

where φ_s is the friction angle for a stationary spherical interfacial grain.

EXAMPLE PROBLEM III-6-6

FIND:

The critical shear velocity u_{*cr} and critical bottom shear stress τ_{cr} for initiation of motion.

GIVEN:

Sediment is quartz sand, $\rho_s = 2,650$ kg/m^3, of diameter $D = 0.2$ mm. Fluid is seawater ($\rho = 1,025$ kg-/m^3, $\nu \approx 1.0 \times 10^{-6}$ m^2/s).

PROCEDURE:

1) Compute value of the sediment-fluid parameter S_* defined by Equation 6-47.

2) *If $S_* < 0.8$ obtain ψ_{cr} from Equation 6-49 (be concerned about sediment being cohesive since formula is limited to cohesionless sediments).*

3) If $S_* > 300$, $\psi_{cr} \approx 0.06$.

4) If $0.8 < S_* < 300$ obtain ψ_{cr} from Figure III-6-8.

5) Solve Equation 6-45 for u_{*cr} with ψ_{cr} obtained in step 2, 3, or 4.

6) Obtain $\tau_{cr} = z\rho\, u_{*cr}^2$ from definition of shear velocity.

SOLUTION:

The sediment-fluid parameter defined by Equation 6-47 is first computed with $s = \rho_s/\rho = 2.59$ and $g = 9.80$ m/s^2

$$S_* = \frac{D}{4\nu}\sqrt{(s - 1)gD} = 2.79$$

With this value of S_* the critical Shields parameter is obtained from Figure III-6-8

$$\psi_{cr} \approx 0.052$$

and using the definition of the Shields parameter given by Equation 6-45

$$u_{*cr} = \sqrt{(s - 1)gD\, \psi_{cr}} = 1.27 \times 10^{-2} \text{ m/s} = 1.27 \text{ cm/s}$$

or $\tau_{cr} = \rho(u_{*cr})^2 = 0.165$ N/m$^2 \approx 0.17$ Pa

If $S_* < 0.8$ had been obtained, Equation 6-49 and *not* Figure III-6-8 should be used to find ψ_{cr}.

d.   Bottom roughness and ripple generation. When the flow intensity (expressed in terms of the Shields parameter ψ) exceeds the critical condition for initiation of motion ψ_{cr}, sediment grains will start to move virtually instantaneously (Madsen 1991). However, for ψ exceeding ψ_{cr} by a modest amount (ψ > (1.1 to 1.2)ψ_{cr}), the plane bottom will no longer remain plane. It becomes unstable, deforms, and exhibits bed forms and ripples.

(1)   The skin friction concept.

(a)   The appearance of bed forms on the sediment-fluid interface changes the appearance of the bottom to the flow above. Rather than giving rise to a flow resistance associated with drag forces on individual sediment grains, i.e., a roughness scaled by the sediment diameter, the flow will separate at the crest of the bed forms and flow resistance will primarily be composed of pressure drag forces on bottom bed forms, i.e., a roughness scaled by the bed form geometry. The appearance of ripples on the bottom changes the physical bottom roughness by orders of magnitude.

(b)   Despite the increased bottom roughness associated with a rippled bed, the *drag force acting on individual sediment grains and not the drag force on the bed forms* is responsible for the sediment motion caused by the flow. In the terminology of bottom shear stresses, it is the average shear stress acting on the sediment grains, *the skin friction*, that moves sediment and not the total shear stress comprising skin friction and form drag. The concept of partitioning bottom shear stress into a skin friction and a form drag component, i.e., taking

$$\tau_b = \tau_b' + \tau_b'' \tag{III-6-52}$$

in which τ_b' denotes the skin friction component and τ_b'' the form drag component, illustrated in Figure III-6-10, has received considerable attention in the context of sediment transport mechanics in steady turbulent flows (Raudkivi 1976).

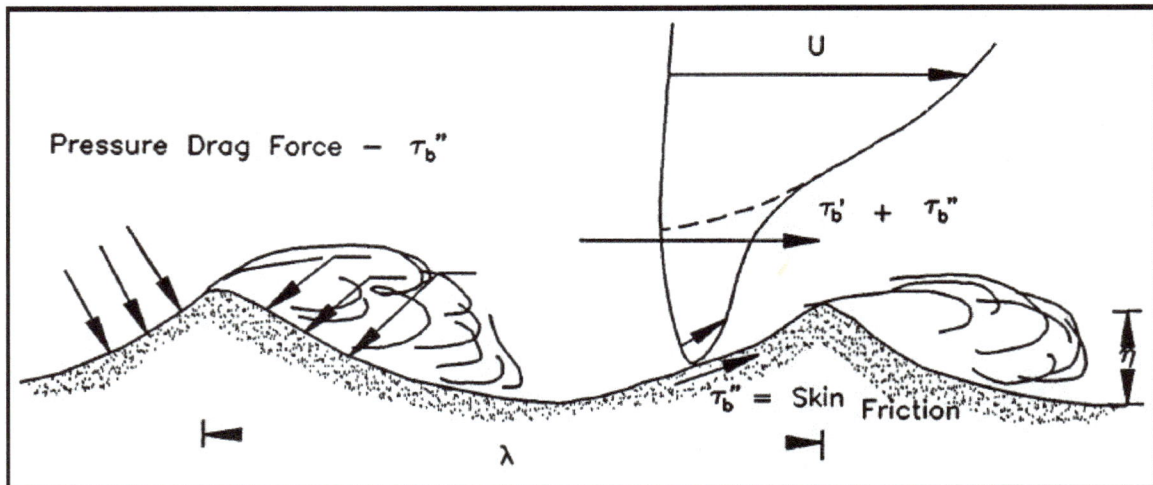

Figure III-6-10.   Conceptualization of pressure drag τ_b'', skin friction τ_b', total shear stress $\tau_b'+\tau_b''$, for turbulent flow over a rippled bed

(c)   For a pure wave motion, it is assumed that the skin friction bottom shear stress is obtained quite simply (Madsen and Grant 1976) by evaluating the bottom shear stress from the given wave condition, u_{bm} and ω, as if the bottom roughness scale were the sediment grain diameter D. Thus, for a wave motion, the skin friction bottom shear stress is obtained from

$$\tau'_{wm} = \frac{1}{2}\rho f'_w u_{bm}^2 \qquad \text{(III-6-53)}$$

where f'_w is the wave friction factor obtained from Equation 6-18, Equation 6-19, or Equation 6-21 for a roughness

$$k_n = k'_n = D \qquad \text{(III-6-54)}$$

Laboratory data on wave-generated ripples as well as ripple geometry for spectral waves is discussed (Madsen 1993) in some detail.

(d)   For combined wave-current flows it has been proposed (Glenn and Grant 1987) to evaluate the skin friction bottom shear stress in a similar manner, i.e., using the wave-current interaction model presented in Part III-b(2)d with a roughness specified by Equation 6-54. However, if a combined wave-current boundary layer flow is specified by a current velocity outside the wave boundary layer, i.e., $u_c(z_r)$ with $z_r > \delta_{cw}$, the direct use of this information in the formulas given in Part III-6(2)d with $k_n = k'_n = D$ would lead to an ambiguous determination of the skin friction, since this approach presumes the variation of the current velocity outside the wave boundary layer to be governed by the grain size roughness, which is true only for a flat bed, i.e., in the absence of form drag.

(e)   To overcome this problem, the combined wave-current theory presented in Part III-6(2)d is first used with k_n = total equivalent Nikuradse sand grain roughness, accounting for the presence of bed forms, to predict the current velocity at the outer edge of the wave boundary layer, i.e.,

$$u_c(z_r) \quad \text{with} \quad z_r = \delta_{cw} \qquad \text{(III-6-55)}$$

is obtained from a model considering total bottom shear stresses.

(f)   With the current specified by Equation 6-55, the skin friction shear stress is now computed from the general wave-current interaction theory using the grain size roughness $k_n = k'_n = D$. In this manner (Wikramanayake and Madsen 1994a), the ambiguity of the prediction of combined wave-current skin friction shear stresses is avoided.

(2)   Field data on geometry of wave-generated ripples.

(a)   When field data on ripple geometry are plotted against the skin friction Shields parameter,

$$\psi'_m = \frac{\tau'_{wm}}{(s-1)\rho gD} = \frac{(u'_{*wm})^2}{(s-1)gD} \qquad \text{(III-6-56)}$$

they exhibit an extensive scatter. By trial and error it was found (Wikramanayake and Madsen 1994a) that the parameter providing the best correlation of field data on ripple geometry was

$$Z = \frac{\psi_m'}{S_*} \qquad \text{(III-6-57)}$$

in which S_* is the sediment-fluid parameter defined by Equation 6-47.

(b)   The equations for the empirical relationships for ripple geometry in the field are (Madsen 1993).

$$\frac{\eta}{A_{bm}} = \begin{cases} 1.8 \times 10^{-2} \, Z^{-0.5} & 0.0016 < Z < 0.012 \\\\ 7.0 \times 10^{-4} \, Z^{-1.23} & 0.012 < Z < 0.18 \end{cases} \qquad \text{(III-6-58)}$$

and

$$\frac{\eta}{\lambda} = \begin{cases} 1.5 \times 10^{-1} \, Z^{-0.009} & 0.0016 < Z < 0.012 \\\\ 1.05 \times 10^{-2} \, Z^{-0.65} & 0.012 < Z < 0.18 \end{cases} \qquad \text{(III-6-59)}$$

where A_{bm} is defined by Equation 6-17, η is ripple height, λ is ripple length, and the lowest and highest range of validity indicate the range covered by the experimental data.

(3)   Prediction of ripple geometry under field conditions.

(a)   From knowledge of the representative periodic wave (recall that this is defined here as the rms wave and not the significant wave) and sediment characteristics, the skin friction shear stress τ_{wm}' is obtained from Equation 6-53.

(b)   With τ_{wm}' known, the value of ψ_m' is obtained from Equation 6-56 and if $2\psi_m' > \psi_{cr}$, obtained from the modified Shields curve in Figure III-6-8, sediment motion is assumed to take place and the parameter Z is obtained from Equation 6-57.

With Z known, the empirical relationship given by Equations 6-58 and 6-59 is used to obtain the ripple geometry. If ψ_m' exceeds a value of 0.35 or if $Z > 0.18$, the bed is assumed flat corresponding to sheet flow conditions.

e.   Moveable bed roughness.

(1)   The determination of <u>moveable bed roughness</u> over naturally rippled beds is limited by the lack of reliable field measurements over a range of wave conditions. The limited amount of available data on wave-generated ripple geometry and the associated rate of energy dissipation in the wave bottom boundary layer above the rippled bed have been analyzed (Wikramanayake and Madsen 1994a). Their analysis results suggest the adoption of the following simple relationship for moveable bed roughness associated with rippled beds

$$k_n = 4\eta \qquad \text{(III-6-60)}$$

where η is the ripple height. However, the translation of ripple geometry to equivalent roughness, Equation 6-60, is based entirely on laboratory data for periodic waves and is of dubious value for field roughness determinations. Therefore, the following procedure is suggested (Madsen 1993).

(2)   For a wave condition specified in terms of the representative periodic wave, the ripple height is obtained from Equation 6-58, and the movable bed roughness k_n is obtained from Equation 6-60.

(3)   If ψ_m' suggests no sediment motion, $2\psi_m' < \psi_{cr}$, the movable bed roughness is taken as $k_n = k_n' = D$ unless information on bottom roughness associated with bioturbation of the bottom sediments is available.

(4)   If $\psi_m' > \approx 0.35$ or if $Z > 0.18$, the bed is expected to be flat, i.e., corresponding to sheet flow conditions. Recent results from field experiments (Madsen et al. 1993) suggest a value of $k_n = 15D$ to be appropriate for sheet flow conditions. Since the conditions from which this result was obtained correspond to values of $\psi_m' \approx 1$ and since it is physically reasonable to assume the movable bed roughness for sheet flow conditions depends on the intensity of sediment transport over the flat bed, which in turn is related to the magnitude of ψ_m', it is *tentatively proposed* that

$$k_n = 15 \psi_m' D \qquad\qquad (III\text{-}6\text{-}61)$$

be adopted for sheet flow conditions. This expression agrees in form with an expression proposed for steady flow (Wilson 1989) although the coefficients differ.

III-6-4. Bed-load Transport

a. Introduction.

(1)   When flow and sediment characteristics combine to produce a Shields parameter greater than the critical value, sediment is set in motion. One result of this is the generation of bed forms, discussed in Part III-6-3; another is, of course, the initiation of a non-zero transport of sediment.

(2)   For values of the Shields parameter slightly above critical or, more specifically, for low transport rates, the predominant mode of sediment transport takes place as individual grains rolling, sliding, and/or jumping (saltating) along the bed. This mode of sediment transport is referred to as *bed load* and, since it takes place in close proximity of the bed, it is dependent on *skin friction*.

(3)   Several empirical bed-load transport formulas have been proposed for steady turbulent boundary layer flows (Raudkivi 1976). One of these, the Meyer-Peter and Müller formula (Meyer-Peter and Müller 1948), enjoys particular popularity in the engineering literature. However, since it was originally developed from data obtained for steady flows in rivers and channels of negligible slope, its adoption for unsteady wave or combined wave-current turbulent boundary layer flows over an inclined bed is not straightforward. A conceptual model for bed-load transport of sediment grains rolling or sliding along an inclined bottom has been proposed (Madsen 1993) as a physical interpretation of the purely empirical Meyer-Peter and Müller formula. The resulting formulation is time-averaged and solved under the assumptions that: waves dominate

EXAMPLE PROBLEM III-6-7

FIND:

The height η, length λ, and movable bed roughness k_n, of wave-generated ripples in the field.

GIVEN:

The representative periodic wave motion is specified as in Example Problem III-6-3. Bottom sediment is quartz, $\rho_s = 2{,}650$ kg/m^3, of diameter $D = 0.2$ mm. Fluid is seawater ($\rho = 1{,}025$ kg/m^3, $\nu \approx 1.0 \times 10^{-6}$ m^2/s).

PROCEDURE:

1) Obtain the wave shear velocity u'_{*wm} assuming grain size roughness $k'_n = d =$ median sediment diameter, following the procedure described in Example Problem III-6-2. (Result is denoted here by *prime* to signify *skin friction* shear velocity.)

2) With u'_{*wm} from step 1 compute ψ'_m from Equation 6-56.

3) Evaluate S_* as defined by Equation 6-47.

4) Determine ψ_{cr} following procedure of Example Problem III-6-6.

5) If $\psi'_m < \frac{1}{2}\psi_{cr}$ there will be no sediment motion and no ripples, i.e., $\eta = \lambda = 0$. Bottom roughness *is* the sediment grain diameter ($k_n = k'_n = D$) unless other information is available, e.g., photos showing relict ripples or other roughness features.

6) If $\psi'_m > 0.35$, sheet flow is assumed. The bed is flat, i.e., $\eta = \lambda = 0$, and the movable bed roughness k_n is obtained from Equation 6-61.

7) If $\frac{1}{2}\psi_{cr} < \psi'_m < 0.35$, the parameter Z defined by Equation 6-57 is computed.

8) If Z falls within the range indicated for Equations 6-58 and 6-59, these are used to compute η and λ. The movable bed roughness k_n is obtained from Equation 6-60.

9) If $Z < 0.0016$ and $D < 0.6$ mm, it is *recommended* that η, λ, and k_n be obtained as if $Z = 0.0016$.

10) If $Z > 0.18$ and $D > 0.08$ mm, sheet flow is assumed (i.e., $\eta = \lambda = 0$), and Equation 6-61 is used to obtain k_n.

(Continued)

Example Problem III-6-7 (Concluded)

SOLUTION:

In Example Problem III-6-3, the initial solution for the wave friction factor neglected the presence of a current ($\mu = 0$, $C_\mu = 1$) and gave $f_w = 7.86 \times 10^{-3}$. Since the wave motion and the bottom roughness ($k_n = k_n' = D = 0.2$ mm) are the same here, the maximum skin friction shear velocity is given by

$$u_{*wm}' = \sqrt{f_w'/2} \; u_{bm} = 2.19 \times 10^{-2} \text{ m/s} = 2.19 \text{ cm/s}$$

From the definition of the Shields parameter, Equation 6-56, it follows that

$$\psi_m' = \frac{\left(u_{*wm}'\right)^2}{(s - 1)gD} = 0.154$$

The sediment being the same as specified in Example Problem III-6-5 where $\psi_{cr} = 0.052$ was obtained, it is seen that the sediment will be moving, since $\psi_m' > \psi_{cr}$. Since $\psi_m' < 0.35$, flat bed (sheet flow) is not predicted and bed forms are therefore expected. The geometry of these bed forms is obtained from the field relationships, Equations 6-58 and 6-59. With the parameter

$$Z = \frac{\psi_m'}{S_*} = 0.055$$

since $S_* = 2.79$ in Example Problem III-6-6, which is within the range $0.012 < Z < 0.18$, one therefore obtains from Equation 6-58

$$\frac{\eta}{A_{bm}} = 7 \times 10^{-4} \; Z^{-1.23} = 2.47 \times 10^{-2}$$

or, with $A_{bm} = 44.6$ cm (Example Problem III-6-3)

$$\eta = \frac{\eta}{A_{bm}} A_{bm} = 1.1 \text{ cm}$$

The ripple steepness is obtained from Equation 6-59

$$\frac{\eta}{\lambda} = 1.05 \times 10^{-2} \; Z^{-0.65} = 6.91 \times 10^{-2} \rightarrow \lambda = \eta/(\eta/\lambda) = 15.9 \text{ cm} \approx 16 \text{ cm}$$

From Equation 6-60, one obtains the movable bed roughness

$$k_n = 4\eta = 4.4 \text{ cm}$$

EXAMPLE PROBLEM III-6-8

FIND:

The skin friction shear velocities associated with the current u'_{*c}, the wave u'_{*wm}, and the maximum combined u'_{*m}, as well as the phase angle of the wave-associated shear stress φ', for a combined wave-current flow over a movable bed in the field.

GIVEN:

The representative periodic wave is specified by its near-bottom maximum orbital velocity $u_{bm} = u_{br} = 0.35$ m/s and period $T = 8.0$ s. The current is specified by its magnitude $u_c(z_r) = 0.35$ m/s at $z_r = 1.00$ m and direction $\varphi_{wc} = 45°$ relative to the direction of wave propagation. The bed consists of uniform quartz sand, $\rho_s = 2,650$ kg/m³, of diameter $D = 0.2$ mm. The fluid is seawater ($\rho = 1,025$ kg/m³, $\nu \approx 1.0 \times 10^{-6}$ m²/s).

PROCEDURE:

1) Considering *only* the wave motion, the movable bed roughness k_n is obtained following the procedure given in Example Problem III-6-7.

2) With k_n from step 1, the combined wave-current problem is solved following the procedure described in Example Problem III-6-4.

3) From the combined wave-current flow solution obtained in step 2, the current velocity $u_c(z = \delta_{cw})$ is computed by the procedure of Example Problem III-6-5.

4) With the current specification $u_c = u_c(\delta_{cr})$ at $z = \delta_{cw}$ from step 3, the wave-current interaction problem is solved for $k'_n = D$ following the procedure described in Example Problem III-6-3.

5) The skin friction phase angle φ' is obtained from Equation 6-12 with $u_{*wm} = u'_{*m}$.

SOLUTION:

The *first step* to follow in determining the skin friction components for a combined wave-current boundary layer flow is to determine the movable bed roughness. Since the presence of a current is neglected in performing this step, this was done in Example Problem III-6-7 for the wave and sediment characteristics considered here and resulted in $k_n = 4.4$ cm.

The *second step* is to determine the current velocity at the edge of the boundary layer, $u_c(\delta_{cw})$ from the wave-current interaction model using the movable bed roughness obtained in step 1. This was done in Example Problems III-6-4 and III-6-5 where the bottom roughness was specified as $k_n = 4.4$ cm, i.e., exactly the movable bed roughness determined in step 1 for the given wave and sediment characteristics.

(Continued)

Example Problem III-6-8 (Concluded)

The *third step* is to determine the shear velocities and stresses associated with the combined wave-current boundary layer flow over a bottom of roughness $k_n = k_n' = D$ = sediment diameter with the current specified as obtained in step 2. From Example Problem III-6-5, the current specification is therefore

$$u_c(z_r) = 10.2 \text{ cm/s} \quad \text{at} \quad z_r = \delta_{cw} = 3.40 \text{ cm}$$

at an angle $\varphi_{wc} = 45$ deg to the direction of wave propagation. With this current specification, $k_n = k_n' = D = 0.2$ mm, and wave characteristics $u_{bm} = 35$ cm/s and $\omega = 2\pi/T = 0.785$ s^{-1} ($A_{bm} = u_{bm}/\omega = 44.6$ cm) solution of the wave-current interaction problem follows Example Problem III-6-3.

For $\mu = 0$ and $C_\mu = 1$, the first iteration is identical to that in Example Problem III-6-3, resulting in $u_{*wm}' = u_{*m}' = 2.19$ cm/s and $\delta_{cw}' = 1.12$ cm. With these values and using the current specifications given above, Equation 6-40 gives

$$u_{*c}' = 0.96 \text{ cm/s}$$

Second iteration is therefore based on $\mu = \left(u*c'/u_{*wm}'\right)^2 = 0.194$ and, from Equation 6-33, $C_\mu = 1.15$. For this value of C_μ solving iteratively for $x = 4\sqrt{f_{cw}'/C_\mu}$ in the manner outlined in Example Problem III-6-3 results in $x^{(3)} = 0.349$, or

$$f_{cw}' = 8.73 \times 10^{-3}$$

and therefore

$$u_{*wm}' = 2.31 \text{ cm/s} \; ; \quad u_{*m}' = 2.48 \text{ cm/s} \quad \text{and} \quad \delta_{cw}' = 1.26 \text{ cm}$$

with which Equation 6-40 gives

$$u_{*c}' = 1.01 \text{ cm/s}$$

The next value of $\mu = \left(u_{*w}/u_{*wm}'\right)^2 = 0.191$ is virtually identical to the previous value, so no further iteration is necessary.

The solution is therefore

$$u_{*c}' = 1.01 \text{ cm/s} \; , \quad u_{*wm}' = 2.31 \text{ cm/s} \; , \quad u_{*m}' = 2.48 \text{ cm/s}$$

and the phase angle of the wave-associated skin friction bottom shear stress is obtained from Equation 6-12 with $u_{*wm} = u_{*m}'$ and $z_0 = d/30$

$$\tan \varphi' = 0.25 \rightarrow \varphi' = 13.8° \approx 14°$$

currents $\mu = \tau'_c/\tau'_{wm} << 1$; maximum shear stress is much larger than critical $\mu_{CR} = \tau_{CR}/\tau'_{wm} << 1$; and bottom slope is relatively small

$$\mu_b = \frac{\tan \beta}{\tan \varphi_m} << 1 \qquad \text{(III-6-62)}$$

where $\beta > 0$ is bottom slope upward relative to direction of flow and φ_m is friction angle (Madsen 1993). Results of this analysis provide the following formulations for the time-averaged bed-load transport $\overline{q_{SB}}$.

b. Pure waves on sloping bottom. For a non-zero bottom slope, i.e., $\mu_b \neq 0$, but still without any current, the time-averaged net bed-load transport is

$$\left(\frac{\overline{q}_{sB}}{\sqrt{(s-1)gD}\,D} \right)_\beta = 8 \left(\psi'_{wm} \right)^{\frac{3}{2}} \overline{|\cos \theta|^{\frac{3}{2}}} \{-\mu_b, 0\} = 4.5 \left(\psi'_{wm} \right)^{\frac{3}{2}} \{-\mu_b, 0\} \qquad \text{(III-6-63)}$$

in the direction of the wave propagation, $+x$, if $-\mu_b > 0$ and opposite the wave direction if $-\mu_b < 0$. Since $\mu_b > 0$ if the wave propagates up a slope, $\beta_w > 0$, and $\mu_b < 0$ if the wave is directed towards deeper water ($\beta_w < 0$), this simply states that the net transport is down-slope as expected.

c. Combined wave-current flows. Since the approximate bed-load transport formula is linear in the small parameters, the net bed-load transport associated with combined wave-current flows can be evaluated by time averaging for $\mu_b = 0$. The resulting net bed-load transport is given by

$$\left(\frac{\overline{q}_{sB}}{\sqrt{(s-1)gD}\,D} \right)_{wc} = 6 \left(\psi'_c \right)^{\frac{3}{2}} \frac{u'_{*wm}}{u'_{*c}} \left\{ \frac{3}{2} \cos \varphi_{wc}, \sin \varphi_{wc} \right\} \qquad \text{(III-6-64)}$$

where ψ'_c denotes the Shields parameter based on the current skin friction shear stress. It should be noted that the direction of sediment transport is *not* in the direction of the current but is obtained from

$$\tan \varphi_{ws} = \frac{2}{3} \tan \varphi_{wc} \qquad \text{(III-6-65)}$$

This expression shows that φ_{ws}, the direction of the net bed-load transport measured counterclockwise from the x-axis, deviates from the current direction towards the direction of wave orbital velocities.

d. Combined current and bottom slope effect. It is interesting to note that the superposition of a current in the direction of wave propagation, i.e., $\varphi_{wc} = 0$, up a slope may lead to zero net transport. Thus, the sum of transport given by Equations 6-63 and 6-64 is zero if

$$\mu_b = 2\mu \qquad \text{(III-6-66)}$$

Although the upslope transport here is associated with an assumed current, it is reasonable to anticipate a similar upslope effect associated with nonlinear wave effects, i.e., a larger upslope shear stress under the crest than the downslope stress under the trough. The balance expressed in Equation 6-66 may therefore be thought of as a simulation of an equilibrium profile condition for inner shelf waters.

e.   Extension to spectral waves. To compute the bed load for spectral wave conditions, the procedures outlined in the foregoing sections are followed using the representative periodic wave characteristics defined in Part III 6-2-*b* (4).

f.   Extension to sediment mixtures.

(1)   For a sediment mixture consisting of several distinct size classes of diameter D_n with the volume fraction of the nth size class being f_n in the bottom sediments, the total bed-load transport of the sediment mixture is given by

$$\overline{q}_{sB} = \sum f_n \overline{q}_{sB,n} \qquad \text{(III-6-67)}$$

(2)   This extension to sediment mixtures follows directly from the conceptual bed-load transport model (Madsen 1993) when it is assumed that the fraction of excess skin friction shear stress carried by moving grains of the nth size class is $f_n \left(\tau_b' - \tau_{cr,n} \right)$.

(3)   For each size class, the procedure is therefore exactly as outlined in the preceding section with D replaced by D_n, the diameter of the nth size class. The skin friction bottom shear stress is, however, the same for all size classes and is computed for a single representative grain diameter ($D = D_r = D_{50}$ = the median diameter of the sediment mixture in the bed).

III-6-5. Suspended Load Transport

a.   Introduction.

(1)   For low transport rates, the predominant mode of transport is sediment grains moving along the bottom, i.e., in the form of bed load. As the flow intensity and, hence, the bed-load transport, is increased individual grains moving along the bottom take off from the bottom with increasing frequency, e.g., due to impact with stationary grains protruding above their neighbors. Thus, as the flow intensity increases, the mode of transport changes from one of rolling and sliding along the bottom to one of jumping along the bottom. For this mode of transport, moving grains are in direct contact with immobile bottom grains only for a fraction of the time they spend in transport, and the basic assumptions behind the conceptual transport model presented in the preceding section are no longer valid.

(2)   Sediment transport associated with individual sediment grains making isolated or a series of jumps along the bottom, referred to as saltation, provides a transition from bed-load transport that takes place immediately above the bottom to suspended load transport that takes place in the overlying water column. From this distinction between the bed-load and suspended load transport modes, it follows that suspended load is governed by the turbulence associated with the *total bottom shear stress*, i.e., the shear stress predicted from the movable bed roughness estimate obtained from Part III-6-3, whereas the bed load is governed by the *skin friction shear stress*.

EXAMPLE PROBLEM III-6-9

FIND:

The bed-load transport rate for a combined wave-current flow over a sloping bottom.

GIVEN:

The wave is specified by its maximum near-bottom orbital velocity $u_{bm} = 0.35$ m/s, period $T = 8.0$ s, and angle of incidence $\alpha = 45$ deg. The current is at an angle of $\varphi_{wc} = 45$ deg to the wave direction and has a magnitude of $u_c(z_r) = 0.35$ m/s at $z_r = 1.00$ m. The bottom has a uniform slope of $\beta = 1$ deg and consists of uniform quartz sand $\rho_s = 2,650$ kg/m³, with diameter $D = 0.2$ mm. The fluid is seawater ($\rho = 1,025$ kg/m³, $\nu \approx 1.0 \times 10^{-6}$ m²/s).

PROCEDURE:

1) The skin friction shear velocities and stresses are computed using the procedure of Example Problem III-6-8.

2) The skin friction Shields parameters, ψ_c' and ψ_{wm}', are obtained from Equation 6-43 with $u_* = u_{*c}'$ and u_{*wm}', respectively, as obtained in step 1.

3) The time-averaged combined wave-current bed-load transport vector is obtained from Equation 6-64 using u_{*c}' and u_{*wm}' from step 1 and ψ_c' from step 2.

4) The bottom slope parameter μ_b is calculated from Equation 6-62, where β is bottom slope in the direction of wave propagation (positive if wave is traveling towards shallower waters) and $\varphi_m = 30$ deg.

5) The time-averaged bottom slope bed-load transport vector is obtained from Equation 6-63 with ψ_{wm}' from step 2.

6) The total time-averaged bed-load transport vector is obtained by vector addition of the wave-current and slope bed-load transport vectors obtained in step 3 and step 5, respectively. (The x component denotes the transport in the direction of wave propagation, the y component is transport 90 deg counterclockwise to the direction of wave propagation.)

(Continued)

Example Problem III-6-9 (Concluded)

SOLUTION:

To calculate bed-load transport, one needs to use skin friction shear stresses. The problem specification is here identical to that of Example Problem III-6-8, so the skin friction problem was solved there, i.e., $u'_{*c} = 1.01$ cm/s and $u'_{*wm} = 2.31$ cm/s. The corresponding skin friction Shields parameters are obtained from Equation 6-56 with $s = \rho_s/\rho = 2.59$ and $D = 0.2$ mm

$$\psi'_c = 0.033 \quad \text{and} \quad \psi'_{wm} = 0.17$$

With these values, the bed-load transport associated with the combined wave-current flow is obtained in nondimensional form from Equation 6-64

$$\left(\frac{\bar{q}_{sB}}{\sqrt{(s-1)gDD}} \right)_{wc} = 0.082 \left\{ \frac{3}{2} \cos \varphi_{wc}, \quad \sin \varphi_{wc} \right\}$$

or in dimensional form with $\varphi_{wc} = 45$ deg

$$\left(\bar{q}_{sB} \right)_{wc} = \{9,7,6.5\} \times 10^{-3} \text{ cm}^3/(\text{cm s})$$

To obtain the contribution due to bottom slope, the slope β_w in the direction of wave propagation is needed. Since the bottom slope is $\beta = 1$ deg and the angle of incidence is $\alpha = 45$ deg, the bottom slope in the direction of wave propagation is $\tan \beta_w = \tan \beta \cos \alpha = 0.012$. Equation 6-62 therefore gives, with $\varphi_m = 30$ deg, $\mu_b = \tan \beta_w / \tan \varphi_m = 0.021$. Bed-load transport due to bottom slope is obtained from Equation 6-63 in nondimensional form

$$\left(\frac{\bar{q}_{sB}}{\sqrt{(s-1)gDD}} \right)_{\beta} = 0.32 \{ -\mu_b, 0 \}$$

or in dimensional form

$$\left(\bar{q}_{sB} \right)_{\beta} = \{ -7.5, 0 \} \times 10^{-4} \text{ cm}^3/(\text{cm s})$$

The total bed-load transport rate is obtained as the sum of the wave-current and bottom slope contributions, i.e.,

$$\bar{q}_{sB} = \left(\bar{q}_{sB} \right)_{wc} + \left(\bar{q}_{sB} \right)_{\beta} \approx \{9.0, 6.5\} \times 10^{-3} \text{ cm}^3/(\text{cm s})$$

with the x component being in the direction of wave propagation and the y component being at 90 deg counterclockwise from the wave direction.

(3) Distribution of suspended sediment in the water column is governed by the fall velocity of the sediment w_f, the sediment diffusion coefficient v_s, and the volume concentration of sediment in suspension c. Assuming the sediment-fluid mixture behaves as a clear fluid, a close analogy can be drawn between the turbulent diffusion of momentum and sediment diffusion such that

$$v_s = v_t = k u_* z \qquad \text{(III-6-68)}$$

For combined wave-current bottom boundary layer flows, the concentration of suspended sediments c may be resolved into its mean $\bar{c}$ and time-varying wave-associated c_w components.

b. Sediment fall velocity.

(1) In order to solve the equations governing the distribution of suspended sediment concentration in the vertical, it is necessary to obtain the value of the sediment fall velocity w_f.

(2) Assuming a spherical sediment grain, the force balance of submerged weight and fluid drag on a grain falling through an otherwise quiescent fluid gives

$$(\rho_s - \rho) \, g \left(\frac{\pi}{6} D^3 \right) = \frac{1}{2} \rho \, C_D \left(\frac{\pi}{4} D^2 \right) w_f^2 \qquad \text{(III-6-69)}$$

or

$$\frac{w_f}{\sqrt{(s - 1) g D}} = \sqrt{\frac{4}{3 C_D}} \qquad \text{(III-6-70)}$$

(3) The drag coefficient C_D is a function of the Reynolds number $Re_D = D \, w_f / v$, which is a function of S_*, the sediment-fluid parameter defined by Equation 6-47.

(4) From the empirical relationship of C_D versus Re_D (Schlichting 1960), C_D is obtained for a specified value of Re_d. With this value of C_D the nondimensional fall velocity is obtained from Equation 6-70, and this value is used with the specified value of Re_D to obtain the corresponding value of S_*. In this manner (Madsen and Grant 1976), the graph of nondimensional fall velocity as a function of the sediment fluid parameter, shown in Figure III-6-11, is obtained.

(5) Extending the graph to values of S_* above the upper limit of Figure III-6-11 is accomplished by using

$$\frac{w_f}{\sqrt{(s - 1) g D}} = 1.82 \quad \text{for} \quad S_* > 300 \qquad \text{(III-6-71)}$$

(6) For values of S_* below unity, i.e., for quartz grains of $d < 0.1$ mm in seawater, the fall velocity is obtained from Stokes Law, which in the present notation may be written

$$\frac{w_f}{\sqrt{(s - 1) g D}} = \frac{2}{9} S_* \quad \text{for} \quad S_* < 0.8 \qquad \text{(III-6-72)}$$

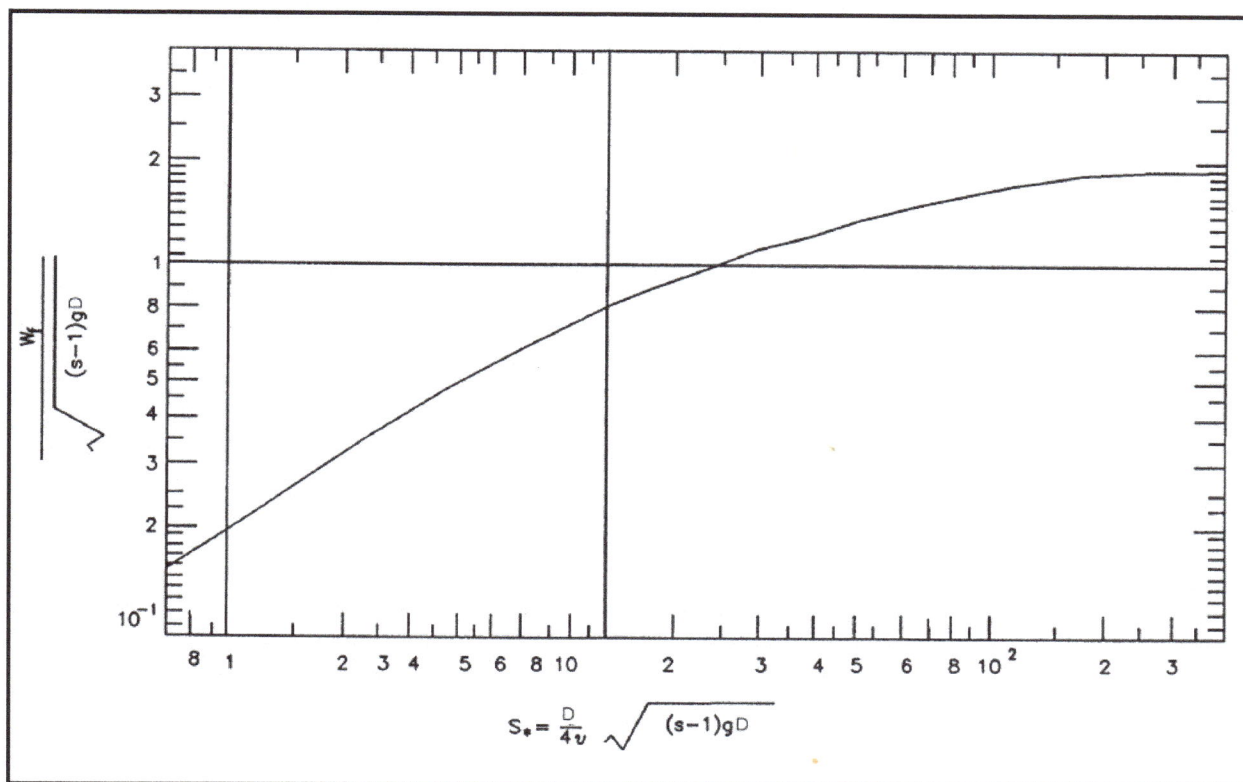

Figure III-6-11. Nondimensional fall velocity for spherical particles versus the sediment fluid parameter (Madsen and Grant 1976)

(7) Although Figure III-6-11 is obtained for sediment assumed to be spherical, it yields reasonably accurate fall velocities for natural cohesionless granular sediments (Dietrich 1982).

c. *Reference concentration for suspended sediments.*

(1) Introduction.

(a) With the fall velocity determined, all parameters in the equation governing the suspended sediment concentration distribution are in principle known. To solve the equation it is, however, necessary to specify boundary conditions. One boundary condition is simply that no sediment is transported through the water surface or that the suspended sediment concentration vanishes at large distances above the bottom, if the water depth is sufficiently large. Another is the boundary condition that expresses the amount of sediment available for entrainment immediately above the bottom. Whereas the former is universally agreed upon the latter boundary condition is the subject of considerable controversy. It is beyond the scope of this presentation to get into the subtleties of this controversy, so the most commonly accepted form of the bottom boundary condition for suspended sediment concentration, the *specification of a reference concentration*, is adopted here.

(b) The *reference concentration* is given as (Madsen 1993)

$$c_R = \gamma C_b \left(\frac{|\tau_b'(t)|}{\tau_{cr,\beta}} - 1 \right)$$
(III-6-73)

EXAMPLE PROBLEM III-6-10

FIND:
Find the sediment fall velocity

GIVEN:
Sediment is quartz sand, $\rho_s = 2,650$ kg/m³, of diameter $D = 0.2$ mm.
Fluid is seawater ($\rho = 1,025$ kg/m³, $v \approx 1.0 \times 10^{-6}$ m²/s).

PROCEDURE:
1) The sediment-fluid parameter S_*, defined by Equation 6-47, is calculated.

2) If $S_* > 300$, the sediment fall velocity w_f is obtained from Equation 6-71.

3) If $S_* < 0.8$, the sediment fall velocity w_f is obtained from Equation 6-72.

4) If $0.8 < S_* < 300$, w_f is obtained from Figure III-6-11 using S_* from step 1 as entry.

SOLUTION:
The sediment-fluid parameter is calculated from

$$S_* = \frac{D}{4v}\sqrt{(s - 1) g D} = 2.79$$

with $s = \rho_s/\rho = 2.59$.

Figure III-6-11 then gives the nondimensional fall velocity

$$\frac{w_f}{\sqrt{(s - 1) g D}} \approx 0.40$$

and therefore
$$w_f = 2.23 \text{ cm/s}$$

---

in which γ is the so-called resuspension parameter and C_b is the volume concentration of sediment in the bed, generally taken as 0.65 (Smith and McLean 1977) for a bed consisting of cohesionless sediment.

(c) It is emphasized that the resuspension parameter γ in Equation 6-73 is intimately related to the choice of reference elevation z_r, which explains at least part of the considerable variability of γ values reported in the literature. Similarly, γ values reported in the literature can be used only in conjunction with their particular z_r values. This being said, the numerical values

$$\gamma = \begin{cases} 2 \times 10^{-3} & \text{for rippled bed} \\ \\ 2 \times 10^{-4} & \text{for flat bed} \end{cases}$$
(III-6-74)

EXAMPLE PROBLEM III-6-11

FIND:
The reference concentration for suspended sediment computations for combined wave-current boundary layer flows.

GIVEN:
Wave, current, sediment, and bottom slope specifications are identical to those of Example Problem III-6-9.

PROCEDURE:

1) Skin friction shear velocities and stresses are computed following the procedure of Example Problem III-6-8.

2) The time-averaged (mean) reference concentration $\bar{c}_R$ is obtained from Equation 6-75 with $C_b = 0.65$, τ'_{wm}, and τ_{cr} as obtained in step 1, and the appropriate choice of γ from Equation 6-74.

3) The periodic (wave) reference concentration c_{Rw} is obtained from Equation 6-76 using γ and C_b as in step 2 and shear velocities obtained in step 1. The bottom slope in the direction of wave propagation β_w is positive if the wave is traveling up-slope, and $\varphi_m = 30$ deg.

SOLUTION:
The reference concentration for suspended load computations depends on skin friction shear velocities, obtained in Example Problem III-6-8 for the same problem specification as here, and may be separated into a mean and a periodic component.

The mean reference concentration is given by Equation 6-75, which may alternatively be written

$$\bar{c}_R = \gamma \, C_b \left\{ \frac{2}{\pi} \left(\frac{u'_{*wm}}{u_{*cr}} \right)^2 - 1 \right\}$$

where $u'_{*wm} = 2.31$ cm/s (Example Problem III-6-8) and $u_{*cr} = 1.27$ cm/s (Example Problem III-6-6). The resuspension parameter is given by Equation 6-74. Since the bed is rippled for these problem specifications, cf. Example Problem III-6-7, $\gamma = 2 \times 10^{-3}$ is chosen here. With $C_b = 0.65$ the *mean reference concentration* is

$$\bar{c}_R = 1.44 \times 10^{-3} \; (\text{cm}^3/\text{cm}^3)$$

(Continued)

Example Problem III-6-11 (Concluded)

The periodic reference concentration is given by Equation 6-76. The maximum value may be written

$$c_{Rwm} = \gamma \, C_b \left\{ \frac{4}{\pi} \left(\frac{u'_{*c}}{u_{*cr}} \right)^2 \cos \varphi_{wc} - \left(\frac{u'_{*wm}}{u_{*cr}} \right)^2 \frac{\tan \beta_w}{\tan \varphi_m} \right\}$$

which is evaluated using $u'_{*c} = 1.01$ cm/s (Example Problem III-6-8), $\varphi_{wc} = 45$ deg, and $\tan \beta_w / \tan \varphi_m = \mu_b = 0.021$ (Example Problem III-6-9) to give the *wave-associated maximum reference concentration*

$$c_{Rwm} = 0.65 \times 10^{-3} \text{ (cm}^3/\text{cm}^3)$$

with a temporal variation $\cos(\omega t + \varphi')$ with $\varphi' = 14$ deg (Example Problem III-6-8). Both mean and periodic reference concentrations are specified at

$$z_R = 7D = 0.14 \text{ cm}$$

---

obtained for and from a model closely resembling the present model, with reference elevation $z_R = 7D$, are adopted here (Wikramanayake and Madsen 1994b). It should, however, be emphasized that there is considerable uncertainty associated with the adoption of these γ values, or any other values for that matter.

(d) To evaluate the general expression for the reference concentration given by Equation 6-73 for combined wave-current flows, the assumption of wave dominance is introduced in the context of evaluation of bed-load transport as previously done in Part III-6-4. Subject to the limitations stated there, the reference concentration is valid for dominant wave conditions ($\tau'_c << \tau'_{wm}$) exceeding critical conditions ($\tau_{cr} << \tau'_{wm}$) over a gently sloping bottom ($\tan\beta_w << \tan\varphi_m$).

(2) Mean reference concentration. From the general expression for the reference concentration (Madsen 1993), the time-invariant, mean reference concentration is obtained as

$$\bar{c}_R = \gamma \, C_b \left(\frac{2}{\pi} \frac{\tau'_{wm}}{\tau_{cr}} - 1 \right) \tag{III-6-75}$$

and is applied at $z = z_r = 7D$.

(3) Wave reference concentration. The periodic component of the reference concentration (Madsen 1993), the wave reference concentration, is given by

$$c_{Rw} = \gamma \, C_b \left(\frac{4}{\pi} \frac{\tau_c'}{\tau_{cr}} \cos \varphi_{wc} - \frac{\tau_{wm}'}{\tau_{cr}} \frac{\tan \beta_w}{\tan \varphi_m} \right) \cos \theta = c_{Rwm} \cos \theta \qquad \text{(III-6-76)}$$

at $z = z_r = 7D$. In this expression it is recalled that the wave phase is given by

$$\theta = \omega t + \varphi' \qquad \text{(III-6-77)}$$

where φ' is the phase angle given by Equation 6-14 for skin friction conditions, i.e., with $u_{*wm} = u_{*m}'$.

   d.   Concentration distribution of suspended sediment. For a combined wave-current flow, the eddy viscosity or eddy diffusivity is given by

$$\nu_s = \begin{cases} \kappa u_{*m} z & z < \delta_{cw} \\ \kappa u_{*c} z & z > \delta_{cw} \end{cases} \qquad \text{(III-6-78)}$$

with

$$\delta_{cw} = \frac{\kappa u_{*m}}{\omega} \qquad \text{(III-6-79)}$$

and the shear velocities being those determined from the wave-current interaction model described in Part III-6-2-c using the equivalent Nikruradse roughness corresponding to the movable bed roughness obtained by the procedures outlined in Part III-6(3)e.

   (1)   Mean concentration distribution. Using the equation governing the distribution of mean concentration (Madsen 1993), and introducing the eddy diffusivity from Equation 6-78 yields

$$\bar{c} = \bar{c}_R \left(\frac{z}{z_r} \right)^{\left(-\frac{w_f}{\kappa u_{*m}} \right)} \qquad \text{for} \quad z < \delta_{cw} \qquad \text{(III-6-80)}$$

and

$$\bar{c} = \bar{c}_R \left(\frac{\delta_{cw}}{z_r} \right)^{\left(-\frac{w_f}{\kappa u_{*m}} \right)} \left(\frac{z}{\delta_{cw}} \right)^{\left(-\frac{w_f}{\kappa u_{*c}} \right)} \qquad \text{for} \quad z > \delta_{cw} \qquad \text{(III-6-81)}$$

where $\bar{c}_R$ is given by Equation 6-75, and the mean concentrations are matched at $z = \delta_{cw}$.

   (2)   Wave-associated concentration distribution. The wave-generated suspended sediment concentration can be approximated as (Madsen 1993)

$$c_w \approx c_{Rwm} \cos(\omega t + \varphi') - c_{Rwm} \frac{2}{\pi} \sin \varphi_s \, \ln \frac{z}{z_r} \cos(\omega t + \varphi' + \varphi_s) \qquad \text{(III-6-82)}$$

where

$$\tan \varphi_s = \cfrac{\cfrac{\pi}{2}}{\ln \cfrac{\kappa u_{*m}}{z_r \omega} - 1.15} \qquad (III\text{-}6\text{-}83)$$

This is a highly simplified solution for the wave-associated suspended sediment concentration within the wave boundary layer, i.e., for $z < \delta_{cw}$. For a more accurate representation of the wave-associated concentration profile, reference is made to Wikramanayake and Madsen (1994b).

    e.   Suspended load transport. The suspended load transport is obtained from the product of the velocity vector, and the concentration of suspended sediment followed by integration over depth is denoted by h. Time averaging the instantaneous transport rate produces a net transport rate, which is the quantity of interest. Indicating time-averaging by an overbar, the total net suspended load transport rate is obtained from

$$\overline{\boldsymbol{q}}_{sS,T} = \int_{z_r}^{h} \boldsymbol{u}_c \, \overline{c} \, dz + \int_{z_r}^{h} \overline{\boldsymbol{u}_w \, c_w} \, dz \qquad (III\text{-}6\text{-}84)$$

This equation identifies the contribution to the total net suspended load transport as one associated entirely with *mean suspended load transport*

$$\overline{q}_{sS} = \int_{z_r}^{h} u_c \, \overline{c} \, dz \qquad (III\text{-}6\text{-}85)$$

which takes place in the direction of the current ,i.e., at an angle φ_{wc} to the direction of wave propagation, and *mean wave-associated suspended load transport*

$$\overline{q}_{sSw} = \int_{z_r}^{h} \overline{u_w \, c_w} \, dz \qquad (III\text{-}6\text{-}86)$$

which is in the direction of wave propagation when positive.

    (1)   Mean suspended load transport. The mean suspended load transport, given by Equation 6-85, must be considered for two specific conditions.

    (a)   When $z_r > z_0$, the current velocity profile is valid for $z \geq z_r$ and the mean suspended load transport is obtained from

$$\overline{q}_{sS} = \frac{u_{*c}}{\kappa} \, \overline{c}_R \left(\frac{\delta_{cw}}{z_r} \right)^{\left(\frac{-w_f}{\kappa u_{*m}} \right)} \delta_{cw} \, (I_1 + I_2) \quad \text{for} \quad z_r > z_0 \qquad (III\text{-}6\text{-}87)$$

where

$$I_1 = \frac{\kappa u_{*c}}{\kappa u_{*m} - w_f}\left\{\ln\frac{\delta_{cw}}{z_0} - \frac{\kappa u_{*m}}{\kappa u_{*m} - w_f} - \left(\frac{z_r}{\delta_{cw}}\right)^{\frac{\kappa u_{*m} - w_f}{\kappa u_{*m}}}\left(\ln\frac{z_r}{z_0} - \frac{\kappa u_{*m}}{\kappa u_{*m} - w_f}\right)\right\}$$ (III-6-88)

represents the contribution from inside the wave boundary layer; and

$$I_2 = \frac{\kappa u_{*c}}{\kappa u_{*c} - w_f}\left\{\left(\frac{h}{\delta_{cw}}\right)^{\frac{\kappa u_{*c} - w_f}{\kappa u_{*c}}}\left(\ln\frac{h}{z_{0a}} - \frac{\kappa u_{*c}}{\kappa u_{*c} - w_f}\right) - \left(\ln\frac{\delta_{cw}}{z_{0a}} - \frac{\kappa u_{*c}}{\kappa u_{*c} - w_f}\right)\right\}$$ (III-6-89)

expresses the transport above the wave boundary layer.

(b) When $z_r < z_0$, the current velocity profile is not defined for $z_r < z < z_0$. To remedy this physically unrealistic situation, the current velocity profile given by Equation 6-27 is modified to read

$$u_c = \frac{u_{*c}}{\kappa}\frac{u_{*c}}{u_{*m}}\frac{\ln\frac{\delta_{cw}}{z_0}}{\ln\frac{\delta_{cw}}{z_r}}\ln\frac{z}{z_r} \quad \text{for} \quad z < \delta_{cw}$$ (III-6-90)

for the purpose of evaluating the transport within the wave boundary layer. The modification retains the matching condition with the outer solution at $z = \delta_{cw}$ and extends the velocity profile down to $z = z_R$.

(c) Introducing Equation 6-90 in the integration leads to

$$\overline{q}_{sS} = \frac{u_{*c}}{\kappa}\overline{c}_R\left(\frac{\delta_{cw}}{z_r}\right)^{-\frac{w_f}{\kappa u_{*m}}}\delta_{cw}(I_3 + I_2) \quad \text{for} \quad z_r < z_0$$ (III-6-91)

where

$$I_3 = \frac{\ln\frac{\delta_{cw}}{z_0}}{\ln\frac{\delta_{cw}}{z_r}}\frac{\kappa u_{*c}}{\kappa u_{*m} - w_f}\left\{\ln\frac{\delta_{cw}}{z_r} - \frac{\kappa u_{*m}}{\kappa u_{*m} - w_f}\left[1 - \left(\frac{z_r}{\delta_{cw}}\right)^{\frac{\kappa u_{*m} - w_f}{\kappa u_{*m}}}\right]\right\}$$ (III-6-92)

and I_2 is given by Equation 6-89.

(2) Mean wave-associated suspended load transport. The mean wave-associated suspended load transport is obtained from Equation 6-86. Since u_w and c_w, given by Equations 6-11 and 6-82, respectively, are approximations valid only for $z < \delta_{cw}/\pi$, the integration is extended only to this upper limit. Similar to the computation of the mean suspended load transport, it is necessary here to distinguish between the two cases of $z_0 < z_r$ and $z_r < z_0$. If $z_r > \delta_{cw}/\pi$, wave-associated transport is considered negligibly small.

EXAMPLE PROBLEM III-6-12

FIND:

The mean suspended load transport rate for a combined wave-current boundary layer flow.

GIVEN:

Wave, current, sediment, and bottom slope as given in Example Problem III-6-9. Water depth $h = 5.0$ m.

PROCEDURE:

1) Movable bed roughness k_n is determined as described in Example Problem III-6-7.

2) Total shear velocities u_{*c}, u_{*wm}, and u_{*m}, and wave boundary layer thickness δ_{cw} are obtained by solving the wave-current interaction following the procedure of Example Problem III-6-4 with k_n from step 1.

3) Apparent bottom roughness z_{0a} and shear stress phase angle φ are obtained as described in Example Problem III-6-5.

4) Skin friction shear velocities u'_{*c}, u'_{*wm}, and u'_{*m}, and phase angle φ' are obtained as in Example Problem III-6-8.

5) Example Problem III-6-10 is followed to obtain the fall velocity w_f.

6) Mean reference concentration $\overline{c}_R$ is obtained as in Example Problem III-6-11.

7) If $z_r = 7D > z_0 = k_n/30$, the mean suspended sediment transport $\overline{q}_{sS}$ is obtained from Equations 6-87 through 6-89.

8) If $z_r = 7D < z_0 = k_n/30$, the mean suspended sediment transport $\overline{q}_{sS}$ is obtained from Equations 6-91, III-6-92, and 6-89.

9) The mean suspended sediment transport obtained in step 7 or step 8 is in the direction of the current, which in turn is at an angle of φ_{wc} to the direction of wave propagation. In a coordinate system with x in the direction of wave propagation, the mean suspended sediment transport vector is $\overline{q}_{sS} = \overline{q}_{sS}\{\cos\varphi_{wc}, \sin\varphi_{wc}\}$.

(Continued)

Example Problem III-6-12 (Concluded)

SOLUTION:

Since quantities needed for the computation of suspended load transport have been obtained in previous example problems, these are summarized here.

Reference concentration (Example Problem III-6-11): $\bar{c}_R = 1.44 \times 10^{-3}$ at $z_r = 0.14$

Fall velocity (Example Problem III-6-10): $w_f = 2.23$ cm/s

Shear velocities and *wave boundary layer thickness* based on total movable bed roughness (Example Problem III-6-4): $u_{*c} = 2.94$ cm/s, $u_{*m} = 6.67$ cm/s, and $\delta_{cw} = 3.40$ cm

The *movable bed roughness* and the *apparent bottom roughness* (Example Problems III-6-7 and III-6-5): $z_0 = k_n/30 = 0.15$ cm and $z_{0a} = 0.85$ cm

From the input values summarized above it is seen that $z_r = 0.14$ cm $\approx z_0 = 0.15$ cm. One could therefore choose either Equation 6-87 or Equation 6-91 to obtain the mean suspended load transport. Since $z_r < z_0$, Equation 6-91 is chosen here.

The leading term in Equation 6-91 is computed first

$$\frac{u_{*c}}{\kappa} \bar{c}_R \left(\frac{\delta_{cw}}{z_r} \right)^{-\frac{w_f}{\kappa u_{*m}}} \delta_{cw} = 2.5 \times 10^{-3} \text{ cm}^3/(\text{cm s})$$

The integral I_3 is evaluated from Equation 6-92

$$I_3 = \frac{3.12}{3.19} \times 2.68\{3.19 - 6.08[1 - 0.59]\} = 1.83$$

and I_2 is obtained from Equation 6-89

$$I_2 = -1.12[0.012(6.38 + 1.12) - (1.39 + 1.12)] = 2.71$$

The mean suspended load transport is therefore obtained from Equation 6-91 as

$$\bar{q}_{sS} = 2.5 \times 10^{-3}(I_3 + I_2) = 1.1 \times 10^{-2} \text{ cm}^3/(\text{cm s})$$

and is in the direction of the current, i.e., at an angle $\varphi_{wc} = 45$ deg to the direction of wave propagation.

(a)   For $z_0 < z_r < \delta_{cw}/\pi$, Equations 6-11 and 6-82 may be introduced directly into Equation 6-86 to give

$$\bar{q}_{sSw} = \frac{1}{\pi^2} \sin\varphi \; u_{bm} \, c_{Rwm} \, \delta_{cw}$$
$$\times \left(\cos(\varphi - \varphi')I_4 - \frac{2}{\pi} \sin\varphi_s \cos(\varphi - \varphi' - \varphi_s)I_5 \right) \quad \text{for} \quad z_0 < z_r \tag{III-6-93}$$

in which

$$I_4 = \ln\frac{\delta_{cw}}{\pi z_0} - 1 - \frac{\pi z_r}{\delta_{cw}}\left(\ln\frac{z_r}{z_0} - 1 \right) \tag{III-6-94}$$

and

$$I_5 = \ln\frac{z_r}{z_0}\left(\ln\frac{\delta_{cw}}{\pi z_r} - 1 + \frac{\pi z_r}{\delta_{cw}} \right) + \left(\ln\frac{\delta_{cw}}{\pi z_r} \right)^2 - 2\ln\frac{\delta_{cw}}{\pi z_r} + 2\left(1 - \frac{\pi z_r}{\delta_{cw}} \right) \tag{III-6-95}$$

(b)   For $z_r < z_0$ the same modification of the logarithmic velocity profile introduced in the context of mean suspended load transport yields, for $z_r < z_0$

$$\bar{q}_{sSw} = \frac{1}{\pi^2} \sin\varphi \; u_{bm} \; c_{Rwm} \; \delta_{cw} \frac{\ln\dfrac{\delta_{cw}}{\pi z_0}}{\ln\dfrac{\delta_{cw}}{\pi z_r}}$$
$$\times \left(\cos(\varphi - \varphi')I_6 - \frac{2}{\pi} \sin\varphi_s\cos(\varphi - \varphi' - \varphi_s)I_7 \right) \tag{III-6-96}$$

where

$$I_6 = \ln\frac{\delta_{cw}}{\pi z_r} - 1 + \frac{\pi z_r}{\delta_{cw}} \tag{III-6-97}$$

and

$$I_7 = \left(\ln\frac{\delta_{cw}}{\pi z_r} \right)^2 - 2\ln\frac{\delta_{cw}}{\pi z_r} + 2\left(1 - \frac{\pi z_r}{\delta_{cw}} \right) \tag{III-6-98}$$

EXAMPLE PROBLEM III-6-13

FIND:

The mean wave-associated suspended load transport rate for a combined wave-current boundary layer flow.

GIVEN:

Same problem specifications as in Example Problem III-6-12.

PROCEDURE:

1) First steps are identical to steps 1 and 2 in Example Problem III-6-12.

2) If $z_r = 7D > \delta_{cw}/\pi$, wave-associated suspended sediment transport is negligibly small.

3) If $z_r = 7D < \delta_{cw}/\pi$, steps 3 through 5 of Example Problem III-6-12 are followed.

4) The wave-associated reference concentration c_{Rw} is obtained as in Example Problem III-6-11.

5) The sediment suspension phase angle φ_s is obtained from Equation 6-83 with $z_r = 7D$.

6) If $z_r = 7D > z_0 = k_n/30$, the mean wave-associated suspended sediment transport $\bar{q}_{sSw}$ is obtained from Equations 6-93 through 6-95.

7) If $z_r = 7D < z_0 = k_n/30$, the mean wave-associated suspended sediment transport $\bar{q}_{sSw}$ is obtained from Equations 6-96 through 6-98.

8) The suspended sediment transport obtained in step 6 or step 7 is in the direction of wave propagation, i.e., $uu\bar{q}_{sSw} = \bar{q}_{sSw}\{1, 0\}$ with x in the direction of wave propagation.

(Continued)

Example Problem III-6-13 (Concluded)

SOLUTION:

In addition to the quantities given in Example Problem III-6-12, the computation of the mean wave-associated suspended load transport requires knowledge of *phase angles*

$$\varphi = 38° \quad \text{and} \quad \varphi' = 14°$$

obtained from Example Problems III-6-5 and 8. The sediment suspension phase angle φ_s is obtained from Equation 6-83 with $z_r = 7D = 0.14$ cm

$$\tan \varphi_s = 0.77 \rightarrow \varphi_s = 37.6° \approx 38°$$

and the *maximum wave-associated reference concentration* (Example Problem III-6-11) is

$$c_{Rwm} = 0.65 \times 10^{-3}$$

Since $z_r = 0.14$ cm $< z_0 = 0.15$ cm, the wave-associated mean transport is obtained from Equation 6-96.

The leading term of this equation becomes

$$\frac{1}{\pi^2} \sin \varphi \; u_{bm} \; c_{Rwm} \; \delta_{cw} \; \frac{\ln \dfrac{\delta_{cw}}{\pi z_0}}{\ln \dfrac{\delta_{cw}}{\pi z_r}} = 5.0 \times 10^{-3} \text{ cm}^3/(\text{cm s})$$

and the integral I_6 is obtained from Equation 6-97

$$I_6 = 2.05 - 1 + 0.13 = 1.18$$

and I_7 is given by Equation 6-98

$$I_7 = 4.18 - 4.09 + 2(1 - 0.13) = 1.83$$

Introducing these quantities in Equation 6-96, the *mean wave-associated suspended load transport* is obtained as

$$\overline{q}_{sSw} = 5.0 \times 10^{-3} \left(1.18 \cos (38° - 14°) - 1.83\frac{2}{\pi} \sin 38° \cos (38° - 14° - 38°) \right) = 1.9 \times 10^{-3} \text{ cm}^3/(\text{cm s})$$

The direction of this transport is in the direction of wave propagation.

(3) Computation of total suspended load transport for combined wave-current flows.

(a) With the wave and current components of the skin friction bottom shear stress including the phase angle φ' obtained following the procedures given in Part III-6-3-c, the mean $\overline{c}_R$ and wave-associated c_{Rwm} reference concentrations are obtained from Equations 6-75 and 6-76, respectively, and the phase angle for the wave-associated concentration φ_s is obtained from Equation 6-83.

(b) From the hydrodynamic wave-current interaction model corresponding to movable bed roughness, u_{*c}, u_{*wm}, u_{*m}, and φ are known and w_f is obtained from Part III-6-5-a. Thus, with δ_{cw} given by Equation 6-79, all quantities necessary to evaluate Equations 6-87 and 6-93, for $z_0 < z_r = 7D$, or Equations 6-91 and 6-96, for $z_r = 7D < z_0$, are available.

f. *Extensions of methodology for the computation of suspended load.*

(1) Extension to spectral waves. For a wave motion described by its near-bottom orbital velocity spectrum, the representative periodic wave characteristics discussed in Part III-6-2-b (4) are used. Once again it is emphasized that this representative wave corresponds to the use of the rms wave height and *not* the significant wave height.

(2) Extension to sediment mixtures. For a sediment mixture, the suspended load transport may be calculated as described above for a single grain size of diameter D.

(a) The skin friction $\tau_b'(t)$ and movable bed roughness k_n are computed using the median grain size $D = D_{50}$.

(b) For each grain size class of diameter D_n, the reference concentration is obtained from the formulas given in Part III-6(5)c with τ_{cr}, the critical shear stress, replaced by $\tau_{cr,n}$, the critical shear stress for the nth size class, and C_b, the sediment concentration in the bed, being replaced by $f_n C_b$ where f_n, is the volume fraction in the bed of sediment of diameter D_n. This reference concentration is assumed to be specified at $z = z_r = 7D_{50}$ for all size classes.

(c) For each size class, the suspended load transport is now computed as detailed in Part III-6(5)e using the fall velocity $w_{f,n}$ appropriate for the nth size class.

(d) When suspended load transport has been obtained for each size class, the total suspended load transport for the sediment mixture is obtained by adding the contribution of individual size classes.

III-6-6. Summary of Computational Procedures

a. *Problem specification.* A properly posed problem requires the following specifications:

(1) *The waves* are specified by near-bottom orbital velocity u_{bm}, radian frequency ω ($A_{bm} = u_{bm}/\omega$), and their angle of incidence α ($|\alpha| < \pi/2$) if waves are traveling towards shallower water). For spectral waves, a representative periodic wave is defined in Part III-6(2)c(5). If the significant wave characteristics H_s and T_s are given, the equivalent periodic wave has a height $H_{rms} = h_s/\sqrt{2}$ and period $T = T_s$.

(2) *The current* is specified by a reference current $u_c(z_r)$ at $z = z_r$ above the bottom and its direction φ_{wc} measured counterclockwise from the direction of wave propagation; or by the average bottom shear stress τ_c and its direction φ_{wc}.

(3) *The fluid* is specified by its density, $\rho \approx 1{,}025$ km/m^3 for seawater, and kinematic viscosity, $\nu \approx 10^{-6}$ m^2/s for seawater.

(4) *The sediment* is specified by its diameter D and density ρ_s ($s = \rho_s/\rho$). The sediment must be characterized as cohesionless.

(5) *The bottom slope* is specified normal to depth contours by $\tan\beta$ ($\beta > 0$). The slope in direction of wave propagation is then $\tan\beta_w = \tan\beta\cos\alpha > 0$ if bottom is sloping upwards in wave direction ($|\alpha| < \pi/2$).

b. Model parameters.

(1) A number of model parameters and constants have been introduced and must be specified to proceed towards a solution

κ = von Karman's constant = 0.4

φ_s = static friction angle of sediment = $50°$

φ_m = moving friction angle of sediment = $30°$

C_b = volume concentration of sediment in the bed = 0.65

γ = resuspension parameter = 2×10^{-3} for rippled beds, 2×10^{-4} for flat beds (sheet flow)

(2) Figure III-6-12 specifies coordinate, angle, and bottom slope definitions used in this chapter.

c. Computational procedures. The following steps should be followed to obtain the total sediment transport rate at a point:

(1) Movable bottom roughness is determined from the wave and sediment characteristics. The maximum wave skin friction shear stress τ'_{wm} is obtained following the procedures described in Part III-6(2)c using $k_n = k'_n$ = D. Sediment motion is checked using the modified Shields diagram, Part III-6(3)c. If sediment *is not* moving, $k_n = D$ *is* the movable bed roughness. If sediment is found to move, the movable bed roughness k_n is obtained following the procedures outlined in Part III-6(3)e using $\tau'_{wm} = \tau'_m$.

(2) With the movable bed roughness k_n known, the wave-current interaction model (Part III-6(2)d) is used to obtain total bottom shear stresses, shear velocities, and phase angles so that wave orbital and current velocity profiles may be evaluated.

(3) Using the current at the edge of the wave boundary layer obtained from the movable bed roughness model as the current specification, the wave-current interaction model (Part III-6(2)d) is used with $k'_n = D$ to obtain the combined wave-current skin friction shear stress $\tau'_b(t)$.

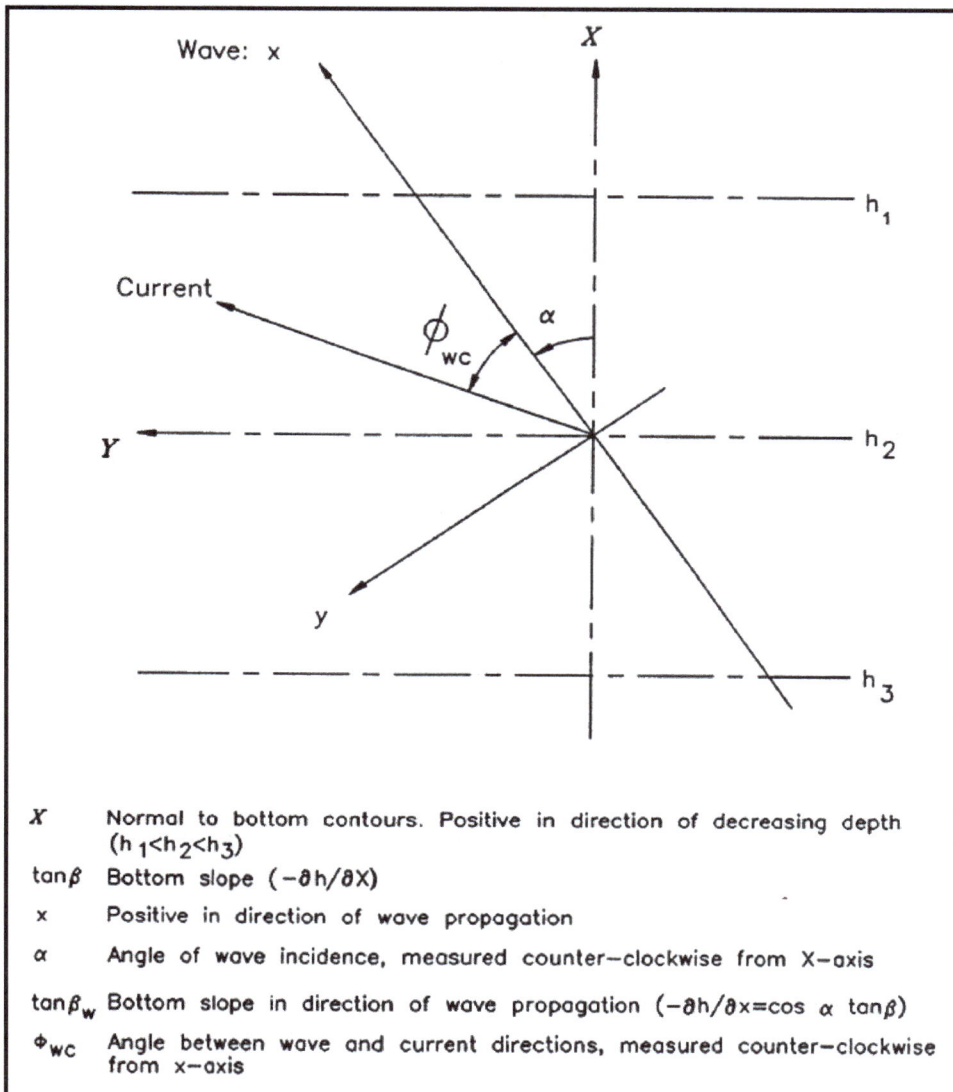

X   Normal to bottom contours. Positive in direction of decreasing depth $(h_1 < h_2 < h_3)$

$\tan\beta$ Bottom slope $(-\partial h/\partial X)$

x   Positive in direction of wave propagation

α Angle of wave incidence, measured counter-clockwise from X-axis

$\tan\beta_w$ Bottom slope in direction of wave propagation $(-\partial h/\partial x = \cos \alpha \ \tan\beta)$

ϕ_{wc} Angle between wave and current directions, measured counter-clockwise from x-axis

Figure III-6-12. Definition sketch of coordinates, angles, and bottom slope

(4) Bed-load transport can now be computed following the methodology presented in Part III-6(4).

(5) The reference concentration for suspended sediment is next obtained from Part III-6(5)*c*, and the total suspended load transport is obtained by vector addition of the mean and the mean wave-associated contributions that are obtained from Part III-6(5)*e*. The former is in the direction of the current, whereas the latter is in the wave direction.

(6) Adding bed-load and suspended load transport provides a prediction of the total sediment transport rate, expressed as a vector.

EXAMPLE PROBLEM III-6-14

FIND:
The total sediment transport vector for a combined wave-current boundary layer flow.

GIVEN:
All problem specifications are identical to those given in Example Problems III-6-9, 12, and 13.

PROCEDURE:
1) Bed-load sediment transport vector is computed following Example Problem III-6-9.

2) Mean suspended sediment transport vector is obtained following the procedure of Example Problem III-6-12.

3) Mean wave-associated suspended sediment transport vector is obtained from Example Problem III-6-13.

4) Total sediment transport vector is obtained as the vector sum of steps 1, 2, and 3.

SOLUTION:
With x being in the direction of wave propagation and y perpendicular to x, the total sediment transport in the x direction is obtained from Example Problems III-6-9, 12, and 13.

x-component of bed load = 9.0×10^{-3} cm³/(cm s)
x-component of mean suspended load = 1.1×10^{-2} cos φ_{wc} = 7.8×10^{-3} cm³/(cm s)
x-component of mean wave-associated transport = 1.9×10^{-3} cm³/(cm s)
Total transport in wave direction = $\bar{q}_{sTx}$ = 1.9×10^{-2} cm³/(cm s)

y-component of bed load = 6.5×10^{-3} cm³/(cm s)
y-component of mean suspended load = 1.1×10^{-2} sin φ_{wc} = 7.8×10^{-3} cm³/(cm s)
y-component of mean wave-associated transport = 0 cm³/(cm s)
Total transport perpendicular to wave direction = $\bar{q}_{sTy}$ = 1.4×10^{-2} cm³/(cm s)

Thus, the magnitude of the total sediment transport vector is

$$|\bar{q}_{sT}| = \sqrt{\bar{q}_{sTx}^2 + \bar{q}_{sTy}^2} = 2.4 \times 10^{-2} \text{ cm}^3/(\text{cm s})$$

directed at an angle $\varphi_{wT} \approx 37°$ to the wave direction.

III-6-7. References

Bakker and van Doorn 1978
Bakker, W. T., and Van Doorn, T. 1978. "Near-Bottom Velocities in Waves with a Current," *Proceedings Sixteenth International Coastal Engineering Conference*, American Society of Civil Engineers, Vol 2, pp 1394–1413.

Bijker 1967
Bijker, E.W. 1967. "Some Considerations about Scales for Coastal Models with Movable Bed," Technical Report, Delft Hydraulics Lab, The Netherlands.

Davies, Soulsby, and King 1988
Davies, A. G., Soulsby, R. L., and King, H. L. 1988. "A Numerical Model of the Combined Wave and Current Bottom Boundary Layer," *Journal of Geophysical Research*, Vol 93, No. C1, pp 491 - 508.

Dietrich 1982
Dietrich, W. E. 1982. "Settling Velocity of Natural Particles," *Water Resources Research*, Vol 18, No. 6, pp 1615–1626.

Dyer and Soulsby 1988
Dyer, K. R., and Soulsby, R.L. 1988. "Sand Transport on the Continental Shelf," *Annual Review of Fluid Mechanics*, Vol 20, pp 295-324.

Fredsoe 1984
Fredsoe, J. 1984. "Sediment Transport in Current and Waves," *Series Paper 35*, Institute of Hydrodynamics and Hydrologic Engineering, Technical University of Denmark, Lyngby.

Glenn and Grant 1987
Glenn, S. M., and Grant, W. D. 1987. "A Suspended Sediment Correction for Combined Wave and Current Flows," *Journal of Geophysical Research*, Vol 92, No. C8, pp 8244–8264.

Grant and Madsen 1979
Grant, W. D., and Madsen, O. S. 1979. "Combined Wave and Current Interaction with a Rough Bottom," *Journal of Geophysical Research*, Vol 85, No. C4, pp 1797–1808.

Grant and Madsen 1982
Grant, W. D., and Madsen, O. S. 1982. "Movable Bed Roughness in Unsteady Oscillatory Flow," *Journal of Geophysical Research*, Vol 87, No. C1, pp 469–481.

Grant and Madsen 1986
Grant, W. D., and Madsen, O. S. 1986. "The Continental Shelf Bottom Boundary Layer," M. Van Dyke, ed., *Annual Review of Fluid Mechanics*, Vol 18, pp 265–305.

Heathershaw 1981
Heathershaw, A.D. 1981. "Comparisons of Measured and Predicted Sediment Transport Rates in Tidal Currents," *Marine Geology*, Vol 42, pp 75-104.

Jonsson 1966
Jonsson, I. G. 1966. "Wave Boundary Layers and Friction Factors," *Proceedings of the Tenth International Coastal Engineering Conference*, American Society of Civil Engineers, Vol 1, pp 127–148.

Jonsson and Carlson 1976
Jonsson, I. G., and Carlsen, N. A. 1976. "Experimental and Theoretical Investigation in an Oscillatory Turbulent Boundary Layer," *Journal of Hydraulic Research*, Vol 14, No. 1, pp 45–60.

Kajiura 1968
Kajiura, K. 1968. "A Model of the Bottom Boundary Layer in Water Waves," *Bulletin of the Earthquake Research Institute*, Disaster Prevention Research Institute, Tokyo University, Vol 46, pp 75–123.

Madsen 1991
Madsen, O. S. 1991. "Mechanics of Cohesionless Sediment Transport in Coastal Waters," *Proceedings, Coastal Sediments '91*, American Society of Civil Engineers, Vol 1, pp 15–27.

Madsen 1992
Madsen, O. S. 1992. "Spectral Wave-Current Bottom Boundary Layer Flows," *Abstracts, Twenty-Third International Coastal Engineering Conference*, American Society of Civil Engineers, pp 197–198.

Madsen 1993
Madsen, O.S. 1993. "Sediment Transport Outside the Surf Zone," unpublished Technical Report, U.S. Army Engineer Waterways Experiment Station, Vicksburg, MS.

Madsen and Grant 1976
Madsen, O. S., and Grant, W. D. 1976. "Quantitative Description of Sediment Transport by Waves," *Proceedings, Fifteenth International Coastal Engineering Conference*, American Society of Civil Engineers, Vol 2, pp 1093–1112.

Madsen and Wikramanayake 1991
Madsen, O. S., and Wikramanayake, P. N. 1991. "Simple Models for Turbulent Wave-Current Bottom Boundary Layer Flow," Contract Report DRP-91-1, U.S. Army Engineer Waterways Experiment Station, Vicksburg, MS.

Madsen et al., in preparation
Madsen, O. S., Wright, L. D., Boon, J. D., and Chisholm, T. A. "Wind Stress, Bottom Roughness and Sediment Suspension on the Inner Shelf during an Extreme Storm Event," in preparation, *Continental Shelf Research*.

Madsen, Poon, and Graber 1988
Madsen, O. S., Poon, Y.-K., and Graber, H. C. 1988. "Spectral Wave Attenuation by Bottom Friction: Theory," *Proceedings, Twenty-First International Coastal Engineering Conference*, American Society of Civil Engineers, Vol 1, pp 492–504.

Mathisen 1993
Mathisen, P.P. 1993. "Bottom Roughness for Wave and Current Boundary Layer Flows over a Rippled bed," Ph.D. diss., Department of Civil and Environmental Engineering, Massachusetts Institute of Technology, Cambridge, MA.

Meyer-Peter and Müller 1948
Meyer-Peter, E., and Müller, R. 1948. "Formulas for Bed-Load Transport," *Report on Second Meeting of International Association for Hydraulic Research*, pp 39–64.

Nikuradse 1933
Nikuradse, J. 1933. "Strömungsgesetze in rauhen Rohren," VDI Forschungsheft No. 361 (English translation NACA Technical Memorandum No. 1292).

Raudkivi 1976
Raudkivi, A. J. 1976. *Loose Boundary Hydraulics* (2nd ed.), Pergamon Press, Oxford, England.

Schlichting 1960
Schlichting, H. 1960. *Boundary Layer Theory* (fourth ed.), McGraw-Hill, New York.

Shields 1936
Shields, A. 1936. "Application of Similarity Principles and Turbulent Research to Bed-Load Movement," (translation of original in German by W. P. Ott and J. C. van Uchelen, California Institute of Technology), *Mitteilungen der Preussischen Versuchsanstalt für Wasserbau und Schiffbau.*

Smith 1977
Smith, J. D. 1977. "Modeling of Sediment Transport on Continental Shelves," *The Sea*, E. D. Goldberg, ed., Wiley Interscience, New York, pp 539–577.

Smith and McLean 1977
Smith, J. D., and McLean, S. R. 1977. "Spatially Averaged Flow Over a Wavy Surface," *Journal of Geophysical Research*, Vol 82, No. 12, pp 1735–1745.

Trowbridge and Madsen 1984
Trowbridge, J. H., and Madsen, O. S. 1984. "Turbulent Wave Boundary Layers," *Journal of Geophysical Research*, Vol 89, No. C5, pp 7989–8007.

White, Milli, and Crabbe 1975
White, W.R., Milli, H., and Crabbe, H.O. 1975. "Sediment Transport Theories: A Review," *Proceedings, Institute of Civil Engineering,* Vol 59, pp 265-292.

Wikramanayake and Madsen 1994a
Wikramanayake, P. N., and Madsen, O. S. 1994a. "Calculation of Movable Bed Friction Factors," Contract Report DRP-94-5, U.S. Army Engineer Waterways Experiment Station, Vicksburg, MS.

Wikramanayake and Madsen 1994b
Wikramanayake, P. N., and Madsen, O. S. 1994b. "Calculation of Suspended Sediment Transport by Combined Wave-Current Flows," Contract Report DRP-94-7, U.S. Army Engineer Waterways Experiment Station, Vicksburg, MS.

Wilson 1989
Wilson, K. C. 1989. "Friction of Wave Induced Sheet Flow," *Coastal Engineering*, Vol 12, pp 371–379.

III-6-8. Definition of Symbols

β	Bottom slope [deg]
β_w	Bottom slope in the direction of wave propagation [deg]
γ	Resuspension parameter [dimensionless]
δ_{cw}	Boundary layer thickness for a combined wave-current turbulent boundary layer [length]
δ_w	Boundary layer thickness [length]
η	Ripple height [length]
κ	von Karman's constant (= 0.4)
λ	Ripple length [length]
μ	Ratio of Shear stress due to currents (τ_c) to Maximum bottom shear stress (τ_{wm})
μ_b	Bottom slope parameter
v	Kinematic viscosity [length^2/time]
v_s	Sediment diffusion coefficient
v_T	Turbulent eddy viscosity [length^2/time]
ρ	Mass density of water (salt water = 1,025 kg/m^3 or 2.0 slugs/ft^3; fresh water = 1,000kg/m^3 or 1.94 slugs/ft^3) [force-time^2/length^4]
ρ_s	Mass density of sediment grains [force-time^2/length^4]
τ_b	Bottom shear stress [force/length^2]
τ_c	Shear stress due to currents [force/length^2]
τ_{cr}	Critical bottom shear stress for initiation of motion [force/length^2]
τ_m	Maximum combined bottom shear stress [force/length^2]
τ_w	Bottom shear stress [force/length^2]
τ_{wm}	Maximum bottom shear stress [force/length^2]
τ'_b	Bottom skin friction shear stress [force/length^2]
τ'_{wm}	Skin friction bottom shear stress for wave motion [force/length^2]
τ''_b	Bottom drag shear stress [force/length^2]
φ	Phase lead of near-bottom wave orbital velocity (bottom shear stress phase angle) [deg]
φ_m	Friction angle [deg]
φ_s	Friction angle for a stationary interfacial sediment grain [deg]

Ψ	Shields parameter (Equation III-6-43) [dimensionless]
Ψ_{cr}	Critical Shields parameter (Equation III-6-45) [dimensionless]
$\Psi_{cr,\beta}$	Critical Shields parameter for flow over a sloping bottom (Equation III-6-51) [dimensionless]
Ψ'_C	Shields parameter based on the current skin friction shear stress [dimensionless]
Ψ'_m	Skin friction Shields parameter [dimensionless]
ω	Wave angular or radian frequency (= $2\pi/T$) [time^{-1}]
A_{bm}	Bottom excursion amplitude predicted by linear wave theory [length]
c	Volume concentration of sediment in suspension [length^3/length^3]
$\bar{c}$	Mean volume concentration of sediment in suspension [length^3/length^3]
$\bar{c}_R$	Mean reference concentration [length^3/length^3]
C_μ	Factor relating maximum wave and maximum combined bottom shear stresses (equation III-6-33)
C_b	Volume concentration of sediment in the bed
c_R	Reference concentration [length^3/length^3]
c_{Rwm}	Wave-associated maximum reference concentration [length^3/length^3]
c_w	Wave-associated volume concentration of sediment in suspension [length^3/length^3]
D	Sediment grain diameter [length - generally millimeters]
D_p	Time-averaged rate of energy dissipation in the wave bottom boundary layer (Equation III-6-25)
f_c	Current friction factor [dimensionless]
f_{cw}	Wave friction factor in the presence of currents [dimensionless]
f_w	Wave friction factor [dimensionless]
f'_w	Wave friction factor [dimensionless]
g	Gravitational acceleration (32.17 ft/sec^2, 9.807m/sec^2) [length/time^2]
h	Water depth [length]
H_{rms}	Root-mean-square wave height [length]
H_s	Significant wave height [length]
k_n	Equivalent Nikuradse sand grain roughness [length]
k_n	Movable bed roughness [length]
q	Sediment transport rate [length^3/time]

$\overline{q}_{sB}$	Total bed-load transport of a sediment mixture [length³/length-time]
RE	Wave Reynolds number [dimensionless]
Re_*	Boundary Reynolds number [dimensionless]
S_*	Sediment-fluid parameter (Equation III-6-47) [dimensionless]
T	Significant wave period [time]
T_s	Significant wave period [time]
u	Horizontal particle velocity [length/time]
u_*	Shear velocity [length/time]
u_{*c}	Current shear velocity [length/time]
u_{*cr}	Critical shear velocity [length/time]
u_{*m}	Maximum combined shear velocity [length/time]
u_{*wm}	Maximum wave shear velocity [length/time]
u_{bm}	Maximum near-bottom wave orbital velocity [length/time]
u_w	Wave orbital velocity [length/time]
w_f	Sediment fall velocity [length/time]
W_{grain}	Submerged weight of individual sediment grains
z	Elevation from bottom [length]
Z	Parameter providing a correlation of field data on ripple geometry [dimensionless]
z_{0a}	Arbitrary constant of integration (apparent bottom roughness)

III-6-9. Acknowledgments

Authors of Chapter III-6, "Sediment Transport Outside the Surf Zone:"

Ole S. Madsen Ph.D., Dept. of Civil and Environmental Engineering, Massachusetts Institute of Technology, Cambridge, Massachusetts.
William Wood, Ph.D. (Deceased).

Reviewers:

Scott M. Glenn, Ph.D., Institute of Marine and Coastal Sciences, Rutgers University, New Brunswick, New Jersey.
James R. Houston, Ph.D., Engineer Research and Development Center, Vicksburg, Mississippi.
Stephen McLean, Ph.D., Department of Mechanical and Environmental Engineering, University of California, Santa Barbara, Santa Barbara, California.
Todd L. Walton, Ph.D., Coastal and Hydraulics Laboratory, Engineer Research and Development Center (ERDC), Vicksburg, Mississippi.

www.ingramcontent.com/pod-product-compliance
Lightning Source LLC
Chambersburg PA
CBHW070157240326
41458CB00127B/6089